Irrigation Show and Education Conference 2018

Long Beach, California, USA
3 - 7 December 2018

ISBN: 978-1-5108-7677-4

Printed from e-media with permission by:

Curran Associates, Inc.
57 Morehouse Lane
Red Hook, NY 12571

Some format issues inherent in the e-media version may also appear in this print version.

Printed by Curran Associates, Inc. (2019)

Irrigation Association (IA)
8280 Willow Oaks Corporate Drive
Suite 400
Fairfax, Virginia 22031
USA

Phone: 703.536.7080
Fax: 703.536.7019

info@irrigationshow.org

Additional copies of this publication are available from:

Curran Associates, Inc.
57 Morehouse Lane
Red Hook, NY 12571 USA
Phone: 845-758-0400
Fax: 845-758-2633
Email: curran@proceedings.com
Web: www.proceedings.com

Agriculture Track – Table of Contents

Landscape Track – Table of Contents

Water in the Cloud:
A new system for field water monitoring with Cloud data access

Steven R. Evett, Research Soil Scientist
USDA ARS Conservation & Production Research Laboratory, Bushland, TX, steve.evett@ars.usda.gov

Harry S. Schomberg, Research Ecologist
USDA ARS Sustainable Agricultural Systems Laboratory, Beltsville, MD

Alondra Thompson, Agricultural Engineer
USDA ARS Sustainable Agricultural Systems Laboratory, Beltsville, MD

Robert C. Schwartz, Research Soil Scientist
USDA ARS Conservation & Production Research Laboratory, Bushland, TX

Susan A. O'Shaughnessy, Research Agricultural Engineer
USDA ARS Conservation & Production Research Laboratory, Bushland, TX

Manuel A. Andrade, Research Agricultural Engineer
Oak Ridge Institute for Science and Education, Sponsored by USDA ARS, Bushland, TX

Abstract. Agricultural field water monitoring for irrigation management is still problematic despite the advent of automatic data collecting systems for electromagnetic (EM) soil water sensors. Annual fees for data access can be large, equipment can be expensive, data may not be easily available, and data may not reflect meaningful soil water content values that can be the basis for irrigation decisions. A new system based on open source software and the Arduino hardware platform for datalogging, telemetry and transmission of data to the Internet Cloud has been coupled with accurate, low power EM soil water sensors based on TDR technology to form a complete system for accurate, low cost field soil water content monitoring with Cloud data access. The new wireless sensor network (WSN) system consists of two devices, a wireless node and a wireless gateway, that are installed in weather-proof enclosures in the field. The node collects data from soil water sensors and transmits it wirelessly using the LoRa protocol to the gateway. The gateway collects data from multiple nodes and then retransmits it to a URL on the Internet using a cellular modem. The ARS Irrigation Scheduling Supervisory Control and Data Acquisition (ISSCADA) system for center pivot variable rate irrigation control was updated to receive soil water data wirelessly and to display it in a graphical user interface. The ISSCADA system also can use the soil water data to inform decisions that are automatically presented daily to the irrigator as a prescription map that can be accepted and downloaded to a center pivot control panel with a mouse click. Both the WSN and ISSCADA systems are scheduled for release as commercial products in 2019.

Keywords: Internet of things, wireless sensor network, Internet Cloud, node, gateway, datalogger

INTRODUCTION

The irrigation scheduling supervisory control and data acquisition (ISSCADA) system patented by USDA ARS (Evett et al., 2014) employs wireless sensor networks (WSN) to gather data on crop canopy temperature using infrared thermometers (IRT) mounted on a moving center pivot lateral and in the field, soil water sensors at select locations in the field, and weather instruments (relative humidity, air temperature, wind speed and solar irradiance). It calculates the crop water stress index (CWSI) on a 1-min or 5-min interval using the weather and canopy temperature data. The CWSI data for daylight hours between approximately 2 h after sunrise to 2 h before sunset are integrated into a single integrated crop water stress index (iCWSI) for each measurement zone in the field. Maps of canopy temperature and iCWSI are made available for visual inspection, and algorithms are applied to the iCWSI data combined with the soil water data to determine where in the field to irrigate and how much. With a mouse click, the irrigation manager may upload the resulting prescription map to the center pivot control panel to direct irrigation.

An effective and relatively inexpensive IRT wireless sensor network was conceived and tested by O'Shaughnessy et al. (2013) and later commercialized through a cooperative research and development agreement (CRADA) (model SAP-IP-IRT, Dynamax, Inc., Houston, TX). And, accurate, low-power and relatively low cost soil water content sensors based on time domain reflectometry (TDR) were developed and commercialized through another CRADA (model TDR-315L and TDR-310S, Acclima, Inc., Meridian, ID), becoming an ASABE 50 Award winner. However, the soil water sensor data were telemetered using rather expensive wireless dataloggers, making the soil water WSN costly and thus effectively limiting the number of field locations that could be monitored.

In 2017, Schomberg et al. (2017) demonstrated a low-cost wireless node and gateway system based on the Arduino platform. The node functioned as a datalogger capable of acquiring data from sensors that employed the SDI-12 wired communications protocol, then storing the data in an on-board flash drive. The data were periodically transmitted to the gateway using the LoRa long-range, low power radio protocol. A gateway could accept data from multiple nodes. In field situations a single gateway at the edge of a field would acquire data from multiple nodes with each node capable of handling eight sensors. Data were downloaded from the gateway to a mobile phone, tablet or other Bluetooth capable device using Bluetooth wireless communications. This system was successfully deployed in 57 fields in four Southeastern seaboard states (Fig. 1). The system's low cost and effectiveness appeared ideal for use with the ISSCADA system. Schomberg developed a CRADA to transfer the technology to a commercial manufacturer, while simultaneously providing for gateway communications to the Internet Cloud using a cellular modem in place of the Bluetooth radio.

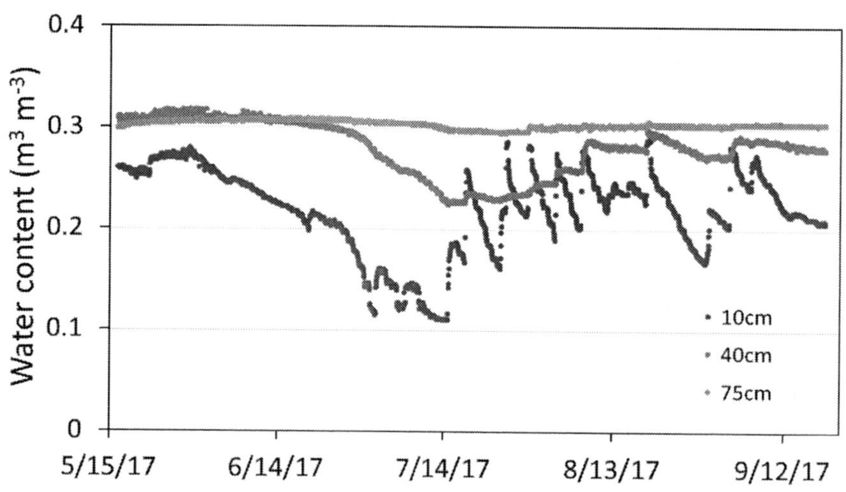

Figure 1. Example soil volumetric water content data from 2017 field experiment with no cover crop (Schomberg et al.). Sensors were placed at 10-, 40- and 75-cm depths.

METHODS and MATERIALS

The main hardware components of the experimental Arduino-based node and gateway system cost approximately $190.00 for the gateway and $65.00 for a node (Table 1) (Thompson et al., 2018).

Table 1. Main hardware components.

Part	Description	Manufacturer
MoteinoMEGA with LoRa transceiver (915 MHz)	Microcontroller	LowPowerLab (Canton, MI)
microSD Transflash Breakout	microSD slot	SparkFun Electronics (Niwot, CO)
PCF8523 Real-Time Clock (RTC) Breakout Board	Clock (node)	Adafruit Industries (New York, NY)
DS3231 Precision RTC Breakout Board	Clock (gateway)	Adafruit Industries (New York, NY)
FONA 3G Cellular + GPS Breakout	Cellular module	Adafruit Industries (New York, NY)

The 2018 version of the Arduino-based node and gateway are illustrated in Figure 2. The cellular modem transmits data to an FTP site using an Internet-of-Things (IOT) platform from MathWorks, Inc. known as ThingSpeak (https://thingspeak.com/).

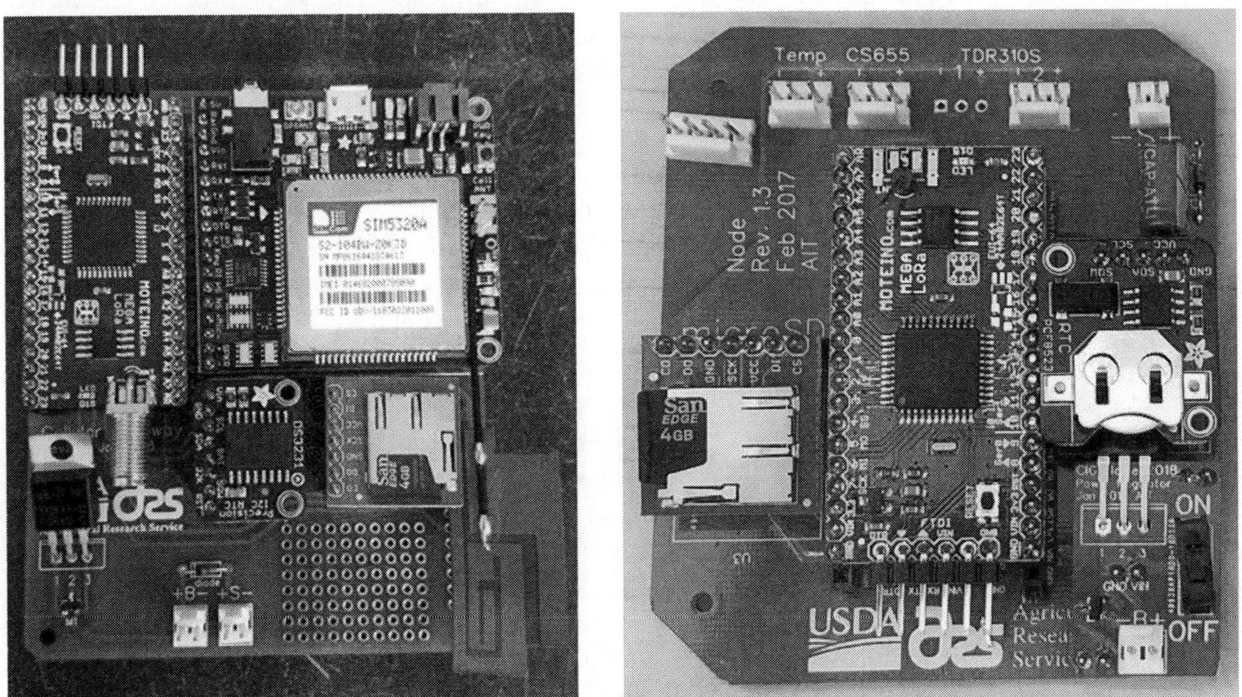

Figure 2. The Arduino-based gateway with cellular modem and LoRa transceiver (left) and node with LoRa transceiver (right). (Photos by A. Thompson)

Data can be pulled from the MathWorks URL and visualized. Results were not available at the time of writing.

RESULTS and FUTURE GOALS

The node and gateway system with cellular modem uplink to the Cloud was tested by Thompson et al. (2018), and was deployed in 2018 at four field locations in Uzbekistan (data not shown). A prototype commercial node has been developed (Fig. 3) as well as a prototype commercial gateway. The design includes a watertight plastic enclosure with integrated solar panel for a stand-alone product.

Figure 3. A prototype commercial node with lithium-ion battery showing front and back sides of the circuit board (left) and the enclosure with integrated solar panel (right). (Photos by A. Thompson)

The upcoming commercial products are intended for use with the model TDR-315, TDR-315L and model TDR-310S TrueTDR soil water sensors from Acclima, Inc. (Fig. 4). These sensors utilize TDR to capture a TDR waveform (Evett, 2003) and analyze it for pulse travel time, which is related nonlinearly to the soil apparent permittivity and practically linearly to the soil volumetric water content. As implemented in these sensors, the TDR method is not strongly influenced by soil temperature and bulk electrical conductivity (BEC) changes, including those related to water content (BEC increases with water content), temperature (BEC increases with temperature) and clay content and type (BEC increases with soil content of most clay types). Testing in clayey soil has verified the efficacy of the TDR method implemented in these sensors (Schwartz et al., 2016). Plans for field testing of the commercial node-gateway system include locations in several U.S. states and a watershed in Jordan where the cellular data uplink has potential to greatly reduce travel time and increase data availability to researchers and watershed managers.

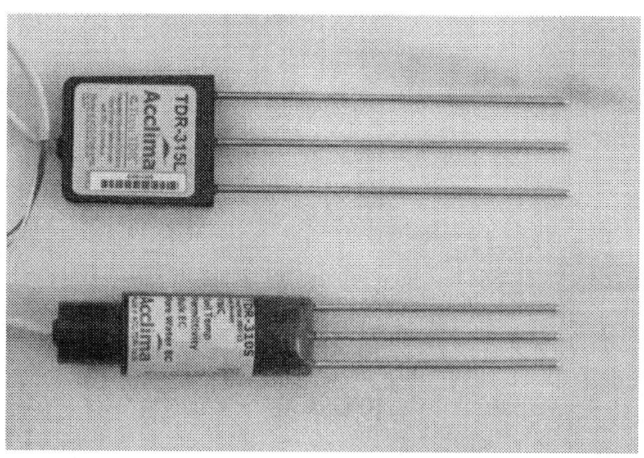

Figure 4. Acclima model TDR-315L (top) and model TDR-310S (bottom) sensors for soil apparent permittivity, bulk electrical conductivity, temperature and volumetric water content.

The commercial node-gateway system will also be tested with the ARS ISSCADA system for center pivot irrigation decision support. The ISSCADA system utilizes a client-server software architecture (Andrade et al., 2016, 2017) that was modified to include a wireless soil water sensing network. The GIS-based graphical user interface displays icons showing the locations in the field of the soil water sensor nodes and the relative depletion levels at each location are indicated by icon color (Fig. 5). Clicking on a particular icon brings up a display of the soil water content for all sensors connected to the corresponding node (Fig. 6). This was used in 2017 experiments in an irrigation decision support algorithm that included the iCWSI and soil water content (O'Shaughnessy et al., 2018, this proceedings). In beta tests at a humid and two sub-humid locations and a semi-arid location, crop yields for cotton and corn managed with the modified ISSCADA system were similar to those obtained using local irrigation scheduling practices. Also, corn water use efficiencies were similar among irrigation management methods. The irrigation water use efficiency was greater for cotton produced with the ISSCADA system in a sub-humid climate.

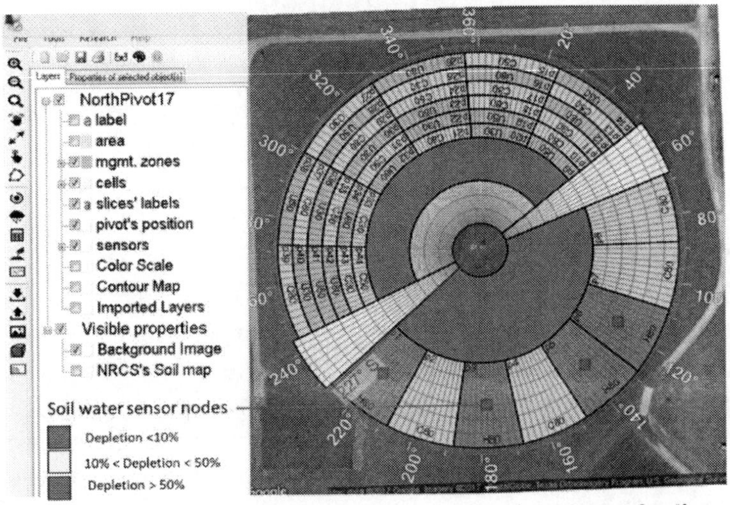

Figure 5. A screenshot of the ISSCADA software interface showing locations of soil water sensor nodes and the relative level of soil water depletion. Also shown are management zones for an experiment contrasting VRI zone control (upper left half of pivot) with sector control (lower right half of pivot).

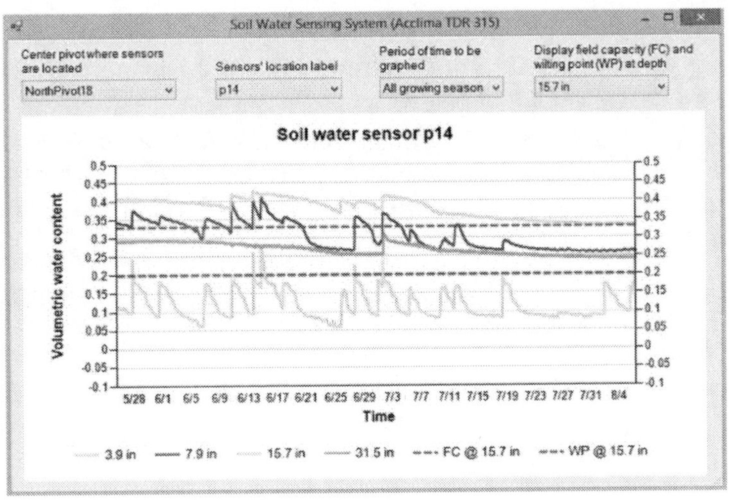

Figure 6. A screenshot of the ISSCADA software interface showing the time series of volumetric soil water content for each sensor at that node and the depth of the sensor.

SUMMARY

The Internet-of-Things (IoT) approach so sensor systems is already being applied in many industrial settings and increasingly for agricultural field operations (e.g., Kohanbash et al., 2013). The LoRa based node and gateway system for soil water sensor data acquisition and wireless telemetry described here provides an effective, low-cost, solar-powered solution for delivering data to the Internet Cloud at a URL that can be accessed by anyone with access rights. For irrigation decision support systems such as ISSCADA, this provides a data access solution that fits well with the underlying wireless in-field and multiple field communications concept that allows user interaction with a data-laden interface on a remote cellular telephone, tablet or other computer that communicates with a single or with multiple irrigation systems for both control and data acquisition.

REFERENCES

Andrade, M.A., O'Shaughnessy, S.A., and Evett, S.R. (2016). A GIS-based Decision Support Tool for Center Pivot Irrigation Systems. *In Proc. 2016 ASABE Annual International Meeting*, St. Joseph, MI.

Andrade, M.A., O'Shaughnessy, S.A., and Evett, S.R. (2017). ARSPivot- A sensor based decision support tool for the integrated irrigation management of VRI center pivot systems. In Proceedings of the 28th Annual Central Plains Irrigation Conference, Burlington, CO. February 21-22, 2017.

Evett, S.R. Soil Water Measurement by Time Domain Reflectometry. 2003. In B.A. Stewart and Terry A. Howell (editors). Encyclopedia of Water Science, Marcel Dekker, Inc. New York. Pp. 894-898.

Evett, S.R., S.A. O'Shaughnessy and R.T. Peters. 2014. Irrigation Scheduling and Supervisory Control Data Acquisition System for Moving and Static Irrigation Systems. United States Patent No.: US 8,924,031 B1. Issued December 30, 2014.

Kohanbash, D., G. Kantor, T. Martin and L. Crawford. 2013. Wireless sensor system design for monitoring and irrigation control: User-centric hardware and software development. HorrTechnology 23(6): 725-734.

O'Shaughnessy, S.A., S.R. Evett, P.D. Colaizzi and T.A. Howell. 2013. Wireless sensor network effectively controls center pivot irrigation of sorghum. *Appl. Engr. Agric.* 29 (6), 853-864. doi 10.13031/aea.29.9921

O'Shaughnessy, S.A., M.A. Andrade, K.C. Stone, E.D. Vories, R. Sui, and S.R. Evett. 2018. Adapting a VRI irrigation scheduling system for different climates and fields. Irrigation Association Technical Conference, Dec 7-10, 2018, Long Beach, CA.

Schomberg, H.S., A. Thompson, S.R. Evett and D.K. Fisher. 2017. Arduino based gateway-node system for soil water measurement in on-farm research. Presented at the 2017 ASA-CSSA-SSSA Meeting – Managing Global Resources for a Secure Future. Oct. 22-25, 2017, Tampa, FL. https://scisoc.confex.com/crops/2017am/recordingredirect.cgi/id/31692

Schwartz, R.C., S.R. Evett, S. Anderson and D. Anderson. 2016. Evaluation of a direct-coupled TDR for determination of soil water content and bulk electrical conductivity. Vadose Zone J. 15(1)2016. doi: 10.2136/vzj2015.08.0115.

Thompson, A., H. Schomberg, S. Mirsky, C. Reberg-Horton, B. Davis and S. Evett. 2018. On-farm gateway-node system for monitoring soil water content and soil temperature. Poster presented at 2018 ASA-CSSA-CSA Meeting - Enhancing Productivity in a Changing Climate, Nov. 4-7, 2018, Baltimore, MD.

How to Determine the Type of VRI Best Suited for My Field

Jake LaRue P.E., Valmont Irrigation, jlarue@valmont.com

Abstract.
The center pivot has been used by a farmer/operator to apply a selected depth of water uniformly across the entire field. Currently several options are available from OEMs and third party's providing the grower the ability to apply different amounts of water and crop production products to defined management zones along the pivot's path and/or sectors around the field. The paper will discuss ways how to evaluate a field and determine which type of VRI, if any, will provide the best economic return for the field. Considered will be Speed Control, Zone Control and Individual Sprinkler Control. The paper will also review potential payback of several fields. It will close with a discussion of future needs for variable rate irrigation

Keywords. Center pivot, precision irrigation, VRI, variable rate irrigation, site specific irrigation

Introduction.

Since the introduction of center pivot irrigation in the mid 1960's, the goal has been to achieve uniform application of water both along the center pivot and in the direction of travel. In a field with uniform soil type, topography, tillage and crop health, this offers a good solution to the producer to maximize profitability. But, the agriculture community has since recognized that fields and crops are not necessarily uniform spatially or temporally. Machines to variably apply fertilizer, seed and other crop production products by management zones have become common in use. The major center pivot / linear irrigation manufacturers have followed to offer a range of hardware / software packages to meet the grower's interest in VRI. In addition some third parties offer VRI hardware and software. The available VRI packages fall into two basic categories:

- VRI speed: Varying the speed of the center pivot around the field
    - Application depth can be varied in the direction of travel by changing the speed of the center pivot
    - Application depth is uniform along the length of the center pivot
    - Overall flow remains constant through the center pivot (excluding the endgun)
- VRI zone: Controlling individual or groups of sprinklers.
    - Application depth can be varied in the direction of travel.
    - Application depth also can be varied along the length of the center pivot.
    - Overall flow varies as sprinklers are pulsed on and off in the direction of travel or along the center pivot.

It was stated in 2011 documented and proven water conservation strategies using variable rate irrigation are quite limited, and its cost-effectiveness has not been demonstrated by researchers (Evans 2011) in the last six years this has changed some. However many growers continue to look closely at VRI technology to determine the value it could bring to their operation. The potential values of VRI commonly considered are saving water, saving energy and improving yield but really the focus needs to be on overall field profitability. The questions still remains what tools are available to help me determine if VRI is a good fit for a particular field and the type of VRI which is best suited?

Discussion

As suggested by LaRue 2012, the solution for VRI decision making is a process. Based on recent experience this still seems to be true in 2018. The following is a process to help with the decision making by helping to understand the variability in the field. However if looking at adding VRI to an existing center pivot it is strongly recommended before you start to check the age and performance of your sprinkler package. You must start from a point of uniformity.

The first step is for the grower is to determine what he wants by answering a question – what does he want to accomplish? This is often done working with his irrigation dealer or farm consultant.

- Vary the application depth by management zone?
- Shut off irrigation completely for particular non-crop areas, such as ponds, drainage ditches, roads, etc.?

The second step is if the grower wants to vary the application depth by management zone is to determine sources of available data. The following is a manual way of doing. Your irrigation dealer or consultant can help with this. The data may include but is not limited to data from sources such as USDA NRCS soil maps, electro conductivity (EC_a) field surveys, yield maps and/or aerial images.

In the United States one of the easiest sources of data is NRCS USDA Web Soil Survey with an example represented in a field (Fig 1).

Mason County, Illinois (IL125)

Map Unit Symbol	Map Unit Name	Acres in AOI	Percent of AOI
53B	Bloomfield sand, 1 to 7 percent slopes	2.9	0.7%
54B	Plainfield sand, 1 to 7percent slopes	4.0	1.0%
54D	Plainfield sand, 7 to 15 percent slopes	3.8	0.9%
54E	Plainfield sand, 15 to 30 percent slopes	13.3	3.3%
88B	Sparta loamy sand, 1 to 7 percent slopes	14.1	3.5%
88D	Sparta loamy sand, 7 to 15 percent slopes	56.6	14.1%
98B	Ade loamy fine sand, 1 to 7 percent slopes	6.3	1.6%
125	Selma clay loam	116.4	29.0%
131C2	Alvin fine sandy loam, 5 to 10 percent slopes, eroded	1.9	0.5%
150B	Onarga sandy loam, 2 to 5	4.9	1.2%

Figure 1. Graphical and tabular representation of the different soil types and area present in one field.

Continuing the second step, using the soil data, texture and area, one then can start to evaluate What needs to be done is to look at the soils and the area covered for both the finest textured and then the coarsest. These typically are the clays, silts and the sands and record their name, texture and area. One way to determine potential yields is to use crop productivity information. Since this field is in Illinois can use the Optimum Crop Productivity Ratings for Illinois Soil, (Olson 2000) (Figure 2). Other examples are Iowa Corn Suitability Rating (CSR/CSR2), Productivity Index (PI) in North Dakota, South Dakota and Minnesota and Nebraska Soil Rating for Pant Growth (SRPG) now replaced by NCCPI, National Commodity Crop Productivity Index. Other states have or are adding types of rating systems. Then look up the soil textures in fig 2, in the guide for soils with optimum management for the crop of interest, in this case corn and record the yield. One must assume most of the yield difference is attributable to lack of soil moisture.

Table S2. Productivity of Illinois Soils Under Optimum Management, Slightly Eroded, 0% to 2% Slopes

IL map symbol	Soil type name	Subsoil rooting[a]	Corn bu/ac	Soybeans bu/ac	Wheat bu/ac	Oats[b] bu/ac	Sorghum[c] bu/ac	Alfalfa[d] hay ton/ac	Grass-legume[e] hay ton/ac	Crop productivity index for optimum management	IL map symbol
109	Racoon silt loam	FAV	130	41	51	0	103	3.50	0.00	106	109
111	Rubio silt loam	FAV	139	44	57	70	0	0.00	4.29	114	111
112	Cowden silt loam	FAV	143	45	57	0	107	0.00	4.41	117	112
113	Oconee silt loam	FAV	148	45	57	0	107	0.00	4.75	119	113
115	Dockery silt loam	FAV	156	51	62	77	0	0.00	4.52	128	115
116	Whitson silt loam	FAV	142	45	54	68	0	0.00	4.29	116	116
119	Elco silt loam	FAV	136	45	53	68	0	3.84	0.00	112	119
120	Huey silt loam	UNF	98	38	38	0	86	0.00	3.16	89	120
122	Colp silt loam	UNF	121	38	51	64	0	0.00	3.84	98	122
123	Riverwash	Crop yield data not available									123
125	Selma loam	FAV	157	51	62	80	0	0.00	4.75	129	125

Table S2. Productivity of Illinois Soils Under Optimum Management, Slightly Eroded, 0% to 2% Slopes

IL map symbol	Soil type name	Subsoil rooting[a]	Corn bu/ac	Soybeans bu/ac	Wheat bu/ac	Oats[b] bu/ac	Sorghum[c] bu/ac	Alfalfa[d] hay ton/ac	Grass-legume[e] hay ton/ac	Crop productivity index for optimum management	IL map symbol
764	Coyne fine sandy loam	FAV	128	42	53	63	0	3.28	0.00	105	764
765	Trempealeau silt loam	FAV	136	46	54	70	0	3.28	0.00	113	765
767	Prophetstown silt loam	FAV	171	53	63	85	0	0.00	4.75	138	767
768	Backbone loamy sand	FAV	103	36	43	49	0	0.00	3.28	87	768
769	Edmund silt loam	UNF	106	37	50	58	0	2.60	0.00	89	769
770	Udolpho loam	FAV	124	41	50	66	0	0.00	3.73	103	770

Figure 2. Information from the Optimum Crop Productivity Ratings

Step three - For map symbol 125, the soil is Selma clay loam; from this, the estimated corn yield is 157 bu/ac. For map symbol 770, Udolpho fine sandy loam is the soil type and the estimated corn yield is 124 bu/ac. Knowing the area of each soil type, the potential impact of VRI to minimize under and/or overwatering can be estimated.

If the assumption is the lower yield of the Udolpho soil is primarily caused by under watering, and the available water for irrigation is not limited, there is a potential yield increase between Udolpho soils and others of 33 bushel on approximately 78.2 acres for a total of 2,737 bushels. For example, if the farmer uses VRI zone or individual sprinkler control, and if the price per bushel is $3.50, then his potential increase in income could be about $9,032. If the farmer uses VRI speed control, and he may experience a field corn yield increase and his potential increase in income could be $1,689.

By looking at the soil map, a "guess" can be made as to whether VRI speed will provide much benefit. Speed control does not appear to offer much potential. VRI zone or VRI individual sprinkler control would be the better choice for the field. Using a quote from you irrigation dealer you are in a position to estimate the ROI.

VRI Control Type		Years for Payback
Speed	assumes have to replace panel	2.9
Zone	assumes only need to add VRI hardware	3.3
Individual Sprinkler	assumes only need to add VRI hardware	4.6

These numbers reflect an assumption water is the main limiting factor, which is a simplistic approach.

As mentioned previously, the agriculture community recognizes that fields and crops are not uniform. Yield maps provide information on both a field's yield and the location in the field where yield variability occur. Figure 3 is the same field as shown in Figure 1. Yield maps help confirm soil information, as well as help to make decisions regarding the type of VRI package in which to invest, particularly if it is believed an area of the field may be improved and not under or overwatered.

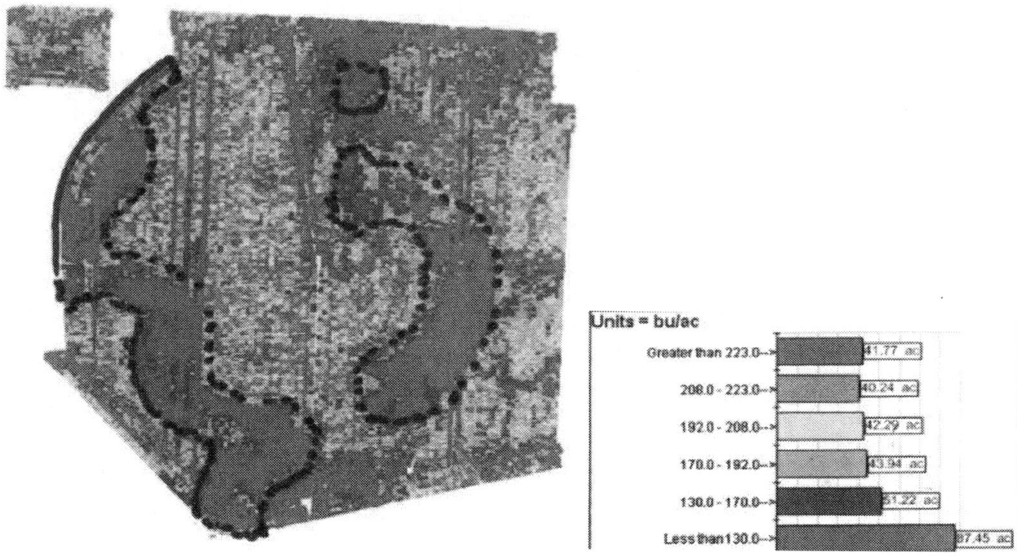

Figure 3. Yield map for field in figure 1

The yield map seems to confirm the assumptions and selections both the area more than estimated and the yields.

Figure 4 illustrates another example where the determination for VRI can be made visually via soil data.

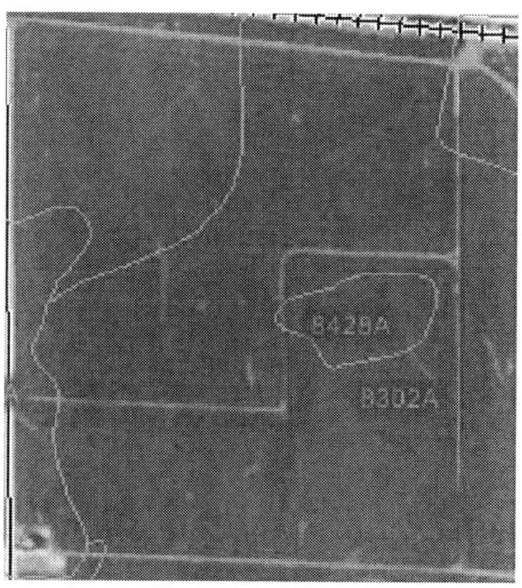

Figure 4 Soil Map

The primary soils in this field are 8428A and 8070A, which are Ambraw clay loam and Beaucoup silty clay loam, respectively. The soils map does not show significant variability. Without a crop productivity guide or a yield map to help show specific variability or particular topographic features and the grower indicates only an 8 bu/ac variation in the Ambraw clay loam , the conclusion can be drawn that VRI may not be a good fit for this field. The ROI could be ten years or more, and, in most cases, would be hard to justify based on current economics. About 50% of the time it has been found the ROI for VRI does not meet the grower's requirements

Another approach is using automated generated prescriptions based on USDA SSURGO data as shown in (fig 5). If the assumption is the lower yield of the coarser texture soil is primarily caused by under watering, and the available water for irrigation is not limited, there is a potential yield increase between the coarser and fine textured soils. This is automatically calculated to be 15 bushel/acre on approximately 64.3 acres for a total of 965 bushels of corn. For example, if the farmer uses VRI zone it is estimated to be an additional $3,376 income or if individual sprinkler control an additional $3,554 assuming the price per bushel is $3.50. For this case due to the way the soils are aligned in the field VRI speed control is a good choice and may experience a corn yield increase and his potential income could be $2,701. The same cost assumptions apply has in the first example.

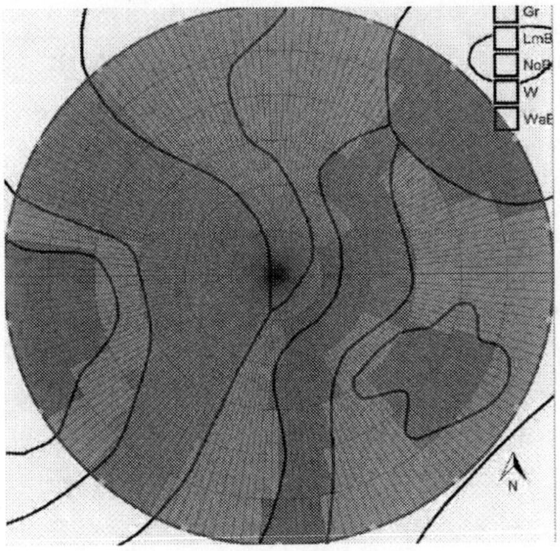

Figure 5 Automated Prescription

VRI Control Type	Years for Payback
Speed	1.8
Zone	8.8
Individual Sprinkler	11.8

A number of other options are available which automatically quantitatively evaluate a field's response to VRI. Consulting groups, such as CropMetrics™ and others offer tools using EC_a data, topographic and other sources of data to develop an estimate of the variability in the field. The assumption is if water can be applied to match the field's variability, then crop yields can be optimized and profits maximized.

Step four is to determine if the grower wants to, or should, proceed with a VRI package, and which VRI package is suited for his operation. This determination is evaluated via the information that has been collected, as well as the potential ROI.

If the farmer is water constrained, then the analysis will be different and the focus will be applying the optimum amount of irrigation on the management zones depending on the yield potential. In this case, more observation needs to be completed, so that better tools are made available to help evaluation the situation.

At the beginning of step one, the decision was made if the grower was looking to:

- Vary the application depth
- Shut off irrigation for particular areas, such as ponds, drainage ditches and other non-crop areas

During to the discussion and decisions for varying application depth, it has been determined that the grower should shut off irrigation for non-crop areas.

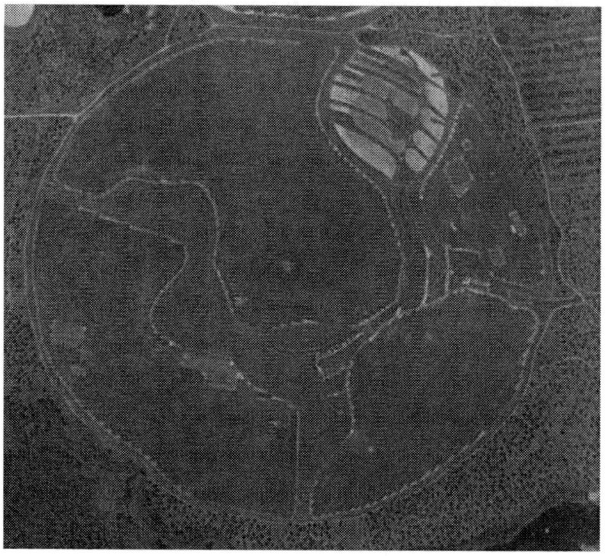

Figure 6. Aerial image of field

For cases such as this, an aerial image can typically be used to estimate the type of VRI package that is most suitable for the field. Either conventional aerial photographs, soil maps or other (Fig 6) are generally sufficient to make a determination of how to proceed along with the farmer experience. This may be utilization of VRI zone control, individual sprinkler control or other center pivot's existing features, if the pivot uses a computerized control panel. For these types of fields, VRI speed control will not generally meet the requirements. In this particular example to ensure no irrigation for non-crop areas individual sprinkler or zone control is best suited

The potential yield improvement is different for every field, as well as for every crop. Therefore, it is important to consider using your irrigation dealer, local consultant or advisor (assuming they have VRI experience) to help analyze each particular situation.

Conclusion

The use of VRI by farmers seems to have plateaued. Farmers who are considering VRI should follow a process to help determine the ROI of the various VRI options or if VRI is even the optimum solution. Farmers can use data to make manual estimates and/or work with their irrigation dealer and/or a consultant who can prepare a quantitative report on variability based on EC_a, topographic, soils, yield and/or other data. With either, ROI estimates can be made. The steps to consider are:

- Determine whether want to vary the application depth for management zones scattered across the field or shutoff the irrigation for particular non crop areas.
- Decide to use either a manual or a service with an automated process to determine the type of VRI or use a consultant and software to utilize geo-referenced data
- Collect field information such as (but not limited to) soil, yield and EC_a maps and make analysis of ROI
- Make a determination of what type of VRI, if any, will best meet your particular field situation

Experience indicates the ROI of using VRI speed control can range from one to five years. Many of the center pivot control panels sold since 2010 include speed control as a function in the panel. The ROI of VRI zone control can range from three to twenty years. Also it is critical to remember the base prescription is a starting point and if varying water by management zone will need to be changed during the growing season.

Future work suggested is for additional information is needed to be supplied by states/regions to help estimate yields based on soil types and how limiting is water to yield. More observation needs to be completed to better characterize the impact of VRI with various soils, topographies, climates and other factors. Considerations also need to be made around limitations to available irrigation water. With this data and these tools, a better decision as to whether a farmer should use VRI speed, zone or individual sprinkler control can be made, as well as an estimate of the potential return on investment.

References

CropMetrics, 2011, North Bend Nebraska, http://cropmetrics.com

Evans, Robert; LaRue, Jacob; Stone, Kenneth; King, Bradley, 2011, Adoption of Site-Specific Variable Rate Sprinkler Irrigation Systems, Irrigation Association 2011 Expo

LaRue, Jake, 2012, Decision Process for the Application of Variable Rate Irritation, ASABE 12-1337451, ASABE Annual meeting, Dallas Texas

LaRue, Jake; Evans, Robert. 2012 Considerations for Variable Rate Irrigation for Central Plains Conference. Proceedings of the 24th Annual Central Plains Irrigation Conference. Colby KS

NRCS USDA. Web Soil Survey. Washington, D.C.: USDA. http://websoilsurvey.nrcs.usda.gov

Olson, K.R. Optimum Crop Productivity Ratings for Illinois Soil: 2000. University of Illinois. Urbana-Champaign, Illinois

Adapting a VRI Irrigation Scheduling System for Different Climates

S.A. O'Shaughnessy, M.A. Andrade, K.C. Stone, E.D. Vories, R. Sui, and S.R. Evett

Abstract. An irrigation scheduling supervisory control and data acquisition (ISSCADA) system was developed by scientists at USDA-ARS in Bushland, Texas. The system was comprised of hardware and software components for control and decision support for center pivot variable rate irrigation (VRI) management and designed for use in a semi-arid region. Multiple years of field studies have demonstrated that the ISSCADA system produced yields and water use efficiencies similar to or better than irrigation scheduling based on weekly neutron probe readings. However, the system had not been tested in other climates. Beta tests were conducted at USDA-ARS laboratories located in sub-humid and humid climates by outfitting VRI center pivots with the ISSCADA system. The goals of the beta tests were to demonstrate that irrigation management with the ISSCADA system produced crop yields and water use efficiency (WUE) or irrigation water use efficiencies (IWUE) generally comparable to or better than those obtained when scheduling irrigations according to best local management practices. Results demonstrated that crop yields for corn and cotton managed with the ISSCADA system were similar to those produced using local irrigation scheduling methods. Water use efficiencies for corn were also similar among irrigation management methods, while IWUE was greater for cotton produced with the ISSCADA system in a sub-humid climate.

Keywords: *Center pivot, crop water stress index, GIS-database, plant feedback, soil water sensing feedback, wireless sensor network systems*

Introduction

The irrigation scheduling supervisory control and data acquisition (ISSCADA) system has a hybrid architecture of hardware and software components for control and decision support for center pivot variable rate irrigation (VRI) management (Andrade et al., 2015; 2016 and 2017). The system was patented at Bushland, Texas (Evett et al., 2014) and uses plant feedback as the primary algorithm for irrigation scheduling. The plant feedback method is based on the theoretical crop water stress index (CWSI) derived by Jackson et al. (1981). The ISSCADA system uses varying levels of pre-established thresholds of an integrated CWSI (iCWSI) to determine irrigation timing and amount (O'Shaughnessy et al., 2013 and 2015). The iCWSI is the summation of the CWSI at one-minute or five-minute time steps during pre-selected daylight hours. The ISSCADA system can also incorporate feedback from soil water sensing stations to ensure closed-loop control and prevent over- and under-irrigation (Figure 1).

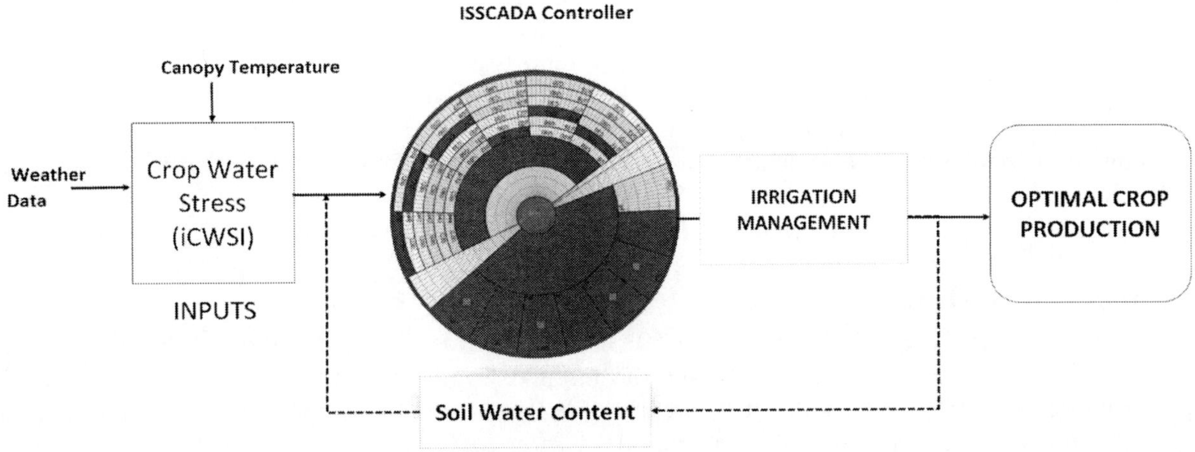

Figure 1. Graphical depiction of the Irrigation Scheduling Supervisory Control and Data Acquisition system showing roles of wireless sensor networks in feedback for optimal crop production.

The core hardware components of the ISSCADA system included wireless sensor network systems for monitoring canopy temperature (infrared thermometers- IRTs, Dynamax Inc., Houston, TX), weather, and soil water content (optional). Data from the wireless network systems and the center pivot are collected at an embedded computer at the pivot point for centralized data assembly, information processing, dynamic prescription map building and site-specific sprinkler irrigation control. The ISSCADA software is of a server-client architecture. The server software is GIS-based and was developed by USDA-ARS (Andrade et al., 2017). Because the ISSCADA system was developed in a semi-arid climate, at the Conservation and Production Research Laboratory in Bushland, TX. The initial operation of the system and some of the ancillary hardware were adapted for VRI center pivot sprinkler systems designed to reduce water losses due to evaporation. For example, the sprinkler spacing on these center pivots is typically 5 ft apart and the application method is low elevation spray application (LESA) with nozzles approximately 1.5 ft above the ground or low energy precision application (LEPA) drag socks (Lyle and Bordovsky, 1983), where the double open-ended socks are dragged in the furrows. The radii of throw for LESA and LEPA sprinklers is < 12 ft. enabling light-weight brackets to be mounted on the center pivot truss rods to hold the wireless network of IRTs forward of the irrigation spray, viewing a dry canopy. The LESA and LEPA application methods allow prescription maps to be built from data collected while the ISSCADA system is irrigating as well as when it runs dry. Furthermore, the two wireless IRTs required in each control zone were located at the borders, facing inwards towards the center of the VRI management zone (MZ) at an oblique angle. The number of MZs and sensors in each MZ was determined by the user.

Beta testing of the ISSCADA system began at two USDA-ARS sites in addition to the ARS Conservation and Production Research Laboratory (CPRL) in Bushland in 2016. During this first year, only the plant feedback method was used to schedule irrigations. In 2017, soil water sensing stations were installed and used in the ISSCADA system in Florence and Bushland. Soil water sensors were installed in 2017 in Portageville but not incorporated into the ISSCADA system for irrigation management. At Stoneville, the ISSCADA system was delayed from being fully operational until the diesel-powered system was converted to electrical and the CAMS panel was powered continuously from the grid. All three of the locations were located east of Bushland, Texas in climates that are more humid than a semi-arid climate. One of the major concerns of using the ISSCADA system to schedule irrigations in humid climates is that high moisture content in the air can reduce crop transpiration and increase canopy temperature independently of soil moisture (Wanjura and Upchurch, 1997) resulting in false positive irrigation signals leading to over-irrigation. This paper addresses the adaptations made in the operation and architectural design of the ISSCADA system to accommodate the differences in center pivot sprinkler design, irrigation practices, and climate and crop choice, and reports preliminary results in crop yield and irrigation water use efficiency.

The goals of the beta tests were to demonstrate that irrigation management with the ISSCADA system produced crop yields and water use or irrigation water use efficiencies generally comparable to or better than those obtained when scheduling irrigations according to best local management practices.

Materials and Methods

A mid-maturing corn hybrid (Dekalb DKC64-69 GENVT3P) was managed at the Coastal Plain Soil, Water and Plant Conservation Research Center in Florence, South Carolina (a humid climate). Treatments were irrigation scheduling with soil water potential (SWP) measurements, the ISSCADA system and rainfed treatment plots, each with eight replications (Figure 2a). Cotton, variety PHY 333WRF, was grown at the University of Missouri Fischer Delta Research Center Marsh Farm at Portageville, Missouri (a sub-humid climate) under a 3-tower VRI center pivot using the Arkansas Irrigation Scheduler (AIS) method, the ISSCADA system, and rainfed treatment plots, each with four replications (Figure 2b). At Florence and Portageville, the goal was to maximum crop yields. At Bushland, a mid-season corn hybrid (Pioneer P1151AM) and a short season corn hybrid (P0157AM) were managed under a three- and a six-span VRI center pivot system, respectively. Yields and water use efficiencies from the ISSCADA treatment plots were compared with treatment plots irrigated using weekly neutron probe readings and refilling a percentage of soil water depletion to field capacity. At this location, the ISSCADA system was used to manage both speed-control VRI to maximize yields and zone-control VRI to control irrigation levels.

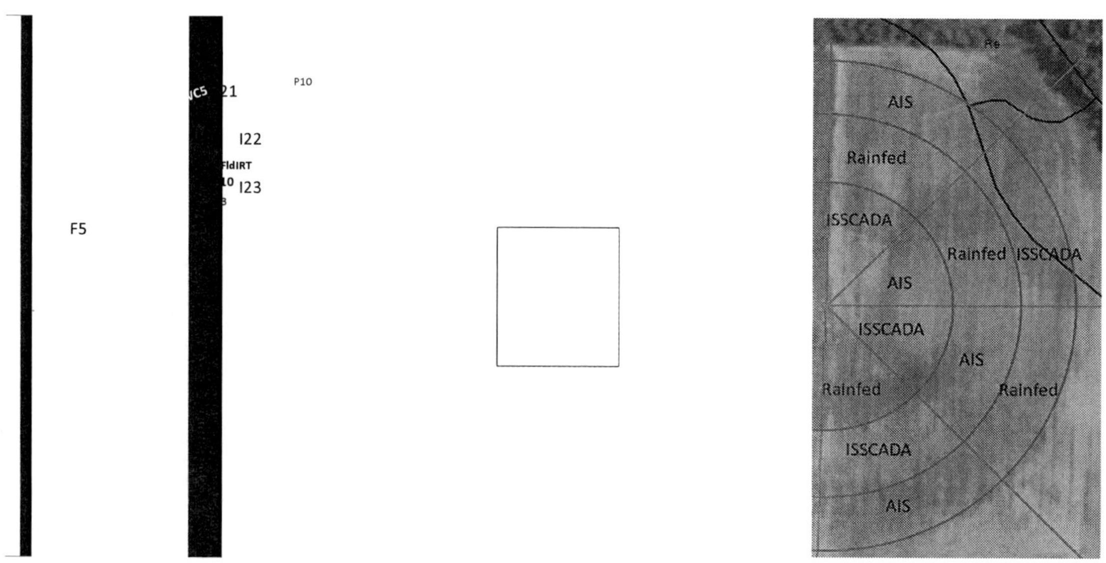

(a) (b)

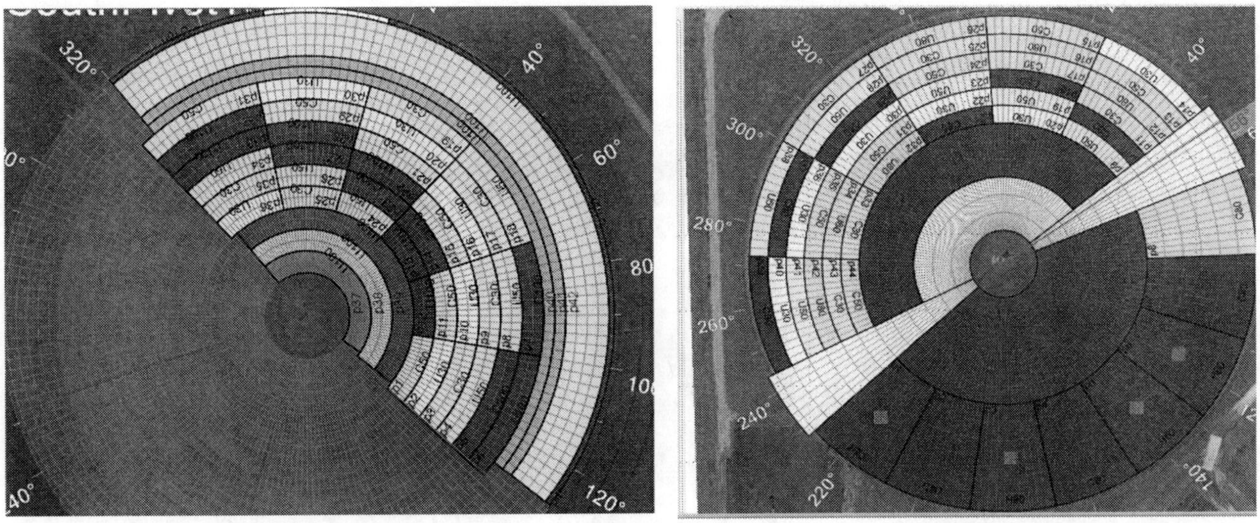

(c) (d)

Figure 2. Plot plans for the beta testing performed on the ISSCADA system at three USDA-ARS laboratories: (a)Florence, South Carolina in 2017, (b) Portageville, Missouri in 2016, (c) six-span at Bushland in 2016, and (d) three-span at Bushland, Texas in 2017.

Each ISSCADA system had a minimum of 2 IRTs located in each control zone, and two IRTs located in a well-watered area of the field. The IRTs on the center pivot pipelines in Florence and Portageville were mounted on steel brackets fabricated by Valmont Industries that extended approximately 15 ft beyond the pipeline and clamped the IRTs at an angle 45° from nadir and a 45° horizontal angle relative to the direction of pivot travel. Standalone weather stations using sensors (to measure air temperature, RH, solar irradiance, and windspeed) and dataloggers from Campbell Scientific (Logan, Utah) were in each field in Florence and Bushland. In Portageville, weather data was automatically downloaded from the University of Missouri website before midnight, using a feature in the ARS client-server program. In all locations, each ISSCADA MZ was managed independently of every other ISSCADA MZ within each beta test field.

In the second year (2017), the Portageville study was modified to include 5 replications. In addition to the IRTs, soil water sensors were installed in Florence, Portageville and under the 3-span center pivot in Bushland. Irrigation scheduling with the ISSCADA system in Florence was based on plant feedback and soil water sensing feedback. In Portageville, the ISSCADA method was based only on plant feedback, while in Bushland, treatment methods by the ISSCADA system included both plant feedback and plant and soil water sensing feedback. When soil water sensing information was used by the ISSCADA system, soil water depletion levels were assessed before the calculated iCWSI values were compared with pre-established thresholds. If soil water depletion was less than the minimum threshold, no irrigation was applied. If the maximum soil water depletion threshold was exceeded, a pre-determined irrigation amount was applied. If soil water depletion was between the minimum and maximum threshold values, irrigation was applied based on the plant feedback thresholds. Three iCWSI thresholds with corresponding irrigation amounts were established for each ISSCADA system. An example of the iCWSI and soil water sensing thresholds are shown in Table 1 for Bushland. Pre-determined thermal stress thresholds were crop specific and pre-determined irrigation amounts were region specific and determined after interviewing the users.

Table 1. Integrated CWSI thresholds for VRI zone control irrigation management and irrigation amounts (inches) (top half) and soil water depletion thresholds for VRI speed control for irrigation management of corn hybrid (P1151AM) grown at Bushland, TX.

	2017			
	Northwest half of field			
Zone	Treatment Type	Irrigation Threshold1	Irrigation Threshold2	Irrigation Threshold3
	Plant feedback	100<iCWSI<= 150	150<iCWSI <=250	iCWSI > 250
	C80	0.60	0.80	1.2
	C50	0.38	0.50	0.75
	C30	0.23	0.30	0.45
	Southeast half of the field			
Speed	Hybrid Plots- 1,3,5,6			
	Soil + Plant feedback	If soil water depletion in 1.5 m < 0.10 (in the root zone)	If 0.10 < soil water depletion < 0.5 (in the root zone)	If soil water depletion in 1.5 m > 0.50 (in the root zone)
	Irrigation amount	0.0 inch	Use iCWSI Threshold Levels shown for Plant feedback for C80 Treatment type	1.0 inch
	Plant Feedback Plots – 2,4,7,8			
	C80	100<iCWSI<= 150	150<iCWSI <=250	iCWSI > 250
		0.60	0.80	1.2

Besides adapting irrigation amounts, it was necessary to adjust the operation of the ISSCADA system outfitted onto the VRI center pivots in Florence, Portageville and Stoneville. Sprinkler spacing for the center pivot in Florence is 5ft, while sprinkler spacing for the center pivots in Portageville and Stoneville were 9 ft. apart. Nozzle height for all three sprinklers was at mid-elevation height (approximately 5 to 9 ft. above the ground). This configuration resulted in radii of throw > 12 ft. The brackets holding the wireless IRTs did not extend their field-of-view beyond a wet canopy. This required the ISSCADA system to be run dry across the cropped area of the field before building a prescription map. In addition to differences in sprinkler design, irrigation and agronomic practices differed in these humid and sub-humid regions as compared to the semi-arid climate of Bushland. Smaller irrigation amounts per irrigation event were common in the sub-humid and humid regions because water losses due to evaporation were minimal under conditions of high relative humidity (RH) and lower average wind speeds. Often, irrigation management practices in sub-humid and humid regions are such that depleted soil water in the profile is filled to less than field capacity to leave storage for unpredictable precipitation events and reduce surface runoff (Nguyen et al. 2015).

The ISSCADA system was customized for each center pivot system by inputting information that was specific to the climate, design and operation of the center pivot sprinkler, crop type, planting date, and local irrigation practices. Information regarding the location of each IRT sensor on the pipeline, its identity, the location of each soil water sensing station, its identity, and the depth of each TDR at the location was used to automatically build prescription maps for management by the ISSCADA method.

The USDA-ARS Crop Production Systems Research Unit in Stoneville, Mississippi managed soybean (HBK4955 Libertylink) under a four-span VRI center pivot outfitted with the ISSCADA system. The system integrated both IRTs and TDRs for plant and soil water sensing feedback. Irrigation scheduling with the ISSCADA system was compared with a VRI irrigation scheduling method that employed a static prescription for MZs labeled VRI (Figure 3), whereby irrigations were triggered when soil water depletion reached a pre-established threshold level. The goal was to maximize crop water use efficiency. Soybean crops were planted under the pivot on April 20, 2018. The total precipitation and ET_0 were 644mm and 438mm, respectively, in 2018. Using the ISSCADA system, three dry-scans were performed on June7, June 27, and July 25, respectively. Starting from June 9, seven irrigations were conducted according to the ISSCADA outputs and the soil moisture measurements of a sensor-based irrigation scheduling method. 0.75" water was applied in each irrigation event to the MZs at 100% irrigation rate in VRI treatment. Amount of water applied to the ISSCADA treatment was dynamically determined by the ISSCADA system based on the information from the dry-scan. Crops were harvested on Oct. 4, 2018. As of this date, yield data were not processed. This study will be repeated in 2019.

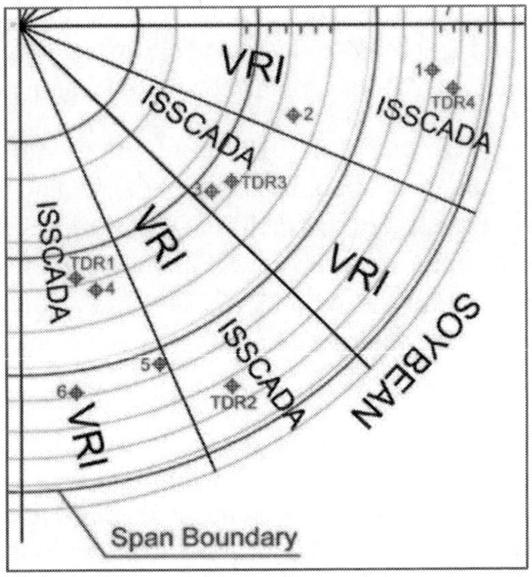

Figure 3. Plot plan and sample prescription map for four-span VRI center pivot located at Stoneville, Mississippi in 2018.

Preliminary Results

Florence, SC

In the first year (2016), the outfitting of the ISSCADA hardware onto the VRI center pivots in Florence and Portageville did not occur until mid-June. Therefore, the system (using only plant feedback) was tested during the reproductive stage of corn in Florence. Approximately of 20 in. of precipitation was received from April through August. Irrigations were terminated after July 8, when the crop was in the dent stage and due to abundant precipitation. Grain yield (15.5% MC), water use efficiency (WUE) [(grain yield/(precipitation + irrigation+ water stored in the soil)] and irrigation water use efficiencies (IWUE) [(irrigated grain yield- rainfed yield)/irrigation] for the SWP (153.9 bu ac^{-1}, 7.04 bu ac^{-1} in^{-1}, 73.9 bu ac^{-1} in^{-1}) and ISSCADA (149.3 bu ac^{-1}, 7.01 bu ac^{-1} in^{-1}, 94.3 bu ac^{-1} in^{-1}) plots were similar and both were significantly greater than the responses from the rainfed treatment.

In 2017, precipitation totaled 17.9 inches, with most of the rainfall occurring in June and July. The beta test in Florence was repeated, with a change in the management of the ISSCADA treatment plots, such that the ISSCADA system used both plant- and soil water sensing feedback for irrigation scheduling in all eight replications. There was a total of seven irrigation events. Mean corn plot yields and WUE for the three irrigation treatments were similar with 172.4, 162.8 and 169.0 bu ac^{-1}; and 2.54, 3.35 and 3.70 bu ac^{-1} in^{-1} for the SWP, ISSCADA and rainfed treatments, respectively. Irrigation water use efficiency was significantly greater for the SWP treatment.

Portageville, MO

In 2016, Portageville received 17.9 inches of precipitation from May through September, while reference evapotranspiration (ET_0) (ASCE, 2005) was 20.9 in. The average maximum and minimum air temperatures for the growing season were 87.7 °F and 68.8 °F, respectively. Two irrigations were applied to the AIS treatment plots totaling 1.5 inches. One of the ISSCADA treatment plots received two irrigations totaling 1.12 inches, and one irrigation totaling 0.87 inches was applied to the other three ISSCADA treatment plots. Regression analysis for seed cotton yield for AIS, ISSCADA and rainfed treatment plots were 2625, 2849 and 2404 lbs seed cotton ac^{-1}; mean irrigation water use efficiency (IWUE) was 197.32 lbs ac^{-1} in^{-1} and 447.24 lbs ac^{-1} in^{-1} for the AIS and ISSCADA plots, respectively.

Portageville received 16.34 in. of precipitation in 2017 during the months of May through September. The average maximum and minimum air temperatures were 86.3 °F and 67.0 °F, respectively. All of the AIS plots received six irrigations for a total of 4.4 inches. Two of the ISSCADA plots received a total of 3.0 inches of irrigation (5 events), while two received 2.5 inches (3 events) and one received 1.9 inches of irrigation (3 events). Seed cotton yield for the AIS, ISSCADA and rainfed plots were 2634, 2704 and 2219 lbs ac^{-1}, respectively. Mean IWUE was significantly greater for the ISSCADA treatment.

Bushland, TX

The 2016 growing season was hot and dry until August. Total precipitation from May through September was 11.26 in. with more than 61% occurring in August, while ET_o was 32.0 in. Under the 3-span VRI center pivot, a total of 19 irrigation events occurred on plots irrigated manually using information from weekly neutron probe readings and from the treatment plots irrigated with the ISSCADA system using both VRI speed and zone control. Under VRI speed and zone control, grain yield and WUE for the P1151AM corn hybrid was 267 bu ac^{-1} and 10.1 bu ac^{-1} in^{-1}; and 274 bu ac^{-1} and 9.9 bu ac^{-1} in^{-1}, respectively.

In 2016, under the 6-span VRI center pivot system, grain yield and WUE for corn hybrid, P0157AM, managed by manual irrigations at the 100%, 50% and 30% treatments were 198, 176 and 150 bu ac^{-1}, and 7.8, 8.3 and 7.5 bu ac^{-1} in^{-1}, respectively. Treatment plots managed by the ISSCADA plant feedback system resulted in grain yield and WUE of 212, 191, and 176 bu ac^{-1}, and 8.3, 8.0, and 8.4 bu ac^{-1} in^{-1}, respectively.

The beta test for VRI speed and zone control were repeated in 2017 under the 3-span VRI center pivot. A mild spring was experienced with 7 inches of precipitation occurring January through mid- May. However, after planting, minimal rainfall occurred for the next 22 days and a total of four 1-inch irrigations were applied uniformly across all plots, approximately every fifth day. A hailstorm hit on July 2 and damaged leaves and corn stalks on every plant. The damage from the storm slowed crop maturity by approximately two weeks and resulted in an estimated 20% yield penalty. Total rainfall from planting to harvest was 18.5 in. Four of the eight plots managed using VRI speed control used plant and soil water sensing feedback for irrigation scheduling, while plots managed with VRI zone control used only plant feedback for irrigation scheduling. There was no significant difference between grain yield and WUE for plots managed with plant feedback only (202.4 bu ac^{-1} and 7.19 bu ac^{-1} in^{-1}) and plots managed with the combination of plant and soil water sensing feedback (204.3 bu ac^{-1} and 7.35 bu ac^{-1} in^{-1}). Grain yields for the VRI zone control plots using the weekly neutron probe readings were 205, 198, and 135 bu ac^{-1}, and 7.50, 7.68, and 5.29 bu ac^{-1} in^{-1}; compared with 207, 202, and 169 bu ac^{-1} and 7.13, 7.42 and 6.75 bu ac^{-1} in^{-1} for the plots managed by the ISSCADA system at the 100%, 50% and 30% iCWSI thresholds, respectively.

Summary and Future Work

Beta tests performed on the ISSCADA system in the humid and sub-humid regions of Florence, SC and Portageville, MO demonstrated that the ISSCADA developed in Bushland required some modifications to the way the system was operated to develop dynamic prescription maps and that irrigations per each event were typically less than practiced in Bushland. Furthermore, the iCWSI values were adjusted in the software when daytime RH was high. In Florence, the ISSCADA performed as well as irrigation management using SWP measurements, however, the ISSCADA system was not used for the entire growing season during 2016 and in both years, irrigation was limited due to plentiful rainfall. In Portageville, MO, seed cotton yields produced with the ISSCADA method were similar to the AIS method in both years, while IWUE for the ISSCADA system was significantly greater compared with the AIS scheduling method. Results reported from Bushland, showed that corn managed with the ISSCADA system under VRI speed and zone control produced grain yields and WUE similar to irrigation scheduling using weekly neutron probe readings in 2016 and 2017. Beta tests in this semi-arid location also demonstrated that the ISSCADA system can control irrigation treatment levels using a combination of iCWSI thresholds and corresponding irrigation amounts.

Future work will include changing the software interface to make the ISSCADA system easier to operate. Also, more beta test sites representing different climates will utilize the ISSCADA system for irrigation management of soybean and grain crops. It is also anticipated that scientists at the Alberta Agriculture and Forestry Research Centre in Lethbridge, Canada will compare the ISSCADA system with irrigation management practices common for farmers in this region using a grain crop.

Acknowledgements

Funding was received as part of a Cooperative Research and Development Agreement between USDA-ARS and Valmont Industries, Inc., Valley, NE (Agreement No.: 58-3K95-0-1455-M). This material is based upon work that is supported by the National Institute of Food and Agriculture, U.S. Department of Agriculture, under award number 2016-67021-24420 and the Ogallala Aquifer Program, a consortium between USDA-Agricultural Research Service, Kansas State University, Texas AgriLife Research, Texas AgriLife Extension Service, Texas Tech University, and West Texas A&M University.

References

Andrade, M.A., O'Shaughnessy, S.A., and Evett, S.R. (2015). ARSmartPivot v.1 - Sensor based management software for center pivot irrigation systems. *In Proc. 2015 ASABE Annual International Meeting*, St. Joseph, MI.

Andrade, M.A., O'Shaughnessy, S.A., and Evett, S.R. (2016). A GIS-based Decision Support Tool for Center Pivot Irrigation Systems. *In Proc. 2016 ASABE Annual International Meeting*, St. Joseph, MI.

Andrade, M.A., O'Shaughnessy, S.A., and Evett, S.R. (2017). ARSPivot- A sensor based decision support tool for the integrated irrigation management of VRI center pivot systems. In Proceedings of the 28th Annual Central Plains Irrigation Conference, Burlington, CO. February 21-22, 2017.

Allen, R.G., Walter, I.A., Elliott, R.L. Howell, T.A., Itenfisu, D., & Jensen, M.E. (2005). The ASCE standardized Reference Evapotranspiration Equation. Report by the Task Committee on Standardization of Reference Evapotranspiration. https://www.kimberly.uidaho.edu/water/asceewri/ascestzdetmain2005.pdf

Evett, S. R., O'Shaughnessy, S.A., and Peters, R.T. (2014). Patent application serial No. 13/403,091 Entitled "Irrigation Scheduling and Supervisory Control and Data Acquisition System for Moving and Static Irrigation Systems". December 31, 2014.

Jackson, R.D., Idso, S.B., Reginato, R.J., and Pinter, P.J. Jr. (1981). Canopy temperature as a crop water stress indicator. *Water Resour. Res.* 17(4):1133-1138.

Lyle, W.M., & Bordovsky, J.P. (1983). LEPA irrigation system evaluation. *Trans. ASAE* 26 (3), 776-781.

Nguyen, A., Thompson, A., Sudduth, K., and Vories, E. (2015). Automating variable rate irrigation management prescription for center pivots from field data maps. Paper No. 152147794. *In Proceedings of Emerging Technologies for Sustainable Irrigation a joint ASABE/IA Irrigation Symposium.* Long Beach, CA. ASABE, St. Joseph, MI.

O'Shaughnessy, S.A., Evett, S.R., Colaizzi, P.D., and Howell, T.A. (2013). Wireless sensor network effectively controls center pivot irrigation of sorghum. *Appl. Engr. Agric.* 29 (6), 853-864. doi 10.13031/aea.29.9921

O'Shaughnessy, S.A., Evett, S.R., and Colaizzi, P.D. (2015). Dynamic prescription maps for site-specific variable rate irrigation of cotton. *Agric. Water Manage*. 159, 123-138. http://dx.doi.org/10.1016/j.agwat.2015.06.001

Wanjura, D.F., & Upchurch, D.R. (1997). Accounting for humidity in canopy-temperature-controlled irrigation scheduling. *Agric. Water Manage*. 34, 217-231.

Grain Sorghum Irrigation in the US Eastern Coastal Plain using Variable Rate Irrigation

Kenneth C. Stone, Research Agricultural Engineer
Gilbert C. Sigua, Research Soil Scientist
Philip J. Bauer, Research Agronomist

USDA-ARS Coastal Plains Soil, Water, and Plant Research Center, Florence, SC

Abstract. Grain sorghum is one of the top five cereal crops and an important grain crop throughout the world. It is generally considered more drought tolerant compared to other grain crops such as maize. Recently, in the US eastern Coastal Plain region, there was an emphasis on increasing regional grain production in which grain sorghum played an important role. The region's soils have low water holding capacities that combined with high rainfall variability cause crops frequently to be exposed to water stress. In this research, an experiment was conducted to evaluate the yield response of two grain sorghum varieties at different supplemental irrigation levels and three nitrogen levels. During our 3-year study, seasonal rainfall was adequate to produce acceptable grain sorghum yields and ranged from 421, 365, and 357 mm in 2012, 2013, and 2014, respectively. These rainfall amounts were greater than the seasonal calculated crop evapotranspiration requirement, but supplemental irrigation was required to maintain soil water potential above -30 kPa. However, irrigation did not increase grain sorghum yields. Additionally, no significant differences were found in grain yield between the two sorghum varieties or for increasing nitrogen applications. Results from this study suggest that there would be little benefit for supplemental irrigation for sorghum production in the US eastern Coastal Plain.

Keyword: Irrigation, grain sorghum, irrigation management

Grain sorghum (*Sorghum bicolor* L.) is an important grain crop that is considered drought tolerant and suitable to be grown in regions drier than those for corn (Swick, 2011). In some areas of the country with low rainfall such as the Southern High Plains, sorghum has to some extent replaced corn (Bordovsky and Lyle, 1996). In many of these areas, irrigation is used to supplement rainfall in grain sorghum fields. In 2013, the average irrigation water applied to grain sorghum was approximately 1100 mm in Arizona, 365 mm in Texas, and 60 mm in North Carolina (USDA-NASS 2017, Quick Stats). In 2012, the US Census of Agriculture reported grain sorghum production at approximately 2.1 million hectares with 0.25 million hectares irrigated (~12%). In much of the Southern High Plains, grain sorghum is deficit irrigated at rates below that of a well-watered crop. In western Kansas, Klocke et al. (2012) studied the impact of deficit irrigation on grain sorghum yields. In their study, grain sorghum yields increased linearly with increased irrigation amounts. Similarly, in northern Texas, O'Shaughnessy et al. (2012) reported on deficit irrigation levels in grain sorghum using automatic and manual irrigation scheduling. They found that optimum deficit irrigated yields and water use efficiencies were obtained when irrigation was from 55 to 80% of well watered crop evapotranspiration rates.

Only limited data are available on irrigating grain sorghum in humid regions. Recently in Mississippi, Bruns (2015) found no significant impact of irrigation on sorghum grain yield and concluded that supplemental irrigation would not be necessary in the Mid-south under normal seasonal conditions.

Grain sorghum yields are also influenced by fertilization. Hibberd and Hall (1990) studied the response of grain sorghum hybrids to nitrogen fertilizer. In a 3-year furrow irrigation study, they reported significant increases in sorghum grain yields by increasing nitrogen applications. In 2 of the 3 years, they observed differences among the evaluated hybrids. They determined the optimum rate of nitrogen application to be approximately 120 to 180 kg N ha^{-1}. They also observed a leveling off of grain yields at higher nitrogen concentrations. Assefa and Staggenborg (2010) investigated the impact of grain sorghum yield and hybrid advancement from 1957 through 2008. In the 52 years of data analyzed, they found hybrid yields increasing nearly 50 kg ha^{-1} per year in dryland sites with nitrogen fertilizer explaining 34% of the yield increases while the remaining yield increase was due to improved hybrids. However, irrigated grain sorghum yields remained unchanged over the same time period. They concluded that hybrid improvement programs had selected hybrids with better drought tolerance characteristics for dryland rather than for irrigated production.

In 2012, Murphy-Brown, Inc. (a division of Smithfield Grain, Rose Hill, NC) initiated a grain sorghum production program in the Carolinas and Virginia to increase local grain supplies for the swine industry (Murphy-Brown, 2012). Even though grain sorghum is generally drought tolerant, supplemental irrigation during drought conditions could impact grain sorghum yield potentials because the agricultural soils of the US south eastern Coastal Plain Region are generally coarse-textured with low water holding capacities (Camp and Sadler, 2002). Additionally, rainfall during the growing season can be highly variable. Sheridan et al. (1979) documented that there was a 50% chance of a 20-day drought during the annual growing season in the southeastern Coastal Plain. In this research, our objective was to evaluate the yield response of two grain sorghum varieties at different supplemental irrigation depths and three nitrogen levels. Results from this study will provide the regions sorghum growers information to make informed irrigation management decisions.

Materials and Methods

Field Experiment

A sorghum experiment was conducted from 2012 through 2014 under a 6-ha variable-rate center-pivot irrigation system (Camp et al., 1998) on a Norfolk loamy sand (Typic Kandiudult) near Florence, South Carolina (34°14'36.94" N, 79 48'34.14" W, elevation 42 m). The site is located in the humid U. S. Eastern Coastal Plain region and has an annual mean rainfall of 1089 mm and mean temperature of 18°C. In 2012, there were four irrigation treatments (0%, 33%, 66%, and 100% irrigation) and two nitrogen fertilization treatments (85 and 170 kg N ha^{-1}) with one cultivar (Dekalb A571). In 2013 and 2014, there were three irrigation treatments ((0%, 50%, and 100% irrigation), three nitrogen fertilization treatments (0, 85 and 170 kg N ha^{-1}), and 2 varieties (Dekalb A571 and Pioneer 84P80, both mid to late maturity varieties, i.e. greater than 70 days to mid-bloom). In all years there were 4 replications. Grain sorghum was planted at the rate of 272,000 seeds per hectare. The planting dates for each year are shown in Table 1.

Table 1. Seasonal rainfall, irrigation for the 100% treatment, crop evapotranspiration (ETc), planting and harvest dates for the 2012 to 2014 sorghum growing seasons.

	Seasonal Rainfall (mm)	Irrigation (mm)	ETc (mm)	Planting Date	N application dates	Harvest Date
2012	421.8	50.8	335.7	6/20/12	7/11, 7/19	10/22-23/12
2013	364.8	38.1	291.7	7/10/13	7/26, 8/26	11/18-19/13
2014	357.3	12.7	313.3	6/18/14	6/30, 8/7	10/1-3/14

The Norfolk loamy sand has a water holding capacity of approximately 24.3 mm in the surface 0.30 m (Peele et al., 1970) (field capacity 8%, 12 kPa; wilting point 2.3%, 1500 kPa; sand/silt/clay 86%,10%, 4%; bulk density 1.3 g/cm^3). A soil water potential (SWP) value of -30 kPa corresponds to approximately 50% depletion of the plant available water holding capacity. A 12.5 mm irrigation was initiated when SWP at the 0.30 m depth was below -30 kPa in the 100% irrigation plot with high N. Irrigations were applied on the following dates: 7/20/12, 7/27/12, 8/2/12, 8/6/12, 9/12/13, 9/16/13, 9/20/13, and 8/27/14. Soil water potentials were measured in all irrigation treatments for the high N rate using tensiometers at two depths (0.30 and 0.60 m). Measurements were recorded at least two times each week. The other irrigation treatments (0%, 33%, and 66%) received an application proportional to the 100% 12.5 mm application. The seasonal evapotranspiration for the sorghum crop was found by calculating the daily reference evapotranspiration from an adjacent weather station using the ASCE standard for grass (Walter et al., 2000) and the dual-crop-coefficient method of Allen et al. (1998) (table 1).

The experimental design in 2012 was a split-plot with irrigation rate as main plots and N levels as subplots. The experimental design in 2013 and 2014 was a split-split-plot with irrigation rate as the main plots and N levels as subplots and cultivars as the sub-subplots. The plot size was approximately 9.1 m wide by 45° of travel for the 137 m center pivot with four replicates (32 plots in 2012 and 72 plots in 2013 and 2014).

Fertilizer Applications

All nitrogen fertilizer was applied via fertigation through the center-pivot system annually in two split applications. The first N application each year was to all plots at the rate of 85 kg N ha^{-1}. A second N application was applied at this rate to only the high N plots. Nitrogen was applied using the center-pivot irrigation system and injecting urea and ammonium nitrate (UAN) 30% into the incoming center pivot water stream. Nitrogen applications were applied with the minimal water application depths in order to minimize irrigation water applications to non-irrigated plots. For this experiment, all nitrogen was delivered with 1.8 mm irrigation depth operating at 100% duty cycle. At the end of the nitrogen application, the system was again run in non-plot area to purge the system of nitrogen. Phosphorus and K were uniformly applied in granular form across all plots each spring based on soil testing and recommendations of the Clemson University Extension Agricultural Service Laboratory. Fertilizer applied was 30 kg ha^{-1} P_2O_5 and 50 kg ha^{-1} K_2O in 2012, 30 kg ha^{-1} P_2O_5 and 80 kg ha^{-1} K_2O in 2013, and 25 kg ha^{-1} P_2O_5 and 60 kg ha^{-1} K_2O in 2014.

Harvest

The grain sorghum was harvested when the grain moisture was below 20%. Grain sorghum yields were determined by weighing the grain harvested from a 6.1-m length of two rows near the center of each plot using an Almaco plot combine (Almaco, Nevada, Iowa). Whole plot samples were weighed and a sub-sample from each plot was collected and dried at 60° C for three days to determine grain moisture concentration. Grain sorghum yields were reported as dry grain yields. After yields and total water applied to each treatment were determined, the water use efficiency (WUE) was calculated by dividing the mean plot yield by the total water applied (irrigation + rainfall). The WUE values were reported in units of kg grain ha^{-1} mm^{-1} of water applied.

STATISTICAL ANALYSES

Yield and WUE data were statistically analyzed for the treatment effects and year using analysis of variance (ANOVA). Means were separated by calculating the least significant difference (LSD) (SAS version 9.4, Proc GLM, Statistical Analysis System, SAS Institute, Cary, N.C.). All significant differences were evaluated at the 0.05 level.

RESULTS AND DISCUSSION

The growing season rainfall totals for each year were lower than the long term average total of approximately 430 mm. However, in each year, the rainfall totals exceeded the calculated crop evapotranspiration (table 1). In 2012, irrigations were required to maintain SWP's greater than -30 kPa (at the 30 cm depth) during the early part of the growing season 30 to 50 DAP (growth stage 3, 8-leaf to stage 5, boot) and rainfall was adequate throughout the remainder of the growing season. In both 2013 and 2014, rainfall was generally adequate to maintain adequate SWP's except during the midpoint of the season from approximately 60 to 70 days after planting (stage 6, mid-bloom to stage 7, soft dough). During this time period, 3 irrigations in 2013 and one irrigation in 2014 were needed to maintain the SWP's above -30 kPa. The reduced rate irrigation during these time periods did have SWP values that were below -30 kPa levels, but after subsequent rainfalls, their SWP values increased and typically were above the -30 kPa threshold. Soil water potentials for the 60-cm depth followed similar trends to those of the 30-cm depth.

SORGHUM YIELD

In 2012, the overall mean grain sorghum yield was 2.8 Mg ha^{-1}. The mean treatment yields for the four irrigation and two nitrogen application rates are shown in table 2. An analysis of variance for the 2012 year indicated no significant differences in grain yield for irrigation application, nitrogen application rates, or the interactions between the irrigation and nitrogen rates. In 2013, the individual mean grain sorghum yield was 3.3 Mg ha^{-1} while in 2014, the mean yield was 3.1 Mg ha^{-1} (Table 3). In both 2013 and 2014, there were no significant differences in sorghum grain yields for the three irrigation or three nitrogen application rates. Additionally, we combined data from 2012-2014 for the Dekalb variety, fertilized at 85 and 170 kg N ha^{-1} for the fully irrigated (100%) and non-irrigated treatments and reanalyzed the data for the entire 3-year study (Table 7). No significant differences among the irrigation or nitrogen treatments for the sorghum grain yields occurred.

Table 2. The 2012 mean sorghum grain yields and water use efficiencies for the four irrigation and two nitrogen treatments.

	Nitrogen Rate			Nitrogen Rate		
	85	170	Mean	85	170	Mean
	Yield ($Mg\ ha^{-1}$)			Water Use Efficiency ($kg\ ha^{-1}\ mm^{-1}$)		
Water						
0	3.0	2.8	2.9	6.4	6.0	6.1
33	2.8	2.9	2.9	6.0	6.1	6.0
66	2.4	2.6	2.5	5.0	5.6	5.2
100	3.3	2.8	3.1	7.0	6.0	6.5
Mean	2.9	2.8	2.8	6.1	5.9	6.0

Overall, our sorghum grain yields were below those reported in the literature for Kansas (~7 Mg ha^{-1}, Klocke et al., 2012) and for Texas (5 to 7 Mg ha^{-1}, O'Shaughnessy et al., 2012) where supplemental irrigation is often necessary for production. A study in a more humid region by Bruns (2015) in Mississippi reported grain sorghum yields of 5.2 to 5.3 Mg ha^{-1}. Yang et al. (2001), in south Texas, reported sorghum grain yields similar to ours with yields ranging from 2.3 to 4.7 Mg ha^{-1}. Our study yields may have been impacted due to late season bird damage and delayed harvest until the crop was dry enough to harvest with a plot combine.

Unlike other reported studies, we observed no increase in yield response due to increasing nitrogen applications. Abunyewa et al. (2017) reported a quadratic relationship of increasing grain sorghum yields with nitrogen fertilizer applications. Our study site had been in a 4-year highly fertilized bermudagrass production study (Stone et al., 2012) and a 1-year flax study in the previous 5 years, and the soil in this experiment may have had adequate holdover nitrogen to produce an adequate sorghum crop.

Water Use Efficiency

In 2012, the overall WUE was 6.0 kg ha^{-1} mm^{-1} of water received (rainfall + irrigation) and varied among the treatments from 5 to 7 kg ha^{-1} mm^{-1} (table 2). In 2013 and 2014, the overall WUE was 8.7 $kgha^{-1}$ mm^{-1} and varied from 7.1 to 10.8 kg ha^{-1} mm^{-1} (table 4). In both the 2012 and 2013-2014 studies, there were no significant differences in WUE between irrigation or nitrogen treatments. However, in the combined 2012-2014 dataset, there was a significant WUE difference for the year of the study, with the 2012 year having significantly lower WUE of 6.7 kg ha^{-1} mm^{-1} compared to the 2013 and 2014 WUE's of 8.2 and 7.7 kg ha^{-1}/mm^{-1}, respectively (table 5). The 2012 WUE values were lower because of higher total seasonal rainfall. Our calculated WUE values for the grain sorghum were intermediate between those published by O'Shaughnessy et al. (2012) which ranged from 4.5 for non-irrigated to 17 kg ha^{-1} mm^{-1} for an 80% irrigated study in Texas.

Table 3. Mean sorghum grain yields for the 2013 and 2014 variety, irrigation, and nitrogen treatments.

			Water			
			0	50	100	Overall Mean
			Yield ($Mg\ ha^{-1}$)			
			Mean	Mean	Mean	Mean
year	Variety	Nitrogen				
2013	DeKalb	0	3.3	3.0	3.5	3.3
		85	2.8	3.3	3.3	3.1
		170	3.3	3.0	3.2	3.2
		Mean	3.1	3.1	3.4	3.2
	Pioneer	Nitrogen				
		0	3.2	3.5	3.4	3.4
		85	4.0	3.8	3.5	3.8
		170	3.4	3.1	3.5	3.3
		Mean	3.5	3.5	3.5	3.5
2014	DeKalb	Nitrogen				
		0	3.5	2.6	2.8	3.0
		85	3.4	3.0	3.2	3.2
		170	3.1	2.9	2.8	2.9
		Mean	3.3	2.8	2.9	3.0
	Pioneer	Nitrogen				
		0	3.0	3.1	2.8	3.0
		85	3.5	3.1	3.6	3.4
		170	3.1	3.5	3.1	3.2
		Mean	3.2	3.2	3.2	3.2
	Overall Mean		3.3	3.2	3.2	3.2

Table 4. Mean water use efficiencies for the 2013 and 2014 variety, irrigation, and nitrogen treatments.

			Water			
			0	50	100	Overall Mean
			Water Use Efficiency ($kg\ ha^{-1}\ mm^{-1}$)			
year	Variety	Nitrogen				
2013	DeKalb	0	9.1	7.9	8.8	8.6
		85	7.6	8.6	8.3	8.2
		170	9.0	7.9	8.0	8.3
		Mean	8.6	8.1	8.4	8.3
		Nitrogen				
	Pioneer	0	8.8	9.1	8.4	8.8
		85	10.8	9.9	8.7	9.8
		170	9.3	8.0	8.7	8.7
		Mean	9.7	9.0	8.6	9.1
		Nitrogen				
2014	DeKalb	0	9.8	7.1	7.5	8.1
		85	9.6	8.2	8.5	8.8
		170	8.7	7.9	7.7	8.1
		Mean	9.4	7.7	7.9	8.3
		Nitrogen				
	Pioneer	0	8.5	8.5	7.6	8.2
		85	9.8	8.5	9.8	9.4
		170	8.8	9.6	8.3	8.9
		Mean	9.0	8.9	2.6	2.3
Overall Mean			9.2	8.4	8.4	8.7

Table 5. Mean sorghum grain yields and water use efficiencies for the 2012-2014 common variety, irrigation, and nitrogen treatments.

			Water			Water		
			0	100	Mean	0	100	Mean
			Yield ($Mg\ ha^{-1}$)			Water Use Efficiency ($kg\ ha^{-1}\ mm^{-1}$)		
year	Variety	Nitrogen						
2012	DeKalb	85	3.0	3.3	3.2	7.1	7.0	7.1
		170	2.8	2.8	2.8	6.7	6.0	6.3
		Mean	2.9	3.1	3.0	6.9	6.5	6.7
		Nitrogen						
2013	DeKalb	85	2.8	3.3	3.1	7.6	8.3	7.9
		170	3.3	3.2	3.2	9.0	8.0	8.5
		Mean	3.0	3.3	3.1	8.3	8.1	8.2
		Nitrogen						
2014	DeKalb	85	3.4	3.2	3.3	9.6	8.5	9.0
		170	3.1	2.8	3.0	8.7	7.7	8.2
		Mean	3.3	3.0	3.1	9.1	8.1	8.6
Overall Mean			3.0	3.1	3.1	8.0	7.5	7.7

Summary and Conclusions

Grain sorghum was grown in 2012, 2013, and 2014 to investigate the potential for irrigation to increase yields in the humid SE US. Supplemental irrigation was required for 4, 3, and 1 irrigation events in 2012, 2013, and 2014 respectively, to maintain soil water potential above -30 kPa in the 100% treatments. However, these water applications did not increase grain sorghum yields. Additionally, we found no significant difference in grain yield for increasing nitrogen application. This lack of response to nitrogen applications may have been related to adequate supplies in the soil from previous crops. Our results are similar to previous work from the mid-South USA in that irrigation did not increase sorghum grain yields above those of rainfed and suggest that there would be little benefit for supplemental irrigation for sorghum production in the US eastern Coastal Plain.

References

Abunyewa, A.A., Ferguson, R.B., Wortmann, C.S., Mason, S. C. 2017. Grain sorghum nitrogen use as affected by planting practice and nitrogen rate. *Journal of Soil Science and Plant Nutrition 17*(1):155-166.

Allen, R. G., Pereira, L.S., Smith, M., and Raes, D. 1998. Crop evapotranspiration. FAO Irrigation and Drainage Paper No. 56. Rome, Italy: Food and Agriculture Organization of the United Nations

Assefa, Y. and Staggenborg, S. A. 2010. Grain Sorghum Yield with Hybrid advancement and Changes in Agronomic Practices from 1957 through 2008. *Agronomy Journal 102*(2):703-706.

Bruns, H. A. 2015. Irrigation, seeding rates, and row type effects on grain sorghum in the midsouth. Agronomy Journal, 107(1), 9-12.

Bordovsky, J. P., and Lyle, W. M. 1996. LEPA irrigation of grain sorghum with varying water supplies. *Transactions of the ASAE, 39*(6), 2033-2038

Camp, C. R., Sadler, E. J., Evans, D. E., Usrey, L. J., and Omary, M. (1998). Modified center-pivot system for precision management of water and nutrients. *Appl. Eng. Agric., 14*(1), 23-31. http://dx.doi.org/10.13031/2013.19362

Camp, C.R., and Sadler, E. J. 2002. Irrigation, deep tillage, and nitrogen management for a corn-soybean rotation. *Transactions of the ASAE 45*(3):601-608.

Hibberd, D. D. and Hall, B. D. 1990. The responses of maize and grain sorghum hybrids to nitrogen fertilizer in South East Queensland. *Australian Journal of Experimental Agriculture* 30:825-31.

Klocke, N. L., Currie, R. S., Tomsicek, D. J., and Koehn, J. W. 2012. Sorghum yield response to deficit irrigation. *Transactions of the ASABE, 55*(3), 947-955.

Murphy-Brown. 2012. 2012 Sorghum Program. http://mbgrain.com/wp-content/uploads/2012/03/MurphyBrown_SorghumProgram_Draftv3.pdf. Accessed 8/8/2017.

Peele, T. C., Beale, O. W., and Lesesne, F. F. 1970. The physical properties of some South Carolina soils. Technical Bulletin 1037. South Carolina Agricultural Experiment Station, Clemson University, Clemson, SC.

O'Shaughnessy, S. A., Evett, S. R., Colaizzi, P. D., & Howell, T. A. 2012. Grain sorghum response to irrigation scheduling with the time-temperature threshold method and deficit irrigation levels. *Transactions of the ASABE, 55*(2), 451-461.

Rhodes, J. D. 1996. Salinity: Electrical conductivity and total dissolved solids. In: D. Sparks, editor, Methods of soil analysis. Part 3. SSSA Book Ser. 5. SSSA, Madison, WI. P. 417-435.

Sheridan, J. M., Knisel, W. G., Woody, T. K., & Asmussen, L. E. 1979. Seasonal variation in rainfall and rainfall-deficit periods in the southern coastal plain and flatwoods regions of Georgia. Bulletin 243. Athens, Ga.: University Georgia Agricultural Experiment Station.

Stone, K.C., Bauer, P.J., Andrae, J., Busscher, W.J., Millen, J.A., Strickland E.E., and Evans, D.E. 2012. Irrigation and nitrogen impact on bermudagrass yield response in the southeastern Coastal Plain. *Transactions of the ASABE 55*(3):969-978.

Swick, R.A. 2011. Global feed supply and demand. Recent Advances in Animal Nutrition Australia, University of New England, Armidale, Australia 18:1–8.

USDA-NASS. 2017. USDA Quick Stats. U.S. Department of Agriculture, National Agricultural Statistics Service, Washington, DC. (https://quickstats.nass.usda.gov/) Accessed 8/8/2017.

Walter, I. A., Allen, R. G., Elliott, R., Jensen, M. E., Itenfisu, D., Howell, T. A., Snyder, R., Brown, P., Echings, S., Spofford, T., Hattendorf, M., Cuenca, R. H., Wright, J. L., and Martin, D. 2000. ASCE's standardized reference evapotranspiration equation. In *Proc. 4th National Irrigation Symposium,* 209-215. R. G. Evans, B. L. Benham, and T. P. Trooien, eds. St. Joseph, Mich.: ASAE.

Yang, C., Everitt, J. H., and Bradford, J. M. 2001. Comparisons of uniform and variable rate nitrogen and phosphorus fertilizer applications for grain sorghum. *Transactions of ASAE 44*(2):201-209.

TILLAGE MANAGEMENT AND SPRINKLER-IRRIGATED CORN PRODUCTION

Freddie R. Lamm
Research Irrigation Engineer
Northwest Research-Extension Center
Kansas State University
Colby, Kansas
Voice: 785-462-6281
Fax: 785-462-2315
flamm@ksu.edu

INTRODUCTION

Tillage management strategies that leave greater amounts of residue on the soil surface are beneficial in sprinkler-irrigated corn production in terms of improving infiltration of both irrigation and rainfall, and in reduction of soil evaporative losses early in the growing season. Additionally, sometimes early season crop growth is delayed under higher residue conditions and this can result in the shifting of crop evapotranspiration to later in the season for higher residue treatments. This paper will discuss 4 years (2004-2007) of sprinkler-irrigated corn research under conventional, strip-tillage, and no tillage.

BRIEF DESCRIPTION OF PROCEDURES

The study was conducted under a center pivot sprinkler at the KSU Northwest Research-Extension Center at Colby, Kansas during the years 2004 to 2007. Corn was also grown on the field site in 2003 to establish baseline residue levels for the three tillage treatments. The study area had conventional tillage in 2003. The soil type was a medium textured, deep, well drained, silt loam soil. The region has an average annual precipitation of 19 inches with a summer pattern resulting in an average corn cropping season precipitation of 12 inches. The average seasonal total crop evapotranspiration (ETc) for corn in this region is 21 inches.

A corn hybrid of approximately 110-day relative maturity (Dekalb DCK60-191 in 2004 and DCK60-18 in 2005 through 2007) was planted in 30-inch spaced circular rows on May 8, 2004, April 27, 2005, April 20, 2006 and May 8, 2007, respectively. The two hybrids differ only slightly with the latter hybrid having an additional genetic modification of corn rootworm control. Three target seeding rates (26,000, 30,000 and 34,000 seeds/acre) were superimposed onto each tillage treatment in a complete randomized block design.

Irrigation was scheduled with a weather-based water budget, but was limited to the 3 treatment capacities of 1 inch every 4, 6, or 8 days (IC-4, IC-6 and IC-8, respectively). The weather-based water budget was constructed using data collected from a NOAA weather station located approximately 1800 ft. northeast of the study site.

The three tillage treatments [Conventional tillage (CT), Strip Tillage (ST) and No Tillage (NT)] were replicated in a Latin-Square type arrangement widths at three different radii (Figure 1). Planting was in the approximate same row location each year for the Conventional Tillage treatment to the extent that good farming practices allowed. The Strip Tillage and No-Tillage treatments were planted between corn rows from the previous year.

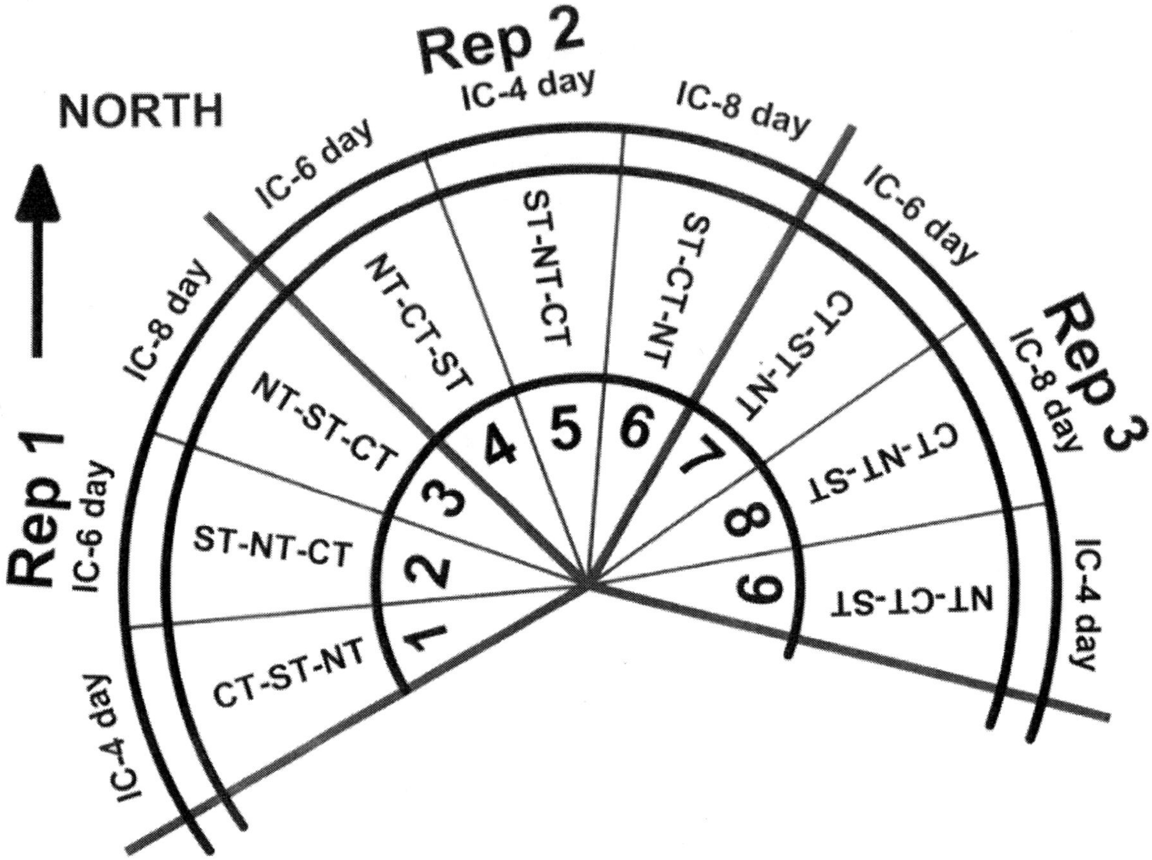

Figure 1. Physical arrangement of the irrigation capacity (IC-4 IC-6, or IC-8) for the nine different pie sectors and tillage treatments (CT, ST or NT) randomized within the outer sprinkler span.

Fertilizer N for all 3 treatments was applied at a rate of 200 lbs/acre. Phosphorus was applied with the starter fertilizer at planting at the rate of 45 lbs/acre P_2O_5. Weekly to bi-weekly soil water measurements were made in 1 ft. increments to an 8 ft. depth with a neutron probe. All measured data was taken near the center of each plot. Corn yield was measured in each of the 81 subplots at the end of the season by hand harvesting the ears from a 20 ft. section of one corn row near the center of each plot. Water use and water productivity (i.e., grain yield divided by seasonal water use) were calculated for each subplot using the soil water data, precipitation, applied irrigation and crop yield.

RESULTS AND DISCUSSION

Weather Conditions and Irrigation Requirements

In general, conventional tillage treatments were observed to emerge earlier and have improved growth during May and June as compared to the strip and no-tillage treatments, probably because of warmer soil temperatures. However, by about mid-summer in most of the years the conventional tillage treatments began to show greater water stress, particularly for the reduced irrigation capacities, as evidenced by some observed mid-day wilting. The conventional tillage plots also tended to senesce earlier in most years with the exception of 2004. Summer seasonal precipitation was approximately 2 inches below normal in 2004, near normal in 2005, nearly 3 inches below normal in 2006, and approximately 2.5 inches below normal in 2007. Calculated well-watered corn evapotranspiration was near normal in 2004, 2005, and 2006, but was approximately 3 inches below normal in 2007. Full irrigation needs varied between years with greatest needs in 2005 and 2006 at approximately 15 inches and was lower in 2004 and 2007 at approximately 12 inches.

Corn Yields, Yield Components, Water Use, and Water Productivity

Average (2004-2007) study wide corn yield, yield components, water use and water productivity are summarized in Table 1 and Figure 2 and 3.

Greatest corn yield was obtained by the fully irrigated treatment (Table 1 and Figure 2 top panel) where irrigation was supplied as needed at a capacity limited to 1 inch/4 days. Average yields decreased approximately 20 bushels/acre for the deficit irrigated treatments (1 inch/6 days and 1 inch/8 days). The greater irrigation capacity was positively reflected in both greater number of kernels/ear and kernel mass (Table 1 and Figure 2 middle and bottom panel), but the number of kernels/ear had the greatest effect (Figure 4 top panel).

Table 1. Summary of average corn yield and water use parameters from a sprinkler-irrigated research study, KSU Northwest Research-Extension Center, 2004-2007.

Treatment	Irrigation (inches)	Yield (bu/a)	Plant Density (p/a)	Ears /Plant	Kernels /Ear	Kernel Mass (mg)	Water Use (inches)	Water Productivity (lbs/acre-in)
1 in/4 day	13.8	243	30129	1.00	575	361	25.99	528
1 in/ 6 day	12.3	224	30137	1.00	546	350	25.05	503
1 in/ 8 day	10.3	221	30016	1.01	545	347	23.56	526
Conventional	12.1	221	30016	1.00	534	35.4	24.56	505
Strip Tillage	12.1	237	30226	1.01	567	35.3	25.15	530
No-Tillage	12.1	231	30040	1.00	565	35.0	24.88	522
PD 26,000 p/a	12.1	221	26822	1.01	577	36.2	24.85	501
PD 30,000 p/a	12.1	232	30145	1.00	556	35.3	24.87	524
PD 34,000 p/a	12.1	235	33315	1.00	532	34.2	24.88	531

The strip tillage and no-tillage treatment had greater yields (≈ 10 bushels/acre) than the conventional tillage treatment (Table 1 and Figure 2 top panel). The improved tillage schemes were positively reflected in a greater number of kernels/ear (≈ 32, Table 1 and Figure 2 middle panel). This nearly 6% greater number of kernels/ear for the two treatments resulted in an approximately 4.5 to 7% greater yield (Figure 4 middle panel).

Increasing plant density from 26,000 to 30,000 or 34,000 plants/acres increased yield approximately 12 bushels/acre (Table 1 and Figure 2 top panel). Both kernels/ear and kernel mass were reduced for the greater plant densities (Table 1 and Figure 2 middle and bottom panel), but the greater plant densities were able to fully compensate for these lower values by the increased number of plants (Table 1, Figure 4 bottom panel).

Water use was greatest for the fully-irrigated irrigation treatment (1 inch/4 days) as might be anticipated due to greater irrigation amounts (Table 1 and Figure 3 top panel). However it had the numerically greatest water productivity, although all three treatments had relatively similar values (Table 1 and Figure 3 bottom panel).

There were very little differences in overall crop water use as affected by tillage treatment (Table 1 and Figure 3 top panel). This is to be anticipated as with good irrigation management all treatments either used the provided irrigation or added it to soil water storage. Due to greater yields, the strip and no-tillage treatments had greater water productivity (Table 1 and Figure 3, bottom panel).

The range of plant densities from 26,000 to 34,000 plants/acre had no effect on crop water use (Table 1 and Figure 3, top panel) with the range of average values differing by less than 0.03 inches. Much greater water productivity was obtained by the greater plant densities (Table 1 and Figure 3, bottom panel).

CONCLUSIONS

The greatest irrigation capacity (1 inch/4days) had approximately 9% greater yield than the deficit-irrigated treatments (1 inch/6 days or 1 inch/8 days) due to a greater number of kernels/ear and to a lesser extent greater kernel mass. All irrigation treatments had relatively similar and high water productivity.

Strip tillage and no-tillage were superior to conventional tillage in terms of grain yield and water productivity. The number of kernels/ear was greater for the two reduced tillage schemes.

Increasing plant density from 26,000 to 30,000 or 34,000 plants/acre increased grain yield and water productivity although the number of kernels/ear and kernel mass were decreased by the increase in plant density. Crop water use was not affected over the entire range of plant density.

ACKNOWLEDGEMENTS

This paper is part of a center pivot irrigation technology transfer effort is supported by the Ogallala Aquifer Program, a consortium between USDA Agricultural Research Service, Kansas State University, Texas A&M AgriLife Research, Texas A&M AgriLife Extension Service, Texas Tech University, and West Texas A&M University.

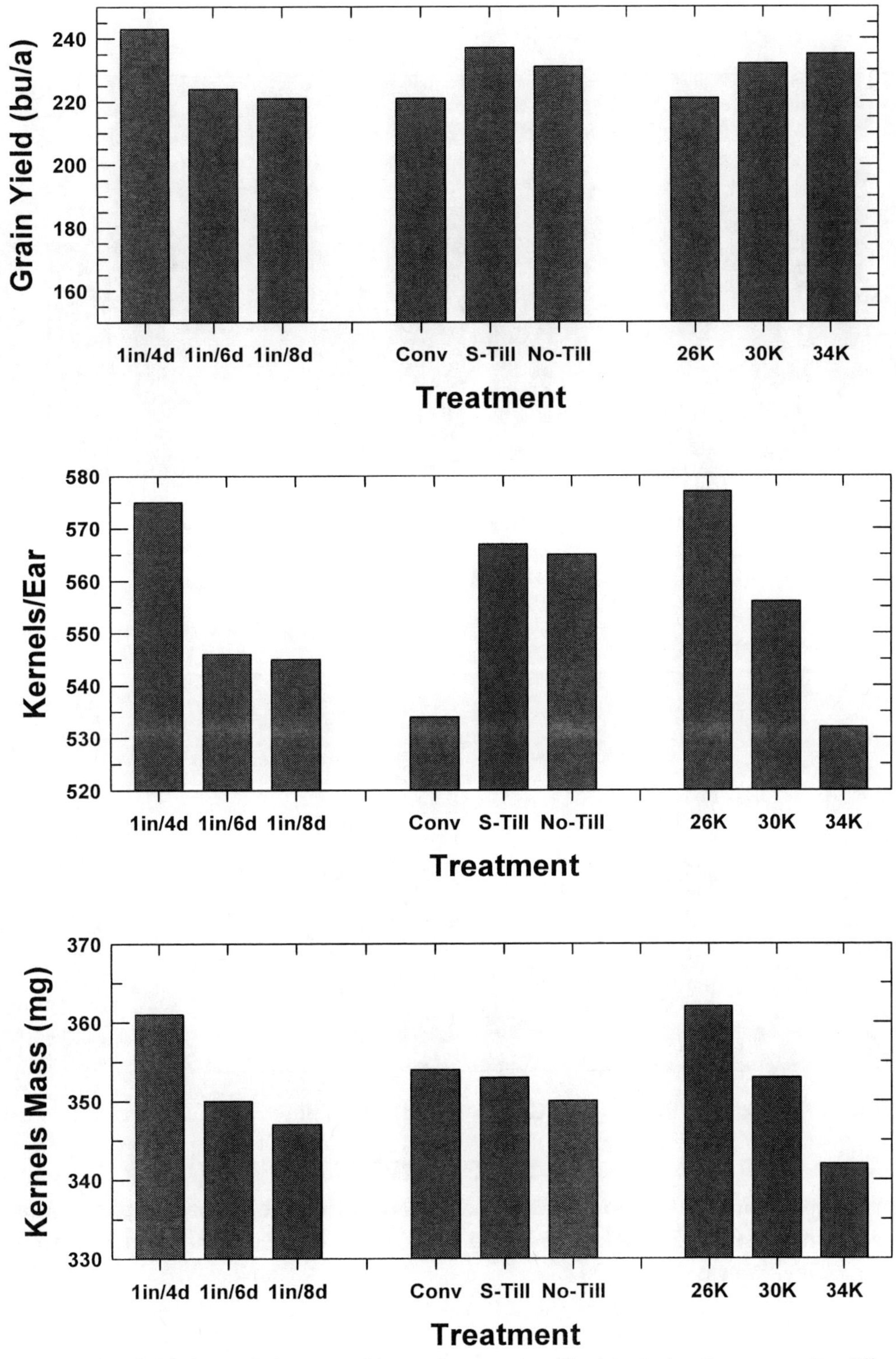

Figure 2. Corn grain yield, kernels/ear, and kernel mass as affect by irrigation capacity, tillage treatment and target plant density in a sprinkler-irrigated research study, KSU Northwest Research-Extension Center, Colby, Kansas, 2004-2007.

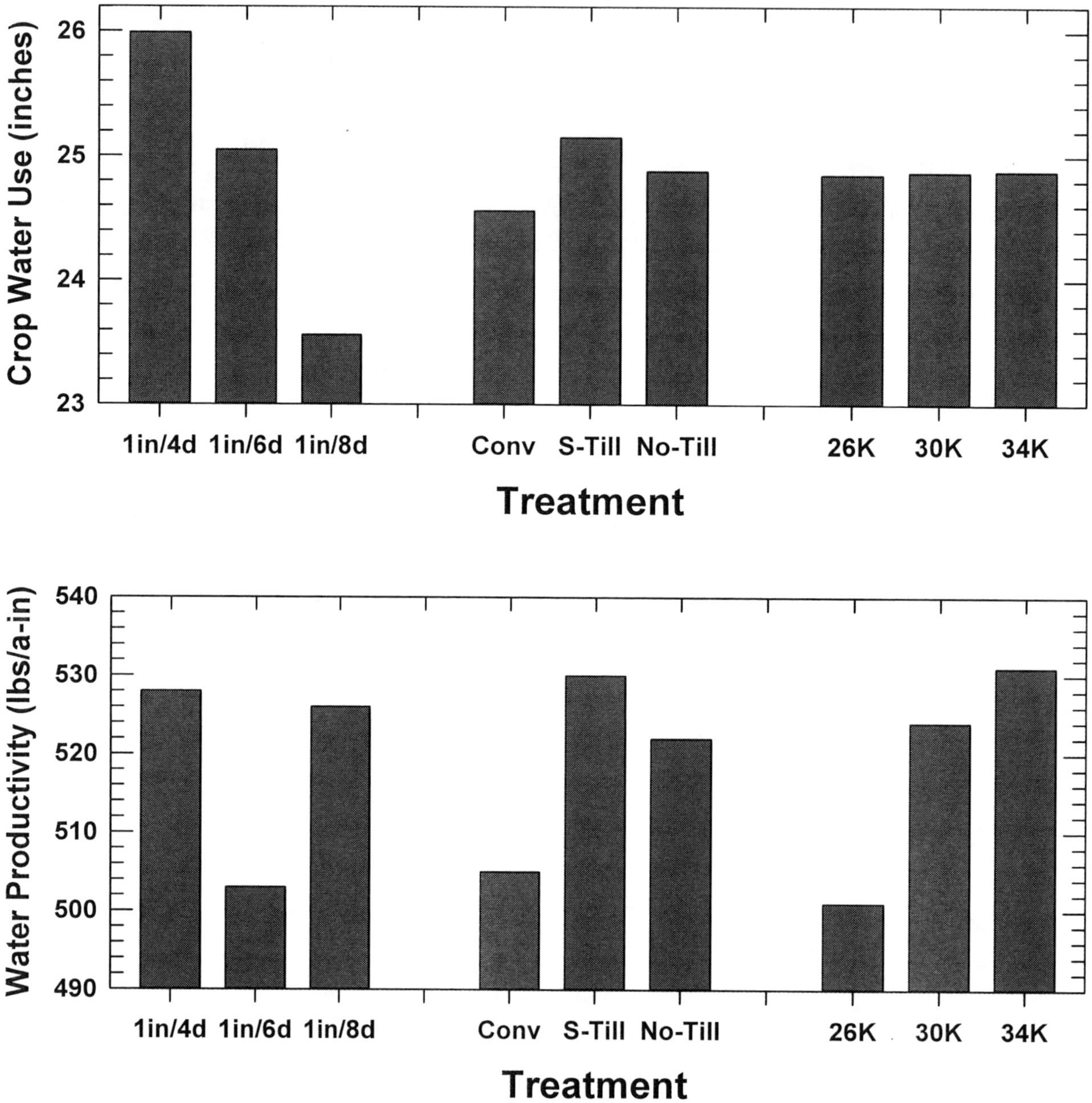

Figure 3. Corn water use and water productivity as affected by irrigation capacity, tillage treatment and target plant density in a sprinkler-irrigated research study, KSU Northwest Research-Extension Center, Colby, Kansas, 2004-2007.

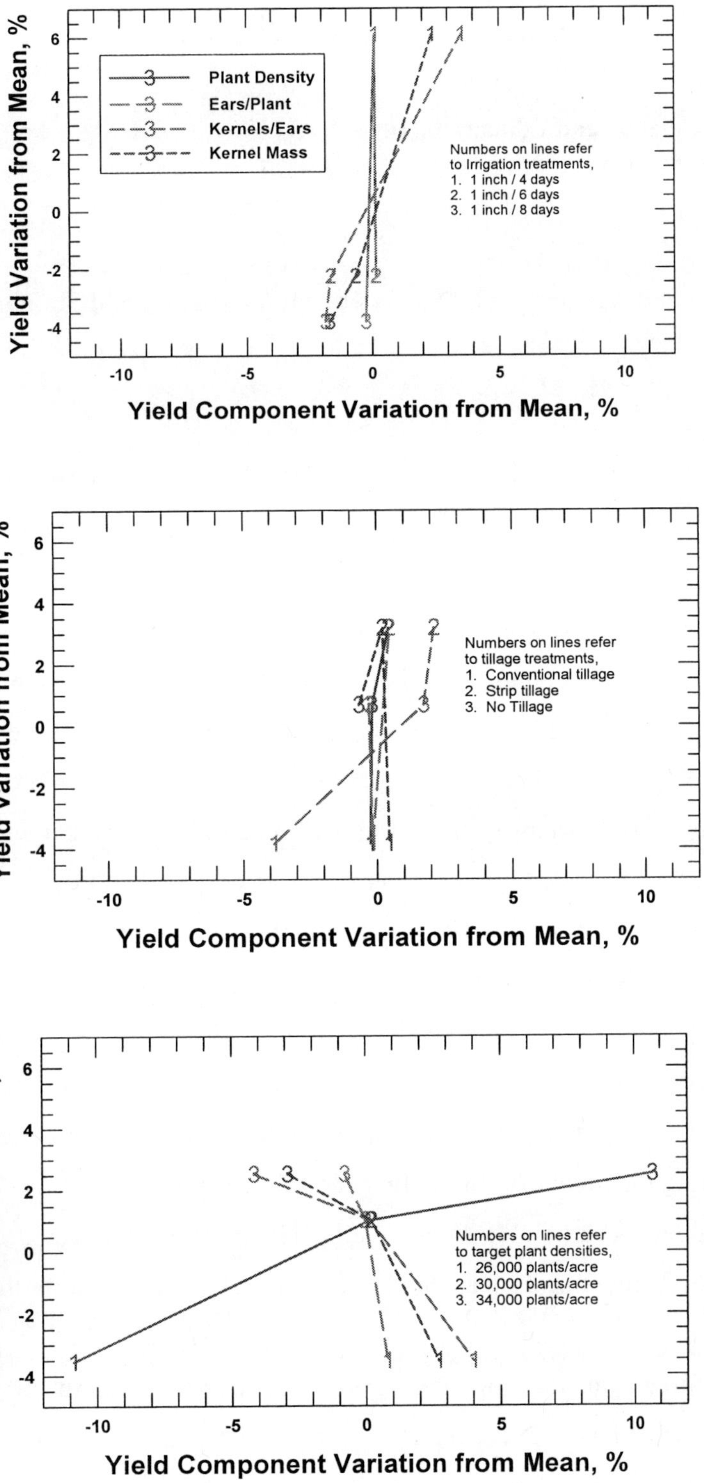

Figure 4. Contribution of various yield components to the yield variation for the three irrigation capacities (top panel), the three tillage systems (middle panel) and the three target plant densities (bottom panel). *Note: Upward sloping lines to the right, such as Kernel/Ear, indicate that treatment heavily affected the yield component and subsequently affected the grain yield, while vertical lines with little or no yield component variation from zero indicate little effect.*

Machine learning algorithms applied to the forecasting of crop water stress indicators

Manuel A. Andrade
Oak Ridge Institute for Science and Education, 100 ORAU Way, Oak Ridge, TN 37830, USA, Alejandro.Andrade@ars.usda.gov

Steven R. Evett
USDA-Agricultural Research Service, P.O. Drawer 10, 2300 Experimental Station Road, Bushland, TX, 79102, USA, Steve.Evett@ars.usda.gov

Susan A. O'Shaughnessy
USDA-Agricultural Research Service, P.O. Drawer 10, 2300 Experimental Station Road, Bushland, TX 79102, USA, Susan.O'Shaughnessy@ars.usda.gov

Abstract. Recent advances made by scientists with the USDA-Agricultural Research Service (ARS) can provide farmers with irrigation scheduling tools based on crop stress indicators to assist the management of Variable Rate Irrigation (VRI) center pivot systems. These tools were integrated into an Irrigation Scheduling Supervisory Control and Data Acquisition System (ISCCADAS). The ISSCADAS automates the collection of data from a network of wireless infrared thermometers (IRTs) distributed on a center pivot's lateral and in the field irrigated by the center pivot, as well as data from a wireless soil water sensor network in the field and a microclimate weather station located at the pivot point. This study analyzes the use of Artificial Neural Networks, one of the most popular machine learning algorithms, for the forecasting of canopy temperatures obtained by a wireless network of IRTs mounted on a three-span VRI center pivot irrigating corn near Bushland, TX, during the summer of 2017. Two case studies were conducted for this purpose using data collected from periodic scans of the field performed during the growing season by running the pivot dry. In the first case, data from the first three scans were used to train an Artificial Neural Network (ANN). Canopy temperatures estimated using the ANN were then compared against canopy temperatures measured by the network of IRTs during the fourth scan. In the second case, data from the first six scans were used to train ANNs. Then ANN predicted canopy temperatures were compared against canopy temperatures measured during the seventh scan. The Root of the Mean Squared Error (RMSE) of ANN predictions in the first case ranged from 1.04 °C to 2.49 °C, whereas the RMSE of ANN predictions in the second case ranged from 2.14 °C to 2.77 °C. To assess the impact of ANN accuracy on irrigation management, estimated canopy temperatures were fed to a plant-stress based irrigation scheduling method and the resulting prescription maps were compared against prescription maps obtained by the same method using the canopy temperatures measured by the network of IRTs. In the first case no difference was found between both prescription maps. In the second case only one plot (out of 26) was assigned a different prescription. Results of this study suggest that machine learning techniques can be used to assist the ISSCADAS in situations where canopy temperatures cannot be measured by the network of IRTs due to poor visibility conditions, or because the center pivot cannot traverse the field within a reasonable amount of time.

Keywords: machine learning, metamodeling, center pivot irrigation, variable rate irrigation, irrigation scheduling, sensors.

Introduction

Agriculture, like many other economic sectors, is facing a rapid transformation resulting from the use of data analytics (Pham and Stack, 2018). Machine learning is a data analysis technique that can be used to solve a wide range of problems, since it uses previous data to train computational 'machines' that can make predictions on new data with no human intervention. One of the most popular types of machine learning algorithms is the Artificial Neural Network (ANN), which is capable of handling complex associations between inputs and outputs. This characteristic makes the ANN particularly well fitted for solving agricultural problems where no known deterministic model may be available to simulate the complex interactions occurring between the many components involved in an agricultural system. In agricultural engineering, ANNs have been used to, among other applications, predict soil properties (Schaap et al., 1998; Farifteh et al., 2007), estimate crop yield (Kaul et al., 2005), model greenhouse environmental variables (Ferreira et al., 2002; He and Ma, 2010), and analyze images for the discrimination of crop and weeds (Yang et al., 2000; Kavdir, 2004).

This study uses ANNs to forecast canopy temperatures obtained by a wireless network of infrared thermometers (IRTs) mounted on a three-span VRI-center pivot irrigating corn near Bushland, TX, during the summer of 2017. The network of IRTs is part of an Irrigation Scheduling Supervisory Control and Data Acquisition System (ISSCADAS) patented by scientists with the USDA-Agricultural Research Service (ARS) (Evett et al., 2014). The ISSCADAS collects data from soil water, plant, and weather sensing systems and feeds those data to computerized irrigation scheduling algorithms based on plant stress to generate site-specific prescription maps. The ISSCADAS also integrates hardware functions to manage the submission of prescription maps for their application by VRI center pivot systems operated with a Pro2 control panel (Valmont Industries Inc., Valley NE). A software package, named ARS-Pivot (ARSP), was developed to simplify the operation of the ISSCADAS (Andrade et al., 2015, 2017). The main objective of this study is to analyze the feasibility of using ANNs to predict future canopy temperatures based on those previously measured by the aforementioned network of IRTs as the center pivot moves through the field. Results obtained from this study can provide guidelines for the future integration of canopy temperature forecasting using ANNs as one of the tools available in ARSP. The availability of such a tool can add redundancy to the ISSCADAS so that site-specific prescription maps can be generated even if a direct measurement of canopy temperatures is not possible due to poor visibility conditions, or because the center pivot cannot traverse the entire field within a reasonable amount of time.

Methodology

In the summer of 2017, the ISSCADAS and the ARSP software were used for the integrated irrigation management of a three-span center pivot (131 m) located at the USDA-ARS Conservation and Production Research Laboratory, near Bushland, TX. The center pivot was equipped with a Pro2 control panel and a commercial VRI system (Valmont Industries Inc., Valley NE). A midseason corn hybrid, Dupont Pioneer P1151AM, was planted on May 15, day of year (DOY) 135. Experimental plots used in this study were located within the six outermost sprinkler zones in the field (Fig. 1).

VRI zone control was used for the North-Northwest (NNW) side of the field, which was divided into six control sectors of 28° each and six concentric control zones with a width of 9.14 m (30 ft) each, for a total of 36 management zones, each of which was considered an experimental plot. Plots were organized using a Latin square design (Fig. 1). VRI speed control was used for the South-Southeast (SSE) side of the field, which was divided into eight control sectors of 20° each and a single concentric control zone with a width of 54.9 m, for a total of 8 management zones, each of which was considered an experimental plot

(Fig. 1). The irrigation of plots in the NNW side was triggered by either the integrated Crop Water Stress Index (iCWSI) method described by O'Shaughnessy et al. (2017) or by weekly neutron probe (NP) (model 503DR1.5, Instrotek, Campbell Pacific Nuclear, Concord, CA) measurements. Each of these plots was assigned one of the following irrigation levels: 80%, 50%, or 30% of full irrigation. Full irrigation was defined as the irrigation required to return soil water content in the root zone to field capacity. The combination of irrigation scheduling methods (2) and irrigation levels (3) resulted in six treatments with six replicates per treatment (Fig. 1). Plots irrigated with the iCWSI method are labeled in Fig. 1 as C80, C50, or C30, where 'C' stands for iCWSI-based control and numbers correspond to irrigation levels. Similarly, plots irrigated with the NP method are labeled in Fig. 1 as U80, U50, or U30, where 'U' indicates that irrigation scheduling is controlled by the user.

Plots in the SSE side were all assigned a single irrigation level of 80%; their irrigation was triggered by either the iCWSI method, or by a hybrid method using the iCWSI method and an average soil water depletion in the root zone (SWDr) calculated using sets of three time domain reflectometer (TDR) sensors (model 315, Acclima, Meridian, ID) buried at depths of 15 cm, 30 cm, and 45 cm. The hybrid method used a two-step approach for irrigation scheduling. During the first step, the SWDr was compared against pre-determined lower and upper SWDr thresholds. No irrigation was assigned if the SWDr was lower than 0.1 (lower threshold) and an irrigation depth of 30.5 mm (1.2 in) was assigned if the SWDr was higher than 0.5 (upper threshold). If the SWDr fell between these values, the iCWSI method was used during a second step to determine its prescription. Plots irrigated with the hybrid method are labeled in Fig. 1 as H80.

The iCWSI method is based on calculation of the theoretical Crop Water Stress Index (CWSI) (Jackson et al., 1981) at discrete intervals during daylight hours. CWSI values were calculated for each location *x* in the field at time interval *t* using the normalized difference between the crop canopy temperature in the location and the air temperature at time *t*. Additional details of the iCWSI method and the formulas used for its calculation can be found in O'Shaughnessy et al. (2013) and O'Shaughnessy et al. (2017). Air temperature and other relevant weather parameters (relative humidity, solar irradiance, wind speed, and wind direction) were sampled every 5 s and averaged and stored every minute at a weather station (Campbell Scientific, Logan, UT) located next to the pivot point. Crop canopy temperatures were measured at two fixed locations in the field using wireless infrared thermometers (IRTs) (model SapIP-IRT, Dynamax Inc., Houston, TX) to provide a reference canopy temperature for a well-watered crop (Fig. 1). A network of 12 wireless infrared thermometers (IRTs) (model SapIP-IRT, Dynamax Inc., Houston, TX) was mounted on the center pivot to measure canopy temperatures inside the experimental area (Fig. 1). The IRTs were located forward of the drop hoses, at an oblique angle from nadir. The average of data collected from two IRTs with opposing views of a sprinkler control zone was the primary datum every minute for each sprinkler zone. Pairs of IRTs arranged in such a way are referred hereinafter as IRT groups.

Scans of the field were performed periodically through the growing season by running the center pivot dry. Weather data and canopy temperatures—measured by the network of stationary IRTs in the field and on the center pivot—collected during scans were used to train ANNs to estimate average canopy temperatures obtained by a given IRT group, i.e., by a pair of IRTs with opposing views of a sprinkler zone. Two case studies were conducted to analyze the feasibility of using ANNs for this purpose. In the first case, six types of ANNs (one for each of the six IRT groups located on the center pivot) were trained using data collected during the first three scans that took place on June 26 (DOY 177), July 7 (DOY 188), and July 11 (DOY 192). Since the training of ANNs is a semi-random process that yields different results every time, 50 ANNs were trained for each ANN type and the best performing ANN among them was then selected to be used for the forecasting of average canopy temperatures that would be measured by the corresponding IRT group during the following scan (July 12, DOY 193). The accuracy of the best ANN selected for ANN type n was then assessed by predicting average canopy temperatures that would be measured by IRT group n on this date. In the second case, six types of ANNs were trained using data collected during the first six scans that, in addition to the previous dates, took place on July 17 (DOY 198), and July 20 (DOY 201). 50 ANNs were also trained for each ANN type and the best performing ANN was selected to be used for the forecasting of average canopy temperatures that would be measured by the corresponding IRT group during the following scan (July 24, DOY 205).

The typical structure of an ANN, also known as architecture, is composed of at least three layers of nodes (usually referred to as neurons) and the links between these layers (Fig. 2). The first layer is the input layer, the last one is the output layer, and all others are hidden layers. Nodes in these layers are referred to as input neurons, output neurons, and hidden neurons, respectively. ANNs used in this study were three-layered feed-forward networks consisting of sigmoid hidden neurons and linear output neurons. The number of input neurons was 10, corresponding to the number of variables that were considered relevant for the estimation of average crop canopy temperatures estimated by a given IRT group n mounted on the center pivot (Fig. 1). These variables were: (1) air temperature measured at time t during a scan, (2) relative humidity at time t, (3) solar irradiance at time t, (4) wind direction at time t, (5) wind speed at time t, (6) average canopy temperature measured by stationary IRTs at time t, (7) irrigation level (%) assigned to the experimental plot p being scanned by IRT group n at time t, (8) irrigation scheduling method assigned to plot p, (9) number of days passed since planting at the time of the scan, and (10) cumulative irrigation (including precipitation) received by experimental plot p.

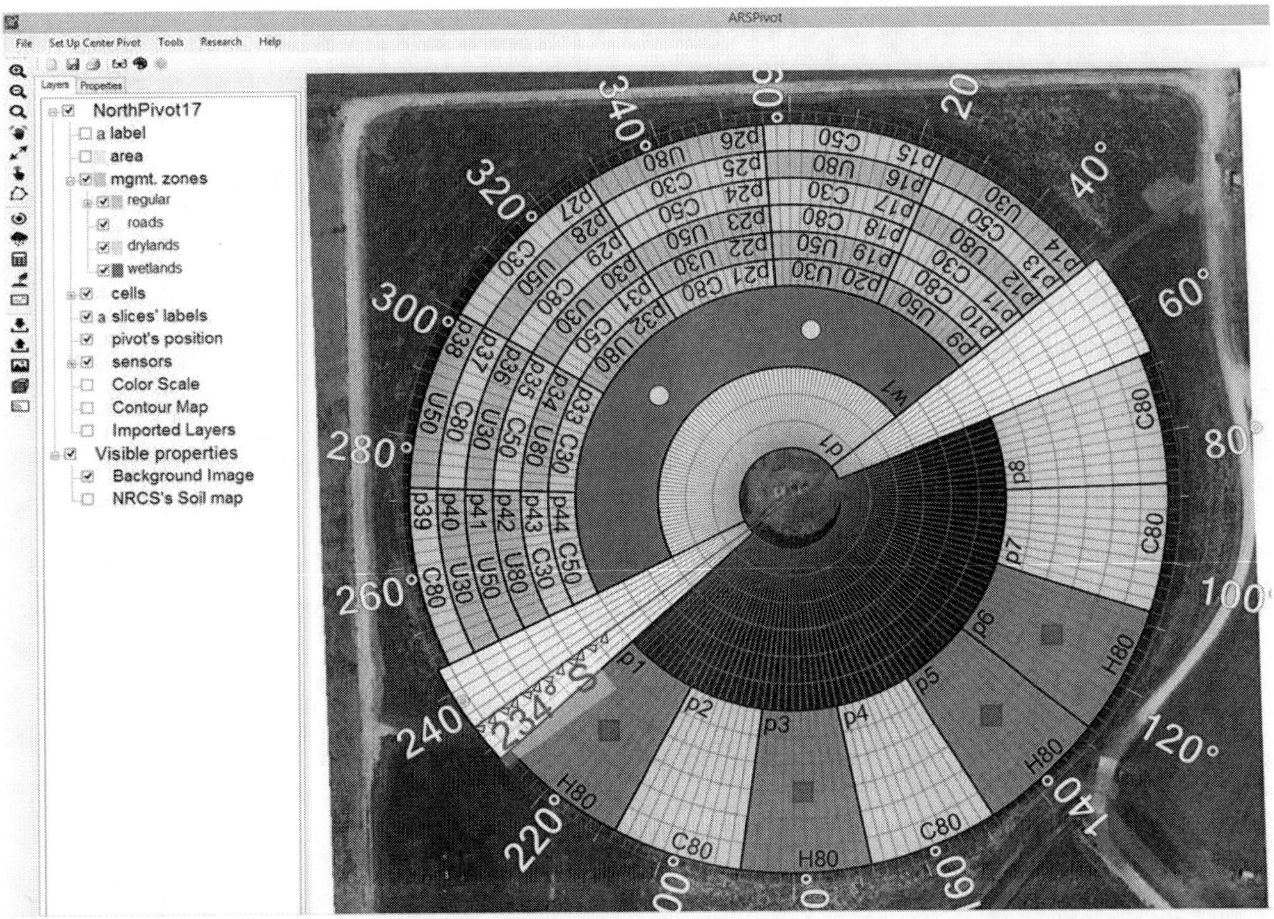

Fig. 1. Experimental setup as displayed in the ARSPivot software. Numbers inside of plots preceded by the letter 'p' indicate the numbers used to identify plots. Squares represent the approximate location of soil water sensors (TDRs). Two-small circles inside a well irrigated area (w1) indicate the approximate location of field IRTs. A red line represents the position of the center pivot and small triangles next to this line indicate the location of IRTs mounted on the center pivot.

ANNs with a single output neuron are expected to be better estimators than ANNs with multiple output neurons (Andrade et al., 2016) and thus a single output neuron was selected for ANNs used in this study. The only output neuron allocated the average canopy temperature measured by a given IRT group mounted on the center pivot (Fig. 2). Using a single output neuron for ANNs in this study offers the additional advantage of allowing ANNs to account for conditions that may be exclusive to a single IRT group, such as scanning a sprinkler zone with a clogged nozzle. The number of neurons in the hidden layer was selected as 12 after running a series of preliminary tests with 2, 4, 6, …, 18 hidden neurons. These numbers were tested following rules of thumb for the selection of the number of hidden neurons (Heaton, 2008). Additional preliminary tests were performed to determine if the addition of a second hidden layer improved the accuracy of ANNs, but their accuracy did not show to be consistently better than ANNs with a single hidden layer. The training of ANNs was performed using Matlab's neural network toolbox software (Beale et al., 2011). The Resilient Backpropagation method was selected for the training of ANNs, since it performed consistently better than other training algorithms tested during preliminary runs. The other algorithms tested were Levenberg-Marquardt, Bayesian Regularization, and Conjugate Gradient.

Datasets used for the training of ANNs in the first case study can be represented by an input matrix with dimensions M by N, and an output vector with M elements, where M is the total number of one-minute intervals occurring during the first three scans performed in the growing season, and N is the number of input variables in the ANNs, i.e., 10 (Fig. 2). The first row in the input matrix contained the values recorded for each input variable during the first one-minute interval, the second row contained the values recorded during the second interval, and so on. The output vector, on the other hand, contained the average canopy temperatures measured by an IRT group at each one-minute interval. Data contained by the training datasets obtained in this way were normalized using a Z-score normalization to account for the differences in the magnitudes of input variables. The percentages of data in these datasets allocated for the training, validation, validation, and testing of ANNs were 70%, 15%, and 15%, respectively.

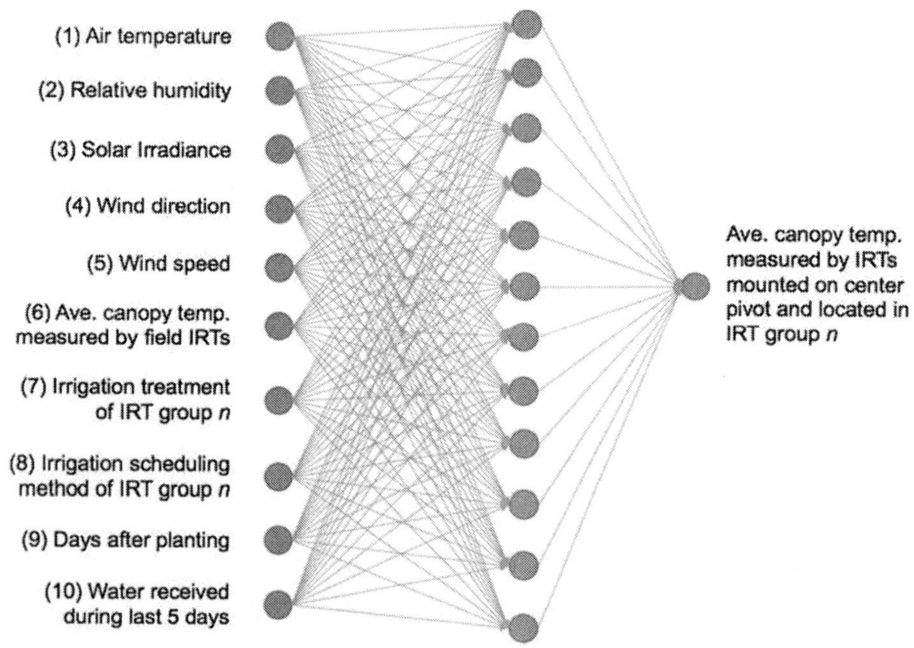

Fig. 2. Arrangement of elements in ANNs used in this study. Elements in the input, hidden, and output layers allocate 10, 12, and 1 neurons, respectively. The number of input neurons correspond to the number of variables that were considered relevant for the estimation of average canopy temperatures estimated by a given IRT group mounted on the center pivot. The only output neuron in the ANN allocates such average temperatures.

Results

Time series of average crop canopy temperatures estimated by ANNs and measured by IRT groups mounted on the center pivot are displayed for the first and second case studies in Fig. 3 and Fig. 4, respectively. On July 12, the scan started at 11.3 h at an angle of 227°. The center pivot then advanced in a counter-clockwise direction through the SSE side of the field and entered the NNW side at approximately 13 h. The scan was completed at 14.2 h when the pivot reached 248°. Since all IRT groups scanned experimental plots in the SSE side (where the highest irrigation level was assigned to all plots) before 13 h, measured canopy temperatures before this time tended to be smaller than temperatures obtained in the NNW side (where irrigation levels varied) after this time (Fig. 3). Nevertheless, ANNs were capable of approximating the oscillating pattern displayed by measured canopy temperatures through the scan (Fig. 3), with a Root Mean Squared Error (RMSE) that ranged from 1.04 °C to 2.49 °C (Table 1). To assess the impact of using ANNs for irrigation management, their estimated canopy temperatures were used by the iCWSI and hybrid methods to recalculate the prescriptions of experimental plots using these methods. No difference was found between the prescription map obtained with canopy temperatures estimated by ANNs and the prescription map obtained with canopy temperatures measured by IRTs. Hence, the accuracy of all ANNs tested in the first case study can be deemed as satisfactory.

Regarding the second case study, the scan started on July 24 at 11 h at an angle of 52°. The center pivot then advanced in a counter-clockwise direction through the NNW side of the field and entered the SSE side at approximately 12.5 h. The scan was completed at 13.7 h when the pivot arrived at 68°. Similar to the first case study, measured canopy temperatures tended to be smaller as the center pivot advanced through the SSE side of the field, i.e., after 12.5 h. As in the first case, ANNs were capable of approximating the oscillating pattern displayed by canopy temperatures through the scan (Fig. 4), with a RMSE that ranged from 2.14 °C to 2.77 °C (Table 2). When comparing the prescription maps obtained with canopy temperatures estimated by ANNs and canopy temperatures measured by IRTs, only one plot (out of 26 assigned either the iCWSI or hybrid methods) was assigned a different prescription (Fig. 5). Therefore, the accuracy of all ANNs tested in the second case study can be also deemed as satisfactory.

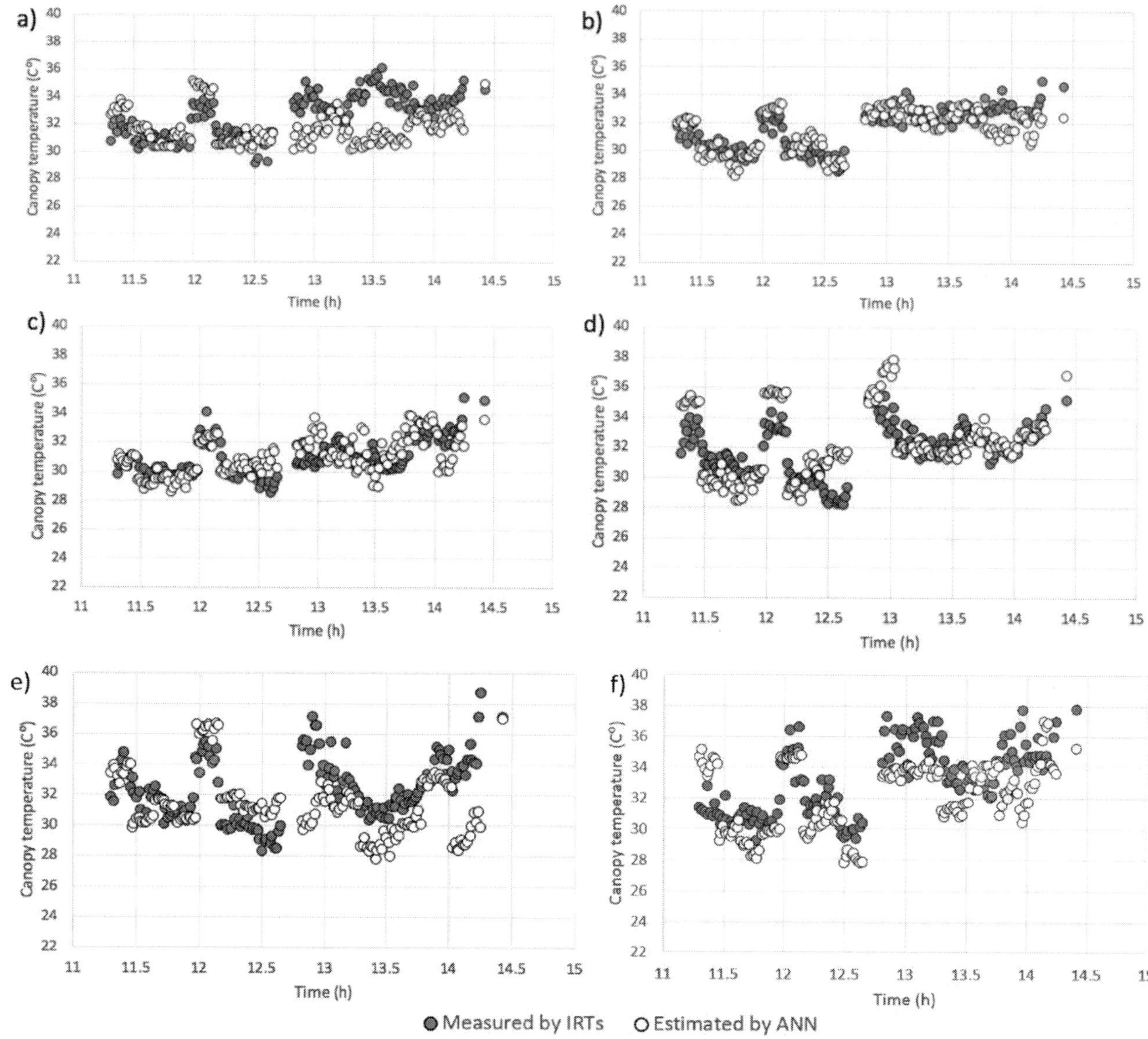

Fig. 3. Time series of canopy temperatures measured by IRTs and estimated by ANNs trained to forecast the average temperatures obtained by IRT groups a) 1, b) 2, c) 3, d) 4, e) 5, and f) 6 during July 12 (DOY 193). IRT group 1 consists of the two IRTs closest to the pivot point and IRT group 6 consists of the two IRTs farthest from the pivot point. ANNs were trained using data collected during the first three scans that took place on June 26, July 7, and July 11.

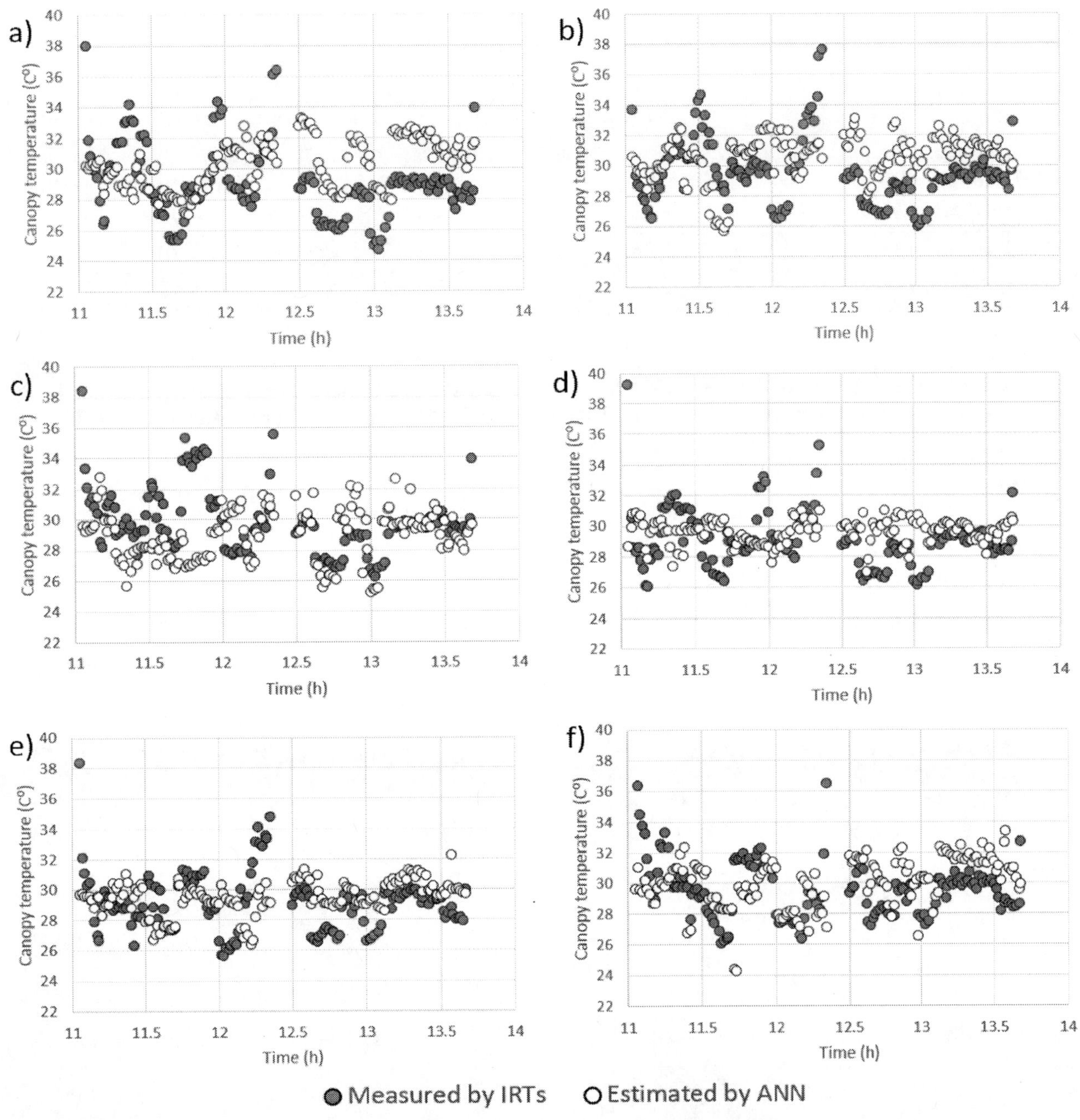

Fig. 4. Time series of canopy temperatures measured by IRTs and estimated by ANNs trained to forecast the average temperatures obtained by IRT groups a) 1, b) 2, c) 3, d) 4, e) 5, and f) 6 during July 24, 2017 (DOY 205). ANNs were trained using data collected during the first six scans that took place on June 26, July 7, July 11, July 12, July 17, and July 20.

Table 1. Root Mean Squared Error (RMSE) of ANNs used in the first case study to forecast average canopy temperatures measured by IRT groups during the scan performed on July 12

Root Mean Squared Error (RMSE)	IRT Group 1	IRT Group 2	IRT Group 3	IRT Group 4	IRT Group 5	IRT Group 6
All irrigation levels	2.06	1.04	1.16	1.52	2.49	2.10
30% irrigation level	1.21	1.07	1.52	0.76	4.02	1.49
50% irrigation level	2.38	0.70	1.18	0.56	3.29	3.11
80% irrigation level	2.14	1.11	1.03	1.84	1.58	1.91

Table 2. Root Mean Squared Error (RMSE) of ANNs used in the second case study to forecast average canopy temperatures measured by IRT groups during the scan performed on July 24

Root Mean Squared Error (RMSE)	IRT Group 1	IRT Group 2	IRT Group 3	IRT Group 4	IRT Group 5	IRT Group 6
All irrigation levels	2.77	2.64	2.72	2.18	2.14	2.42
30% irrigation level	3.20	3.29	5.04	2.30	3.07	3.78
50% irrigation level	2.52	1.34	2.40	2.29	2.12	1.28
80% irrigation level	2.72	2.70	1.67	2.11	1.81	2.15

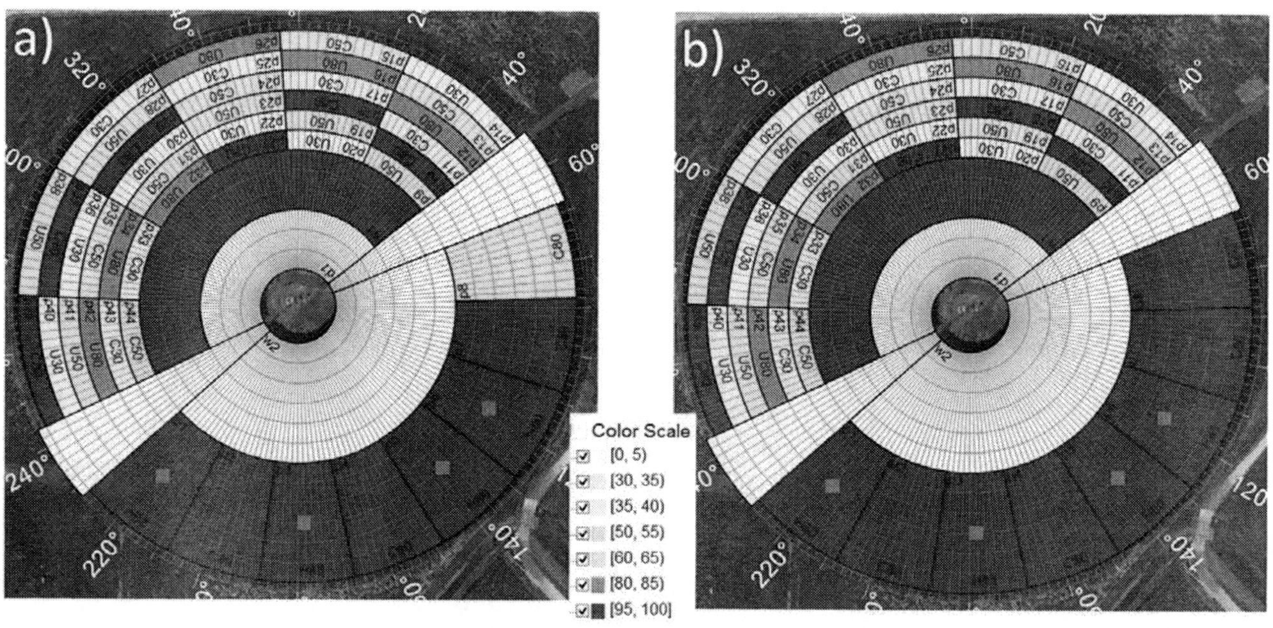

Fig. 5. Prescription maps generated using canopy temperatures a) measured by a network of wireless IRTs mounted on the center pivot and b) estimated by ANNs using data collected on July 24, 2017 (DOY 205). Prescriptions are displayed as percentages of a pre-specified maximum irrigation depth of 30.5 mm (1.2 in). Only one plot (p8) received a different prescription when using the canopy temperatures estimated by ANNs.

Conclusions

Machine learning is a data analysis technique that can be used to solve a wide range of problems, since it uses previous data to train computational machines that can make predictions on new data with no human intervention. This study analyzes the feasibility of using a popular machine learning technique to estimate canopy temperatures in a field irrigated by a Variable Rate Irrigation (VRI) center pivot system. Results indicate that the machine learning technique tested can predict canopy temperatures with a satisfactory accuracy. Machine learning technology can be useful to add redundancy to an Irrigation Scheduling Supervisory Control and Data Acquisiton System (ISSCADAS) patented by scientists from ARS (Bushland, Texas). The addition of machine learning capabilities to the ISSCADAS can assist users when poor visibility conditions prevent the correct estimation of canopy temperatures using the network of sensing systems incorporated by the ISSCADAS.

Acknowledgements

This material is based upon work that is supported by the National Institute of Food and Agriculture, U.S. Department of Agriculture, under award number 2016-67021-24420. Work reported in this paper was also accomplished as part of a Cooperative Research and Development agreement between USDA-ARS and Valmont Industries, Inc., Valley, NE (Agreement No.: 58-3K95-0-1455-M).

References

Andrade, M. A., Shaughnessy, S. A. O., and Evett, S. R. (2015). ARSmartPivot v . 1 – Sensor based management software for center pivot irrigation systems. *In Proc. ASABE Annual International Meeting,* St. Joseph, MI.

Andrade, M. A., Choi, C. Y., Lansey, K., Asce, A. M., and Jung, D. (2016). Enhanced Artificial Neural Networks Estimating Water Quality Constraints for the Optimal Water Distribution Systems Design. *Journal of Water Resources Planning and Management,* (April). https://doi.org/10.1061/(ASCE)WR.1943-5452

Andrade, M. A., Shaughnessy, S. A. O., and Evett, S. R. (2017). ARSPIVOT, A sensor based Decision Support Tool for the Integrated irrigation Management of VRI Center Pivot Systems, Oak Ridge Institute for Science and Education Sponsored by USDA-ARS Email : Alejandro.Andrade@ars.usda.gov USDA-ARS USDA-ARS, 5–14.

Beale, M., Hagan, M., and Demuth, H. (2011). *Neural Network Toolbox User's Guide*. Natick, MA: Mathworks.

Evett, S. R., O'Shaughnessy, S. A., and Peters, R. T. (2014). *8924031 B1*. United States of America.

Farifteh, J., Van der Meer, F., Atzberger, C., and Carranza, E. J. M. (2007). Quantitative analysis of salt-affected soil reflectance spectra: A comparison of two adaptive methods (PLSR and ANN). *Remote Sensing of Environment*, *110*(1), 59–78. https://doi.org/10.1016/j.rse.2007.02.005

Ferreira, P. M., Faria, E. A., and Ruano, A. E. (2002). Neural network models in greenhouse air temperature prediction. *Neurocomputing*, *43*(1–4), 51–75. https://doi.org/10.1016/S0925-2312(01)00620-8

He, F., and Ma, C. (2010). Modeling greenhouse air humidity by means of artificial neural network and principal component analysis. *Computers and Electronics in Agriculture*, *71*(SUPPL. 1). https://doi.org/10.1016/j.compag.2009.07.011

Heaton, J. (2008). *Introduction to Neural Networks with Java* (2nd ed.). st. Luis, MO: Heaton Research, Inc.

Jackson, R. D., Idso, S. B., Reginato, R. J., and Pinter, P. J. (1981). Canopy temperature as a Crop Water Stress Indicator. *Water Resources Research*, *17*(4), 1133–1138. Retrieved from http://onlinelibrary.wiley.com/doi/10.1029/WR017i004p01133/abstract

Kavdir, I. (2004). Discrimination of sunflower, weed and soil by artificial neural networks. *Computers and Electronics in Agriculture*, *44*(2), 153–160. https://doi.org/10.1016/j.compag.2004.03.006

Kaul, M., Hill, R. L., and Walthall, C. (2005). Artificial neural networks for corn and soybean yield prediction. *Agricultural Systems*, *85*(1), 1–18. https://doi.org/10.1016/j.agsy.2004.07.009

O'Shaughnessy, S. A., Evett, S. R., Colaizzi, P. D., and Howell, T. A. (2013). Wireless Sensor Network Effectively Controls Center Pivot Irrigation of Sorghum. *Applied Engineering in Agriculture*, *29*(6), 853–864. https://doi.org/10.13031/aea.29.9921

O'Shaughnessy, S. A., Andrade, M. A., and Evett, S. R. (2017). Using an integrated crop water stress index for irrigation scheduling of two corn hybrids in a semi-arid region. *Irrigation Science*, *31*(5), 451–467. https://doi.org/10.1007/s00271-017-0552-x

Pham, X., and Stack, M. (2018). How data analytics is transforming agriculture. *Business Horizons*, *61*(1), 125–133. https://doi.org/10.1016/j.bushor.2017.09.011

Schaap, M. G., Leij, F. J., and Van Genuchten, M. T. (1998). Neural network analysis for hierarchical prediction of soil hydraulic properties. *Soil Science Society of America Journal*, *62*(4), 847–855. https://doi.org/10.2136/sssaj1998.03615995006200040001x

Yang, C., Prasher, S., Landry, J., Ramaswamy, H., and Ditommaso, A. (2000). Application of artificial neural networks in image recognition and classification of crop of weeds. *Canadian Agricultural Engineering*, *42*(3), 147–152. https://doi.org/10.1007/s00417-014-2568-8

Improving Applications of Center Pivot Irrigation

Dana Porter, Ph.D., P.E., Professor, Extension Program Leader and Associate Department Head, Department of Biological and Agricultural Engineering and Texas A&M AgriLife Research and Extension Service, Lubbock, Texas
d-porter@tamu.edu

Thomas Marek, M.S., P.E., Regents Fellow and Senior Research Engineer
Texas A&M AgriLife Research, Amarillo/Bushland, Texas
tmarek@ag.tamu.edu

Abstract

Low pressure center pivot irrigation systems, offering potential for high energy efficiency and high water application efficiency, are widely used in much of the Great Plains, especially in the Texas High Plains. Further improvements in adoption and management of these systems are possible, and benefits are more likely to be realized with simplified integration of readily available information (such as soil moisture, evapotranspiration, forecast conditions, etc.) for improved irrigation scheduling. Advancements (patents pending) will provide viable solutions to overcome existing barriers to adoption and limitations of commercially available systems by offering platform independent integration of information to simplify and improve irrigation management. These developments simplify integration of information for irrigation decisions using commercially available sensors and components, greatly improving potential benefits of existing technologies (and return on investment for these technologies) for a relatively low cost.

Keywords: Precision irrigation, LEPA, LESA, irrigation management, center pivot control

Background

Low pressure center pivot irrigation was introduced to the Texas High Plains approximately forty years ago with the development of Low Energy Precision Application (LEPA) irrigation (Lyle and Bordovsky, 1981). LEPA and important adaptations of low pressure sprinkler irrigation, including Low Elevation Spray Application (LESA), Mid-Elevation Spray Application (MESA), and Low Pressure In-Canopy Irrigation (LPIC) and other related systems have become the predominant irrigation application platform, comprising approximately 70% of the irrigated acreage in Texas (Wagner, 2012); over 75% of the irrigated acres in the Texas Northern High Plains (Amosson, et al., 2016); 70% of irrigated acres in Oklahoma (Taghvaeian, 2015); and over 50% of irrigated acres in the many other states, including Nebraska and Kansas (USDA-NASS, 2013).

Many factors have contributed to the popularity of low pressure center pivot irrigation systems. Since energy requirement (and hence energy cost) is directly related to system pressure, low pressure center pivot systems require less energy for operation than higher pressure systems. In arid and semi-arid conditions, especially in highly advective environments, low pressure application systems that apply water at or near the soil surface and/or deliver relatively large water droplet sizes reduce evaporation losses during application compared to high pressure systems. While true LEPA systems' applicability is limited to fields with relatively low slope (less than 1% slope), the low pressure spray irrigation systems (LESA, MESA, LPIC) are widely applicable over a range of topographies, soil types, row spacings, and crop types.

During the last forty years, center pivot associated technologies have expanded and have been adopted for a wider range of conditions and crops, as end-users sought - and equipment manufacturers delivered – solutions to practical challenges in the field. From LEPA drag hoses and bubblers, to a wide range of sprinkler options (applicator height and trajectory options, droplet sizes, and chemigation applications), to corrosion-resistant materials, low pressure center pivot irrigation tools are readily available for diverse agricultural irrigation applications.

Cost-share programs, including USDA-NRCS Environment Quality Incentives Program (EQIP) and low interest loan programs offered through groundwater conservation districts have encouraged adoption of higher efficiency irrigation systems to replace lower efficiency systems. These programs have been expanded to incentivize improved irrigation management, including application of soil moisture sensors and weather-based irrigation scheduling, variable frequency drives, and other improvements (USDA-NRCS, 2018).

Challenges

While there are great technologies and tools commercially and readily available, the potential benefits promised by low pressure center pivot irrigation generally are not fully realized. "Implementation gaps" (Lamm and Porter, 2017) often are linked to knowledge gaps related to the technologies. For instance, end-users tend to under-utilize the variable rate irrigation (speed control) capabilities of commercially available center pivot irrigation controllers, either because they are not familiar with their systems' capabilities, or they are not able to obtain, implement, or justify time or expense of updating VRI prescriptions. They may not understand the relationships between design flow rate, pressure, and nozzle packages, and their impacts on distribution uniformity. Hence potential water use efficiency gains afforded by precise placement of water and fertigation are missed. Many knowledge and implementation gaps can (or should) be addressed through additional research and improved communication among - and technology transfer by- research/extension programs, USDA-NRCS, equipment manufacturers and dealers.

It is worth noting that *lack of knowledge* does not necessarily equate to *lack of information*. A great body of research and increasingly available and affordable sensors and telemetry systems make data readily available for weather-based (evapotranspiration-based) irrigation scheduling, soil moisture monitoring, plant water status monitoring, soil and plant fertility monitoring. Even technology savvy and progressive end-users can be overwhelmed with the large quantity of information, and lack time and understanding to interpret it and ability to translate it into their irrigation and crop management decisions. Increasingly, irrigation scheduling tools (software) and irrigation system control technologies are integrating field-based data and research-based recommendations into useful, "user-friendly" packages, whereby information can be processed and interpreted for the user, with minimal effort and expertise required of the user.

Integration of Information and Technologies

Major manufacturers and third party companies offer advanced remote center pivot irrigation system monitoring and control (Mowitz, 2013; Kranz, 2011) and/or field data acquisition and interpretation. Manufacturers, other private (commercial) interests, universities, and USDA-Agricultural Research Service have developed a wide range of irrigation management software (decision support systems), smart phone apps, and other useful resources. Their products range in applicability, technical sophistication, level of technical support, and ease of use. Examples of such decision tools include DIEM (Bordovsky, et al, 2016); SmartIrrigation Apps (Migliaccio, et al, 2016); and many others.

There are a great number of useful advanced irrigation technologies and tools available, with more under development. Maximizing their benefits will require addressing several considerations, including applicability to the specific crop production system (soils, climate, crops/rotations, irrigation system capabilities/constraints, management/labor capabilities/constraints); education and technical support; cost effectiveness; and ease of use. Some of these concerns are addressed in a suite of technologies (patents pending) that include improvements over existing technologies:

- precise center pivot system speed control;
- optimization of the number and placement of soil moisture sensors; and
- irrigation system control with predictive water balance capabilities.

The approach used in these developments include:

- wireless sensors/technologies and automated data acquisition and packaging to minimize data entry requirements by end-users;
- integration of data from multiple sources and appropriate models and machine learning to automate interpretation of field-based information to inform irrigation decision making;
- use and incorporation of commercially available "off-the-shelf" components for cost-effectiveness;
- operational platform independence for transferability and ease of "retrofitting" to existing systems in the field; and
- simple user interface for convenient access via personal computer, tablet or smartphone devices.

Despite the many great advances by industry and public sector programs, knowledge gaps remain. Crop models have their shortcomings; new crop genetics requiring differing management continue to evolve; many site-specific, season/climate-specific, and crop-specific considerations are not easily "simplified" or modeled; and communications and internet connectedness can be limited in rural areas. Agricultural systems are increasingly complex, so ongoing research and development are needed to fill gaps in knowledge and technology applicability.

Summary

Communication, education and collaboration will continue to be important in supporting optimal use of available and emerging irrigation tools and technologies. Improved communication between and among the irrigation industry, service providers, technical advisors, educators, researchers, end-users, water planners and policy makers, and other decision makers will help deliver clearer, consistent messages to support better irrigation decision making. Given that end-users are increasingly busy, and rthat highly qualified technical support professionals are invaluable and often hard to recruit, train and retain, new approaches to education, communication and technical support are needed. Information delivery and technical support must adapt to end-users' preferences (venues, "on-demand" media, concise presentation). Workforce challenges (limited numbers of people interested in irrigation work, exacerbated by limited and declining numbers of academic programs offering relevant irrigation curricula, sometimes attributed to limited external investment in the programs) affect both industry and academic /public sectors. Integration of information and technologies and improvements in technology transfer to support most beneficial use of the information and technologies will be increasingly important as irrigation water resources continue to decline regionally, nationally and globally.

Acknowledgements

The authors greatly appreciate the contributions of Dr. Jiang Hu, Professor, Dr. Justin Sun, Ph.D. graduate, Yanxiang Yang, Ph.D. candidate, and Hongxin Kong, Ph.D. student, Texas A&M University Department of Electrical and Computer Engineering; Dr. Kevin Heflin, Texas A&M AgriLIfe Research – Amarillo/Bushland; and Dr. Gary Marek, USDA-ARS-CPRL, Bushland, TX.

The authors gratefully acknowledge funding and in-kind support for their program provided by the Texas A&M University Water Seed Grant program, USDA-NRCS CIG Program, the USDA-AFRI-CAP Program, High Plains Underground Water Conservation District Research Program, and the USDA-ARS Ogallala Aquifer Program. The Ogallala Aquifer Program (OAP) is a consortium between the USDA Agricultural Research Service, Kansas State University, Texas A&M University, Texas Tech University and West Texas A&M University. Associated projects are included in the USDA REEport system that supports many state and federal Hatch projects.

References

Amosson, Steve, Shyam Nair, Bridget Guerrero, and Thomas Marek. 2015. Panhandle Regional Water Plan Task 5 Report: Agricultural Water Management Strategies. 2016 Panhandle Water Plan Vol II. Panhandle Regional Water Planning Group, Amarillo, TX.

Bordovsky, James, Jim Wall, Dana Porter, Keith Biggers. 2016. DIEM Dashboard for Irrigation Efficiency Management. Texas A&M AgriLife Research and Texas Center for Applied Technology. College Station, TX.

Lamm, Freddie and Dana Porter. 2017. Ogallala Aquifer Program Center Pivot Irrigation Technology Transfer Effort. Proceedings, 2017 Irrigation Association Technical Conference, Orlando, Florida, November 6-10, 2017.

Kranz, William. 2011. Monitoring Irrigation Water Application with Computerized Controllers. Proceedings of the 23rd Annual Central Plains Irrigation Conference, Burlington, CO., February 22-23, 2011.

Lyle, W. M. & Bordovsky, J. P. 1981. Low energy precision application (LEPA) irrigation system. *Trans. ASAE,* 24(5), 1241-1245.

Migliaccio, K.W., K.T. Morgan, G. Vellidis, L. Zotarelli, C. Fraisse, B.A. Zurweller, J.H. Andreis, J.H. Crane, and D.L. Rowland. 2016. Smartphone Apps for Irrigation Scheduling. *Trans. of the ASABE*, Vol. 59(1):291-301. American Society of Agricultural and Biological Engineers.

Taghvaeian, Saleh. 2015. Irrigated Agriculture in Oklahoma. Fact Sheet BAE-1530, Oklahoma Cooperative Extension Service. Stillwater, OK.

Wagner, Kevin. 2012. Status and Trends in Irrigated Agriculture in Texas. Special Report TWRI EM-115. Texas Water Resources Institute, Texas A&M University, College Station, TX.

Mowitz, Dave. 2013. Irrigation Insider: Upgrading a Center Pivot to Remote Monitoring and Control. *Successful Farming* online. Dec.18, 2013. Available at: https://www.agriculture.com/video/irrigation-insider-upgrading-a-center-pivot-to-remote-monitoring-and-control

USDA-NASS. 2013. Set 7. Pressure-sprinkler irrigation statistics by system type. United States Dept. of Agriculture Economic Research Service. Available at: https://www.ers.usda.gov/data-products/irrigated-agriculture-in-the-united-states/summary-of-results/

USDA-NRCS. 2018. Texas Payment Schedules. United States Department of Agriculture Natural Resources Conservation Service. Available at: https://www.nrcs.usda.gov/wps/portal/nrcs/detail/national/programs/financial/?cid=nrcseprd1328414

Effect of Limited Water Supplies on Center Pivot Performance

by
Derrel Martin
Professor Irrigation and Water Resources Engineering
University of Nebraska-Lincoln
Lincoln, NE

Abstract

Performance of a center pivot for a field with a general slope of 1% was simulated for three rotations of the center pivot lateral (upslope, level and downslope). The performance was simulated for a range of inlet pressures to the lateral for center pivots that included pressure regulators and systems without pressure regulators. The operation of the pivot in unison with a pump was simulated for these conditions when the water supplies were high in the spring and consistent with conditions used for design. Conditions were also simulated for low supply conditions that occur because of decreasing flow supplies as water tables drop and or flow is throttled to avoid the supply capacity of the well. Results show that regulators do improve uniformity for high-flow conditions but do not improve the discharge uniformity when inlet pressures—and therefore system inflow—drop below design conditions. The variability of discharge at the distal end of the pivot is less for unregulated conditions than when regulators are used for these conditions. More work is necessary to evaluate a wider range of slope and supply conditions. The approach needs to be expanded to compute a whole-field uniformity. The field map could offer the opportunity to utilize speed control on the pivot to enhance uniformity with rotation around the field. Unified design procedures should also be explored for systems that experience such variations in water supply during the irrigation season.

Introduction

Center pivot systems have been used extensively because of the ability to efficiently apply water for many different soil-crop-management conditions. However, the water supply capacity of aquifers and wells varies throughout the growing season for some locations across the Great Plains. Therefore, center pivots designed to operate efficiently when a full water supply is available may operate during extended portions of the growing season when the water supply is below design requirements. We don't know how many systems face this problem, but the condition is widespread across the Great Plains. The goal of this paper is to describe how pivot system performance is affected when forced to operate under varying supply conditions. The characteristics of pivots for varying inflow pressure/flow conditions will be analyzed. Management strategies and system designs to best operate under varying inflows will be discussed.

Center-pivot irrigation systems utilize sprinkler devices to apply water based on the pressure available at outlets along the pivot lateral. The pressure at an outlet depends on the pressure at the inlet to the pivot lateral, elevation across the field, the original design of sprinkler devices on the pivot and the friction loss along the pivot lateral. Some factors are static based on physical features of the field. Other factors are dynamic based on flow and pressure characteristics across the field and during the irrigation season. The irrigation system must interact with the pumping system. An example of a center pivot supplied by a groundwater well is illustrated in figure 1. The hydraulic components consists of the static water level—the distance from the soil surface to the water table when not pumping—plus the drawdown in the well due to aquifer and pumping characteristics. The static water level and the drawdown represent the lift required to bring water to the pump base. The elevation change from the pump discharge base to the inlet of the center pivot is static but contributes to the total head required for the irrigation system. The friction loss that occurs in the various components on the system is dynamic and depends on the original design and the flow of water. The system also requires pressure to discharge water in a uniform manner across the field. The pressure required for varying system flows compromise what is commonly referred to as the total dynamic head requirement of the irrigation system as illustrated in graph in figure 1.

Matching a pump to a center pivot system involves determining the flow where the head output from the pump equals the requirements for the irrigation system. This is shown in figure 1 where the pump curve intersects the total dynamic head for the system. This analysis focuses on the pressure requirements to operate the center pivot under different water supply conditions. Subtraction of the head requirements for all components except pressure from the pump outpout provides an estimate of the amount of pressure available to operate the center pivot for varying flow rated. The pressure available to the inlet to the center pivot irrigation system can be described expressed as:

$$P_i = \gamma \left(TDH_p - L - s_w - E_L - F_L \right) \tag{1}$$

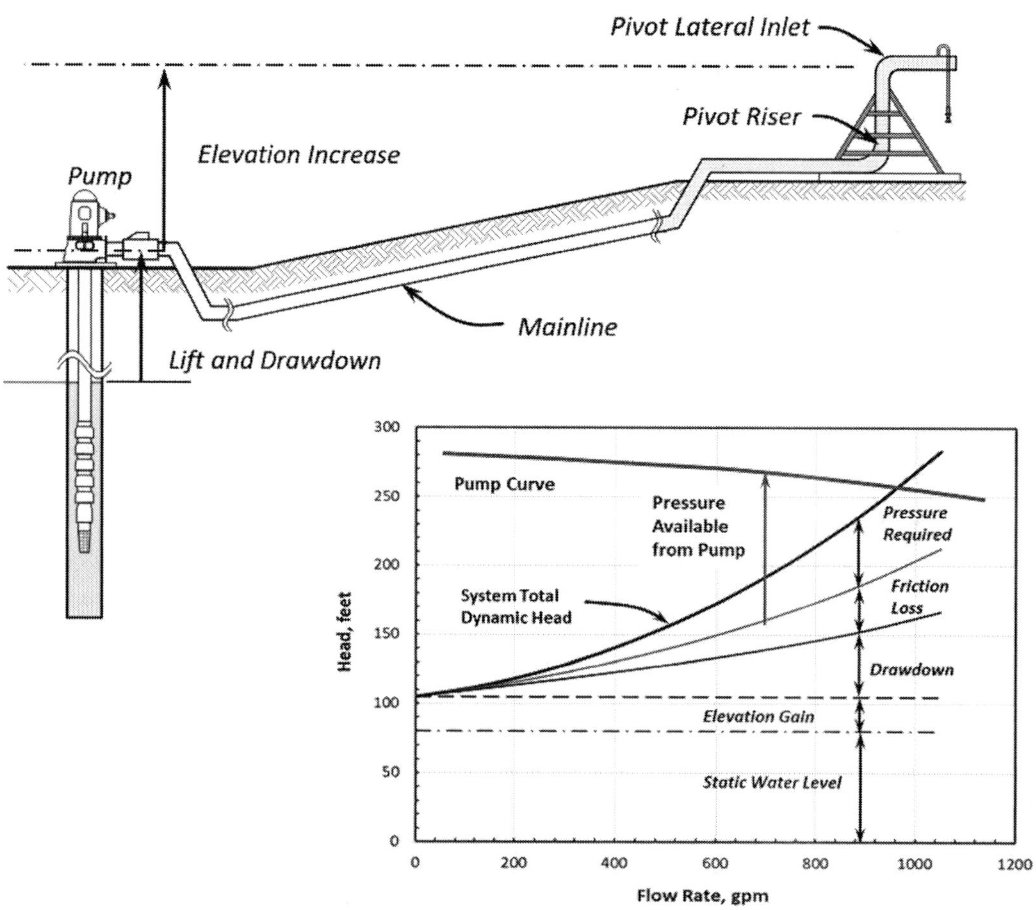

Figure 1. Illustration of the components of a center pivot field and their contribution to the total dynamic head requirement.

Matching a pump to a center pivot system involves determining the flow where the head output from the pump equals the requirements for the irrigation system. This is shown in figure 1 where the pump curve intersects the total dynamic head for the system. This analysis focuses on the pressure requirements to operate the center pivot under different water supply conditions. Subtraction of the head requirements for all components except pressure from the pump outpout provides an estimate of the amount of pressure available to operate the center pivot for varying flow rated. The pressure available to the inlet to the center pivot irrigation system can be described expressed as:

$$P_i = \gamma \left(TDH_p - L - s_w - E_L - F_L \right) \quad (2)$$

where: P_i = the pressure available to the center pivot inlet, TDH_p = the total dynamic head output from the pump, L = the static water level-lift, s_w is the drawdown, E_L is the elevation change from the pump base to the pivot inlet (could be positive or negative), and F_L is the friction loss in the system components.

Some components in equation 1 vary with the flow through the system but may also vary during the irrigations season due to other factors. The depth of the saturated thickness of an aquifer measured in an observation well in southwest Nebraska is shown in Figure 2. The saturated thickness represents the depth of water in the aquifer and reflects the change in the static water level during the year. The saturated thickness is deepest in the spring prior to irrigation during the season and then the water level drops during the pumping season. The saturated thickness drops by 50% some seasons. These measurements were from an observation well that represents general conditions for a region not the level in an irrigation well since water is not pumped from and observation well. The drawdown in an irrigation well is larger when the saturated thickness is smaller. Thus, the pressure available to the inlet of the pivot varies through the growing season due to changing aquifer conditions. In some cases, the depth of water in a well may limit pump discharge as pumps might produce air in the flow stream if pumping capacity exceeds the well yield. Producers often control "pumping air" using throttling valves. Thus, declining water levels which may lead to throttling valves which increases the friction loss. The impact of declining water level is not illustrated in figure 1. These factors illustrate that the pressure available to supply the center pivot can vary throughout the growing season. The goal is to analyze the effect of varying pressure availability on the performance of center pivots.

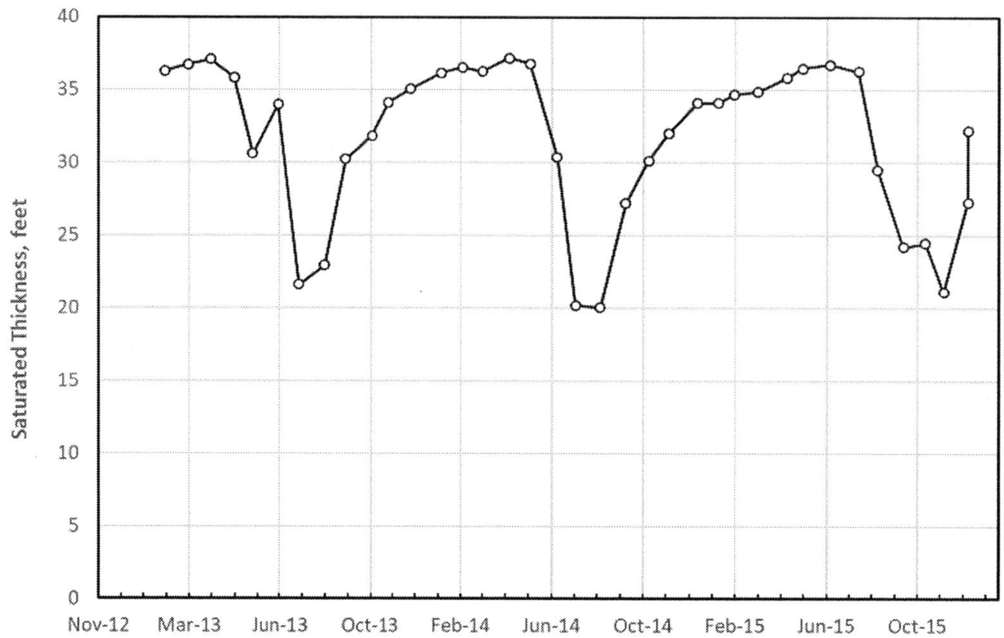

Figure 2. Depth of the saturated thickness of an observation well in southwest Nebraska.

Pivot Performance

The flow rate required from sprinklers along a center pivot lateral must increase in a linear fashion to maintain uniformity. The linear increase is required because sprinklers located at the distal end of the pivot lateral irrigate a much larger area than a sprinkler at the midpoint of the lateral. The discharge required from a sprinkler can be expressed as:

$$q_s = \left(\frac{2\, Q_s\, S}{R_s^{\,2}} \right) R \tag{3}$$

where; q_s = the discharge required from a sprinkler located a distance R from the pivot inlet, Q_s = the flow into the pivot lateral, R_s is the radius of the irrigated field and S = the distance between sprinklers along the lateral at location R. Uniformity requires the slope of the sprinkler discharge included in the parentheses in equation 2.

The actual discharge from sprinklers on a lateral depends on the pressure available to the sprinkler located at radial distance R, and the size of nozzle installed in the sprinkler at location R. The discharge from an individual sprinkler can be represented by:

$$q_s = C_d\, D_n^{\,2} \sqrt{P} \tag{4}$$

where: C_d = the discharge coefficient for a model of sprinkler nozzle, D_n is the effective diameter of the nozzle in the sprinkler and P is the pressure available at the base of the sprinkler. Therefore, if the variation in pressure along the center pivot lateral is known, then the discharge from individual sprinklers can be estimated. The distribution of sprinkler discharge may not follow a linear pattern, and/or the linear slope may not follow the value required for uniformity. Variations in pressure along the pivot lateral that does not follow the pattern that occurs from design will degrade uniformity across the field. Tests are required to determine if the distribution of discharge from sprinklers remains linear—at the proper slope—as the pressure available in the pivot lateral decreases during the growing season. The analysis will focus on whether the slope of linear relationship is adequate to maintain the uniformity of the center pivot.

The performance of center pivot irrigation systems has been simulated with computer models many times. Originally, Bittinger and Longenbaugh (1962) described an overlapping procedure to determine how the water application pattern from successive sprinklers varied along the center pivot lateral. Beccard and Heermann (1981) built a very flexible simulation program to further simulate the effect of sprinkler characteristics, sprinkler spacing, computer movement and terrain on the uniformity of water application for a center pivot oriented in one direction in a field. Other such as James (1982), James (1984), James and Blair (1984), Heermann (1990), Bremond and Molle (1995), and Heermann and Spofford (1998) built on the original concept of Bittenger and Longenbaugh. Numerous other authors have built models and conducted simulation studies of center pivot analysis since these early studies. These approaches rely on a mathematical representation of the pattern of water application about individual sprinklers. These relationships depend on sprinklers operating in a pressure range specified by manufacturers for specific devices. The previous simulation approaches have primarily focused on simulation and testing of sprinkler packages in the design process. The application here requires analysis of an existing sprinkler package operated under conditions quite different than intended in the design process. The product of the analysis has often been the coefficient of uniformity of water application for the sprinkler package.

Heermann (2006) performed an analysis of the impact of lowering water tables on the performance of center pivots considering the interaction of the pumping system. He included limited analysis of the impact of pressure regulators on the water distribution and the increase in operating cost due to pressure loss in the regulators. The analysis by Heermann illustrated how irrigators should modify systems as the water table declines over time; however, the analysis did not consider ramifications to variations during the irrigation season. The analysis conducted here used a spreadsheet model to compute the sprinkler discharge and pressure variation along a center pivot lateral for varying pressure/inflow conditions for several field slopes. The sprinkler package on the pivot was designed to produce the required sprinkler discharge, for the steepest upslope in the field, when the water supply to the pivot was higher in the spring. This is how many systems are currently operated in the Great Plains. The performance for other slopes and inflow pressure conditions was compared to the design conditions at high flows. Altering the inflow pressure—and thus the pivot inflow—changes the pressure and sprinkler discharge distribution along the lateral. The effects of changed operating conditions on uniformity was evaluated by comparing the flow from each sprinkler to the flow for the respective sprinkler when operated as originally designed. If the ratio of sprinkler discharge for the altered inflow to the design requirement is constant along the lateral, then the application uniformity will be the like the original design—even though the depth of water applied will change. If the ratio of discharge is not consistent along the lateral, the uniformity will deviate from that attained through the initial design. The uniformity estimated by Bittinger and Longenbaugh and others requires analysis of the changes in water distribution about individual sprinklers using an overlapping procedure like that in the CPED computer program by Heermann (2006). Analysis would require description of the distribution characteristics of sprinkler devices over a wide range of pressures not anticipated by manufactures. Thus, additional data would be necessary to describe the performance of sprinkler devices operated outside the range they were intended.

It is common for the flow rate from wells to be largest in the spring when the systems are first used. The supply capacity and available pressure may decrease during the growing season leading to smaller water application depths and a reduction in the uniformity of application within the field. The variation in pressure and sprinkler discharge as a function of available pressure to the inlet of the pivot, slope of land in the field, and the impact of pressure regulators on uniformity will be analyzed. Results will illustrate the change in the uniformity and highlight the expected changes in the water supply for the pivot.

The analysis will include an evaluation of the impact of pressure regulators on the operation of the sprinklers on the center pivot. Some experts have noted that pressure regulators may not be advantageous when the amount of pressure available to the pivot decreases during the growing season and therefore the flow decreases. The effect of the pressure regulators on the distribution of water for fields that have mild slopes and for level fields will be evaluated.

Pivot Design

A generic center pivot irrigation system was selected for analysis where the inflow and pressure vary significantly throughout the growing season. For these conditions, it is essential to select sprinkler devices that are operable over a wide range of pressures. It is also likely that the pressure along the system could drop quite low during the latter part of the summer; therefore, the spacing of devices along the lateral should be rather small to maintain local uniformity. Thus, if the throw pattern from the device is negatively affected by pressure drops below design conditions the overlap between closely space devices should provide a more acceptable local uniformity. For example, the variation of the wetted diameter for a sprayhead sprinkler device is illustrated in figure 3. The diameter decreases by as much as 50% for the pressure range shown. The sprayhead can operate over a wider range than shown in figure 3; therefore, sprayheads were chosen for the center pivot sprinkler package. To maintain reasonable uniformity at low pressures the sprayheads were spaced 7.5 feet apart along the lateral.

The center pivot was designed a field that has a general slope of either 1% or 2% across the field. This provides a portion of the field to evaluate the performance when the pivot lateral is generally level. The sprinkler package—including nozzle sizes for each outlet—for the basic design was based upon the water supply conditions at the start of the irrigation season for an inflow of 800 gallons per minute. The nozzles sizes for the sprinkler package was based on pivot lateral aligned along the upslope portion of the field. Therefore, the sprinkler package was designed assuming that the pivot lateral ran directly uphill. That orientation will provide the least pressure available at the far end of the pivot lateral and would be the critical location for selection of sprinkler devices and the estimation of available pressure.

As the center pivot rotates through other angles of the field the pressure at a point in the pipeline will increase reaching the maximum when the pivot lateral points directly downhill. When pressure regulators are utilized to enhance uniformity, the devices are normally selected for the upslope orientation of the pivot as describe. Since that orientation would provide the lowest pressure at the distal end of the lateral, all other areas of the field will experience higher pressures but constant flow when regulators are used. If the inflow pressure drops below that required for the high flows then portions of the field will experience reductions in sprinkler discharge. The impact of pressure regulators for such conditions is uncertain.

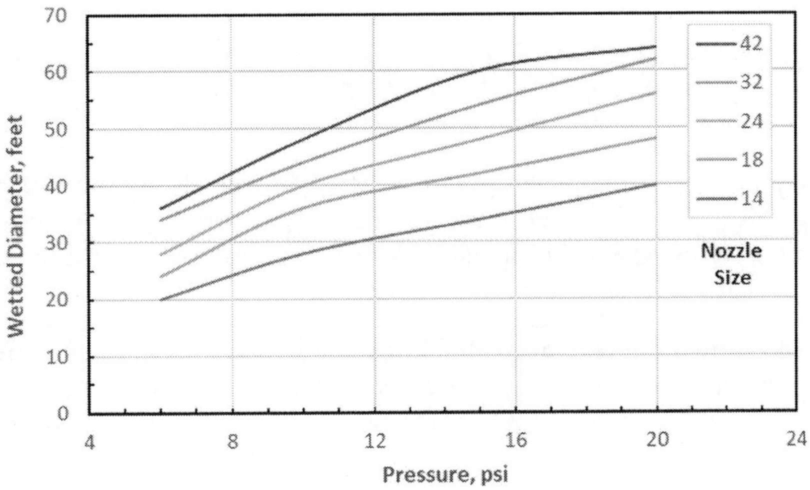

Figure 3. Variation of the wetted diameter of a sprayhead sprinkler device for vary pressure and nozzle size when mounted nine feet high.

We selected 20-psi pressure regulators for the design at 800 gallons/minute (gpm) at the start of the irrigation season. Devices were installed on a 1300-ft center-pivot lateral of 6 ⅝ inch galvanized steel tubing. Sprinkler were spaced 7.5 feet apart along the lateral. The design pressure at the distal end of the pivot lateral when sloping uphill was set to 5 psi above the required pressure for the regulated system. Therefore, pressure at the distal end of the pivot lateral when oriented uphill was 25 psi. If regulators are not used the center pivot was also designed to provide the required flow rate in the uphill position. Since regulators are not utilized the pressure at other angles of rotation would experience higher pressures and larger sprinkler discharges than required for uniformity based on equation 2. The Hazen-Williams equation was used to calculate friction loss within the pivot lateral. A resistance coefficient was set to 140 in Williams equation. The inside diameter of the lateral pipe was 6.407 inches. The pressure distribution for various orientations was computed by starting at the distal end of the pivot and computing the discharge from each sprinkler adding the discharge and calculating the friction loss incrementally from the distal end to the pivot inlet. This provides a relationship between the pressure at the distal end of the pivot lateral and the inflow to the center pivot. The hydraulic distribution along the pivot lateral was computed when the pivot was sloping uphill, downhill and on flat portions of the field.

Pressure Regulator

We modeled the operation of pressure regulators as illustrated in figure 4. Most pressure regulators have a recommended maximum and minute flow rate. The flow variation causes variation in the outlet for the same inlet pressure. The pressure variation about the midrange outlet pressure was approximately ±5% of the rated pressure. Once the inlet pressure reaches the rating of the regulator, the regulator closes to maintain a relatively flat pressure as illustrated in figure 4. The heavy-dark line in figure 4 represents a median flow midway between the outlet pressure at the minimum and maximum rated flows for the regulator. We described the variation about the midpoint flow using a flow coefficient once the pressure exceeded the rated pressure. For example, if the regulator was rated at 50 psi the variation in pressure depending on flow rate with estimated to be 5 psi. We developed the following equation:

$$F_c = \sqrt{\frac{(Q_{hi}^2 - Q_{lo}^2)}{0.1 \times P_{reg}}} \quad thus \quad P_L = \left(\frac{Q}{F_c}\right)^2$$

$$P_o = P_{reg} + \left[\left(\frac{Q}{F_c}\right)^2 - \left(\frac{Q_m}{F_c}\right)^2\right] \tag{5}$$

where; F_c is the flow coefficient for a specific pressure rating, Q_{hi} and Q_{lo} are the maximum and minimum flow rate for the regulator, P_{reg} is the rated pressure of the regulator, P_L is the pressure loss through the regulator for other flow rates, Q_m is the median flow that corresponds to the rated pressure for the regulator and P_O is the outlet pressure from the regulator.

Friction loss also occurs within the regulator when the inlet pressure is below the rated pressure for the regulator. Laboratory research indicated that the slope of the relationship between the outlet pressure and the inlet pressure was approximately 0.91 for the rising limb of the regulator performance curve. The heavy line on the rising segment represents the midrange flow. Therefore, an inlet pressure of 55 psi results would provide an outlet pressure of 50 psi for the midrange flow. The vertex of the sloping segment for a specific flow occurs at the inlet pressure equal to $0.91 \times P_{reg}$ and an outlet computed in equation 4 for that flow rate. The outlet pressure for inlet pressures below the vertex value follows the slope of the line from the origin through the vertex. The results in figure 4 illustrate that the range of pressures for 25-psi regulators is half of that for a 50-psi regulator. This closely matches information available from suppliers and was consistent with descriptions by Rogers et al. (2010) and von Bernuth and Baird (1984). It also is representative of laboratory investigations we conducted. It's been well documented that pressure regulators maintain a hysteresis effect which gives a range of pressures even for constant flows. Given this characteristic this description of the performance of regulators seems adequate simulating performance for varying inflow and inlet pressures.

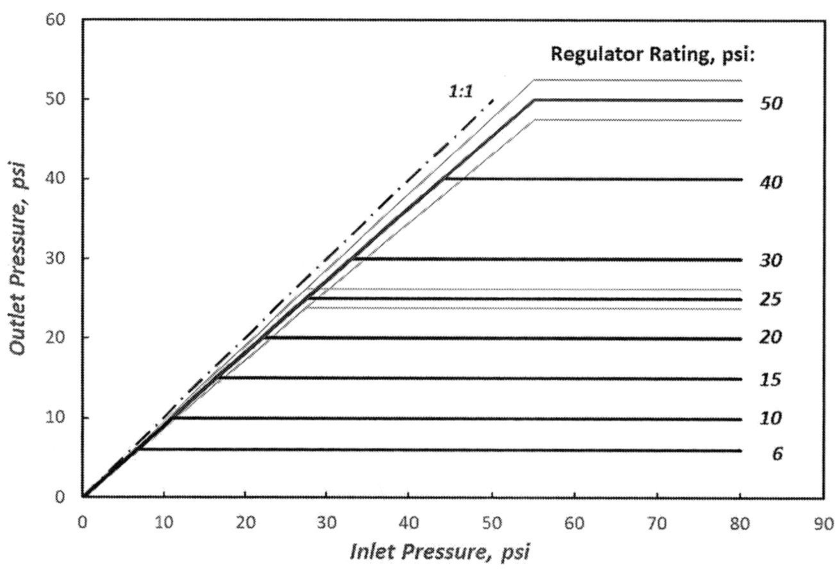

Figure 4. Performance model for pressure regulators.

SIMULATION RESULTS

Simulations were conducted for varying inflow pressures for laterals running up and down slopes of 1% and for level fields. Pivots with and without regulators along the lateral were considered. Variation of the inflow pressure changes the pressure and sprinkler discharge distributions along the lateral. The sum of discharge for all sprinkler equals the inflow to the pivot. The uniformity and effectiveness of water applications can be addressed using the ratio of the sprinkler discharge for a selected inlet pressure relative to the discharge from the same sprinkler from design conditions. The value of the ratio indicates the overall change in application depth per irrigation, while the distribution of the ratio along the lateral demonstrates how the uniformity of application varies relative to design conditions.

Results for pivots without pressure regulators when the lateral was positioned along a 1% upslope from the pivot inlet are presented in figure 5. The distribution of pressure in the lateral from the inlet to the distal end of the pivot is illustrated at the top of figure 5. The shape of the pressure pattern along the lateral is similar for all inlet pressures—just shifted down for smaller inlet pressures. The difference in pressure from the lateral inlet to the distal end of the lateral is larger for higher inlet pressures than for lower inlet pressures because higher pressures result in more sprinkler discharge when regulators are not used. This is expected because the friction loss is larger in the lateral. The pressure is always largest at the inlet and the pressure decreases to approximately 34 psi for the highest inlet pressure and 10 psi for the lowest inlet pressure. The blue line represents the design condition where the pressure was adequate at the distal end of the lateral to provide 25 psi in the lateral. The discharge from individual sprinklers along the lateral are illustrated in the middle of figure 5. The blue line represents the design condition where the pressure was adequate to provide a linear distribution of sprinkler discharge along the lateral. The stairstep discharge pattern occurs because finite nozzle sizes are available; thus, the same nozzle size is used for a small group of sprinklers and then the next larger nozzle is used for the next downstream group of sprinklers. Therefore, the discharge is relatively constant over a small group of sprinklers and then increases for the subsequent group of sprinklers. Since the discharge pattern is nearly linear applications will be uniform along the lateral. The flow rate from individual devices obviously is larger for higher inlet pressures and smaller for lower inlet pressures for unregulated systems. Higher pressures produced nearly linear relationships with distance along the lateral; however, the conditions for lower pressures indicates that the distribution become somewhat curvilinear. The ratio of sprinkler discharge to the design discharge is included at the bottom of figure 5. The results illustrate that the average depth of application decreases as the inlet pressure drops as expected but the depth is reasonably uniform for most inlet pressures. The ratio does decline near the end of the lateral for the lower pressures. This is important as much of the irrigated field is affected by the outer spans of the pivot.

Results for the level portion of the field are presented in figure 6. The sprinkler package design for the upslope condition was used for the level field, *i.e.* the pivot just rotated to the level portion of the field. The pressure distribution along the lateral produces a very linear sprinkler discharge distribution for all inlet pressures; therefore, the discharge ratios are very uniform and the average depth of application declines with overall system inflow. Obviously, regulators are not needed for flat fields but the results show that a sprinkler package designed for uphill orientations performs very well for flat areas of the field.

The results when the pivot lateral rotates to the 1% downslope position are included in figure 7. Higher inflow pressures produce pressure distributions that result is reasonably uniform sprinkler discharge distributions and uniform discharge ratios. For lower pressures the friction loss in the lateral is smaller for the reduced system flow and the downslope produces more pressure gain that lost due to friction. Thus, the pressure distribution gives a discharge ratio trend that increases toward the distal end of the pivot lateral.

Results for the unregulated pivot show that the uniformity along the pivot lateral is reasonably good for all orientations of the pivot lateral when inlet pressures and pivot inflow capacities decrease from the design conditions. The primary impact is that the overall depth of application decreases requiring longer or more frequent irrigation, if possible, to meet the crop needs compared to early in the season.

Results for the same orientations of the lateral are illustrated in figure 8 when a 20-psi pressure regulator was utilized on the center pivot. The pressure distribution along the lateral is similar to the distribution when regulators were not utilized. Expectedly, sprinkler discharge equals design values when the inlet pressure exceeds that for design. However, the discharge pattern in the middle of the figure is more curvilinear when the inlet pressure drops below the design pressure than for the unregulated system. The discharge ratio for any portion of lateral where pressure exceeds the 20-psi rated pressure for the regulator will discharge the design flow and the discharge ratio will equal design values. When the pressure drops below the regulator rating the sprinkler is not experiencing as much pressure as required for the design and the discharge ratio decreases. The impact of reduced depth with the regulated systems occurs at the distal end of the pivot—which represents a large portion of the upslope area. The discharge ratio for smaller inflow pressures that may occur due to decreasing well capacity late in the irrigation season are less uniform for the upslope orientation that when regulators were not used.

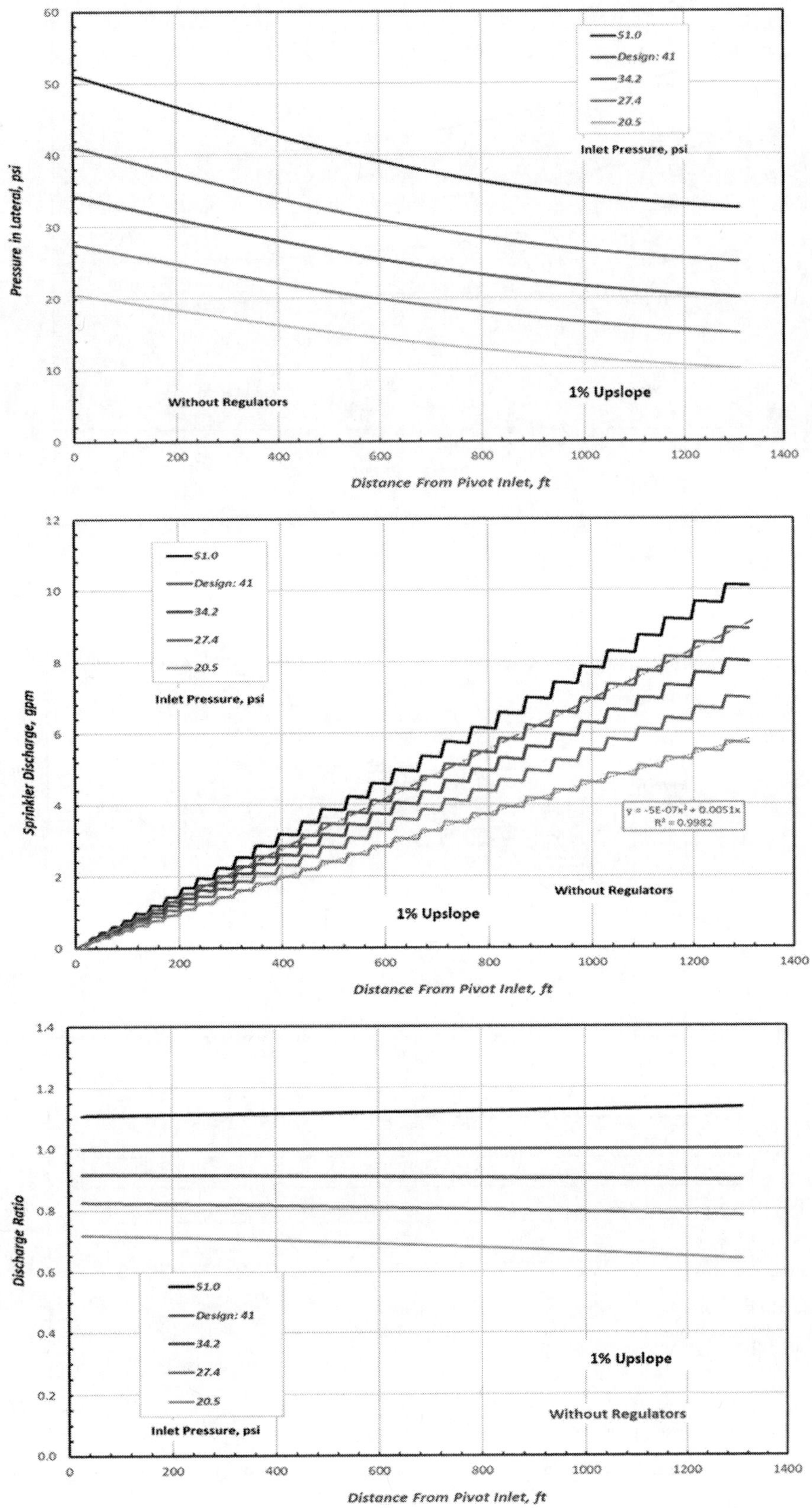

Figure 5. Results of simulations for variable pressures when pressure regulators are not used for a field with a 1% upslope from the inlet.

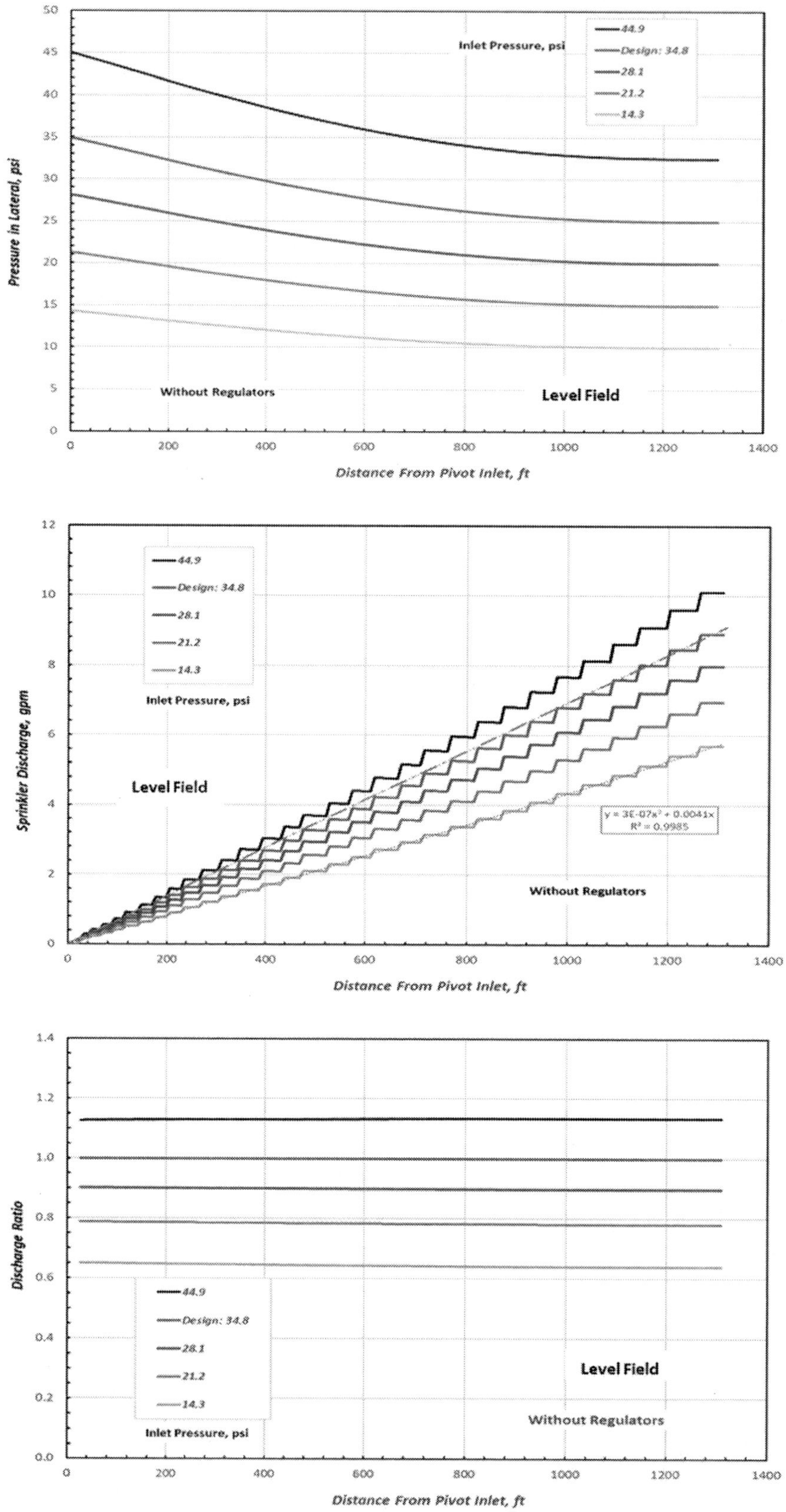

Figure 6. Results of simulations for variable pressures when pressure regulators are not used for a level field.

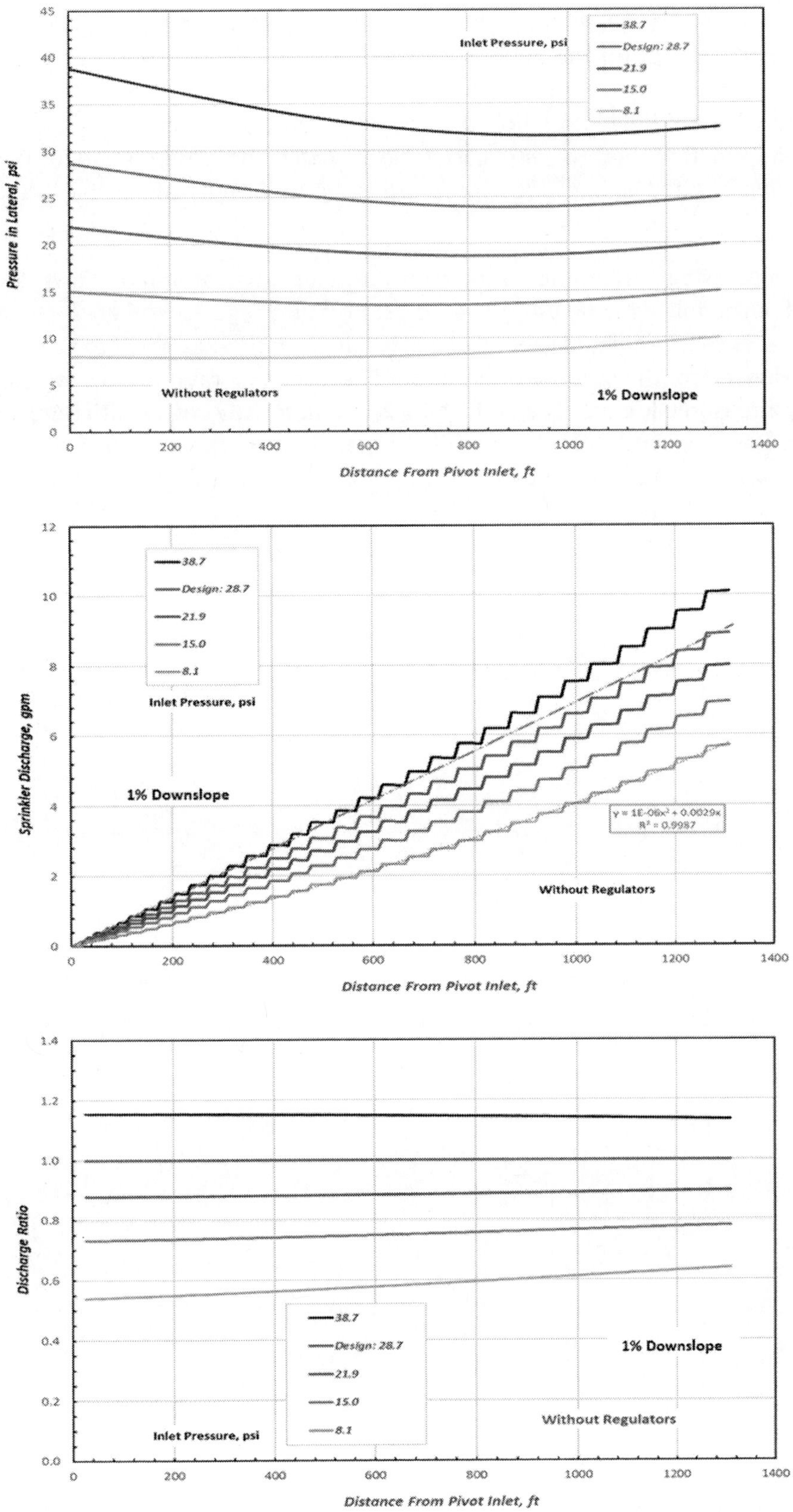

Figure 7. Results of simulations for variable pressures when pressure regulators are not used for a field with a 1% downslope from the inlet.

Simulation results for the level field with 20-psi regulators are illustrated in figure 9. The sprinkler discharge is quite linear except for the lower inlet pressures. Low inlet pressures produce a curvilinear depth of application that falls below that required for uniform application and the outer portion of the field will be under irrigated. Regulators do not provide much better uniformity along the lateral than unregulated system, but the depth of application does match the design goal. Thus, the overall field uniformity would improve for higher inflow pressures. The uniformity along the lateral is poorer with regulation that for unregulated systems when the inflow pressure is well below design values.

Results for the 1% downslope orientation with 20-psi pressure regulators are illustrated in figure 10. at rating is ulcerated in the following figuring. The 20-psi pressure regulators were selected thus locations along the lateral that exceed regulated pressures will produce discharges equal to the design value for the sprinkler package derived for the upslope orientation. The discharge ratio for low inlet slightly curvilinear but the uniformity is reasonably good. The uniformity is not materially better with regulators than without for low inlet pressures. Regulators do throttle discharge when the pressures in the lateral exceed the regulated pressure which consumes some energy but does contribute to field uniformity rather than just lateral uniformity.

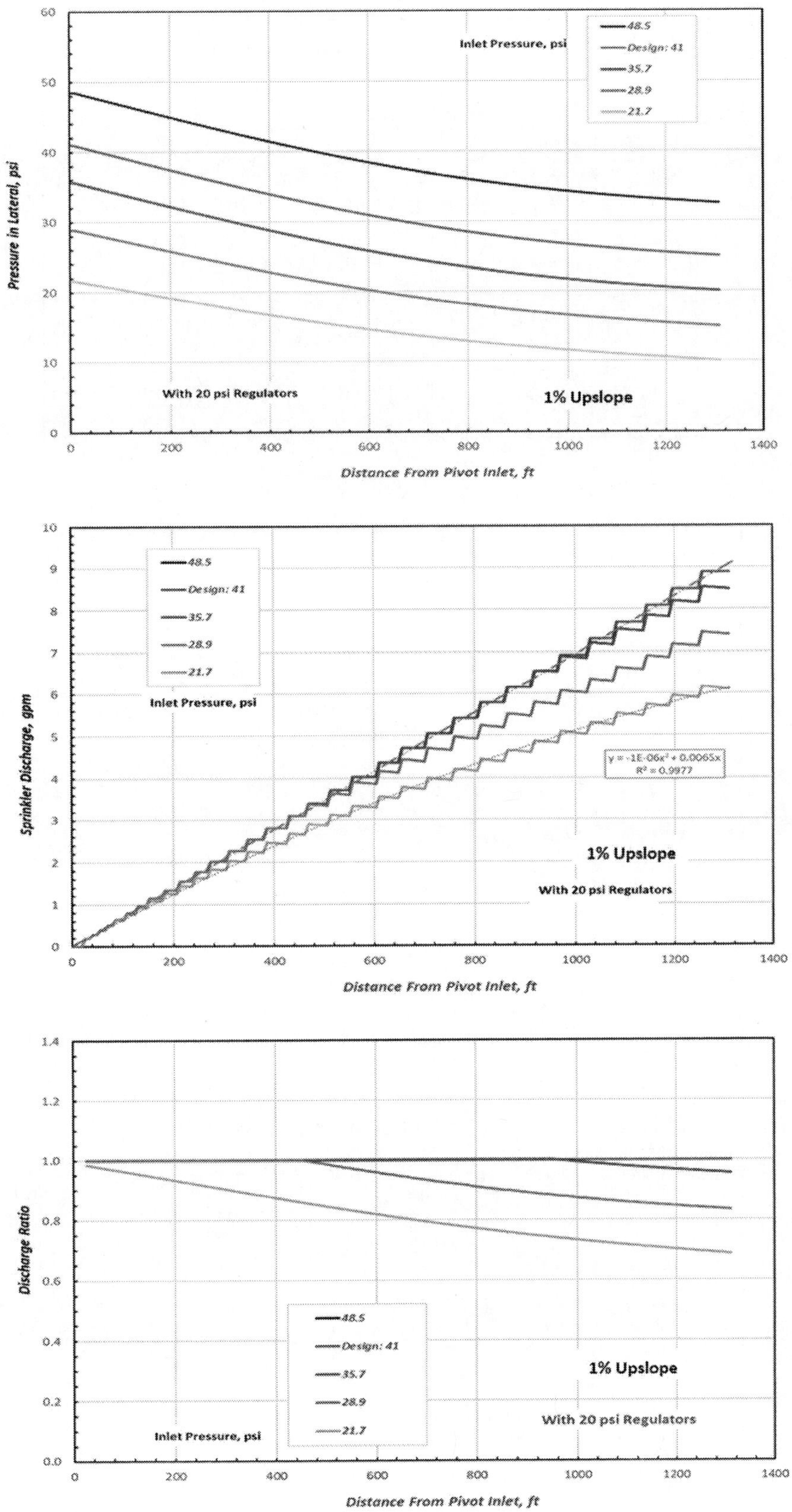

Figure 8. Results of simulations for variable pressures when pressure regulators are used for a field with a 1% upslope from the inlet.

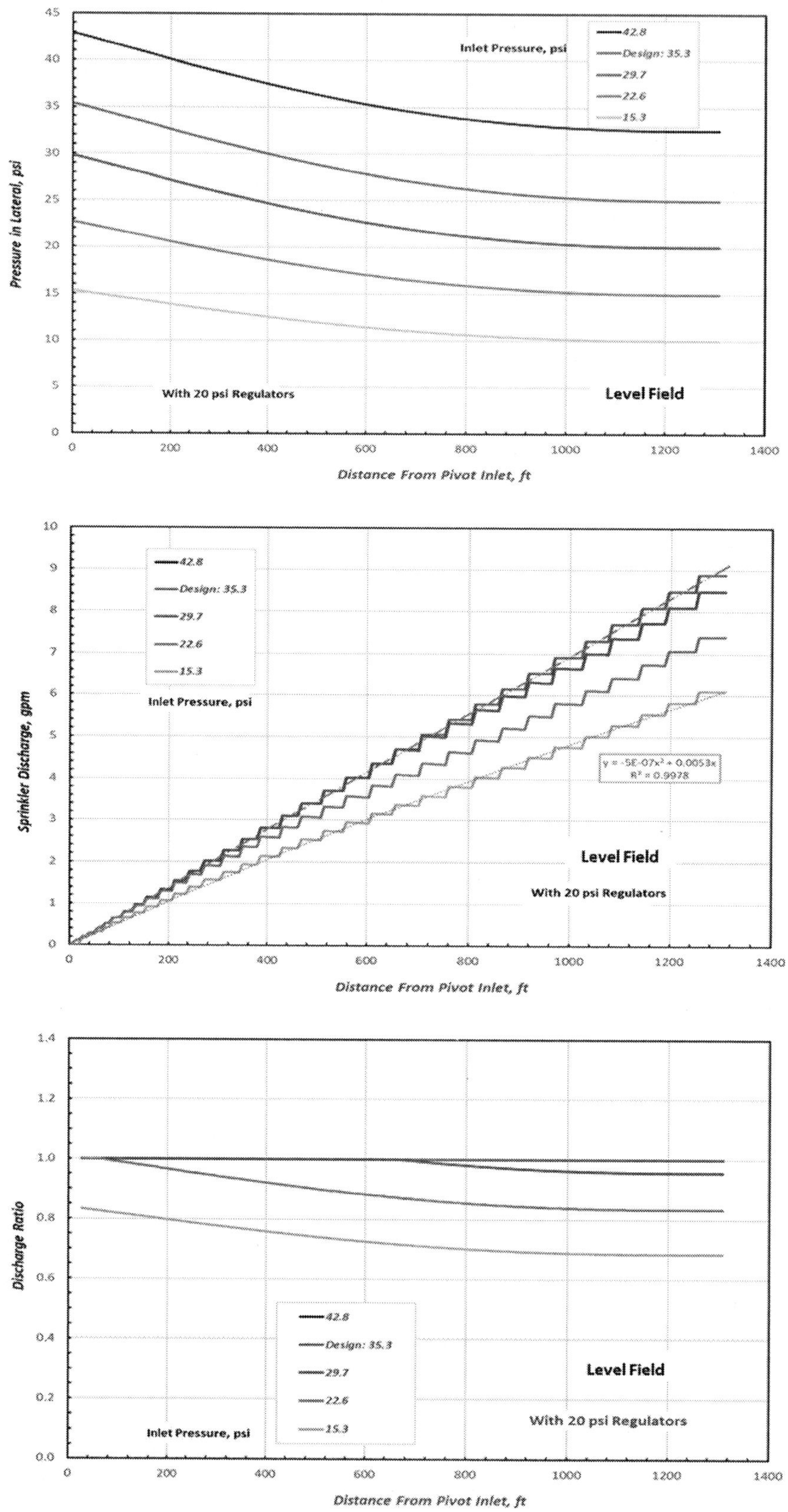

Figure 9. Results of simulations for variable pressures when pressure regulators are used for a level field.

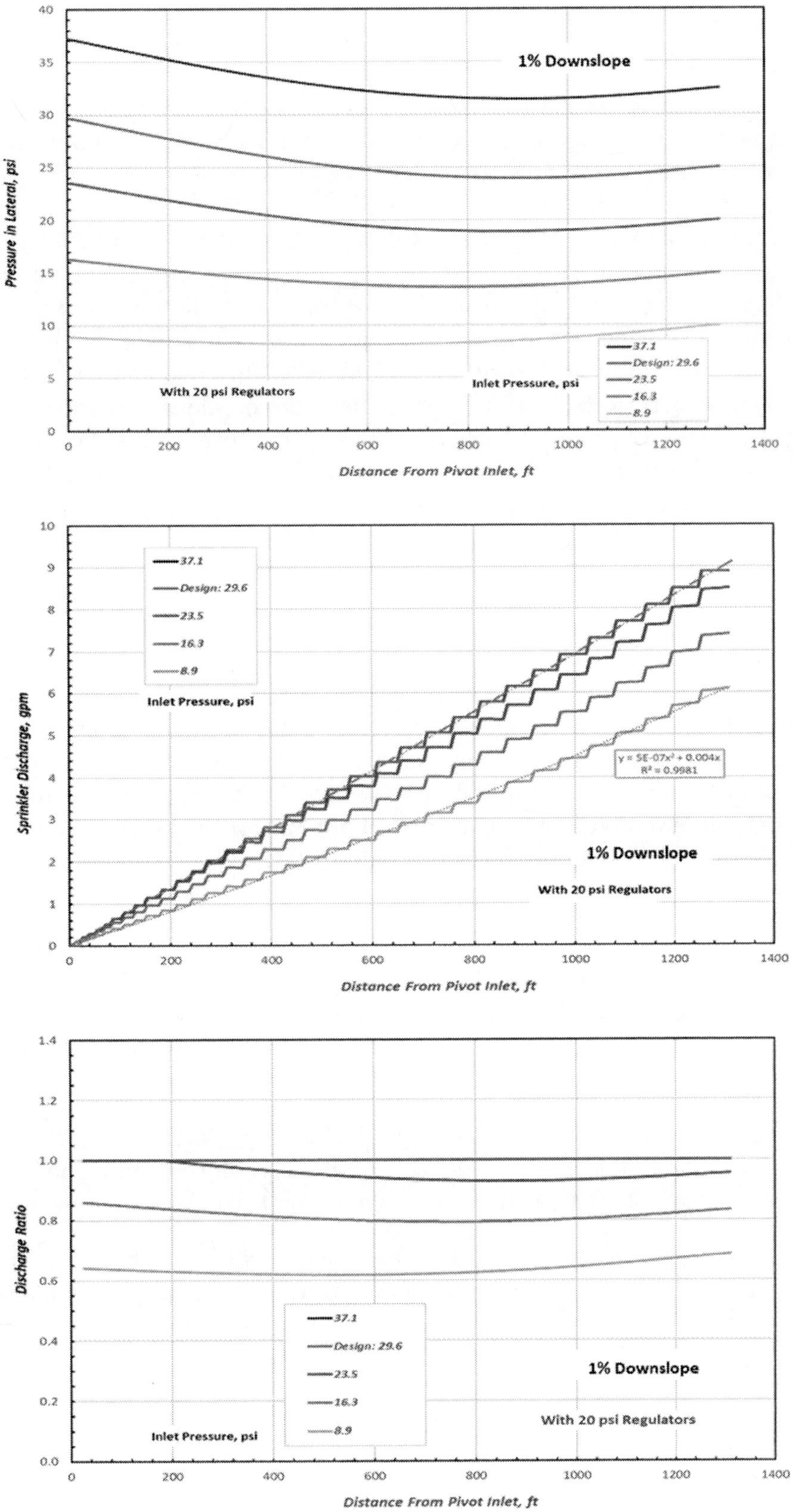

Figure 10. Results of simulations for variable pressures when pressure regulators are used for a field with a 1% downslope from the inlet.

Combining with Pump Curve

The previous results illustrated how laterals perform with different inlet pressures and therefore changing system inflow. It is critical to consider the interaction of the pump with the center pivot to assess the impact of changing water supply conditions. A pump curve for a vertical turbine pump was used to evaluate the effect of varying water supply conditions. The pressure available to the pivot inlet depends on the interaction of the pump performance curve and the system required total dynamic head as illustrated in figure 2. Two water supply conditions were considered as depicted in figure 11. The upper pump curve is labeled high flow which represents water supply conditions early in the irrigation season. The pump cure in figure 11 represents the available inlet pressure from the pump after discounting the factors listed in equation 1. The low flow pump curve represents the available inlet pressure when supply conditions have decreased later in the irrigation season—when the regional water table drops and to throttling inflow may have occurred. The top graph in the figure provides operating points for the three lateral orientations and were 20-psi regulators were utilized. The total system flow rates are essentially equal at about 800 gpm for the high-flow case for all lateral orientations and the inlet pressure is approximately 40 psi for all lateral orientations. The system curve for regulated pivots becomes vertical once the flow rate reaches the design inflow rate for the pivot. Increasing pressure will not increase system flow once all sprinklers have reached the rated pressure of the regulator. Thus, the excess pressure is essentially wasted. Designers may desire some excess pressure to ensure uniformity; therefore, the pump-system match points are essentially the same for all lateral orientations for the high-flow supply condition.

There is considerable variation for the low-flow supply conditions when regulators are utilized. The least discharge will occur when the lateral is pointing upslope and most flow occurs when for the downslope orientation. For this pump and system, the inflow varies by about 50 gallons per minute between the upslope and downslope orientations. The variation is about 10% of the average inflow for the low-flow supply condition. This illustrates that the quantity of water that can be pump produce must be consider for the whole-field uniformity.

The system flow during the low-flow period is approximately 62% of the flow during the high-flow period. This the average depth of application will be less that expected during high-flow periods. The change in flow must be included into the system operation and a new percent timer relationship should be developed. The low-flow system should operate longer to provide the depth expected during high-flow periods.

The lower graph in figure 11 presents the operating points for unregulated pivots. The variation for high-flow conditions is about the same as for low-flow conditions. The variation between upslope and downslope system flows is about 6% for high-flow conditions and 10% for low-flow conditions.

The system flows are higher for the unregulated configuration and for regulated pivots during high-flow periods. The system flows are slightly higher for regulated pivots for the low-flow conditions.

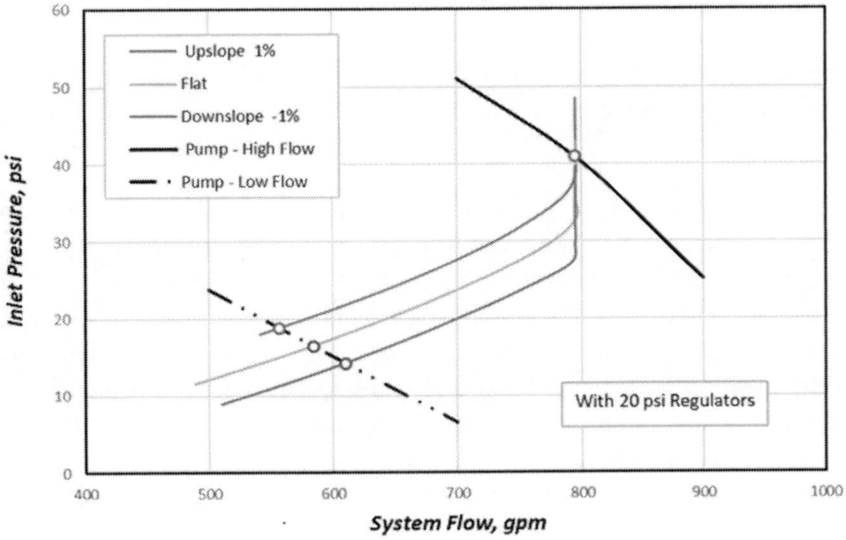

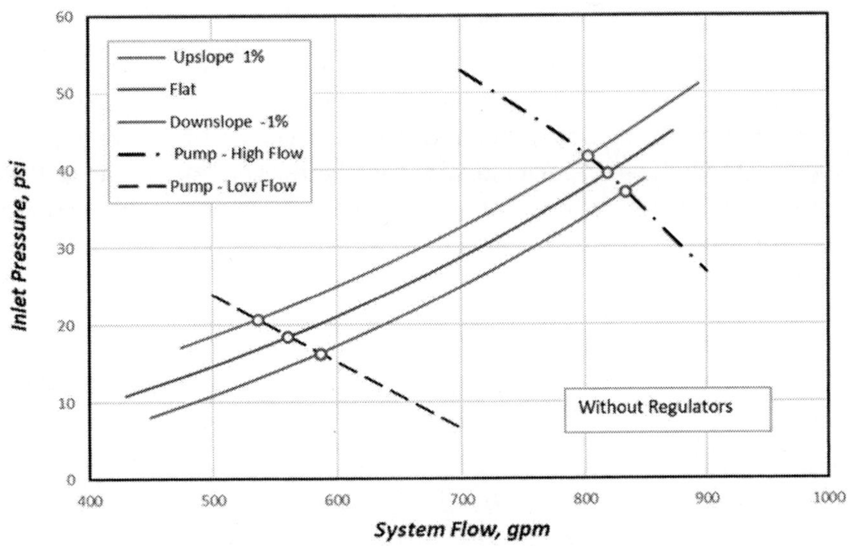

Figure 11. Flow conditions and operating points for high and low flow supply conditions for three lateral orientations for center pivot systems with and without pressure regulators.

The distribution of pressure, discharge and discharge ratio when considering pump interactions are presented in figure 12 for systems using regulators. These results stem from the operating points for the regulated system shown in figure 11. The pressure along the lateral is well above the regulator rating for all lateral orientations for the high-flow supply condition. Therefore, the sprinkler discharge is linear, and the discharge ratio matches the design goal throughout the field. The system works as designed. When the supply capacity drops below the design value the pressure in the lateral is below the regulated pressure for all lateral orientations for the low-flow supply conditions. The discharge distribution is most linear for the downslope condition and thus the discharge ratio is more uniform when the lateral is oriented downhill. The discharge ratio decreases from about 90% near the inlet to about 60% near the distal end when oriented upslope. There is a 20% variation in discharge ratio at the distal end of the lateral between the downslope and upslope orientation.

The results for the pressure, discharge and discharge ratio for the unregulated pivot are presented in figure 13. The pressure distributions for all orientations produces nearly linear discharge patterns and uniform discharge ratios for the high-flow orientation. The depth of application for the downslope orientation appears to exceed the upslope—i.e., the design ratio—by about 13% for the high-flow conditions. The discharge patterns are somewhat curvilinear for the low-flow conditions producing discharge ratios that are less uniform. The discharge ratio for the level and downslope orientations are more uniform that the upslope pattern where the discharge ratio varies from about 7% along the lateral for the upslope orientation. The variation of the discharge ratio is much less for the unregulated system than the regulated system for low-flow conditions, especially for the upslope orientation. The variation of depth at the distal end of the lateral varies by about 15% for the unregulated system and low-flow conditions.

The sprinkler discharge ratios for the regulated and unregulated pivot are summarized for the low-flow supply condition in figure 14. Ratios vary less along the pivot lateral for unregulated pivot than for regulated pivots and the variation is less at the distal end of the lateral is less for the unregulated pivot than when regulated.

A summary of the simulation results for the 1% field is presented in table 1. These results summarize the performance for the range of conditions simulated. Major changes occur for either the regulated or unregulated pivot when the inflow capacity decreases during the irrigation season. The uniformity changes and perhaps most importantly the system inflow changes considerably requiring adjustments to irrigation scheduling procedures. The results also illustrate the variations expected as the pivot rotates around the field. Variation are significant and may merit speed control to improve overall field performance.

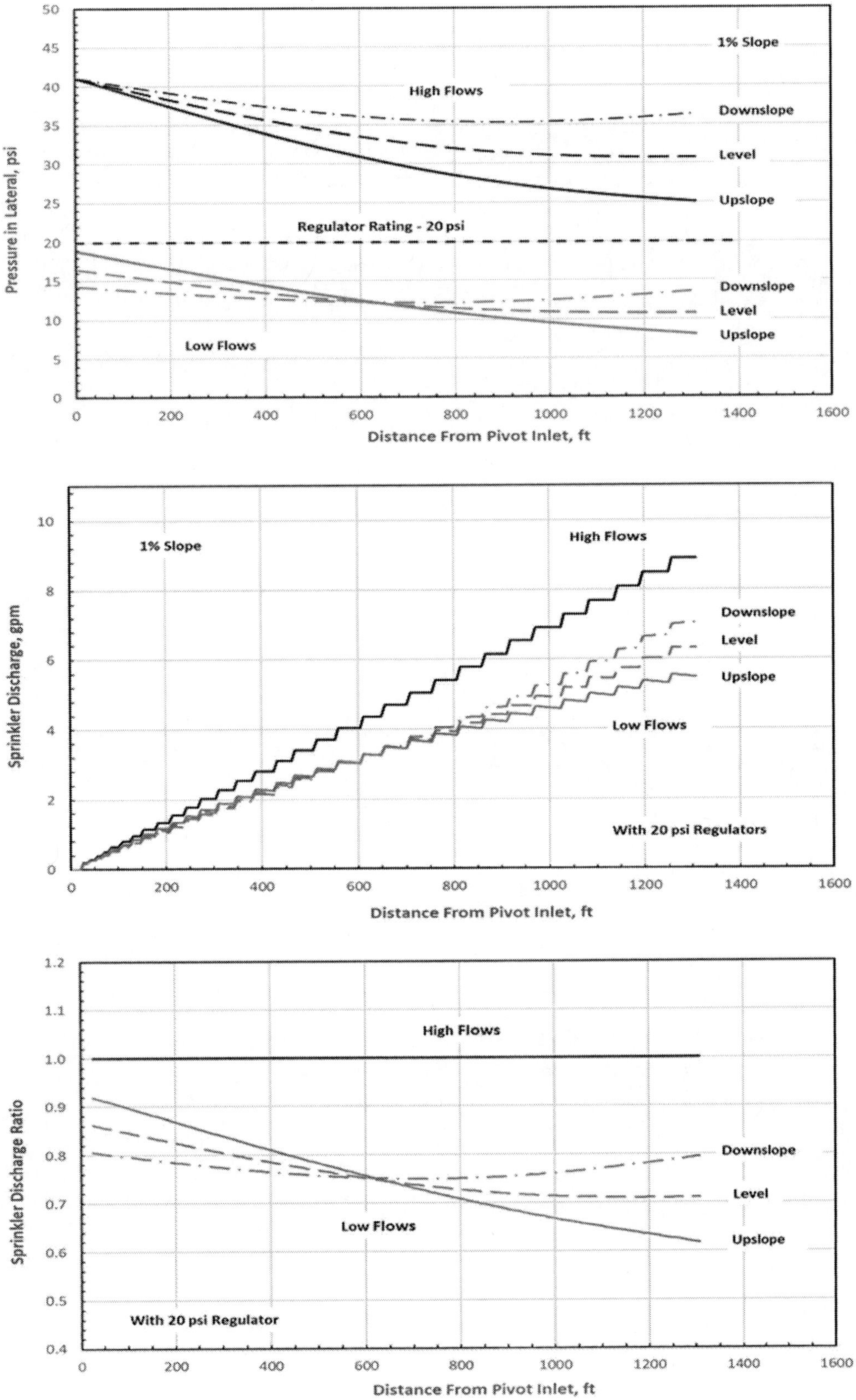

Figure 12. Performance for a center pivot using pressure regulators for a field with a 1% slope and where water inflows vary during the irrigation season.

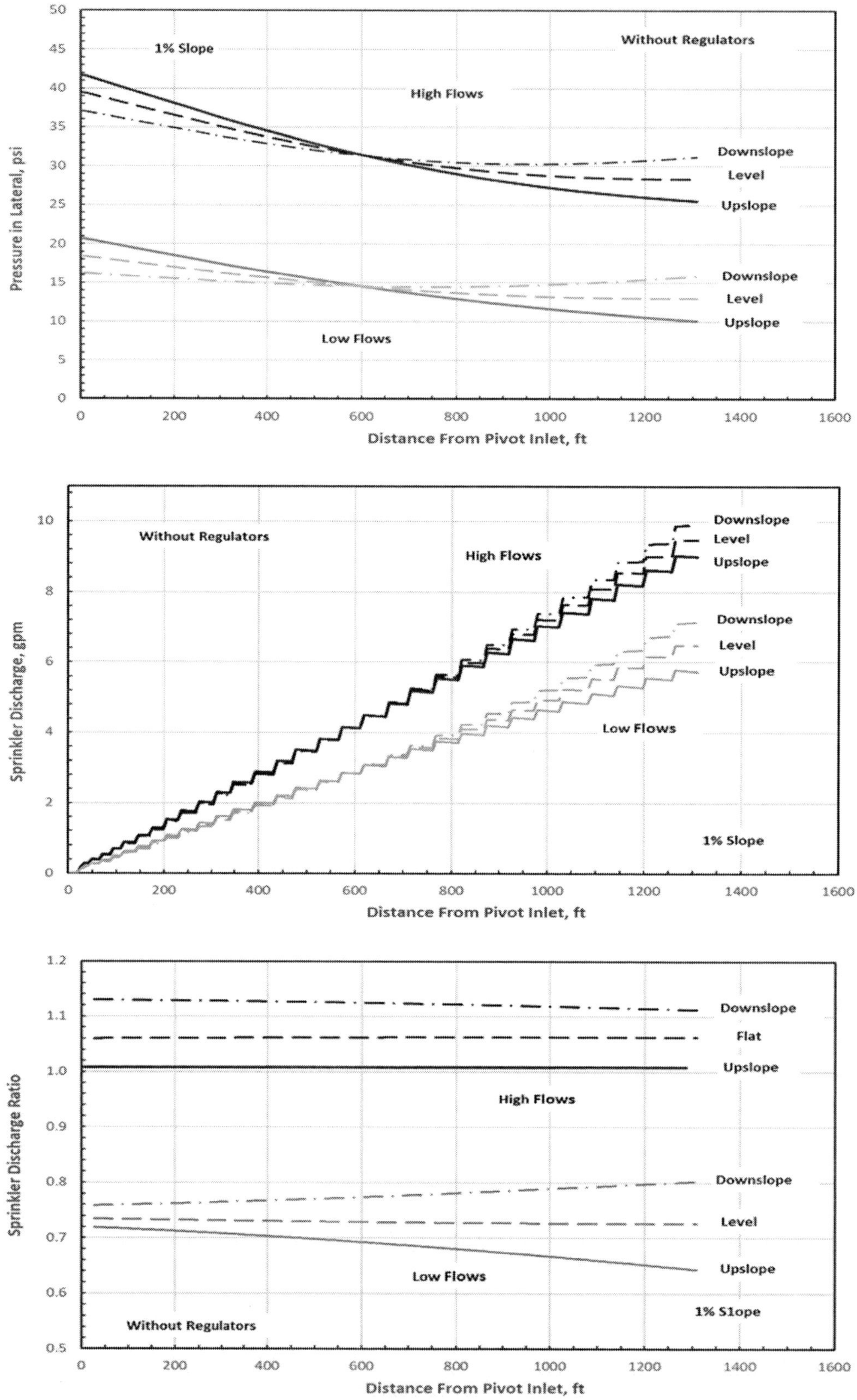

Figure 13. Performance for a center pivot without pressure regulators for a field with a 1% slope and where water inflows vary during the irrigation season.

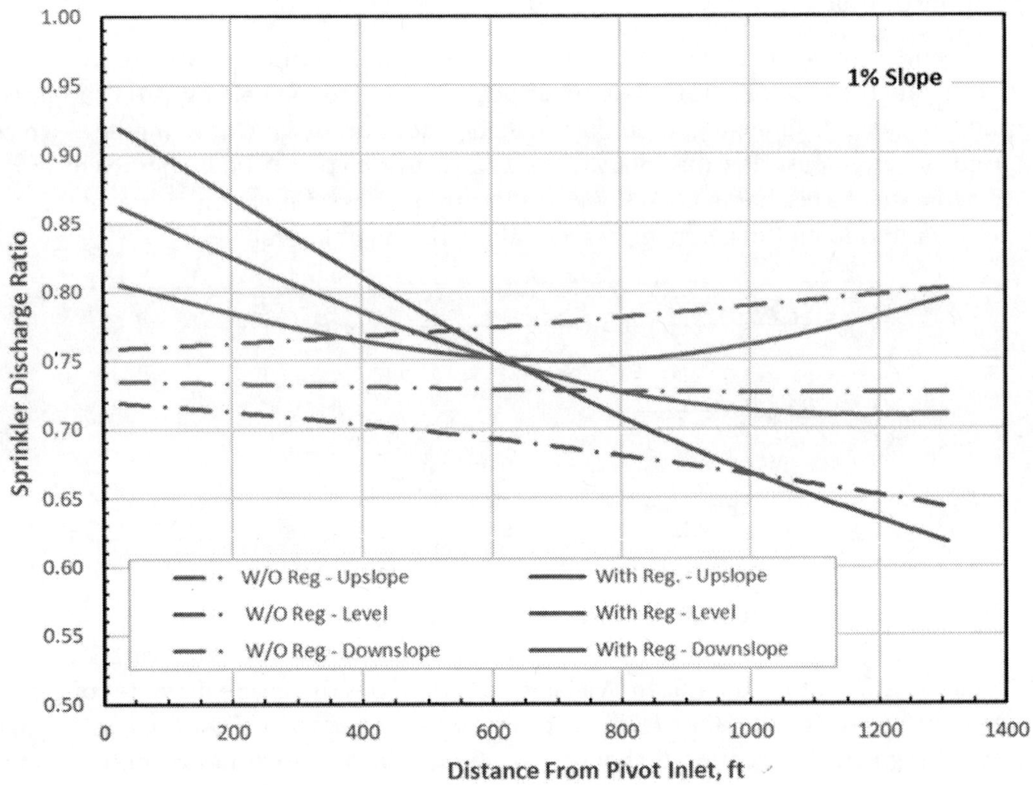

Figure 14. Comparison of the sprinkler discharge ratio for low-flow conditions for regulated and unregulated pivot on a field with a slope of 1%.

Table 1. Simulation results for 1% slope conditions.

	With Regulators					
	High Flow Conditions			Low Flow Conditions		
	Upslope	Level	Downslope	Upslope	Level	Downslope
Flow, gpm	795	795	795	557	583	609
Inlet Pressure, psi	41.0	41.0	41.0	18.8	16.4	14.2

	Without Regulators					
	High Flow Conditions			Low Flow Conditions		
	Upslope	Level	Downslope	Upslope	Level	Downslope
Flow, gpm	802	819	833	536	561	585
Inlet Pressure, psi	41.7	39.4	37.0	20.7	18.4	16.2

Flow Coefficient for Pivot

Irrigation management must consider the uniformity and variability of application in the field. It is also important to estimate the change in total volume of water applied to the field as water supply condi5tions change during the season. Changing conditions can be monitored with flow meters—which are less frequently used—or by measuring the inflow pressure to the pivot. A flow coefficient approach was developed to estimate the discharge from the entire center pivot as a function of the inlet pressure. The flow coefficient was based on the average pressure along the pivot lateral:

$$Q_s = \sqrt{P_i - \frac{E_c\, s\, R_s}{\lambda} - \frac{2}{3}F_p\, 4.52 \frac{Qs^{1.852}}{C^{1.852}} \frac{R_s}{D^{4.866}}} \approx \sqrt{P_i - \frac{E_c\, s\, R_s}{\lambda} - kQ_s^2}$$

$$k = \frac{2}{3}F_p\, 4.52 \frac{1}{C^{1.852}} \frac{R_s}{D^{4.866}} \qquad (6)$$

$$Q_s^2[1-k] = P_i - \frac{E_c\, s\, R_s}{\lambda} \rightarrow Q_s = \frac{\sqrt{P_i - \frac{Ec\, s\, R_s}{\lambda}}}{1-k} \approx F_{cp}\sqrt{P_i - \frac{Ec\, s\, R_s}{\lambda}}$$

Where; Ec is an elevation constant that represents the fraction of the elevation effect occurs at the average pressure, s is the slope, Fp is the friction loss factor for multiple outlets—often taken as 0054—, C is the resistance coefficient for the Hazen-Williams equation, D is the inside diameter of the lateral pipe and other terms are as previously defined. The discharge was capped at the flow for a fully regulated pivot for the case with regulation. Equation 6 shows that about ⅔ of the friction loss occurs at the average pressure. The flow coefficient approach approximates the exponent of the flow factor in the Hazen-Williams equation as 2 rather than 1.852. Solution for the flow rate is then possible. The value of factor k is quite small compared to 1 and is therefore ignored in the final portion of equation 6. The value of 0.3 appears to work well for the unregulated system.

The predicted and simulated flow versus inflow pressure results are illustrated in figure 15. The flow coefficient was applied to the 6 flow conditions: 1% or 2% slopes with 3 different lateral orientations. The function provides very reliable results for the unregulated system in the upper portion of the figure. Similar results are presented in the bottom graph for the regulated system. Results are not good for the regulated system when the lateral runs up or downhill. The solution for the level orientation is reliable; however, the influence of elevation when regulators are used is not well represented. Additional work is needed to predict inflow changes for sloping regulated pivot.

These figures however are very useful in helping a manager decide how to schedule irrigation. The field average depth of water applied is directly proportional to the flow rate of water into the pivot; thus monitoring the inlet pressure to the pivot allows one to estimate the volume flow rate being pumped and to adjust operation as flow supply conditions change during the season.

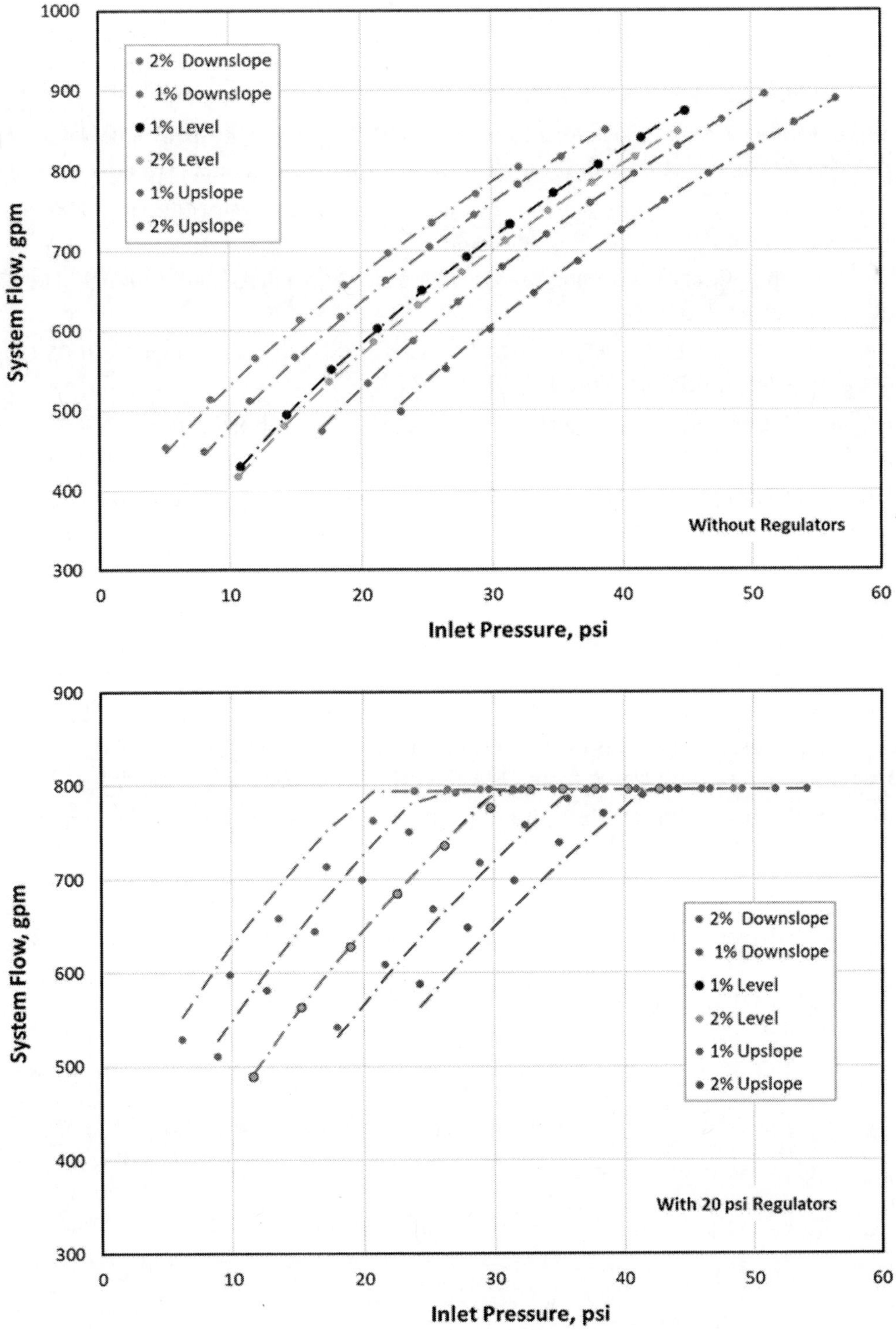

Figure 15. Comparison of simulated relationship between inlet pressure and system inflow and predictions based on a center pivot flow coefficient approach.

REFERENCES

Beccard, R.W. and D.F. Heermann. 1981. Performance of pumping plant-center pivot sprinkler irrigation systems. ASAE Paper 81-2548, St. Joseph MI.

Bremond, B. and B. Molle. 1995. Characterization of rainfall under center pivot: influence of measuring procedure. Journal of Irrigation and Drainage Engineering, ASCE, Vol. 121(5):347-353.

Heermann, D. F. 1990. Center pivot Design and Evaluation. Proceedings of the Third National Irrigation Symposium (ASAE), Oct. 28-Nov. 1. Phoenix, AZ.

Heermann, D. F. 2006. The Effect of Dropping Water Tables on Center Pivot Performance. The 27th Annual International Irrigation Show.

Heermann, D. F. CENTER PIVOT EVALUATION AND DESIGN. 2005. In Proceedings for 2005 Central Plains Irrigation Conference, Sterling, Colorado, Feb 16-17 (p. 137).

Heermann, D.F. and T.L. Spofford. 1998. Evaluating center pivot irrigation systems. ASAE Paper 98-2068. St. Joseph, MI.

James, L.G. 1982. Modeling the performance of center pivot irrigation systems operating on variable topography. Trans. of the ASAE 25(1):143-149.

James, L.G. 1984. Effects of pump selection and terrain of center pivot performance. Trans. of ASAE 27(1):64-68,72.

James, L.G. and S.K. Blair. 1984. Performance of low-pressure center pivot systems. Trans. of the ASAE 27(6):1753-1757, 1762.

Rogers, D.R, L. K. Shaw, S. R. Pragada, and M. Alam. 2010. Evaluation of Pressure Regulators from Center Pivot Nozzle Packages. Proceedings of the 22nd Annual Central Plains Irrigation Conference, Kearney, NE., February 24-25, 2010 Available from CPIA, 760 N. Thompson, Colby, Kansas.

Von Bernuth and Baird. 1990. Characterizing Pressure Regulator Performance. Transactions of the ASAE Vol. 33 (1): 0145-0150.

Effect of collector size on center pivot water depth catch

Danny H. Rogers, P. E., Ph. D.

Professor and Extension Agricultural Engineer, Irrigation, Kansas State University, Department of Biological and Agricultural Engineering, 2018 Seaton Hall, Manhattan, KS. 66506. Phone: 785-532-2933. Email:drogers@ksu.edu

Jonathan Aguilar, Ph. D.

Assistant Professor and Extension Irrigation Specialist, Kansas State University, SWREC, Garden City, KS.

Gary A. Clark, P.E., Ph. D.

Professor and Senior Associate Dean, College of Engineering. Kansas State University, Manhattan, KS.

Scott W. Wiens

Former Graduate Student, Dept. of Biological and Agricultural Engineering, Kansas State University, Manhattan, KS.

Abstract: *The ASABE standard on uniformity testing of a center pivot system contains specifications on the collector size to measure the applied water depth. The standard was developed before many of the current sprinkler application devices were developed. A study comparing the catch depths of five collector sizes for three types of sprinkler application devices: spinning plate, fixed plate, and wobbling plate sprinkler systems using four 5x5 Latin squares. Each collector's water depth was measured and statistically analyzed. Two analysis of variance (ANOVA) tests of the collector size effect were reported. Past experimental results were also compared to this experiment's results.*

Keywords: Center pivot irrigation, uniformity, sprinkler packages collector size

Introduction

Center pivot irrigation systems are used to irrigation over 90 percent of the irrigated land in Kansas (Rogers and Aguilar, 2017) and over 50 percent in the US (USDA, 2013). A variety of reasons related to the adaptability of the systems to a wide variety of field conditions, crops, cropping systems, labor, water supplies and ultimately favorable economics, along with the capability of efficiently and uniformly applying water, have contributed to the conversion to center pivot irrigation systems from surface irrigation systems or the expansion of irrigation into previously unirrigated cropland areas.

Quantification of the uniformity of water application by sprinkler systems was pioneered by Christiansen (1942) which was before the invention of center pivot sprinkler irrigation systems. The adaption of Christiansen's coefficient of uniformity (CU) concept to center pivot systems was accomplished by weighting of the individual application depths along the center pivot lateral relative to the area served at that location (Heermann and Hein, 1968). This adaptation of the

CU equation is the basis for the ASABE standard, *Test Procedure for Determining the Uniformity of Water Distribution of Center Pivot and Lateral Move Irrigation Machines Equipped with Spray or Sprinkler Nozzles,* for quantifying the water application uniformity for center pivot irrigation systems (ASAE S436.1, 2007).

In addition to ASAE S436.1, ASAE S398.1 (2007), *Procedure for Sprinkler Testing and Performance Reporting,* is a standard to quantify the performance of individual sprinkler devices. This standard does not specify the collector size, only that they should be identical and no water can splash in or out. ASAE S436.1 expands the specification to include that height of the collector and the minimum entrance diameter which is to be no less than 60 mm.

Schneider (2000) noted that the use of collectors is the common method of measuring irrigation application depths but cautioned against using small catch cans for use on uniformity studies of spray irrigation systems. Marek and Howell (1987) compared the catches of seven different collectors to a separatory funnel gage used in their previous research and found that the larger diameter collectors had less variability than smaller diameter collectors. They indicated a diameter of 75 mm was needed to measure with +/- 2 percent of the separatory funnel gage.

Clark et al. (2006a) used 100 mm Irrigages (Clark et al., 2004) to measure the application uniformity of three low pressure center pivot nozzle devices and compared the results to a 430 mm pan collector and found the under reported the application depths as compared to the pans. The test was repeated the following year for the same system but a different location within the field, this time the Irrigages collected higher application depths than the pans, leading to the conclusion that the diameter size of the Irrigage used was too small for the low pressure application devices of the irrigation system. Clark et al. (2006b) conducted a lab study to determine the minimum collector diameter for spray systems using diameters from 52 to 780 mm and compared the results to collectors with a 198 by 211 mm opening. The results from the collections from six different spray devices were highly variable, again leading to a conclusion that consistently measuring water application depths is a challenging task.

The goal of this study was to evaluate the accuracy and variability of irrigation water application depths of various collector diameter sizes for three types of water application devices; fixed plate, spinning plate and wobbling plate sprinkler devices. Five collector devices with unique diameters were positioned under center pivot irrigation systems in a 5 x 5 Latin square arrangement to measure the application depth and through the statistical analysis seek to identify the ideal collector size for the three application device types.

Procedures

Field evaluations were conducted at three farms near Garden City, KS, each equipped with one of the target application devices. Site 1 was equipped with spinning plate devices, represented by Nelson A300 Accelerator with Nelson 69 kPa pressure regulators that were positioned approximated 1.5 m above ground surface. At the test zone, the nozzle sizes were two Nelson #28 (5.56 mm; 28/128 in.), and a #29 (5.75 mm; 29/128 in.) with a spacing of 3.05 m. Site 2 was equipped with fixed plate devices, represented by Senninger LDN sprays with D3000 blue plated and #29 (11.5 mm; 29/64 in.) nozzle sizes with Senninger 104 kPa pressure regulators that were positioned approximately 1.8 m above ground surface and spaced at 1.5 m. Site 3 was equipped with wobbling plate devices, represented by Senniger I-Wobs with LA9 pads and 6.55 mm (33/128th in.) nozzles and Senninger 83 kPa pressure regulators that were positioned approximately 2.3 m above ground and at a 2.3 m spacing.

Twenty collectors for each of the five diameters were constructed which were 5.5 cm, 10 cm, 14.8 cm, 20 cm, and 27.4 cm, referenced as C2, C4, C6, C8, and C10, respectfully. All collectors had a height of 20.3 cm and beveled to create a sharp-edge inner lip. Additional construction details are available in Wiens, (2010).

SAS 9.2 was used to generate the design of the collector arrangement in the field. Four Latin square replicates were used for each sprinkler test. Figures 1 and 2 show the details of the generalized test location within the field and example collector position within the Latin square.

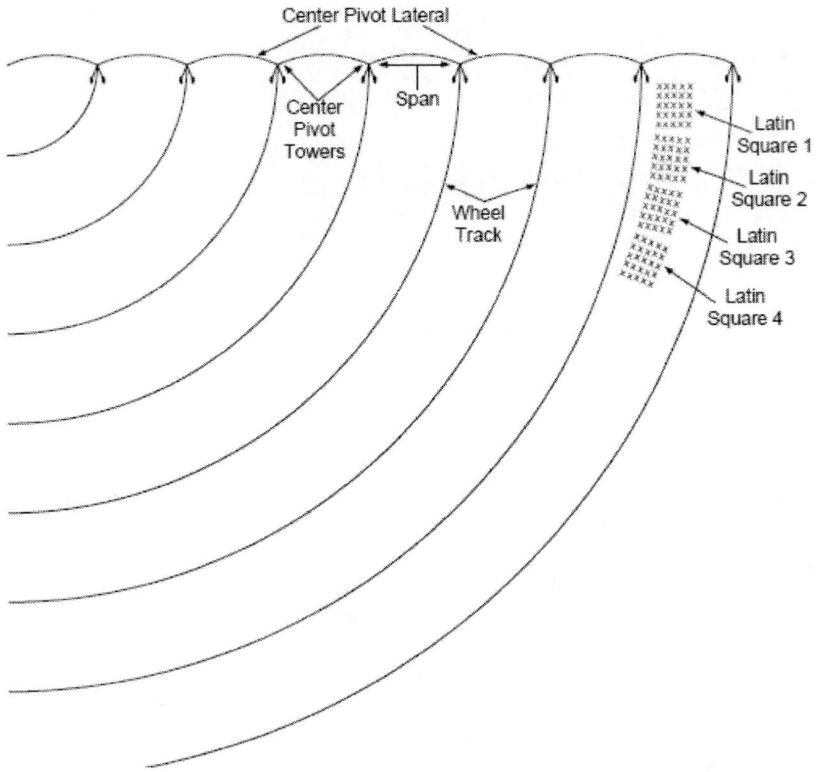

Figure 1. Center pivot irrigation system and location of Latin square field site (Wiens, 2010).

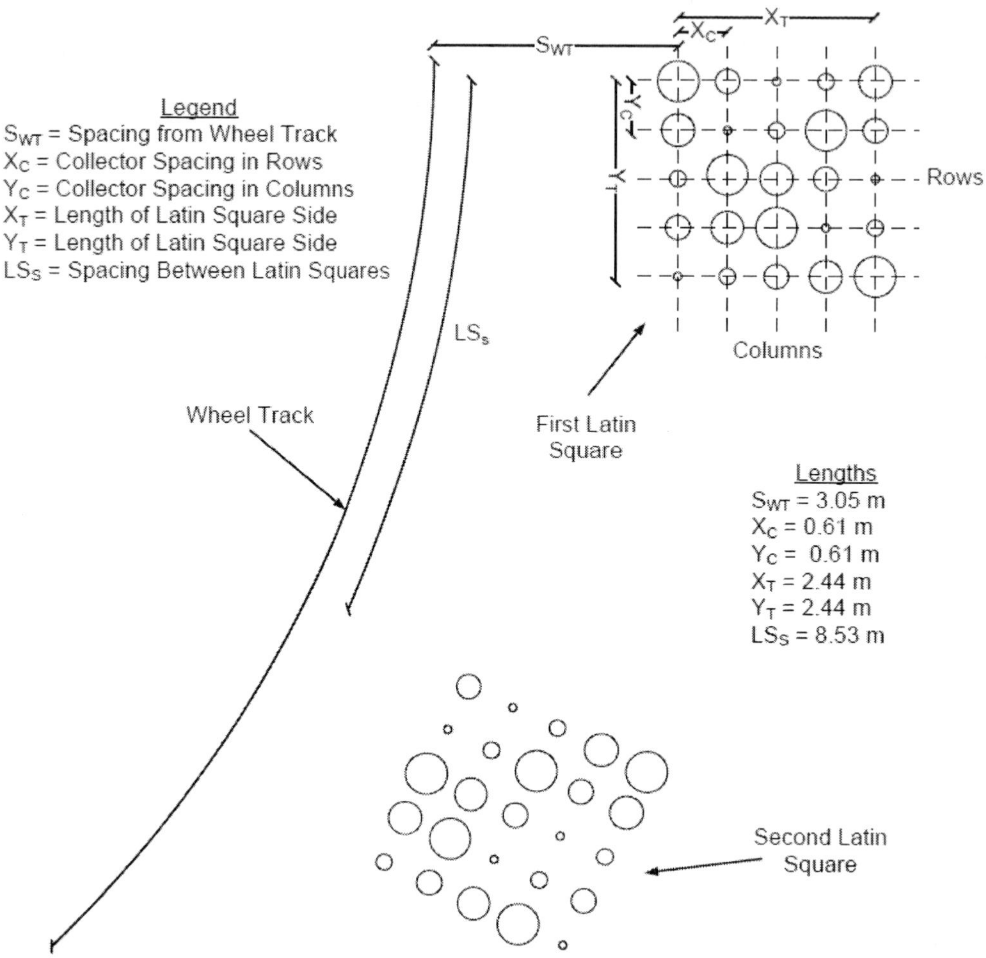

Figure 2. Two Latin squares with the collector spacing, columns, and rows labeled (Wiens, 2010).

Evaporation losses were minimized by measuring the collected volumes as soon as the collectors stopped receiving water from the irrigation event and measured using a graduated cylinder. This volume was converted to an application depth by dividing the volume by the collector surface area.

The SAS 9.2 "Proc Means" command was used to determine the means and standard deviations of the depth measurements for the different collectors, columns and squares. An analysis of variance (ANOVA) of the depths was created using the "Proc Mixed "command for the collector sizes, since this accounted for both fixed (collector size, column and row) and random (Latin square) effect factors in the model. If the ANOVA showed there was a significant effect of collector size on depth, then the collector size measurements were compared.

A second ANOVA was conducted to examine the variability of measured depth of each collector size. The significance of the column effect on measure depth was determined for each test. If there was significance, then the measured depths were adjusted to eliminate the column effect. After adjustments, the standard deviation of the measured depth data was calculated and used to develop an ANOVA of whether collector size had an effect on the variability of the measured depths. If size was significant, then a comparison between collector sizes was made.

In addition, the 95 percent confidence interval was created for each collector size and used to estimate accuracy of a depth measurement for a given number of collectors. These results are unique to the tested systems and should not be generalized to other systems.

Results and Discussion

Various diameter sized catch can depth of application catches for three different types of center pivot irrigation sprinkler packages were analyzed to study the effect of collector size on the measured water depth. The three types of sprinklers packages tested were fixed plate spray (FP), spinning plate (SP), and wobbling plate (WP) sprinklers as described previously. The tests were conducted under acceptable weather conditions and the average application depths meet the minimum depth specified by ASABE S436.1 standard for uniformity of center pivot irrigation systems. The means and standard deviations for the combination of all catches for each system test are shown in Table 1. The applied depth was determined by averaging the measured depth of the catches across all diameter sizes. The FP average applied depth was much larger than SP and WP values, making the standard deviation of FP large but still with a comparable amount of variability due to the higher average applied depth. The coefficients of variability (CV) for the systems were similar with FP being the highest.

Table 1. Means and standard deviation of each sprinkler system combined across collector diameter sizes (Wiens, 2010).

System	Applied Depth (cm)	Nominal Depth (cm)	Lower 95% Mean Confidence Level	Upper 95% Mean Confidence Level	St. Dev. (cm)	CV
Spinning Plate	1.49	1.52	1.46	1.51	0.108	0.07
Fixed Plate	5.11	5.08	5.01	5.21	0.510	0.10
Wobbling Plate	2.27	2.54	2.23	2.31	0.208	0.09

Spinning plate (SP) results are shown in Table 2. The difference between container catches was not statistically different. The largest measurement difference between containers was only 0.05 cm. The standard deviations (SD) of the measurements were similar except for C6 which had a SD of approximately one-half the other containers.

The position of the catch container in each column (see fig 2) were statistically different (Table 2). This is likely due to the position of each column relative to the application depth applied at that position from the combined depth of water from the overlap of the sprinkler device pattern. There were also differences in the water application depth of each square of collectors (Table 2) with a decreasing amount of depth with each successive square. The reason for this change in depth measurement is unclear as the variability of the measurements within the squares was consistent across all squares.

Table 2. Proc Means results for the spinning plate sprinkler (Wiens, 2010).

Treatment	Mean (cm)	Grouping*	St. Dev. (cm)	CV
C2	1.51	A	0.127	0.08
C4	1.48	A	0.108	0.07
C6	1.51	A	0.0598	0.04
C8	1.47	A	0.115	0.08
C10	1.46	A	0.121	0.08
Column	Mean (cm)	Grouping*	St. Dev. (cm)	CV
1 (Inner)	1.45	A	0.0861	0.06
2	1.47	AB	0.0983	0.07
3	1.42	A	0.0850	0.06
4	1.53	BC	0.145	0.09
5 (Outer)	1.55	C	0.0537	0.03
Square	Mean (cm)	Grouping*	St. Dev. (cm)	CV
1	1.54	A	0.101	0.07
2	1.52	AB	0.107	0.07
3	1.47	B	0.0864	0.06
4	1.41	C	0.0925	0.07

* Treatments with the same letter are not different at the 95 percent significance level.

Fixed plate (FP) results are shown in Table 3. There was statistical difference between container catches with C10 having the lowest catch. The CV between containers was similar.

Statistical differences also occur for columns and squares. In general, inner columns catches were less then outer catches as discussed for SP catches. The squares were also statistically different, which would be unexpected.

Wobbling plate (WP) results are shown in Table 4. There were statistical differences between the container sizes with C6 having the lowest catch. The CV of C2 was higher than for the other collectors, which means less confidence in the accuracy of measurement for that collector size as compared to the other options. Higher CV was thought more likely to occur for the FP tests.

Statistical differences also occur for columns and squares. In general, inner columns catches were less then outer catches as discussed for SP catches. The squares were also statistically different, which would be unexpected.

The results of the proc mixed test of measured depths are shown in Table 5. There was statistical significance for each of the sprinkler types with regards to column location as discussed previously. Unexpectedly, the FP collectors had a significant effect due to row location and collector size.. Collector size was not a factor for SP and WP sprinklers in this test.

Table 3. Proc Means results for the fixed plate sprinkler (Wiens, 2010).

Treatment	Mean (cm)	Grouping*	St. Dev. (cm)	CV
C2	5.23	A	0.539	0.10
C4	5.04	AB	0.476	0.09
C6	5.19	A	0.549	0.11
C8	5.22	A	0.486	0.09
C10	4.85	B	0.435	0.09
Column	**Mean (cm)**	**Grouping***	**St. Dev. (cm)**	**CV**
1 (Inner)	4.91	AB	0.340	0.07
2	4.77	A	0.362	0.08
3	5.12	BC	0.486	0.10
4	5.26	CD	0.537	0.10
5 (Outer)	5.48	D	0.501	0.09
Square	**Mean (cm)**	**Grouping***	**St. Dev. (cm)**	**CV**
1	5.45	A	0.541	0.10
2	5.37	A	0.410	0.08
3	4.78	B	0.357	0.07
4	4.82	B	0.311	0.06

* Treatments with the same letter are not different at the 95 percent significance level

Table 4. Proc Means results for the wobbling plate sprinkler (Wiens, 2010).

Treatment	Mean (cm)	Grouping*	St. Dev. (cm)	CV
C2	2.30	AB	0.340	0.15
C4	2.23	A	0.136	0.06
C6	2.35	B	0.214	0.09
C8	2.26	AB	0.134	0.06
C10	2.22	A	0.122	0.06
Column	**Mean (cm)**	**Grouping***	**St. Dev. (cm)**	**CV**
1 (Inner)	2.20	AB	0.365	0.17
2	2.18	A	0.105	0.05
3	2.35	C	0.0687	0.03
4	2.32	BC	0.112	0.05
5 (Outer)	2.31	BC	0.197	0.09
Square	**Mean (cm)**	**Grouping***	**St. Dev. (cm)**	**CV**
1	2.33	A	0.136	0.06
2	2.21	B	0.139	0.06
3	2.39	A	0.300	0.13
4	2.16	B	0.130	0.06

* Treatments with the same letter are not different at the 95 percent significance level.

Table 5. Proc Mixed results for the collector size effect on measured depth for the spinning plate, fixed plate and wobbling plate systems (Wiens, 2010).

Sprinkler Type	Source	Degrees of Freedom	F Value	p-Value	Significance
Spinning Plate	Row	84	1.12	0.3528	NS
	Column	84	8.87	<0.0001	***
	Collector Size	84	1.47	0.2187	NS
Fixed Plate	Row	84	2.62	0.0405	**
	Column	84	18.66	<0.0001	***
	Collector Size	84	6.17	0.0002	***
Wobbling Plate	Row	84	0.34	0.8518	NS
	Column	84	3.63	0.0089	***
	Collector Size	84	1.79	0.1386	NS

* NS indicates collectors are not significantly different, * indicates collectors different at 90 percent significance, ** indicates collectors different at 95 percent significance, and *** indicates collectors different at 99 percent significance.

Table 6 shows the proc mixed test results for the comparison of measured depths for FP since it was the only sprinkler type that showed collector size had a significant effect (from Tables 3-5) on measured depth. C10 had the lowest mean depth and was different than all the other containers. C2 and C4 were also different. It is interesting to note there was not a strong trend of increasing or decreasing depth measurement relative to collector size.

Since the actual water application depth is not known, the ideal collector size cannot be determined from this study. Previously noted C10, the largest collector size had the lowest measured depth result. While large collector diameter is associated with increased accuracy, acceptance of the C10 average depth as the closet to the true actual applied depth, would ignore the that combined estimate from C2, C6 and C8 (sixty observations that were very consistent) which was much higher.

The final statistical test was the proc mix test on the level of variability (SD) of the measured depth for the various collector size for each of the sprinkler type. In this test, the test model adjusted the raw data to eliminate the column effect. The results, shown in Table 7, show only SP had significant differences in the variability of the measured depths. Therefore SP's SD values were proc mix tested for significance. The results are shown in Table 8, for 99 and 90 percent confidence levels. At 99%, C6 had lower variability than C2 or C10. At 90%, C2 had lower variability than C4, C6 was different than C8, and C6 different than C10.

Table 6. Proc Mixed comparison test of the measured depths for the fixed plate sprinkler (Wiens, 2010).

Treatment	Mean (cm)	95% Significance Grouping
C2	5.23	A
C4	5.04	B
C6	5.19	AB
C8	5.22	AB
C10	4.85	C

* Treatments with the same letter are not significantly different at the 95 percent level.

Table 7. Proc Mixed test of the collector size effect on variability of measure depth (Wiens, 2010).

Sprinkler Type	Degrees of Freedom	F Value	p-Value	Significance
Spinning Plate	12	3.43	0.0432	**
Fixed Plate	12	0.10	0.9817	NS
Wobbling Plate	12	1.09	0.4061	NS

* NS indicates collectors are not significantly different, * indicates collectors different at 90 percent significance, ** indicates collectors different at 95 percent significance, and *** indicates collectors different at 99 percent significance.

Table 8. Proc Mixed comparison test of the measured depth variability for the spinning plate system (Wiens, 2010).

Collector Size	Estimated Standard Deviation	99% Significance Grouping	90% Significance Grouping
C2	0.100	A	A
C4	0.0687	AB	BC
C6	0.0477	B	B
C8	0.0785	AB	AC
C10	0.0997	A	A

* Treatments with the same letter are not significantly different at the noted significance level.

Conclusion

Measured depth and variability of measured application depth of three distinct sprinkler application devices were compared for six difference catch can collector using a replicated Latin square study design. The collector sizes were C2, C4, C6, C8, and C10 collectors with diameters of 5.5 cm (2.19 in.), 10.0 cm (3.92 in.), 14.75 cm (5.81 in.), 20.0 cm (7.87 in).

The C6 collector measured the largest water depths for the SP and the WP systems, while the C2 collector measured the greatest depth measurements for the FP system. The C10 collector measured the lowest water depth for all three sprinkler systems. The larger collector sizes measured the same variability of measured depths as the smaller collector sizes for the SP and

FP systems. There was a trend of decreasing measured depth variability as the collector size increased for the WP system, although C4, C8, and C10 collectors had similar levels of measurement variability. The larger collector sizes were expected to record lower measured depth variability for the fixed plate system but the variability was consistent across all collector sizes.

Previous studies had indicated difficulty in accurately measuring the application depth of FP systems using collectors with small diameter with less difficulty of other sprinkler types. An ANOVA analysis of the measured depths showed that the latter expectation was upheld as collector size was not a significant factor in determining the mean water depth for SP and WP systems. However, for the FP system, there was significance at the 95 percent level.

The comparison test of the collector sizes for the FP system indicated that the C2 and C4 collectors measured significantly different depths and the C10 collector was significantly different from all other collectors.

The variability of the measured depths indicated significant differences for the SP system but not for FP and WP systems. It was expected that SP and WP type sprinklers would be similar and the FP more variable. The C6 collector had lower depth measurement variability (90 percent significance) than the C2, C8, and C10 collectors and the C4 collector had measured variability that was significantly lower than the variability of C2 and C10 measured depths.

If the ideal collector size selection was based on minimum variability of depth measurements, then either C6 or C4 collectors could be used. The measurement variability of C6 was numerically lower than the C4 but not significantly different statistically. Though not significant, C4, C8, and C10 had the lowest variability of measured depths for the WP system. A surprising outcome for this study, all collector sizes measured similar levels of variability of depth measurements for the FP system.

Acknowledgement

This study was supported in part by the Ogallala Aquifer Project 58-6209-6-031

References

ASABE S398.1. R2007. Procedure for sprinkler testing and performance reporting. St. Joseph, Mich.: ASABE. ASABE S436.1. R2007. Test procedure for determining the uniformity of water distribution of center pivot and lateral move irrigation machines equipped with spray or sprinkler nozzles. St. Joseph, Mich.: ASABE.

ASABE S436.1. R2007. Test procedure for determining the uniformity of water distribution of center pivot and lateral move irrigation machines equipped with spray or sprinkler nozzles. St. Joseph, Mich.: ASABE

Christiansen, J.E. 1942. Irrigation by sprinkling. Agr. Exp. Sta. Bulletin 670. Berkeley, California.: University of California.

Clark, G.A., E. Dogan, D.H. Rogers, and V.L. Martin. 2006a. Evaluation of IrriGage collectors to measure irrigation depths from low pressure sprinklers. Applied Eng. in Agric. 22(1): 63-72.

Clark, G., D. Rogers, and M. Alam. 2006b. Evaluation of collector size for low pressure, fixedplate sprinklers for center pivots. Contribution No. 07-18-A of the 2006 Technical Conference of the Irrigation Association at San Antonio, TX.

Clark, G.A., D.H. Rogers, E. Dogan, and R. Krueger. 2004. The IrriGage: a non-evaporating infield precipitation gage. Applied Eng. in Agric. 20(4): 463-466.

Heermann, D.F. and P.R. Hein. 1968. Performance characteristics of self-propelled center-pivot sprinkler irrigation system. Trans. ASAE 11: 11-15.

Marek, T.H. and T.A. Howell. 1987. Evaluation of sprinkler application collection devices. ASAE Meeting Paper No. 87-2592. St. Joseph, Mich.: ASAE.

Schneider, A.D. 2000. Efficiency and uniformity of the LEPA and spray sprinkler methods: A review. Trans. ASAE 43(4): 937-944.

Rogers, D.H. and J. Aguilar. 2017. Kansas Irrigation Trends. Kansas State Research and Extension. Irrigation Management Series. MF-2849 (revised). 8 pgs.

USDA. 2013. Farm and Ranch Irrigation Survey. https://www.agcensus.usda.gov/Publications/2012/Online_Resources/Farm_and_Ranch_Irrigation_Survey/fris13_2_028_028.pdf

Wiens, S.W., 2010. Collector Size Effect on the Measurement of Applied Water Depth for Irrigation Systems. Kansas State University, Department of Biological and Agricultural Engineering, Unpublished, Master of Science Thesis.

Kansas center pivot uniformity evaluation overview

Danny H. Rogers, P. E., Ph. D.

Professor and Extension Agricultural Engineer, Irrigation, Kansas State University, Department of Biological and Agricultural Engineering, 2018 Seaton Hall, Manhattan, KS. 66506. Phone: 785-532-2933. Email:drogers@ksu.edu

Jonathan Aguilar, Ph. D.

Assistant Professor and Extension Irrigation Specialist, Kansas State University, SWREC, Garden City, KS.

Abstract: *Center pivot irrigation systems are the most common system type in Kansas for a variety of factors – one of which is the ability to deliver a uniform depth of water application for a variety of crops and field conditions. Uniform applications are dependent on properly designed, installed and operated sprinkler nozzle packages. Uniformity evaluations were conducted as part of the Mobile Irrigation Lab (MIL) project to promote adoption of improved irrigation management practices with an emphasis on ET based irrigation scheduling. Since efficient and uniform water applications are critical to successful irrigation scheduling; MIL included evaluation of sprinkler package performance using a single line catch can test. Catch data was used to calculate the coefficient of uniformity and average application depth. The information was used in extension programs to illustrate the effect of various correctible sprinkler package deficiencies on performance and to encourage irrigation farmers to examine their nozzle packages and operating conditions. A summary of the evaluation results will be presented.*

Keywords: Center pivot irrigation, uniformity, sprinkler packages

Introduction

Center pivot irrigation systems are the dominant irrigation system type in use within Kansas (Rogers and Aguilar, 2017). Irrigation is also the dominant use of water supplies for the state, but in many areas of the state, water supplies are diminishing. However, irrigated agriculture makes significant contributions to the economy so improving irrigation water utility and conservation has long term benefits. Encouraging adoption of improved irrigation management practices is a major goal of the Kansas State Research and Extension (KSRE), including the irrigation scheduling. In the late 1980's and early 1990's, the development of information networks, communication systems and increasing availability of personal computers combined to make ET-based irrigation scheduling an option for irrigation managers to use but lack of familiarity of ET-based irrigation scheduling as well as lack of user friendly scheduling software and limited farmer skills with the operation of PC's remained as barriers to adoption (Rogers et al., 2002a).

On-farm demonstration projects were established in south central and western Kansas to promote ET-based irrigation scheduling using KSU's KanSched scheduling tool. These projects were the forerunners to the Mobile Irrigation Lab (MIL) project which was expanded to include performance evaluate center pivot nozzle packages for uniformity (Rogers et al., 2002a, Rogers et al., 2002b, Clark et al., 2002). One rationale for conducting center pivot nozzle package evaluations was that adoption of improved irrigation management techniques, such as ET-based irrigation scheduling, required a uniform application depth to assure all the crop had equal access to the available water and no areas of the field were either over- or under- watered which would reduce irrigation water productivity. The majority of the tests were conducted using a single line of catch cans of 4 –inch diameter, called Irrigages (Clark et al., 2004), spaced at no more than 80 percent of the sprinkler nozzle spacing.

Catch can evaluations require sufficient clearance of the nozzle above the top of the collector. In a center pivot survey (Rogers et al., 2009), most systems in south central Kansas could be tested using the irrigage catch can evaluation, since over 92 percent have nozzle heights of greater than 4 foot above ground surface. However, in western Kansas, almost 60 per cent of the nozzle packages are mounted at 4 foot or less above the ground surface which is insufficient clearance for an irrigage collector, especially since the top of the irrigage is about 16 inches above ground when installed.

Procedures

The catch can generally used was a 4-inch irrigage which was constructed with a storage bottle attached to the bottom of the collection barrel to which the water drained after capture in the collection barrel. Once in the bottle, evaporation losses were minimal. This allowed data collection without concern for accuracy losses due to evaporation, improved time convenience for collection of data and minimized the on-site labor need for data collection.

The majority of the tests were conducted using a single line of catch cans, spaced at no more than 80 percent of the sprinkler nozzle spacing. The collector spacing was selected so a catch sample would be collected within each nozzle spacing interval but with gradual change in the collection location relative to the nozzle outlet. Although the overall coefficient of uniformity (CU) value could be calculated, another goal was to document the effect of various operational deficiencies on the performance of the sprinkler package. Many of performance issues could have been identified with a visual inspection of the nozzles and/or a comparison of the nozzle package as installed to the sprinkler design package.

The center pivot systems initially evaluated were a part of a demonstration project. Part of the selection criteria for the project field sites included the drive-by visibility of project signage and ease of access for education tours or programs. These systems also thought to be systems with well-maintained and operated at design specifications. Other systems evaluated were at the request of individuals, therefore, the evaluated systems were not randomly selected. The intent was to evaluate as many as systems as possible each year while the MIL program was funded. However many constraints limited the number of evaluations possible , such as winter evaluations were often precluded, spring cultural operations (where a wetted area within the field would not be desirable), scheduling limitations of the operators (we required them to start the systems), crop canopy height limitations, and even water right limitations.

Results and Discussion

Fifty-three center pivot irrigation sprinkler package evaluations were conducted Kansas during the period of 1998 through 2011 using catch cans. These evaluations were conducted on unique systems, expect for tests FI 01A – 99. In this instance, the system was tested in the two modes of operation; with the end gun on and with the end gun off. Both values are included. These results are shown in Table 1 which includes the general classification of the sprinkler type, collector spacing, the CU and slope of the average application depth, pressure regulation, collector diameter, and the measures region of the system.

The sprinkler types were classified as fixed plate, impact, and moving plate sprinklers. Fixed plate sprinklers are primarily spray nozzles with a splash plate that does not move when impacted by the water stream; while a moving plate sprinkler would have a splash plate that spins, oscillates or otherwise moves when impacted by the water stream. The number of each sprinkler type tested and the average CU of the systems are shown in Table 2. The average CU for the three sprinkler types were similar. Only four impact sprinkler packages were tested and were all operated by one producer. In the center pivot survey (Rogers et al., 2009), only about 2 per cent of the survey observations were impact sprinklers. In some instances, tested systems may have had either wider nozzle spacing on the first span and/or a different sprinkler type on the first span but the sprinkler type and later the sprinkler spacing reported reflects the package used on the bulk of the system.

The measured range of the center pivots are included in Table 1 with the majority of the systems being quarter mile systems of approximately 1300 foot in length, although several are longer including one of one half mile in length and one with a corner system (tested with the corner extended). Note that some systems were tested only in the outer spans verses nearly to the pivot point. This range was reflection of whether the test was conducted with the evaluators staying on-site or being able to leave the site to return later for data collection. Graphs of the applied depths of systems often show higher application depths in inner span but including or excluding these values from the CU calculation, since the values are area weighted, have little impact on the overall CU value.

Early tests were conducted using 17-inch diameter pans before the development of the irrigages. The pans nested for easy transportation and storage and they were easy to install since they only needed to be placed on the ground surface. However, they also needed to be read quickly after an irrigation event to minimize pan evaporation losses. The weight of water collected was used as the measurement method. The pans had to be carried to a weigh station which was labor intensive and tedious. While the average CU value of the pan catches was higher than the irrigage catches, the difference was more likely do to the systems selected to tested by the pans rather than the collector size itself. Early systems were demonstration project fields thought to be well maintained and/or relatively new and selected to promote irrigation scheduling; verses later fields that were tested at the request of producers which were field that they suspected may have an issue.

Table 2 also includes the average CU values for pressure regulated (81.67) and non-pressure regulated systems (75.62). In the Kansas center survey (Rogers et al., 2009) about half of the center pivots in SC Kansas were pressure regulated and about 80 percent in western Kansas. In western Kansas, many of the spray systems are close to the ground and therefore not able to be tested with a catch can procedure.

The CU values for the various collector spacings are also summarized in Table 2. Initially, the tests were conducted at about 80 % of the nozzle spacing rounded to the nearest foot. Over time, the tests migrated to being conducted at either 4 foot or 8 foot spacing as a way to streamline the test procedure. There is a tendency for the closely spaced collectors to have higher CU but the data set, especially at wider spacing, is limited.

Figures 1a and 1b are the graph of the same system (FI 01A -99, Table 1) tested with the end gun on and end gun off, respectively. Figure 1a shows an area of good uniformity until the high catch at radius 945 feet. This high catch was due to a leaky tower boot. The next area of catch shows a gradual decrease in catch until radius 1241 when application depth increases dramatically. The area with gradually decreasing application was due to a reversal of the outer two spans nozzles, while the sudden increase was caused by over spray from the end gun onto a portion of the main lateral as the end gun was not ratcheting properly. The area of decreasing application depth due to improper nozzle installation is more visible in figure 1b (Rogers et al., 2008, Rogers, 2012).

The application depth distribution graph for test PR 5-27-99 is shown in figure 2. The CU value for this system is 84.3. The major problem associated with this system was at the outer edge where the application depth dropped to approximately half. This effect was due to an un-installed nozzle and under sizing of the orifices of the next two adjacent nozzles in both directions from the uninstalled nozzle location as compared to the design specifications. This under-watered area covered approximately 9.2 acres. So if the average water application was 12 inches, so this area received around 6 inches of irrigation. A conservative estimate of yield response would be 10 bu/in, resulting in an estimated annual field loss of over 500 bushels which could easily be repaired at minimal cost.

Figure 3 shows the graph for center pivot test SN 7-18-02 which had the lowest CU value of the systems tested (CU = 53.2). The issue associated with this nozzle package was incrustation build-up within the system and on the fixed plate nozzles as shown in figure 4. A regular maintenance requirement for this system included unclogging nozzles at the start of irrigations and the removal of nozzles in the off-season for cleaning of incrustation. Incrustation on the splash plate would interfere with the development of the spokes of water typical for this type of nozzle and prevent proper overlap of the water streams. However, for this very level field, farmed with high residue practices, the applied water was adequately re-distributed on the ground surface as evidenced by the crop appearance (figure 5).

The ASABE standard (ASAE S436.1) describing the test procedure determining the uniformity of water distribution by center pivots has a maximum can spacing of 3 meters (9.84 ft.) for spray devices and 5 meters (16.4 ft.) for impact sprinklers. The MIL tests were conducted using a single line of cans verses two rows for the ASABE test. Never-the-less, the impact of can spacing on CU was examined by calculating the CU values for the base can spacing, then every other can (2 sets) and every third can (3 sets). The results are shown in Table 3, arranged by from lowest can spacing to largest spacing.

The first three systems (PR5-27-99, KI 6-09-99, ED 6-01-99) used a collector spacing of 4 ft. with CU values ranging from 84.3 to 89.9). Recalculating CU values for 2x or 3x spacing values resulted in less than 1.0 change in CU as compared to the base CU. The regression lines through the applied depth of catches were very flat and changed little with the increased spacing. In this case, the 2x catches would have been at a 8 ft. spacing which is still within the ASABE spacing recommendation but results varied little when going to a 12 ft. spacing , which slightly exceeds the ASABE recommendation.

The next two systems (RC-TZ-1998, ED 6-02-99) had CU values of 91.9 and 84 measured at 5 ft can spacing with a flat regression line for the applied depth of application for the first system, and a positive slope for the second, meaning increasingly more water was being applied with distance from the pivot point. The slope of the regression line was not greatly impacted by can spacing and also little impact on the average applied application depth. The CU for RC-TZ-1998 had a maximum CU change of 1.5 for both 2x and 3x spacing. The 2x spacing is 10 ft. or approximately the maximum recommended ASABE spacing, while 3x spacing would exceed the ASABE recommendation. The change in CU value for ED 6-02-99 was only 0.2 at 2x spacing but 5.1 for the 3x spacing.

SN 7-18-02, which was discussed previously and shown in figure 3), had large change in CU calculation estimates with increased spacing, however with the base can spacing at 6 ft, both 2x and 3x catches would exceed the ASABE spacing recommendation. The estimate of applied application depth and the slope of the applied application depth regression line was also impacted by change in spacing.

The next four systems were tested at an 8 ft. spacing and the last system at 10 ft., which would be within or near ASABE guidelines. Two systems (LN 4-21-03, BT 3-27-02) showed spacing had little impact on the CU value. The latter system had a strong slope to the application depth. This was thought to be from improper input operating conditions (The on-site values were accepted at the time since it was a new installation and not independently verified. Test crews returned to the site at later dates twice but a new catch was never successfully completed.). The maximum change in CU value for the other systems ranged from 7.6 to 8.6 with the largest CU change for the 2x spacing.

Conclusion

A series of center pivot uniformity evaluations were conducted over multiple years providing a snapshot of the performance of these systems at the time of the test. A single line test with a catch can spacing of less than the sprinkler spacing was used. The systems tested were not randomly selected. The average CU value of the tested systems was 78.65 with a range of from 91.9 to 53.2. Early tests tended to be on producer fields in a demonstration project and tended to have higher CU values, which indicates that high CU values are achievable. Latter tests, conducted at the request of producers, tended to be systems suspected of having an issue. Many of the sprinkler package deficiencies could have been identified and corrected with a visual inspection and/or a comparison to the sprinkler package design specifications. However, the catch test then documents the impact of a sprinkler package deficiency on the performance.

Information from these tests have been used in meetings and publications to encourage irrigation managers that high CU performance is possible with good package designs and proper operating conditions but also regular sprinkler package maintenance.

Acknowledgements

This study was supported in part by The Mobile Irrigation Lab Project GECG 601490 , funded by the Kansas Water Plan Fund administered by the Kansas Water Office, USDA Project GEGC 601448 and the Ogallala Aquifer Project GEGC 600468.

References

Rogers, D.H. and J. Aguilar. 2017. Kansas Irrigation Trends. Kansas State Research and Extension. Irrigation Management Series. MF-2849 (revised). 8 pgs.

Rogers, D.H., G. A. Clark, M. Alam, R. Stratton, and S. Briggeman. 2002. A Mobile Irrigation Lab for Water Conservation: II Education Programs and Field Data. In proceedings of Irrigation Association International Irrigation Technical Conference, October 24-26, 2002, New Orleans, LA, available from I.A., Falls Church, VA

Rogers, D.H., M. Alam, and L.K. Shaw. 2008. Considerations for Sprinkler Packages on Center Pivots. Kansas State Research and Extension. Irrigation Management Series. L-908 rev.

Rogers, D.H. 2012. Efficient Crop Water Use in Kansas, Chapter 3: Evaluating Center Pivot Nozzle Package Performance. Kansas State Research and Extension. MF-3066. pp. 8-17.

Rogers, D.H., G. A. Clark, M. Alam, and L.K. Shaw. 2005. Field Performance Testing of In-canopy Center Pivot Nozzle Packages in Kansas. In proceedings of Irrigation Association International Irrigation Technical Conference, IA 05-1239. November 6-8, 2005, Phoenix, AZ. pp. 295-301.

Rogers, D.H., M. Alam, and L.K. Shaw. April 2009. Kansas Center Pivot Survey. Kansas State Research and Extension. Irrigation Management Series MF-2870.

Rogers, D.H., M. Alam, G.A. Clark and L.K. Shaw. 2006. MIL Evaluation of Center Pivot Irrigation Systems. Proceedings of the Central Plains Irrigation Conference. Colby, KS February 21-22, 2006. pp 35-43.

Rogers, D.H., G.A. Clark, M. Alam, R. Stratton, D.L. Fjell and S. Briggeman. 2003. Mobile Irrigation Lab (MIL): Center Pivot uniformity Evaluation Procedure and Field Results. Proceeding of the Central Plains Irrigation Short Course, Colby, KS. February 4-5, 2003. Available from CPIA. 760 N. Thompson, Colby, KS. Pp 78-84.

Clark,G.A., D.H. Rogers, and M. Alam. 2006. Evaluation of Collector Size for Low Pressure, Fixed-Plate sprinklers for Center Pivots. In proceedings of Irrigation Association International Irrigation Technical Conference, IA06-1513, November 5-7, 2006. San Antonio, Texas. Pp 368-373.

Clark, G.A., D.H. Rogers, M. Alam, D.L. Fjell, R. Stratton, and S. Briggeman. 2002. A Mobile Irrigation Lab for Water Conservation: I. Physical and Electronic Tools. In proceedings of Irrigation Association International Irrigation Technical Conference, October 24-26, 2002, New Orleans, LA, available from I.A., Falls Church, VA.

Rogers, D.H., G.A. Clark, M. Alam, R. Stratton, and S. Briggeman. 2002. A Mobile Irrigation Lab for Water Conservation: II Education Programs and Field Data. In proceedings of Irrigation Association International Irrigation Technical Conference, October 24-26, 2002, New Orleans, LA, available from I.A., Falls Church, VA.

Rogers, D.H. 2012. Efficient Crop Water Use in Kansas, Chapter 3: Evaluating Center Pivot Nozzle Package Performance. Kansas State Research and Extension. MF-3066. pp. 8-17.

Rogers, D.H., M. Alam, L.K. Shaw, and G.A. Clark. 2009. Impact of collector size and spacing on center pivot uniformity evaluations. ASABE paper no. 09-6522. Available from ASABE, ST. Joesph, MI. 12 pp.

Clark,G.A., D.H. Rogers, and M. Alam. 2006. Evaluation of Collector Size for Low Pressure, Fixed-Plate sprinklers for Center Pivots. In proceedings of Irrigation Association International Irrigation Technical Conference, IA06-1513, November 5-7, 2006. San Antonio, Texas. Pp 368-373.

Clark, G.A., E. Dogan, D.H. Rogers, and V.L. Martin. 2003 .Evaluation of Collector Size for the Measurement of Irrigation Depth. In proceedings of Irrigation Association International Irrigation Technical Conference, November 17-20, 2003, San Diego, Ca. pp. 269-278.

Rogers, D.H. 2016. Performance of Center Pivot Irrigation Systems. In: Proc. 28th annual Central Plains Irrigation Conference, Feb. 23-24, 2016, Kearney, NE. Available from CPIA, 760 N. Thompson, Colby, KS. pp. 12-18.

Rogers, D.H. and J.K. Koelliker. 2011. Evaluating Center Pivot Nozzle Package Performance. In: Proc. 23rd annual Central Plains Irrigation Conference, Burlington, Colorado, Feb. 22-23, 2011. Available from CPIA, 760 N. Thompson, Colby, KS. pp. 122-134.

Rogers, D.H., M. Alam, G.A. Clark and L.K. Shaw. 2006. MIL Evaluation of Center Pivot Irrigation Systems. Proceedings of the Central Plains Irrigation Conference. Colby, KS February 21-22, 2006. Pp 35-43.

Rogers, D.H., G.A. Clark, M. Alam, R. Stratton, D.L. Fjell and S. Briggeman. 2003. Mobile Irrigation Lab (MIL): Center Pivot uniformity Evaluation Procedure and Field Results. Proceeding of the Central Plains Irrigation Short Course, Colby, KS. February 4-5, 2003. Available from CPIA. 760 N. Thompson, Colby, KS. Pp 78-84.

Rogers, D.H., G.A. Clark, and M. Alam. 2001. Field Scale Evaluation of Center Pivot Systems. Proceedings of the 2001 Central Plains Irrigation Short Course, Kearney, NE, February 5-6, 2001. Pp 112-125.

Rogers, D.H., G.A. Clark, M. Alam, R. Stratton, D. L. Fjell, and S. Briggeman. 2004. Mobile Irrigation Lab (MIL): Computer Software and Center Pivot Evaluation. Proceedings of the Bureau of Indians Affairs Tribal Irrigation Workshop. Denver, CO. February 10-12, 2004.

Rogers, D.H., L.K. Shaw, and D.O. Porter. 2011. Field Evaluation of Center Pivot Nozzle Package Application Intensity. Final Report for Ogallala Aquifer Project (OAP web site). 14 pp.

Rogers, D.H., M. Alam, G.A. Clark and L.K. Shaw. 2006. MIL Evaluation of Center Pivot Irrigation Systems. Proceedings of the Central Plains Irrigation Conference. Colby, KS February 21-22, 2006. Pp 35-43.

Rogers, D.H., G.A. Clark, M. Alam, R. Stratton, D.L. Fjell and S. Briggeman. 2003. Mobile Irrigation Lab (MIL): Center Pivot uniformity Evaluation Procedure and Field Results. Proceeding of the Central Plains Irrigation Short Course, Colby, KS. February 4-5, 2003. Available from CPIA. 760 N. Thompson, Colby, KS. Pp 78-84.

Clark,G.A., D.H. Rogers, and M. Alam. 2006. Evaluation of Collector Size for Low Pressure, Fixed-Plate sprinklers for Center Pivots. In proceedings of Irrigation Association International Irrigation Technical Conference, IA06-1513, November 5-7, 2006. San Antonio, Texas. Pp 368-373.

Clark, G.A., D.H. Rogers, M. Alam, D.L. Fjell, R. Stratton, and S. Briggeman. 2002. A Mobile Irrigation Lab for Water Conservation: I. Physical and Electronic Tools. In proceedings of Irrigation Association International Irrigation Technical Conference, October 24-26, 2002, New Orleans, LA, available from I.A., Falls Church, VA.

Rogers, D.H., G.A. Clark, M. Alam, R. Stratton, and S. Briggeman. 2002. A Mobile Irrigation Lab for Water Conservation: II Education Programs and Field Data. In proceedings of Irrigation Association International Irrigation Technical Conference, October 24-26, 2002, New Orleans, LA, available from I.A., Falls Church, VA.

Clark, G.A., D.H. Rogers, E. Dogan, and R. Krueger. 2004. The Irrigage: A Non-Evaporating In-Field Precipitation Gage. Appl. Engr. in Agric. Vol. 20(4): 463-466.

Table 1: Coefficient of Uniformity (CU) and slope of linear regression line of catch depth and selected test information for various center pivot sprinkler packages of fixed plate, moving plate and impact sprinklers.

Test ID	Type of Nozzle	Can Space	CU	Regression Line Slope	Pressure Regulated	Can Dia.	Test Area
		Ft.	%		No or PSI	Ins.	Ft. from pivot point
ED 6-01-99	Fixed	4	86.6	-0.000006	No	17	628 - 1298
FI 01A -99 EG On	Fixed	4	74.8	0.00001	No	17	473 - 1365
FI 01A - 99 EG Off	Fixed	4	78.2	0.00002	No	17	473 - 1313
SV 5-27-99	Fixed	4	73.2	0.0001	No	17	1250 - 2598
FI 5-26-05	Fixed	4	72.8	-0.00005	6	4	266 - 1306
FI 4-17-06 a	Fixed	4	77.6	0.0004	6	4	12 - 1294
HS 8-05-09	Fixed	4	81.7	-0.0004	15	4	20 - 1324
BT 6-28-10	Fixed	4	76.3	-0.00003	No	4	295 - 1470
FI 8 - 12 - 11 (A)	Fixed	4	89.5	0.00002	10	4	8 - 1328
ED 6-02-99	Fixed	5	84.0	0.0001	No	17	660-1352
SN 7-18-02	Fixed	6	53.2	0.0001	No	4	750 - 1290
SV 5-12-05	Fixed	6	79.6	-0.00002	NR*	4	300 - 1296
FI 5-27-05	Fixed	6	87.0	-0.0001	10	4	532 - 1300
FI 7-02-08	Fixed	6	86.6	0.0002	10	4	24 - 1302
FI 7-17-08	Fixed	6	91.1	0.00009	10	4	168 - 1302
FI 3-28-08a	Fixed	6	92.1	0.00006	10	4	184 - 1296
FI 4-16-02	Fixed	8	81.9	0.00003	No	4	16 - 1288
FO 5-16-02	Fixed	8	58.2	-0.0005	No	4	210 - 1322
SN 6-02-02	Fixed	8	86.8	0.0002	10	4	537 - 1249
FI 7-19-05	Fixed	8	75.5	0.000001	No	4	50 - 1298
LN 4-21-03	Fixed	8	71.0	-0.00008	No	4	250 - 1282
RNU01	Fixed	8	68.6	0.0002	No	4	360-1528
FI 6-14-06a	Fixed	8	71.9	0.0003	10	4	24 - 1304
FO 5-27-09	Fixed	8	86.7	-7E-07	10	4	120 - 1392
FI 7-25-05 b	Fixed	8	71.8	-0.0005	10	4	134 - 1286
KI 6-09-99	Fixed	4	89.9	0.00001	No	17	526 - 1326
FO 3-13-06	Impact	8	82.4	-0.0001	No	4	264 - 1352
FO 3-09-06	Impact	8	72.1	-0.00008	No	4	48 - 1336
FO 4-04-07a	Impact	8	82.4	-0.0002	No	4	268 - 1352
FO 3-30-07a	Impact	8	73.5	-0.0003	No	4	270 -1344
PR 5-27-99	Moving	4	84.3	-0.00008	30	4	588 - 1300
RN 5-06-11a	Moving	4	90.9	-0.0002	20	4	8 - 847
MP GS-1998	Moving	5	91.8	-0.0002	No	17	770 - 1290
RC-TZ- 1998	Moving	5	91.9	-0.00003	NR	17	733 - 1213

SG 5-22-02	Moving	6	83.8	-0.0002	No	4	132 - 1212
SD 6-15-05	Moving	6	74.1	0.0002	Yes	4	480 - 1212
FI 7-15-09	Moving	6	90.9	-0.0002	12	4	30 - 1140
GY 4 -01-08 b	Moving	6	73.8	0.0003	10	4	102 - 1338
BT 3-27-02	Moving	8	81.7	0.0003	10	4	326 - 1254
KI 7-8-02	Moving	8	76.4	-0.0003	Yes	4	340 - 1308
MP 8-21-02	Moving	8	76.0	-0.0002	No	4	365 - 1277
MP1 8-21-02	Moving	8	67.0	-0.0003	No	4	486 - 1430
PN 4-01-03	Moving	8	83.1	-0.00007	10	4	350-1278
SW 5-15-03	Moving	8	76.3	-0.0007	10	4	350 - 1278
HV 10-05-11	Moving	8	79.1	-0.0001	20	4	176 - 1253
SG 3-14-03	Moving	8	65.9	0.0002	No	4	148 - 1284
FI 7-25-05	Moving	8	72.2	-0.0003	10	4	62 -1422
RN 6-05-00	Moving	10	74.5	0.0002	NR	4	630 - 1260
RN 7-01-00	Moving	10	88.8	0.0003	No	4	845 - 1335
RC 7-06-00	Moving	10	72.8	-0.0002	No	4	540 - 1230
SF 6-06-00	Moving	10	88	-0.0003	NR	4	624 - 1244
HV 4-10-03	Moving	10	62.6	-0.0002	No	4	383 - 1353
RN 6-08-02	Moving	12	65.3	-0.00005	NR	4	343 - 1311

*NR = not recorded

Table 2: Average CU values for center pivot performance evaluations

Test Summary of CU	**CU**	**Number of Observations**
Overall Average	78.65	53
Type of Sprinkler		
Fixed Plate Average	78.72	26
Impact Sprinkler Average	77.60	4
Moving Plate Average	78.75	23
Size of Catch Can		
4 inch Catch Can Average	77.73	45
17 inch Catch Can Average	83.80	8
Pressure Regulated System		
Pressure Regulated	81.67	23
Non-pressure Regulated	75.62	25
Not Recorded	79.86	5
Catch Can Spacing		
Average 4 ft	81.32	12
Average 5 ft	89.23	3
Average 6 ft	81.22	10

Average 8 ft	75.48	22
Average 10 ft	77.34	5
Average 12 ft	65.30	1

Table 3: Influence of can spacing on CU

Test ID	**Type of Nozzle**	Collector Spacing (Ft.)	CU %	Applied Depth Regression Line Slope	Applied Depth (Ins.)
PR 5-27-99	Moving	**4**	**84.3**	**-0.00008**	**0.3**
		Odd	**83.8**	**-0.00009**	**0.3**
		Even	**84.7**	**-0.00007**	**0.3**
		3.1	**83.3**	**-0.00007**	**0.3**
		3.2	**84.4**	**-0.00007**	**0.3**
		3.3	**85.2**	**-0.0001**	**0.3**
KI 6-09-99	Fixed	**4**	**89.9**	**0.00001**	**0.32**
		Odd	**89.7**	**0.00002**	**0.33**
		Even	**89.9**	**0.000008**	**0.32**
		3.1	**90.8**	**0.000004**	**0.32**
		3.2	**89.4**	**0.00002**	**0.32**
		3.3	**89.2**	**0.00001**	**0.32**
ED 6-01-99	Fixed	**4**	**86.6**	**-0.000006**	**0.54**
		Odd	**87.2**	**0.000008**	**0.54**
		Even	**86**	**-0.00002**	**0.54**
		3.1	**86.1**	**-0.00006**	**0.55**
		3.2	**86.4**	**0.00008**	**0.55**
		3.3	**87.4**	**-0.00004**	**0.53**
RC-TZ- 1998	Moving	**5**	**91.9**	**-0.00003**	**0.81**
		Odd	**91.2**	**0.00005**	**0.82**
		Even	**92.7**	**-0.0001**	**0.81**
		3.1	**91**	**0.0001**	**0.81**
		3.2	**92.5**	**-0.00007**	**0.83**
		3.3	**92.2**	**-0.0001**	**0.8**

ED 6-02-99	Fixed	**5**	**84**	**0.0001**	**0.44**
		Odd	**83.9**	**0.00009**	**0.45**
		Even	**84.1**	**0.0001**	**0.44**
		3.1	**87.2**	**0.0001**	**0.44**
		3.2	**82.1**	**0.00008**	**0.44**
		3.3	**83.1**	**0.0001**	**0.46**
SN 7-18-02	Fixed	**6**	**53.2**	**0.0001**	**0.67**
		Odd	**44.6**	**-0.0004**	**0.68**
		Even	**55.5**	**-0.0003**	**0.66**
		3.1	**44.7**	**-0.0005**	**0.62**
		3.2	**56.2**	**-0.0001**	**0.75**
		3.3	**50.7**	**-0.0004**	**0.64**
PN 4-01-03	Moving	**8**	**83.1**	**-0.00007**	**0.73**
		Odd	**77.9**	**0.0002**	**0.73**
		Even	**86.5**	**-0.0002**	**0.7**
		3.1	**81.3**	**0.000008**	**0.72**
		3.2	**79.2**	**-0.000003**	**0.74**
		3.3	**85.4**	**-0.00001**	**0.68**
LN 4-21-03	Fixed	**8**	**71**	**-0.00008**	**0.56**
		Odd	**70.6**	**0.00008**	**0.57**
		Even	**71.5**	**0.00008**	**0.56**
		3.1	**71.8**	**0.000006**	**0.52**
		3.2	**70**	**0.0002**	**0.61**
		3.3	**71.5**	**-0.000009**	**0.56**
MP 8-21-02	Moving	**8**	**76**	**-0.0002**	**0.69**
		Odd	**78.4**	**-0.0002**	**0.67**
		Even	**74.1**	**-0.0002**	**0.72**
		3.1	**80.5**	**0.00007**	**0.66**
		3.2	**72.2**	**-0.0003**	**0.71**
		3.3	**75.7**	**-0.0003**	**0.71**
BT 3-27-02	Moving	**8**	**81.7**	**0.0003**	**0.63**
		Odd	**82.6**	**0.0003**	**0.62**
		Even	**81**	**0.0003**	**0.65**

		3.1	**82**	**0.0003**	**0.61**
		3.2	**81.9**	**0.0003**	**0.63**
		3.3	**81.4**	**0.0004**	**0.64**
RC 7-06-00	Moving	**10**	**72.8**	**-0.0002**	**0.88**
		Odd	**72.4**	**-0.0001**	**0.89**
		Even	**73.1**	**-0.0003**	**0.88**
		3.1	**70.9**	**0.0001**	**0.85**
		3.2	**70.2**	**-0.0005**	**0.96**
		3.3	**77.8**	**-0.0003**	**0.84**

Figure 1a. Catch can uniformity analysis for Center Pivot FI 01A End Gun On

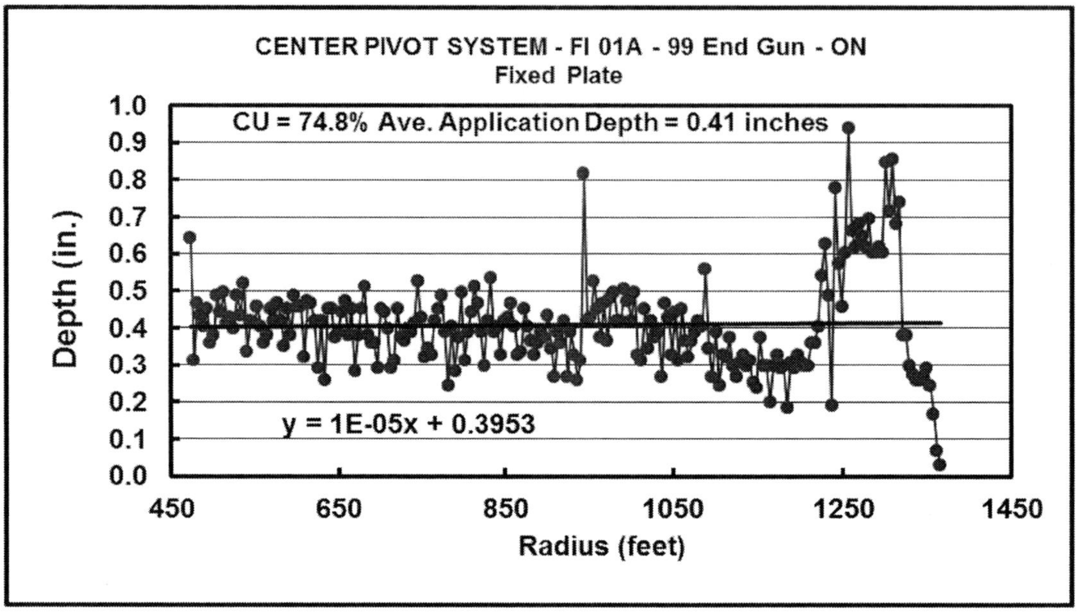

Figure 1b. Catch can uniformity analysis for Center Pivot FI 01A End Gun Off

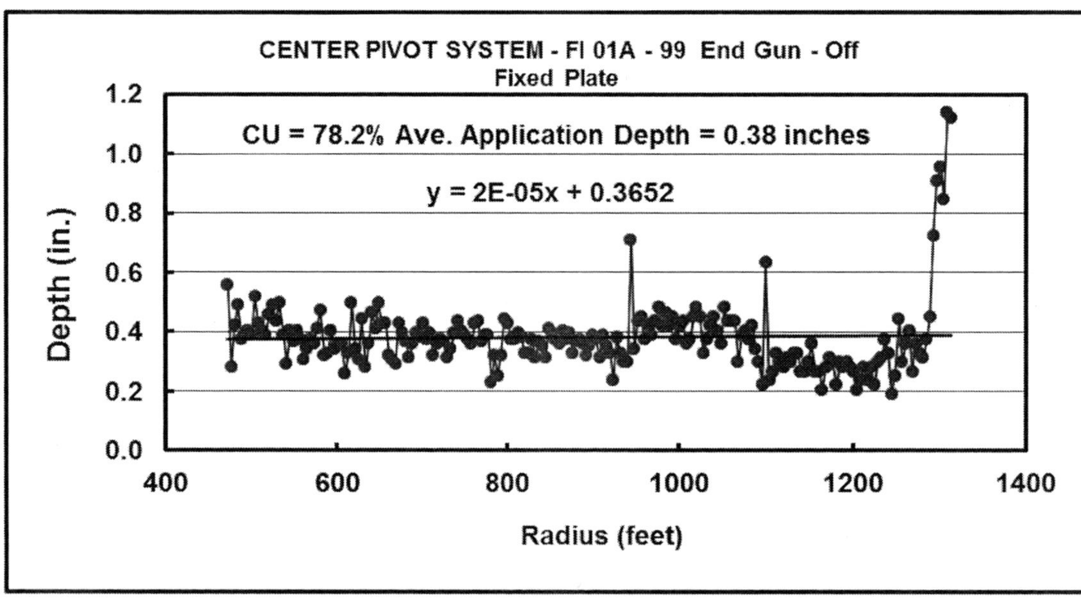

Figure 2: Catch can uniformity analysis for center pivot PR 5-27-99.

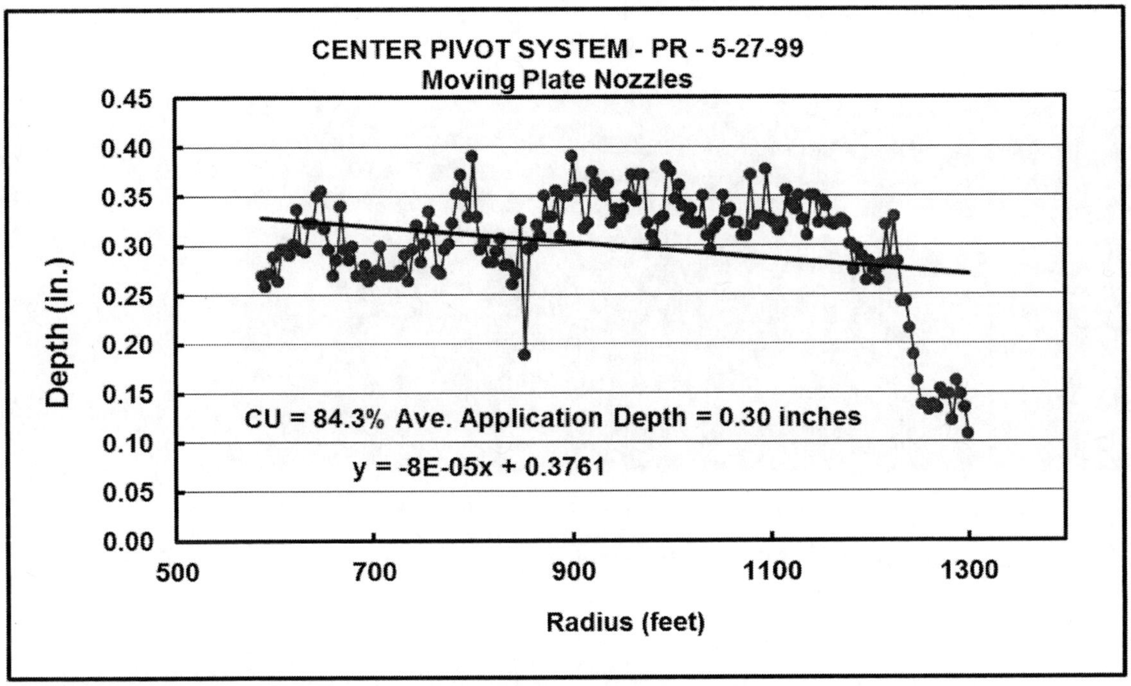

Figure 3: Catch can uniformity analysis for center pivot SN 7-18-02.

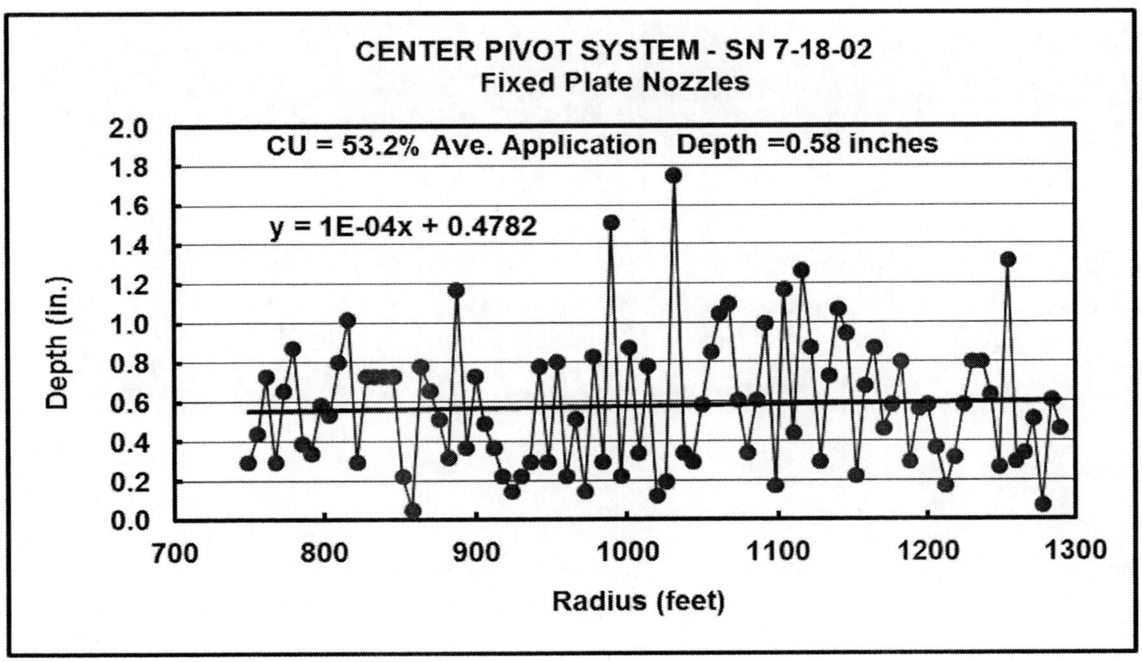

Figure 4: Nozzles incrustation for center pivot SN 7-18-02.

Figure 5: Crop appearance for center pivot SN 7-18-02.

Evaporative loss differences between SDI and sprinkler irrigation – Southern High Plains experience

Steven R. Evett, Research Soil Scientist

USDA-ARS Conservation & Production Research Laboratory, Bushland, Texas, steve.evett@ars.usda.gov

Gary W. Marek, Research Agricultural Engineer

USDA-ARS Conservation & Production Research Laboratory, Bushland, Texas

Paul D. Colaizzi, Research Agricultural Engineer

USDA-ARS Conservation & Production Research Laboratory, Bushland, Texas

David K. Brauer, Research Leader

USDA-ARS Conservation & Production Research Laboratory, Bushland, Texas

Susan A. O'Shaughnessy, Research Agricultural Engineer

USDA-ARS Conservation & Production Research Laboratory, Bushland, Texas

Abstract: Subsurface drip irrigation (SDI) has steadily gained ground despite being considerably more expensive to install than center pivot irrigation. SDI now serves greater than 8% of land irrigated by pressurized irrigation systems (sprinkler, surface and subsurface drip and other microirrigation methods) in the US. In the Texas High Plains, the most southern extension of the High Plains, irrigation water is completely derived from "fossil" aquifers, the most important of which is the declining Ogallala Aquifer, a fossil aquifer that is recharged at most about one inch per year. Well yields are steadily declining, making it difficult in some cases to find adequate capacity to serve a center pivot irrigation system. An SDI system can be zoned to accommodate the smaller well yields. Although center pivot variable rate irrigation (VRI) systems can also accommodate declining well yields, the acceptance of VRI systems is relatively small – although growing. However, other factors influence acceptance of SDI, including larger yields, particularly with cotton, after conversion to SDI. Research has shown warmer soil temperatures obtained with SDI due to the reduced evaporative cooling early in the season, and crop rooting and early growth are improved in the warmer soil, particularly for cotton, which is one reason for larger yields. Not as well established is the degree to which the reduced soil water evaporation in SDI systems affects the soil water balance, water available to the crop, and overall water savings. Grain corn (*Zea mays* L.) and sorghum (*Sorghum bicolor* L. Moench) were grown on four large weighing lysimeters at Bushland, Texas in 2013 (corn), 2014 and 2015 (sorghum) and 2016 (corn). Two of the lysimeters and surrounding fields were irrigated by subsurface drip irrigation (SDI) and the other two were irrigated by mid elevation spray application (MESA). Evaporative losses from SDI fields were two to five inches less than those from sprinkler irrigated fields. Differences were strongly affected by plant height, essentially disappearing when plant height reached the elevation of spray nozzles, indicating that use of LEPA or LESA nozzles could decrease the evaporative losses from sprinkler irrigated fields in this region with its high evaporative demand. Annual weather patterns also influenced the differences in evaporative loss, with differences being exacerbated in dry years.

Keywords: Crop water productivity; Water use efficiency; Irrigation application method; Evaporative loss; Evapotranspiration; Corn; Sorghum; Weighing lysimeter

Introduction

Irrigation application method is known to affect crop performance, including yield and water use efficiency, with subsurface drip irrigation (SDI) having some advantages over spray sprinkler irrigation for corn, cotton and sorghum production. Sprinkler has become the predominant irrigation method for crop production in the USA and Great Plains, with > 80 percent of the irrigated area in the Southern High Plains (SHP) being by center pivot irrigation system (Colaizzi et al., 2009; NASS, 2014). Microirrigation, which includes surface and subsurface drip irrigation, generally results in greater crop water productivity compared with sprinkler, where the greater crop water productivity is due to greater crop production, less consumptive use of irrigation water, or a combination of both (Camp, 1998), less sensitivity to impaired irrigation water (Goldberg and Shmueli, 1970; Berstein and Francois, 1973; Goldberg et al., 1976; Adamsen, 1989; 1992; Wu et al., 2001), and warmer soil temperatures (Wang et al., 2000; Colaizzi et al., 2010). In some cases, this has justified the greater capital costs per unit land area of microirrigation compared with sprinkler (Bordovsky et al., 2000; Bosch et al., 1992; Enciso et al., 2005; O'Brien et al., 1998), leading to > 200,000 acres irrigated by SDI in the SHP, almost entirely for cotton. As cotton, corn and sorghum prices fluctuate, and as other pressures to rotate out of cotton arise, this SDI ground may be planted to corn or even sorghum.

The advantage of SDI is thought to be due in part to decreased loss of water to evaporation (*E*) from the soil surface since the soil surface is directly wetted by spray sprinklers but not with SDI. Also, there are no evaporative losses from wind drift or evaporation from sprinkler-wetted canopies using SDI. A 36% (81 mm) decrease in evaporative loss using SDI vs. surface irrigation (sprinkler or gravity flow) that wets the entire soil surface was estimated using a mechanistic model (Evett et al., 1995), which would mean that more of the applied irrigation water would be available for transpiration (*T*) by plants. Because yield is directly tied to transpiration (Doorenbos and Kassam, 1979), this increase in the *T/E* ratio should result in relatively more yield per unit of water applied with SDI, and a corresponding increase in crop water productivity, also known as crop water use efficiency (WUE) (Howell, 2001).

There are, however, very few direct daily measurements of differences in *E*, *T*, their sum the evapotranspiration (ET), and water and energy balances of crops grown using SDI compared with spray sprinkler irrigation. Instead, these differences and their effect on WUE have been indirectly inferred from numerous crop water productivity studies, particularly those that included limited irrigation (i.e., irrigation rates below the crop ET obtained under full irrigation). Initial studies in the 1960s and 1970s were conducted in Israel (Goldberg and Shmueli, 1970; Goldberg et al., 1976) and California (Berstein and Francois, 1973; Peacock et al., 1977) for bell peppers, melons, cucumbers, tomatoes, and vineyards. Subsequent studies included other common crops, such as alfalfa (Bui and Osgodd, 1990), corn (Adamsen, 1992; Colaizzi et al., 2011), cotton (Bordovsky et al., 2000; Cetin and Bilgel, 2002; Colaizzi et al., 2010), lettuce (Sammis, 1980; Hanson et al., 1997), onion (Al Jamal et al., 2001), peanut (Adamsen, 1989), potato (Sammis, 1980), sorghum (Colaizzi et al., 2004), soybean (Wang et al., 2000; Colaizzi et al., 2010), sugar beet (Tognetti et al., 2003), sunflower (Sezen et al., 2011), tree orchards (Middelton et al., 1979; Bielorai, 1982), and vineyards (Bowen et al., 2012). These studies were conducted under a wide range of climates (e.g., arid, Mediterranean, temperate, humid), soil textures (e.g., loamy sand to clay), water and soil quality (e.g., different pH, salinity, sodicity), and agronomic and irrigation management practices that reflected commercial production in the study region (e.g., full and limited irrigation, irrigation scheduling criteria, tillage, fertilization). The cited studies included numerous variants of designs and configurations for sprinkler (e.g., solid set, center pivot, high pressure impact, low pressure spray) and microirrigation (surface drip, SDI, depth of SDI). Despite the wide range of crops and environmental and technological conditions, in nearly all cases microirrigation resulted in greater crop water productivity compared with sprinkler, which was due to greater crop yield for a given irrigation rate or crop ET, the same crop yield for less irrigation water applied or less crop ET, or a combination of both. Still, direct measurements of evaporative losses and their contribution to the overall water balance are few, limiting understanding of the mechanisms contributing to improved crop WUE and limiting our ability to develop and test crop water use and water use efficiency models that include irrigation application method.

Weighing lysimeters directly measure water losses from the soil (ΔS) due to ET when there is no precipitation (P) or irrigation (I) occurring and when deep flux (F) and runoff (R) are negligible. And, the crop ET can be calculated as the residual of the soil water balance equation

$$\mathrm{ET} = P + I + R + F + \Delta S \qquad (1)$$

for periods during which I, P, F and R are known because the change in soil water storage (ΔS) is known from the lysimeter mass change (Evett et al., 2012b). Energy and water balance modeling tells us that most of the difference in ET from SDI versus spray sprinkler occurs early in the season during pre-irrigation and the period before full cover is established (Evett et al., 1995). The ET difference is due primarily to differences in E, not T from the relatively small plants, and can be determined from weighing lysimeter measurements.

In order to more fully understand water and energy balance and flux differences under SDI compared with spray sprinkler irrigation, we modified the large weighing lysimeter facility at Bushland Texas (Marek et al., 1988) during 2012 and early 2013 so that the eastern two of the four monolithic lysimeters and their surrounding fields could be irrigated using SDI (Evett et al., 2018). Energy and water balances were measured on grain crops grown in 2013 through 2016 to determine the differences in evaporative loss and corresponding differences in yield and water use efficiency, if any.

MATERIALS and METHODS

Site description

Grain corn (*Zea mays* L) was grown in 2013 and 2016 and grain sorghum (*Sorghum bicolor* (L.) Moench) was grown for grain in 2014 and 2015 at the USDA-ARS Conservation and Production Research Laboratory, Bushland, Texas (35° 11′ N, 102° 06′ W, 1170 m elevation above MSL) on a gently sloping (<0.3%) Pullman soil (fine, mixed, superactive, thermic Torrertic Paleustoll). The slowly permeable soil has a dense B22 horizon at 0.3- to 0.5-m depth and a caliche layer at approximately 1.4-m depth that restricts water movement in some seasons. The soil series is common to 1.2 million ha of land and one third of the irrigated area in the Texas Panhandle (Musick et al., 1988). The plant available water holding capacity is approximately 210 mm in the top 1.4 m of the profile. The research location and facilities are situated in the Southern High Plains of the Great Plains and were thoroughly described by Evett et al. (2012a, 2018). Winds are predominantly from the south and southwest during the growing season and often carry advective energy from dryland and rangeland fields and pastures. Additional energy is derived from their passage over the Chihuahuan Desert followed by descent with adiabatic heating along the eastern slope of the southern Rocky Mountains. Mean annual pan evaporation exceeds 2400 mm (Kohler et al., 1959).

Agronomy

Crops were grown in four adjacent, square, 4.4-ha fields, in the center of each of which was a large weighing lysimeter (nominally 3 m × 3 m in surface area and 2.4-m deep) (Evett et al., 2012b; Marek et al., 1988). Fields and the lysimeters within them were designated with reference to the cardinal directions as NE, SE, NW and SW. The crops were managed for high yield using practices common for the northern Texas Panhandle. Fertilizer was applied according to soil tests done by a commercial soil testing laboratory. In 2013, a medium season length corn (109 d) was planted in the four fields at 81,500 seeds ha^{-1}, fertilized with liquid N (32-0-0) and treated with herbicide. The hybrid corn was bred and engineered for water-limited conditions and so a test of this was conducted by irrigating the NW field at 75% of full irrigation.

In 2014, a short-season sorghum (Channel variety 5c35) was planted on 20 June at a rate of 210,000 seeds ha^{-1} on all fields after cotton failed due to heavy rain and hail (>200 mm in five days). Sorghum was fertilized and treated with herbicide following typical production practices in the area. After plant establishment, the SW field was deficit irrigated at 75% of full. The same sorghum variety was planted on June 22, 2015, again after cotton was hailed out, and fertilized and managed using typical methods for the region. Corn was planted May 10-11, 2016 at 87,475 seeds ha^{-1}.

Liquid fertilizer applications on the lysimeters were simulated by digging trenches 10 cm deep and spaced 38 cm apart to simulate the fertilizer applicator, then spraying the liquid in the trench and covering it with soil. Common tillage practices included stubble mulch tillage after harvest to close soil cracks in order to minimize rodent damage to buried drip lines, followed by off season shredding of stalks, and incorporation of residue using a disc plow in the fields and by hand tillage on the lysimeters. Deficit irrigation tests were conducted in NW or SW fields, alternating between fields.

Lysimeter and soil water balance ET measurements

As shown in Eq. 1, crop water use (evapotranspiration, ET) is measured by the soil water balance of a control volume that includes the root zone. In this study, both weighing lysimeters and field soil water balance calculations based on neutron probe measurements were used to determine ET. Weighing lysimeters define the control volume as the depth of the lysimeter (2.3 m at Bushland). The lysimeter mass change is a direct measure of the change in soil water storage and thus of the water lost to evaporation and transpiration (ET) when P, I, R and F are zero. Lysimeter mass changes were converted to a depth of water by dividing the mass change by the density of water at standard atmosphere and pressure and by the effective surface area of the lysimeter (9.15 m^2).

The lysimeters were drained under vacuum equivalent to 1 m of hanging water column into tanks suspended by load cells from the lysimeter soil tanks so that drainage did not change the total mass of the lysimeter. Irrigations were metered, but sprinkler irrigation metered amounts were verified by measuring the change in lysimeter mass caused by each irrigation. Precipitation was measured with rain gages at each lysimeter and again verified (and corrected when precipitation events happened quickly) by observing changes in lysimeter mass (Marek et al., 2014; Evett et al., submitted). The field was furrow diked to inhibit runoff and runon into the lysimeters, and the lysimeter soil boxes had approximately 0.05 m of freeboard that prevented runoff and runon for all irrigation events and almost all precipitation events.

At each of eight neutron probe access tubes in each field, soil profile water content was determined to 2.4-m depth using a neutron probe for measurements centered at 0.10-m depth and at depths in 0.20-m increments below that. The neutron probe was operated and field calibrated to 0.01 $m^3\ m^{-3}$ accuracy using methods described by Evett et al. (2008), and a depth control stand (Evett et al., 2003) was used to ensure repeatedly accurate probe depth placement. The water content as a depth for each 0.20-m thick measured soil layer was calculated by multiplying the volumetric water content by the layer depth. Profile water content as a depth was calculated by summing the water contents for each 0.20-m thick measured soil layer. The change in storage for each period between neutron probe measurements (typically weekly) was calculated as the difference in profile water contents, the precipitation and irrigation amounts were taken as those measured by the lysimeters for each field, the value of R was assumed equal to zero since the fields were furrow diked, and the soil water flux, J_w (m s^{-1}), at the bottom of the control volume for the neutron probe was estimated using Darcy's law:

$$J_w = -K(\Delta H/\Delta z) \quad (2)$$

where soil water contents at the 2.10- and 2.30-m depths were used to estimate the hydraulic conductivity, K (m s^{-1}), and hydraulic gradient, $\Delta H/\Delta z$ (-), for the 2.10- to 2.30-m soil layer using methods described in detail by Evett et al. (2012b).

Eight determinations of field ET by soil water balance were thus made in each field (NW, SW, NE and SE) using the neutron probe. For each of the four fields, these were grouped into four mean ET values for purposes of WUE calculation. For each of the four fields, field ET values were used in calculations of water use efficiency using field combine yields (described in the section on Plant Sampling). Lysimeter-measured ET data were used primarily to illustrate the differences in ET between SDI and MESA irrigation application methods in terms of ET and evaporative losses over time.

Irrigation systems and management

Spray sprinkler irrigation was applied to the NW and SW fields using a ten-span linear irrigation system (Lindsay Manufacturing, Inc., replaced after 2014 with a Valley system, Valmont Industries, Valley, NE)) moving in the E-W direction with spray plates at 1.5-m height (mid-elevation sprays, MESA) on weighted drops with 69 kPa pressure regulators on each drop. Drops were spaced at 1.52-m intervals. Irrigations were typically 19 to 25 mm depth, and occasionally as much as 38 mm. Nozzling was such that a 25-mm irrigation took approximately 12 h. Proximal lateral end pressures were typically 242 kPa and distal lateral end pressures were typically 173 kPa, ensuring that the pressure regulators set and operated correctly after system startup.

The SDI system was installed in the NE and SE fields and lysimeters before the 2013 cropping season using 25-mm diameter tubing (model Typhoon 990, 13 mil wall thickness, Netafim, Inc., Fresno, Calif.) spaced 1.52-m apart and injected at 0.30- to 0.36-m depth in the E-W direction. Emitters were spaced 0.30-m apart and had 0.68 L h^{-1} discharge at the 69 kPa regulated line pressure. Emitter spacing was purposefully chosen to be relatively small so that the drip tape would act more as a line source and more uniformly wet the soil orthogonally to the drip line. The combination of emitter spacing, discharge rate and drip tube spacing was chosen so that application of a given depth of water would take approximately the same time as application of the same amount of water with the linear move irrigation system. Lines were 210-m long and designed for an emission uniformity of 98.6%. The field was divided into 20 zones, with each zone controlled with a separate valve, meter and pressure regulator. The system applied 25 mm of irrigation in approximately 14 h. Water from multiple wells was stored in a reservoir, then pumped through sand filters with automatic flush out (waste stream returned to the reservoir) to remove sediments and algae. A variable frequency drive was used to provide constant supply line pressure downstream of the filters.

The SDI tubing on the lysimeters was buried at the same depth and row spacing as in the field and the same tubing was used as in the field. The two SDI tubes in each lysimeter were plumbed to a buried header at the west side of each lysimeter soil monolith and to a buried flush out line at the east end of the monolith. The number of emitters and emitter spacing relative to the west and east sides of the monoliths was carefully controlled so that the number of emitters per unit area was the same on the lysimeters as in the field. The buried header in the lysimeter was connected to the field SDI supply so that when the field was irrigated the lysimeters were also irrigated. Lysimeter mass gain in the hours just before and after midnight during overnight irrigations, when ET was essentially zero, was examined to verify that irrigation rate on the lysimeters was equal to that in the field. The irrigation rate was verified to be practically constant, which was expected due to the pressure regulation, and ET during irrigations was calculated using the difference between mass added by the constant rate irrigation and mass change of the lysimeter.

Later in 2014, the lysimeter irrigation system was changed so that lysimeter ET could be more clearly differentiated from mass gain due to irrigation (Evett et al., 2018). In each lysimeter, water storage tanks, larger in volume than needed for a 38-mm irrigation depth, were suspended from the monolith using a 4,448 N (1,000-lb) load cell. Charging the tanks required about five minutes, producing a step change in both the 4,448 N load cell and in the lysimeter load cell during which ET was negligible. The mass change determined using the two load cells was compared to verify equality and to verify that the correction irrigation amount was delivered. After the tanks were charged and the solenoid valves were closed, a separate solenoid valve was opened to allow a pressure-regulated pump and pressure tank system to supply water to the lysimeter SDI lines through a pressure regulator set to 69 kPa. Since the water supply tanks and pressure-regulated pump and pressure tank were all suspended from the soil monolith container, the irrigation itself did not change the mass of the lysimeter and any lysimeter mass change could be directly attributed to ET (or precipitation).

After crop emergence, irrigations were applied to replace soil water in the root zone to field capacity based on weekly neutron probe measurements. Pre-plant irrigations were applied in 2013 and 2014 due to dry pre-plant soil conditions that would have prevented uniform germination and emergence. Dry spring conditions are common in the region, but early season precipitation was plentiful in 2015 and 2016, avoiding the need for pre-plant irrigation in those years. The NW and SW fields were managed together and separately from the common management applied to the NE and SE fields. In 2013, the SW field was managed for full (100% replenishment) irrigation, replacing soil water used back to field capacity, while the NW field was irrigated on the same dates but with nozzle size reduced to apply approximately 75% of full irrigation beginning on 6 June. The deficit irrigation treatment in the SW field was applied to provide more data for a longer-term study of deficit sprinkler irrigation at Bushland. There is no similar long-term study using SDI, so a deficit treatment was not applied in the SDI fields. In 2014, the NW field was managed for full irrigation while the SW field was managed for approximately 75% of full irrigation. In 2015, large precipitation events early in the year equalized soil water in the NW and SW fields, so again the SW field was managed for a deficit irrigation of 75% of full, and in 2016 this was reversed with the deficit irrigation treatment in the NW field. In all years, the NE and SE fields were both managed for full irrigation. With a minimum of four crop water use and yield samples in each field, there were sufficient replications for statistical validity.

Plant sampling

Plant counts after emergence were taken in two adjacent rows at each of two separate locations in each of the 10 linear move sprinkler spans, and in each adjacent pair of SDI zones (zones 1 and 2, 3 and 4, etc.). Counts in each row were in a 0.91-m row length. The locations were in opposite halves of each field in the E-W direction. On an approximately biweekly basis as weather allowed, destructive plant samples for leaf area index, LAI (-), and above-ground biomass determination were taken in two adjacent rows, each 0.91-m long in three replicate locations in each field, and leaf area was determined using a calibrated leaf area meter. Specific leaf area index (leaf area per unit leaf mass; m^2 kg^{-1}) was computed as a conservative crop development parameter and an internal data check. Crop height and width were measured in the same locations and on each lysimeter. Both fresh mass and dry mass were determined. Yield sampling was done both by hand and by combine. Yield samples were taken by hand by removing ears or heads in two rows, each 3.28-m long in three replicates for each of the four fields, and for all plants on each lysimeter (four rows, 2.25 m^2 per row). Above-ground biomass was also collected from these sample areas for dry biomass determination. After drying, ear or head mass, and shelled or threshed seed mass were determined. On each replicate sample, the mass of 200 seeds was determined, and after oven drying (24 h at 60°C), dry mass per seed was determined. Combine harvested grain was weighed and moisture content measured separately for each of the 10 linear-move sprinkler spans, and for every two SDI zones, resulting in five yield samples for each of the four lysimeter fields. The combine harvested yields were total yields for each subarea of each field. Reported yields are dry grain yields. Statistical calculations were performed using t-tests assuming unequal variances and by ANOVA using the Holm-Sidak method for means comparisons. Means were considered significantly different at the 5% level. Combine yields and crop water use values calculated from neutron probe data were used for the results presented in this paper.

RESULTS and DISCUSSION

2013 corn

Despite a severe hailstorm during corn emergence, plant stand was 84,000 ha^{-1} in the SDI fields and 96,600 ha^{-1} in the MESA fields. The fully irrigated SW field received 583 mm of MESA irrigation, while the deficit irrigated NW field received 447 mm of MESA irrigation. Corn fully irrigated using MESA yielded 9.38 Mg ha^{-1}, which was significantly greater (30% greater) than the 7.19 Mg ha^{-1} corn harvest resulting from limiting MESA irrigation to 75% of full irrigation. Yield differences translated to differences in water use efficiency. Full MESA irrigation resulted in WUE of 1.29 kg m^{-3}, which was significantly (11%) greater than the 1.16 kg m^{-3} WUE of deficit MESA irrigated corn. Fully irrigated corn grown using SDI yielded on average 11.1 Mg ha^{-1}, which was significantly greater (18% greater) than the yield of MESA fully irrigated corn. The mean WUE (1.66 kg m^{-3}) for SDI corn was significantly greater (29%) than that of MESA fully irrigated corn.

Reasons for the differences in yield and water use are illustrated in Figure 1. For DOY 170 through 189, SDI corn used 48 mm less water than did MESA fully irrigated corn. Once the crop substantially covered the soil, by day of year (DOY) 175, the SW field, which was fully irrigated using MESA, used water at rates much greater than did the deficit irrigated NW field (Fig. 1A). Daily ET exceeded 12 mm several times and exceeded 14 mm once under full MESA irrigation. In contrast, peak water use of deficit irrigated corn was 10 mm d^{-1}. Water use of fully MESA irrigated corn exceeded that of the deficit irrigated crop through seed filling and senescence up until DOY 255. In contrast, the fully irrigated SDI corn exceeded 10 mm daily water use only once, and in fact appeared to use water at daily rates very close to those exhibited by the MESA deficit irrigated corn until DOY 230 when SDI irrigated corn began to use more water than the MESA deficit irrigated corn and began to closely match the water use of the MESA fully irrigated corn (Fig. 1B). This late season water use was likely important for completing grain filling. Overall, fully MESA irrigated corn used the most water (722 mm), deficit MESA irrigated corn used 620 mm, while corn irrigated using SDI used significantly less (649 to 677 mm) than did fully irrigated MESA corn, based on neutron probe readings.

MESA irrigation wetted the soil surface, which resulted in much greater evaporative loss during pre-plant irrigations and in the first 25 days after planting (DAP) when the crop was emerging and not yet covering much of the soil surface (Fig. 1C). As observed from weighing lysimeter data, total MESA irrigation water use in that period was 147 to 161 mm compared with the much smaller 113 to 115 mm water use of SDI irrigated corn. Most of this water was lost to evaporation from the soil surface since the plants were not emerged or were very small. The gross savings in evaporative loss from the use of SDI was 85 mm during this period (Fig. 1C). This is remarkably close to the savings estimated by Evett et al. (1995) who used the ENWATBAL simulation model to estimate an evaporative loss reduction of 81 mm for SDI compared with surface irrigation of corn in a relatively dry year. The greater evaporative losses suffered under MESA irrigation did not end at 25 days after planting, but continued until approximately day of year (DOY) 192 when the corn grew taller than the MESA sprays, totaling another 53 mm more water lost from full MESA irrigation than from full SDI.

Differences in water use did not always translate directly into differences in yield. The fully MESA irrigated corn yield was significantly greater (30%) than that from the deficit MESA irrigated corn. However, the corn irrigated with SDI, which used less water than the Full MESA irrigated treatment, out yielded both significantly with a mean yield of 11.0 Mg ha^{-1}. Overall yields were not as large as expected, partly due to corn earworms that invaded nearly every ear despite the Bt variety grown. The extra yield from the SDI fields was partially due to more water available for transpiration, particularly during grain filling at season's end (Fig. 1C). Water use efficiency for the SDI fields was significantly greater than that for the MESA irrigated fields. And, WUE for the fully MESA irrigated field was significantly greater than that for the deficit MESA irrigated field. Yield loss for the deficit MESA irrigated field occurred even though measured water contents were within the management allowed depletion range (Fig. 2). There are two reasons for this. First, the water content values plotted in Figure 2 are means, and some individual values were less than the management allowed depletion level, which means that some areas of the field were dry enough to cause yield reduction. Second, the mean values in the deficit irrigated field were nearly constantly at or near the management allowed depletion level for a substantial part of the season, including during grain filling, which evidently had a substantial effect on final yield.

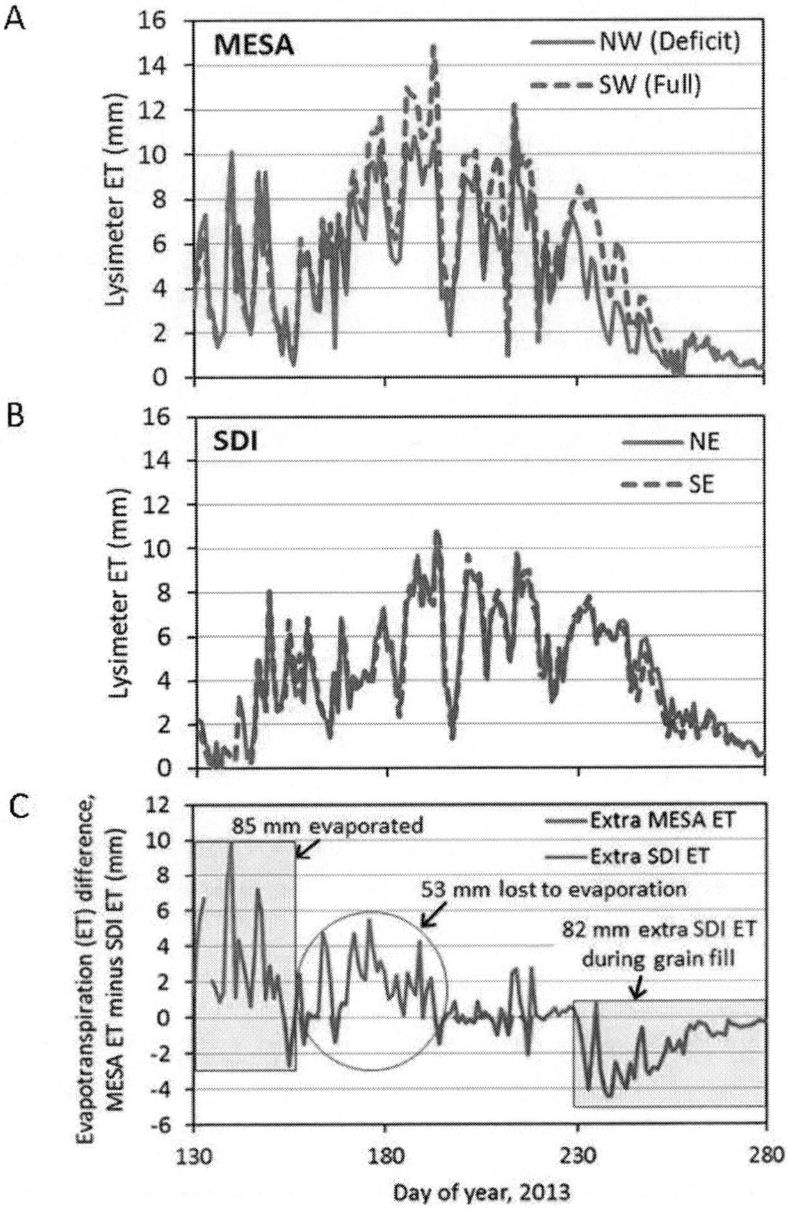

Figure 1. (A) Corn evapotranspiration (ET) in the northwest (NW) and southwest (SW) lysimeter MESA irrigated fields in 2013. (B) Corn ET in the northeast (NE) and southeast (SE) SDI fields. (C) Difference in ET between MESA and SDI irrigation where values >0 are extra ET in the MESA irrigated fields and values <0 are extra ET in the SDI fields.

Increased yield of the SDI irrigated corn may have been somewhat influenced by overall better soil water conditions (Fig. 2). While water storage in the top 1.5-m of soil for the fully irrigated MESA field was always considerably greater than the maximum allowed depletion, water storage in the SDI fields was greater and often near field capacity for the first half of the season (Fig. 2B). This was due to two factors. One is that the SDI fields were heavily irrigated (220 mm) after planting in order to bring water to the surface to germinate the corn. This was enough, in combination with the antecedent water to bring the soil to field capacity. This water remained in the profile and was available to the crop later in the season. There is evidence that corn rooting is enhanced when the Pullman soil is wetter (Tolk and Evett, 2012) because the soil strength (resistance to root penetration) increases greatly as the soil dries. The MESA irrigated fields were irrigated to replenish water content to field capacity, but the large evaporative losses prevented effective use of the irrigation water applied, so water content only increased gradually (Fig. 2A). This is why irrigation to replenish water content to field capacity in the MESA irrigation fields often did not result in measured water content being at field capacity in those fields. The other reason is that neutron probe measurements typically lagged irrigation events by two to three days in the MESA irrigated fields because the fields were too muddy for foot traffic in those fields. In contrast, the neutron probe was used in the SDI fields soon after irrigation because irrigations left the soil surface reasonably dry.

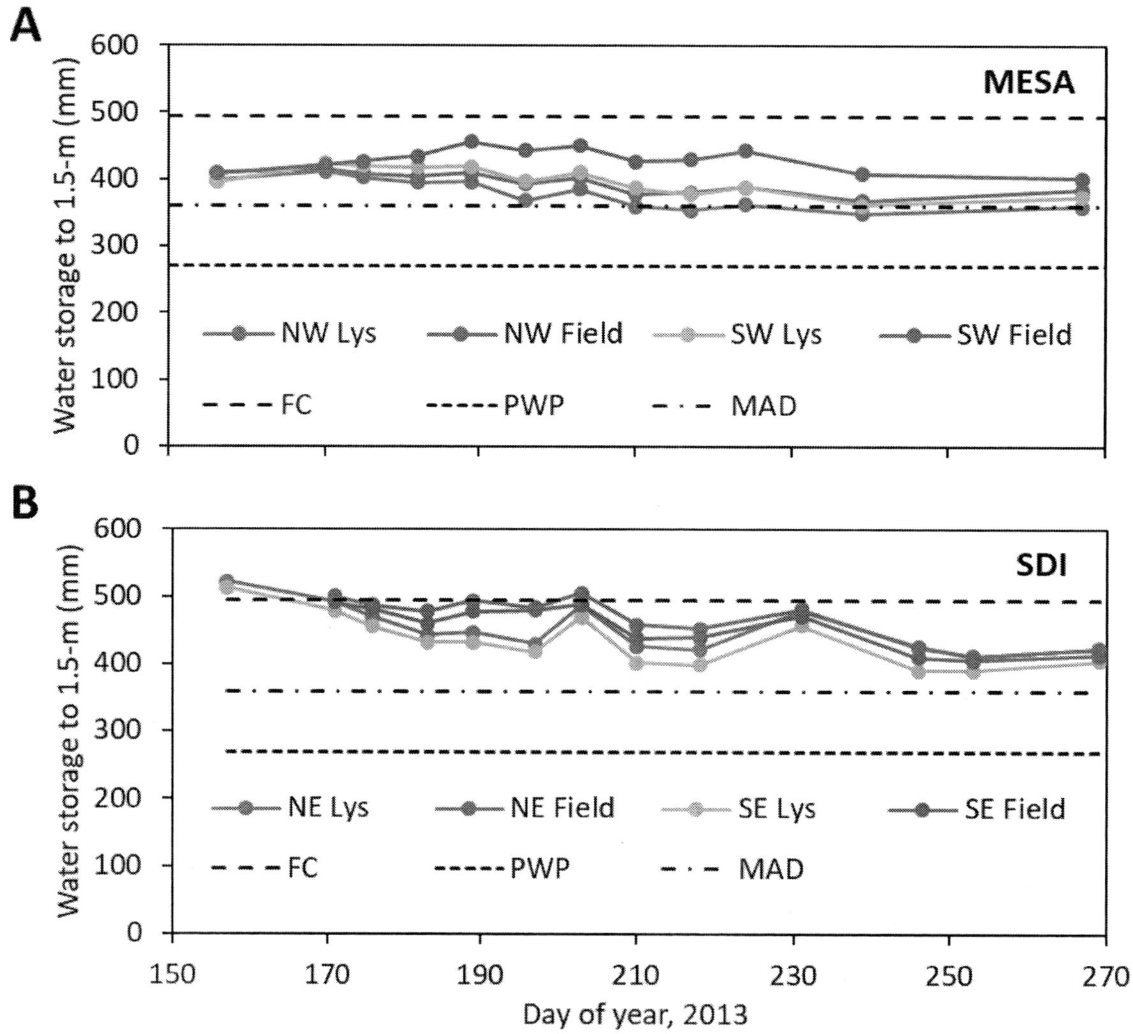

Figure 2. (A) Water storage in the top 1.5 m of soil in 2013 for (A) the northwest (NW) and southwest (SW) MESA irrigated fields and lysimeters (Lys), and (B) the northeast (NE) and southeast (SE) SDI fields and lysimeters. FC is field capacity, PWP is permanent wilting point, and MAD is the maximum allowed depletion.

2014 short season sorghum

Due to a dry spring and dry soil profile, pre-plant irrigations were conducted on all lysimeter fields in 2014. MESA irrigation on the west fields began on May 1, 2014 (DOY 121) and totaled 105 mm before sorghum was planted. SDI applications began on May 12, 2014 (DOY 132) in the east fields and totaled 177 mm before planting due to the drier soil profile in the east fields. Immediately after pre-plant irrigations were ended, precipitation totaling 119 mm occurred over six days (DOY 141-146). Cotton was planted on DOY 154 but failed soon after due to torrential rains. Short season sorghum (Channel 5c35) was planted on June 20 (DOY 171) and emerged five days later. Black layer was noted on DOY 272 and fully developed by DOY 290. The deficit irrigation treatment on the SW field did not begin until August 1 and resulted in only a 73 mm reduction in total irrigation, not enough to much influence crop yield given the full soil profile that resulted from the plentiful rains. Due to the full profile, only 34 mm of MESA irrigation was required from planting to August 1. Hot, dry weather in July and August required 256 mm of irrigation for the fully irrigated treatment in the NW field to finish the season. A large rain (>50 mm) on September 3 finished the irrigation season.

Despite the 73 mm difference in total irrigation, season total sorghum water use (ET) did not differ importantly or significantly between the fully MESA irrigated (694 mm) and deficit irrigated (670 mm) crops (Fig. 3). Even though yield from deficit irrigated sorghum was 4.4% less than that for fully irrigated sorghum, yield was not significantly different (7.31 Mg ha^{-1} for full irrigation and 6.96 Mg ha^{-1} for deficit irrigation). Similar depression of yield when water is limited during grain filling has been reported previously. In the NE and SE fields, yields were smaller for sorghum produced using SDI, 6.76 and 6.19 Mg ha^{-1}, respectively, which did not differ significantly. However, the SDI yields were significantly smaller (9% overall) than those obtained using MESA irrigation. For the same sorghum variety, O'Shaughnessy et al. (2014) reported 8.04 Mg ha^{-1} for 533 mm of ET in 2009, but only an average 6.42 Mg ha^{-1} for an average ET of 648 mm of ET in 2010 and 2011. For one field under SDI, water use efficiency was numerically greater than that for MESA irrigation, but there were no significant differences in WUE amongst the four fields and two irrigation application method treatments.

Water use in the SDI fields was typically less than that in the MESA irrigated fields throughout the season (Fig. 3A,3B). The MESA fully irrigated sorghum used 53 mm more water than did the SDI sorghum during the period from initiation of pre-plant irrigation until 25 days after planting (Fig 3C). Unlike the result for the previous year's tall corn crop, losses to evaporation continued throughout the season for the MESA irrigated sorghum, totaling another 52 mm by season's end for a season-long total of 105 mm lost to evaporation. The likely cause for the greater ET from the MESA irrigated crop through August 1 (DOY 213) is that the sorghum did not reach a height greater than the elevation of the spray plates and so the entire leaf area was wetted by each irrigation.

There are several possible reasons for the smaller sorghum yields with SDI. The large rainfalls may have leached fertilizer from the already full soil profile in the SDI fields, depressing yields. Supporting this idea is the fact that the lysimeters in the SDI fields drained considerably more water than did those in the MESA irrigated fields, indicating larger deep percolation losses in the SDI fields. Pre-plant irrigation with the SDI system was larger than that for the MESA system in order to bring water to the seed bed for cotton germination. This proved unnecessary due to the large rains just after cotton planting, but it did leave the soil profile full of water prior to the large rains.

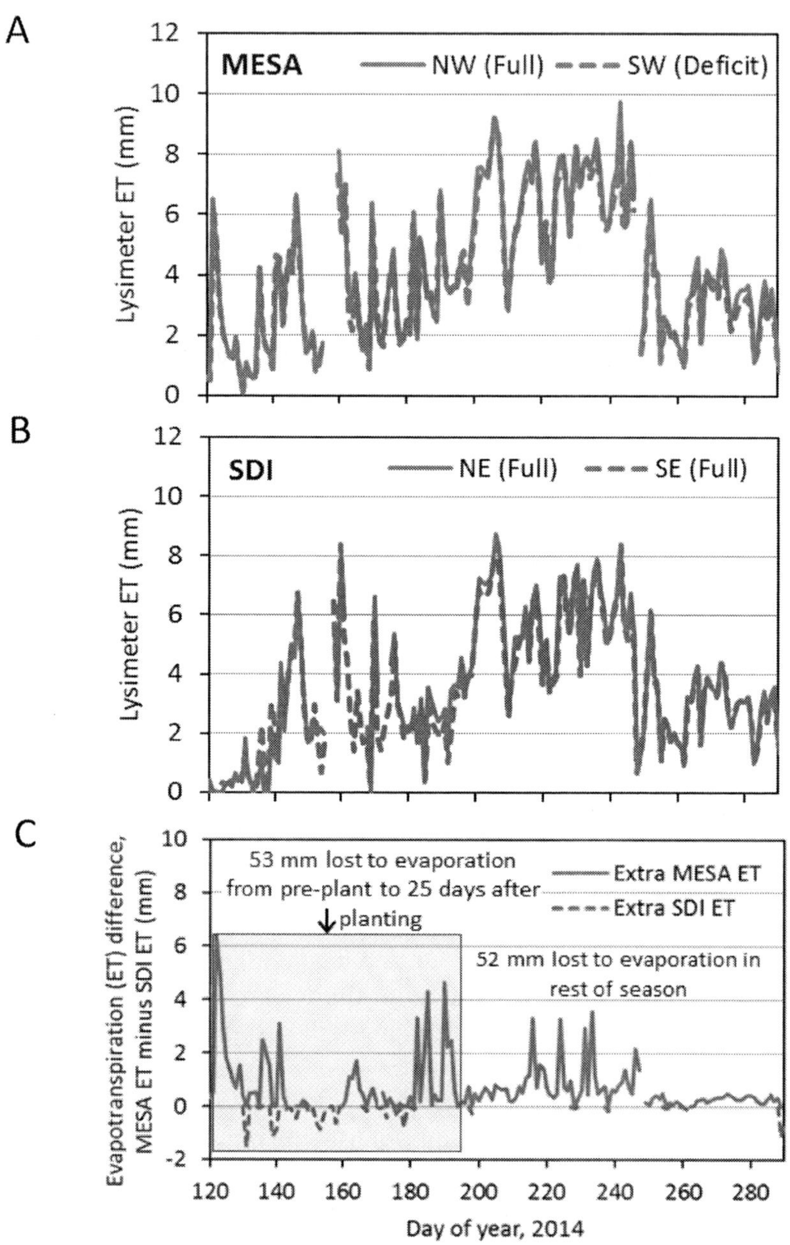

Figure 3. (A) Sorghum evapotranspiration (ET) in the northwest (NW) and southwest (SW) lysimeter fields irrigated with MESA in 2014. (B) Sorghum ET in the NE and SE SDI fields. (C) Difference in ET between MESA and SDI irrigation where values >0 are extra ET in the MESA irrigated fields and values <0 are extra ET in the SDI fields.

Despite the larger pre-plant irrigation, evaporative losses before planting and through 25 DAP were on average 53 mm smaller in the SDI fields compared with the MESA irrigated fields. Season long irrigation using SDI averaged 21 mm more than that for full irrigation using MESA. But, season long water use for SDI sorghum averaged 112 mm less than that for MESA fully irrigated sorghum, mostly due to 84 mm less irrigation in the SDI fields after August 1, which caused water content to decrease to near the management allowed depletion value (Fig. 4). The relatively less irrigation of the SDI fields in August was due to an error in neutron probe readings that went undetected until too late to take correction action.

Although water content did not decrease to less than the management allowed depletion, it is possible that less irrigation in the SDI fields after August 1 combined with loss of nutrients due to deep percolation led to the 9% yield depression in SDI fields. As for the deficit irrigated corn crop in 2013, mean values of soil water content being at the management allowed depletion level indicated that some values were well less than that level and some areas of the field were thus water stressed. It is known from previous studies that deficit irrigation during grain filling can reduce sorghum yields.

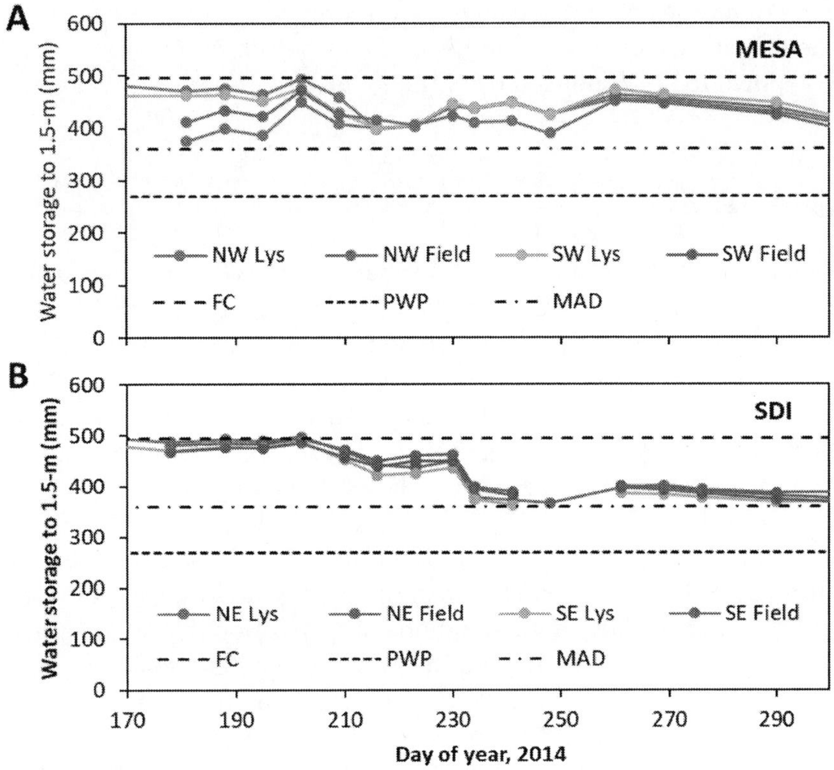

Figure 4. (A) Water storage in the top 1.5 m of soil in 2014 for (A) the northwest (NW) and southwest (SW) MESA irrigated fields and lysimeters (Lys), and (B) the northeast (NE) and southeast (SE) SDI fields and lysimeters. FC is field capacity, PWP is permanent wilting point, and MAD is the maximum allowed depletion.

This result is in line with those of Colaizzi et al. (2004) who reported that yields for both long and short season grain sorghum were on average 12% less for SDI compared with MESA for 75% and 100% of full irrigation rates. They too attributed this yield depression to leaching of nutrients, which was supported by measurements of increasing volumetric water content deep (> 1.8 m) in the soil profile. In their case, leaching was caused by over irrigation due to irrigation management being based on crop coefficients developed for sprinkler irrigation, which did not take into account the reduced crop water use with SDI. Interestingly, for 25% and 50% irrigation rates, Colaizzi et al. (2004) reported that SDI resulted in an average of 36% greater grain yields compared with MESA. This implied that SDI resulted in greater partitioning of water to plant transpiration and less to soil evaporation, especially early in the season. The study reported herein confirms their supposition of reduced evaporative loss with direct measurements, which reinforces their supposition of greater partitioning of water to plant transpiration.

Despite the fact that spray irrigation can lower leaf temperature and in some cases promote greater yields, yield in the SDI fields was likely not adversely affected by greater leaf temperatures. The mean daily air temperatures in 2013 and 2014 were not greatly different, and they varied from 20 to 30°C through the middle part of the growing season when it was warmest, not great enough to incur large yield effects. The under irrigation of SDI fields late in the season did not affect plant height and width, both of which were not noticeable or significantly different between the SDI and MESA irrigated fields. Leaf area index, LAI (-), did, however, reach numerically greater values in the MESA irrigated fields than in the SDI fields, and the difference at maximum LAI (before DOY 230) was statistically significant for the fully MESA irrigated field. Since water content in the SDI fields before DOY 230 was not deficient, and since leaf expansion is known to be influenced by nitrogen fertility, the smaller LAI in the SDI fields was likely due to loss of N fertilizer due to leaching, which resulted in decreased LAI and yield.

2015 short season sorghum

2015 was the wettest year in >70 years of record at Bushland with >870 mm of precipitation. Due to a wet spring, pre-plant irrigations were not conducted in 2015. As in 2014, cotton was planted and hailed out, followed by planting short season sorghum (Channel 5c35) at 210,000 seeds ha^{-1} on June 22-23, 2015 (DOY 173-174). The fully emerged crop was severely damaged by hail on July 8, 2015 (DOY 189), but survived and reached full bloom on August 20 and was harvested on October 20th (DOY 293). Grasshoppers and sugar cane aphids infested the crop in August but were controlled successfully using pesticides. Deficit irrigation of the SW field began on July 1, 2015 (DOY 182).

Irrigation from planting to 25 DAP was not much different between full MESA and SDI irrigation treatments, ranging from 51 to 61 mm. Precipitation in the 60 days before sorghum planting averaged 412 mm, nearly the yearly mean for Bushland. Additional precipitation during the growing season averaged 407 mm. Total season irrigation amounts were correspondingly small, being 286 mm for full MESA irrigation and 262 mm for SDI. ET to 25 DAP was only slight larger (10 mm) for MESA irrigation compared with SDI, and full season ET was 50 mm larger for full MESA irrigation compared with SDI (Fig. 5C). In the wet year, ET rates were less than 10 mm d^{-1} and almost always less than 8 mm d^{-1} (Fig. 5). ET rates for SDI were only slightly less than that for full MESA irrigation throughout the season. Soil water content remained within management allowed depletion in the SDI fields and was similar for the fully irrigated MESA field (Fig. 6). Soil water content in the deficit MESA field approached MAD late in the season, but too late to have appreciable effect on yield. Yields averaged 7.54 Mg ha^{-1} for SDI versus 7.60 ha^{-1} for full MESA and 7.74 ha^{-1} for deficit MESA irrigated fields. Neither yields nor WUE were significantly different across the four fields, although WUE of 1.39 kg m^{-3} was numerically larger than that of 1.28 kg m^{-3} for full MESA and 1.31 kg m^{-3} for deficit MESA irrigation.

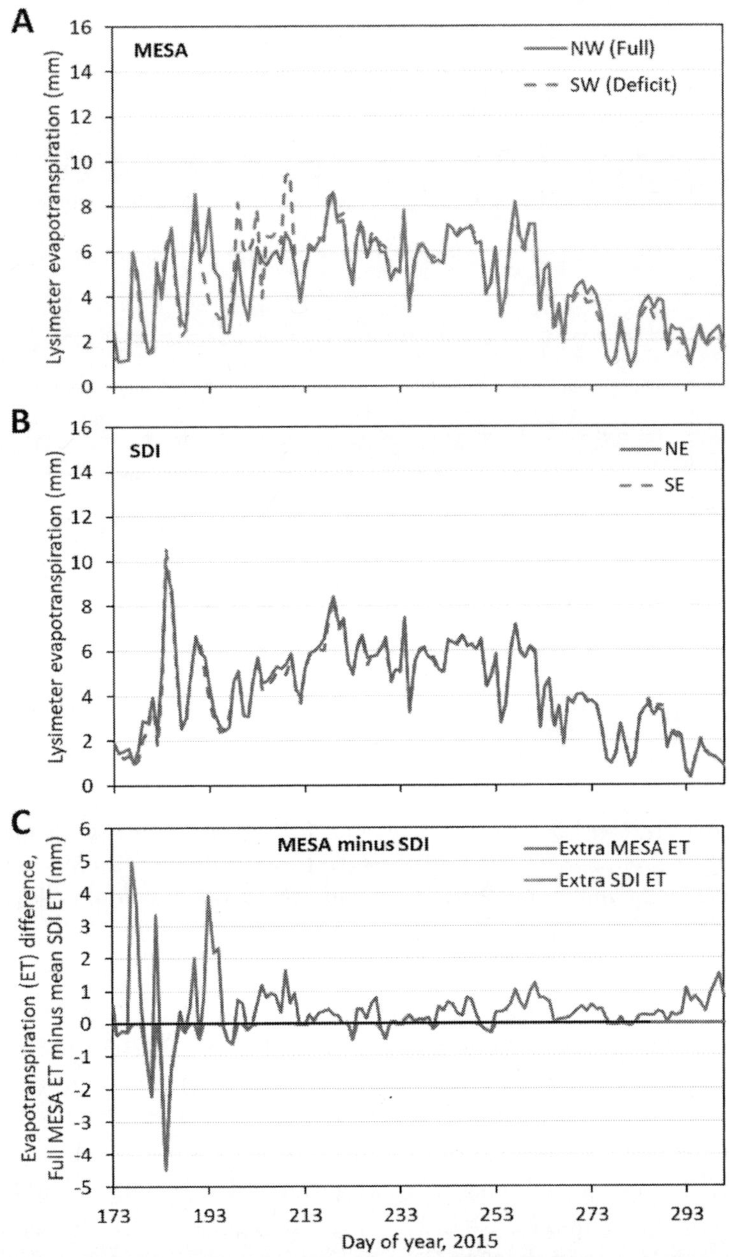

Figure 5. (A) Sorghum evapotranspiration (ET) in the northwest (NW) and southwest (SW) lysimeter fields irrigated with MESA in 2015. (B) Sorghum ET in the NE and SE SDI fields. (C) Difference in ET between MESA and SDI irrigation where values >0 are extra ET in the MESA irrigated fields and values <0 are extra ET in the SDI fields.

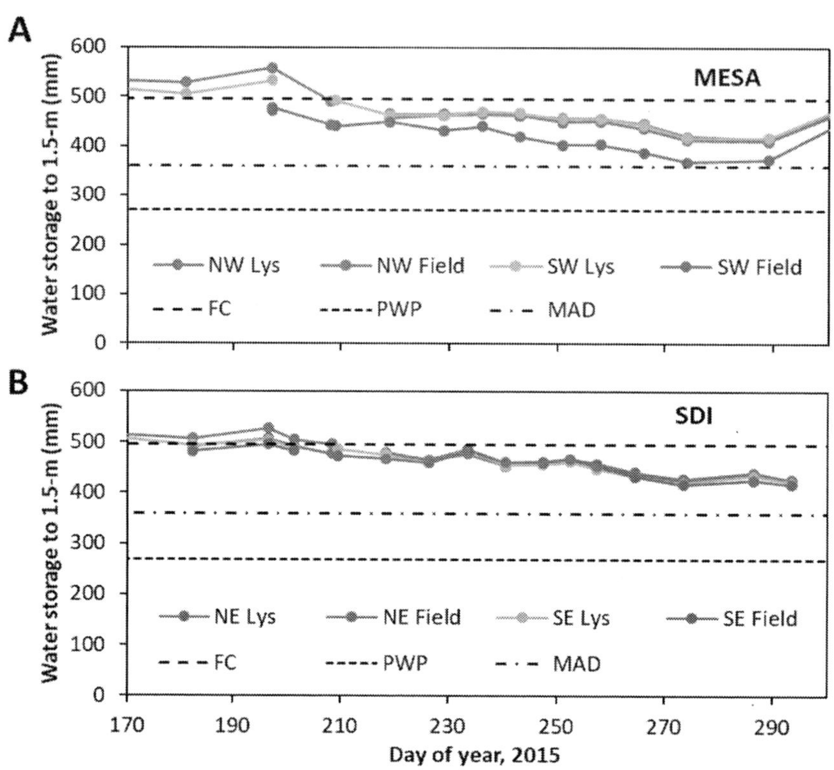

Figure 6. (A) Water storage in the top 1.5 m of soil in 2015 for (A) the northwest (NW) and southwest (SW) MESA irrigated fields and lysimeters (Lys), and (B) the northeast (NE) and southeast (SE) SDI fields and lysimeters. FC is field capacity, PWP is permanent wilting point, and MAD is the maximum allowed depletion.

2016 corn

A drought tolerant corn variety (Pioneer 1151) was planted May 10-11, 2016 (DOY 131) at 87,475 seeds ha^{-1} and had emerged by May 21 (DOY 137). Due to pre-plant precipitation averaging 99 mm, no pre-plant irrigation was required. The 75% deficit irrigation treatment on the NW field began on June 9, 2016 (DOY 161). Along with growing season precipitation that averaged 238 mm and moderate weather, total irrigation averaged 504 mm for SDI. Full MESA irrigation was 18% larger and full MESA ET was 17% larger than for SDI. For deficit irrigated MESA the ET was 8% larger than ET for SDI, even the though total irrigation (392 mm) was 22% less than that for SDI (mean of 504 mm). This result indicates that spray irrigation was relatively inefficient in delivering water to the crop in the deficit regime. Due to timely rains, soil water content under deficit MESA irrigation approached the MAD only twice during the growing season, and the crop ET was relatively larger than that for SDI. Daily ET exceeded 15 mm once and reached 12 mm a few times for fully MESA irrigated corn, but only reached and exceeded 12 mm once for SDI (Fig. 7). Because irrigation before 25 DAP was relatively little (<40 mm), there was not much difference in ET between full MESA and SDI during that period, but ET was consistently larger for full MESA irrigation for the rest of the season (Fig. 7C), resulting in 150 mm more ET over the entire growing season for full MESA irrigation than for SDI. Soil water storage throughout the season indicated a no-stress regime for fully MESA irrigation corn and corn under SDI (Fig. 8A,B). The deficit MESA irrigated corn approached the management allowed depletion level twice during critical growth stages (Fig. 8B).

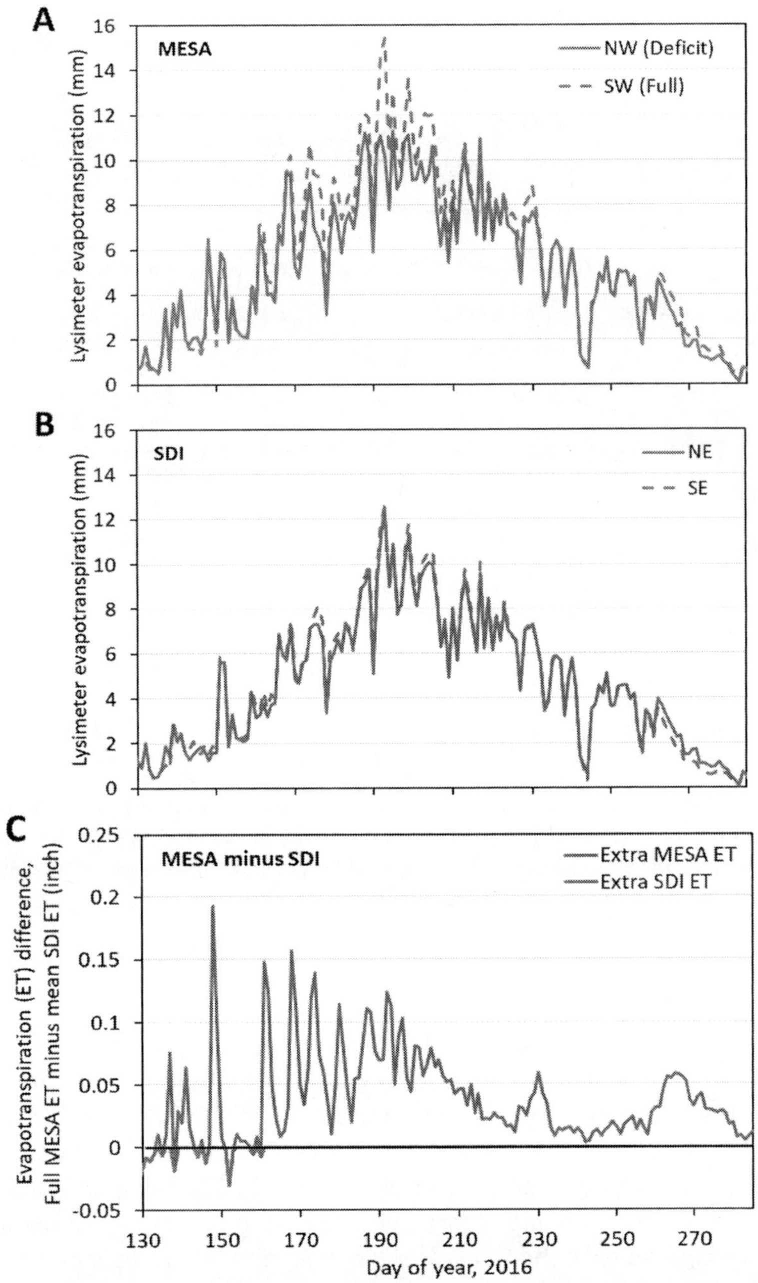

Figure 7. (A) Corn evapotranspiration (ET) in the northwest (NW) and southwest (SW) lysimeter fields irrigated with MESA in 2016. (B) Corn ET in the NE and SE SDI fields. (C) Difference in ET between MESA and SDI irrigation where values >0 are extra ET in the MESA irrigated fields and values <0 are extra ET in the SDI fields.

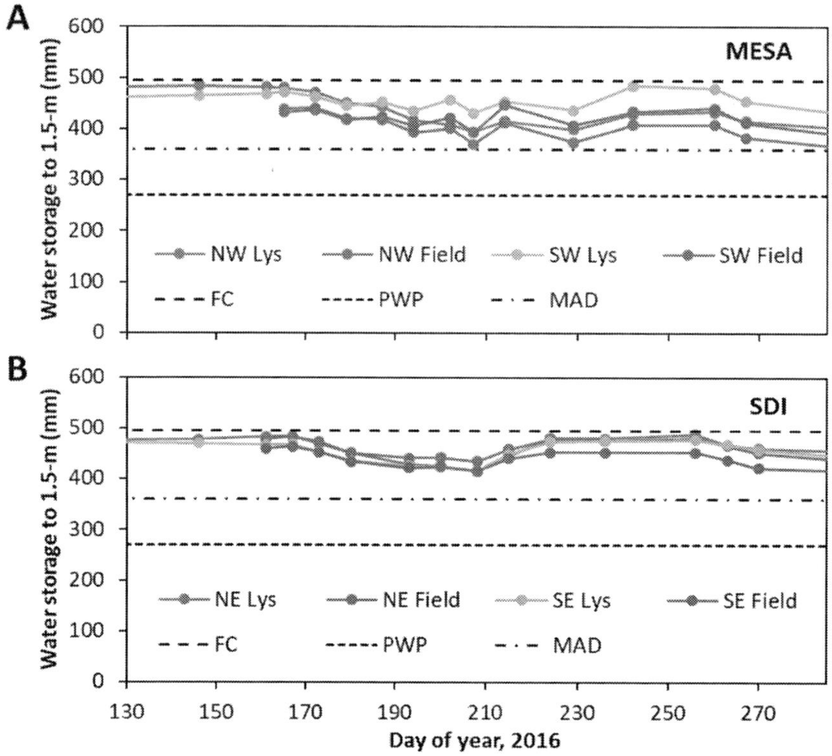

Figure 8. (A) Water storage in the top 1.5 m of soil in 2016 for (A) the northwest (NW) and southwest (SW) MESA irrigated fields and lysimeters (Lys), and (B) the northeast (NE) and southeast (SE) SDI fields and lysimeters. FC is field capacity, PWP is permanent wilting point, and MAD is the maximum allowed depletion.

SUMMARY

The SDI system saved on average from 85 to 53 mm of water that was lost to evaporation early in the season (pre-plant to 25 DAP) from the fully MESA irrigated crop in the 2013 and 2014, respectively, which was consistent with the estimate of evaporative loss reduction for SDI made using the ENWATBAL model. In 2015 and 2016, when spring weather was wetter, the ET savings with SDI to 25 DAP were small (<= 11 mm). Between 25 DAP and mid season, another 53 to 52 mm of water was lost with the MESA irrigation system compared with the SDI system in 2013 and 2014, respectively. In 2015 and 2016, from 25 DAP to harvest there were ET savings of 39 and 139 mm, respectively, using SDI compared with full MESA irrigation. For corn grown in 2013, much of the water saved due to smaller evaporative losses was used during grain fill when SDI corn used 82 mm more water than did MESA fully irrigated corn. In the relatively dry 2013 season, SDI reduced overall corn water use by 147 mm while increasing yields by 1.88 Mg ha^{-1} (20%) and WUE by 0.64 kg m^{-3} (61%) compared with MESA full irrigation. In the relatively wet 2016 season, SDI reduced overall corn water use by 150 mm while increasing WUE by 0.24 kg m^{-3} (17%) compared with MESA full irrigation, although with no significant yield difference. While sorghum, particularly short season sorghum, is not a crop ordinarily considered for SDI, it was grown successfully using SDI with yields and water use efficiencies comparable to others reported for short season sorghum at Bushland. The yield increases in some years and water savings in all years using SDI point to important economic advantages in revenue and reduced pumping costs.

ACKNOWLEDGEMENTS

We gratefully acknowledge support from the USDA-ARS Ogallala Aquifer Program, a consortium between the USDA Agricultural Research Service, Kansas State University, Texas A&M AgriLife Research, Texas A&M AgriLife Extension Service, Texas Tech University, and West Texas A&M University, as well as the dedicated support of Karen Copeland, USDA-ARS Soil Scientist, and Brice Ruthardt, USDA-ARS Biological Scientist.

REFERENCES

Adamsen, F. J. 1989. Irrigation method and water quality effect on peanut yield and grade. Agron. J. 81(4): 589-593.

Adamsen, F. J. 1992. Irrigation method and water quality effects on corn yield in the Mid-Atlantic Coastal Plain. Agron. J. 84(5): 837-843.

Al-Jamal., M. S., S. Ball., and T. W. Sammis. 2001. Comparison of sprinkler, trickle and furrow irrigation efficiencies for onion production. Agric. Water Manage. 46: 253-266.

Bernstein, L., and L. E. Francois. 1973. Comparisons of drip, furrow, and sprinkler irrigation. Soil Sci. 115(1): 73-86.

Bielorai, H. 1982. The effect of partial wetting of the root zone on yield and water use efficiency in a drip- and sprinkler-irrigated mature grapefruit grove. Irrig. Sci. 3: 89-100.

Bordovsky J.P., W. M. Lyle, and E. Segarra. 2000. Economic evaluation of Texas High Plains cotton irrigated by LEPA and subsurface drip. Texas J. Agric. Nat. Resour. 13: 67–73.

Bosch, D. J., N. L. Powell, and F. S. Wright. 1992. An economic comparison of subsurface microirrigation with center pivot sprinkler irrigation. J. Prod. Agric. 5(4): 431-437.

Bowen, P., C. Bogdanoff, and B. Estergaard. 2012. Effects of converting from sprinkler to drip irrigation on water conservation and the performance of merlot grown on a loamy sand. Am. J. Enol. Vitic. 63(3): 385-393.

Bui, W., and R. V. Osgood. 1990. Subsurface irrigation trial for alfalfa in Hawaii. In Proc. 3rd Nat. Irrig. Symp., 658-660. St. Joseph, Mich.: ASAE.

Camp, C. R. 1998. Subsurface drip irrigation: A review. Trans. ASAE 41(5): 1353-1367.

Cetin, O., and L. Bilgel. 2002. Effects of different irrigation methods on shedding and yield of cotton. Agric. Water Manage. 54: 1-15.

Colaizzi, P.D., A. D. Schneider, S. R. Evett, and T. A. Howell. 2004. Comparison of SDI, LEPA, and spray irrigation performance for grain sorghum. Trans. ASAE Vol. 47(5): 1477-1492. Available at http://www.cprl.ars.usda.gov/pdfs/Trans%20ASAE%2047-5-1477-1492.pdf.

Colaizzi, P.D., S.R. Evett, and T.A. Howell. 2005. Comparison of Spray, LEPA, and SDI for Cotton and Grain Sorghum in the Texas Panhandle. Proc. 2005 Central Plains Irrigation Conference, Sterling, Colorado, Feb 16-17. pp. 123-136. Available at http://www.cprl.ars.usda.gov/pdfs/Central%20Plains%20Irrig%20Conf%202005%20Colaizzi.pdf.

Colaizzi, P. D., P. H. Gowda, T. H. Marek, and D. O. Porter. 2009. Irrigation in the Texas High Plains: A brief history and potential reductions in demand. Irrig. and Drain. 58(3): 257-274.

Colaizzi, P.D., S.R. Evett, T.A. Howell, and R.L. Baumhardt. 2010. Crop production comparison with spray, LEPA, and subsurface drip irrigation in the Texas High Plains. 5th Decennial National Irrigation Conf., 5-8 December 2010, Phoenix, Arizona. Paper No. IRR10-9704. ASABE, St. Joseph, Mich. http://www.cprl.ars.usda.gov/wmru/pdfs/Colaizzi%20et%20al%20%20(2010)%20Crop%20Production%20Comparison%20with%20Spray%20LEPA%20and%20SDI_5th%20Dec%20Irrig%20Conf_07-9704.pdf

Colaizzi, P.D., S.R. Evett and T.A. Howell. 2011. Corn production with spray, LEPA and SDI. Pp. 52-67 In Proc. 23rd Annual Central Plains Irrig. Conf., Burlington, CO, 22-23 Feb. 2011. Available at http://www.cprl.ars.usda.gov/wmru/pdfs/Colaizzi%20et%20al%20%20(2011)%20Corn%20production%20with%20spray-LEPA-SDI_CPIC%20meeting.pdf

Doorenbos, J., and A. H. Kassam. 1979. Yield response to water. Irrigation and Drainage Paper No. 33. Rome, Italy: United Nations FAO.

Enciso, J. M., P. D. Colaizzi, and W. L. Multer. 2005. Economic analysis of subsurface drip irrigation lateral spacing and installation depth for cotton. Trans. ASAE 48(1): 197-204.

Evett, S.R., T.A. Howell and A.D. Schneider. 1995. Energy and water balances for surface and subsurface drip irrigated corn. Pp. 135-140. In F. R. Lamm (ed.) Microirrigation for a Changing World: Conserving Resources/Preserving the Environment. Proc. Fifth International Microirrigation Congress. Am. Soc. Agric. Engr., St. Joseph, MI.

Evett, S.R., J.A. Tolk, and T.A. Howell. 2003. A depth control stand for improved accuracy with the neutron probe. Vadose Zone J. 2:642-649. http://www.cprl.ars.usda.gov/pdfs/Evett_Tolk_Howell_VZJ2003.pdf

Evett, S.R., L.K. Heng, P. Moutonnet and M.L. Nguyen (eds). 2008. Field Estimation of Soil Water Content: A Practical Guide to Methods, Instrumentation, and Sensor Technology. IAEA-TCS-30. International Atomic Energy Agency, Vienna, Austria. ISSN 1018–5518. http://www.cprl.ars.usda.gov/wmru/pdfs/Field%20Estimation%20of%20Soil%20Water%20Content%20TCS-30_web.pdf

Evett, S.R., W.P. Kustas, P.H. Gowda, J.H. Prueger, and T.A. Howell. 2012a. Overview of the Bushland Evapotranspiration and Agricultural Remote sensing EXperiment 2008 (BEAREX08): A field experiment evaluating methods quantifying ET at multiple scales. Adv. Water Resourc. 50:4-19. http://dx.doi.org/10.1016/j.advwatres.2012.03.010.

Evett, S.R., R.C. Schwartz, T.A. Howell, R.L. Baumhardt and K.S. Copeland. 2012b. Can weighing lysimeter ET represent surrounding field ET well enough to test flux station measurements of daily and sub-daily ET? Adv. Water Resour. 50:79-90. http://dx.doi.org/10.1016/j.advwatres.2012.07.023.

Evett, S.R., G.W. Marek, P.D. Colaizzi, B.B. Ruthardt and K.S. Copeland. 2018. A subsurface drip irrigation system for weighing lysimetry. Appl. Engineer. Agric. 34(1):213-221. https://dx.doi.org/10.13031/aea.12597

Evett, S.R., G.W. Marek, K.S. Copeland and P.D. Colaizzi. Quality management for research weather data – Bushland, Texas. Submitted to Agrosystems, Geosciences and Environment on Sep 7, 2018.

Goldberg, D., and M. Shmueli. 1970. Drip irrigation—A method used under arid and desert conditions of high water and soil salinity. Trans. ASAE 13(1): 38-41.

Goldberg, D., B. Gornat, and D. Rimon. 1976. Drip Irrigation: Principles, Design and Agricultural practices. Drip Irrigation Scientific Publications, Israel, 296 pp.

Hanson, B.R., L.J. Schwankl, K.F. Schulbach and G.S. Pettygrove. 1997. A comparison of furrow, surface drip, and subsurface drip irrigation on lettuce yield and applied water. Agric. Water Manage. 33: 139-157.

Howell, T.A. 2001. Enhancing water use efficiency in irrigated agriculture. Agron. J. 93(2):281-289. Available at http://www.cprl.ars.usda.gov/pdfs/howell%202001.pdf.

Kohler, M.A., T.J. Nordenson and D.H. Baker. 1959. Evaporation Maps for the United States. Technical Paper No. 37, U.S. Department of the Interior, Weather Bureau, Washington, D.C.

Marek, T.H., A.D. Schneider, T.A. Howell and L.L. Ebeling. 1988. Design and construction of large weighing monolithic lysimeters. Trans ASAE 31: 477-484.

Marek, G.W., S.R. Evett, P.H. Gowda, T.A. Howell, K.S. Copeland, and R.L. Baumhardt. 2014. Post-processing techniques for reducing errors in weighing lysimeter evapotranspiration (ET) datasets. Trans. ASABE 17(2):499-515. http://dx.DOI.org/10.13031/trans.57.10433.

Middleton, J. E., E. L. Proebsting, and S. Roberts. 1979. Apple orchard irrigation by trickle and sprinkler. Trans. ASAE 22: 582-584.

Musick, J.T., F.B. Pringle and J.D. Walker. 1988. Sprinkler and furrow irrigation trends -Texas High Plains. Appl. Engr. Agric. 4(1):46-52. Available at http://www.cprl.ars.usda.gov/pdfs/musick%20pringle%20walker%2088.pdf

NASS, 2014. 2012 Census of Agriculture: Farm and Ranch Irrigation Survey (2013). National Agricultural Statistics Service. Available at: http://www.agcensus.usda.gov/Publications/2012/Online_Resources/Farm_and_Ranch_Irrigation_Survey/fris13.pdf (accessed 26 Aug 2015).

O'Brien, D. M., D. H. Rogers, F. R. Lamm, and G. A. Clark. 1998. An economic comparison of subsurface drip and center pivot irrigation systems. Applied Eng. Agric. 14(4): 391-398.

O'Shaughnessy, S.A.,S.R. Evett, P.D. Colaizzi, J.A. Tolk and T.A. Howell. 2014. Early and late maturing grain sorghum under variable climatic conditions in the Texas High Plains. Trans. ASABE 57(6):1583-1594. http://www.cprl.ars.usda.gov/wmru/pdfs/O'Shaughnessy%20et%20al%20%20(2014)%20Early%20and%20Late%20Maturing%20Sorghum%20in%20THP.pdf

Peacock, W. L., D. E. Rolston, F. K. Aljibury, and R. S. Rauschkolb. 1977. Evaluating drip, flood, and sprinkler irrigation of wine grapes. Am. J. Enol. Vitic. 28(4): 193-195.

Sammis, T. W. 1980. Comparison of sprinkler, trickle, subsurface, and furrow irrigation methods for row crops. Agron. J. 72(5): 701-704.

Sezen, S. M., A Yazar, B. Kapur, and S. Tekin. 2011. Comparison of drip and sprinkler irrigation strategies on sunflower seed and oil yield and quality under Mediterranean climatic conditions. Agric. Water Manage. 98: 1153-1161.

Shalhevet, J., D. Shimshi, and T. Meir. 1983. Potato irrigation requirements in a hot climate using sprinkler and drip methods. Agron. J. 75: 13-16.

Tognetti, R., M. Palladino, A. Minnocci, S. Delfine, and A. Alvino. 2003. The response of sugar beet to drip and low-pressure sprinkler irrigation in southern Italy. Agric. Water Manage. 60: 135-155.

Tolk, J.A. and S.R. Evett. 2012. Lower limits of crop water use in three soil textural classes. Soil Sci. Soc. Am. J. 76(2):607–616.

Wang, D., M. C. Shannon, C. M. Grieve, and S. R. Yates. 2000. Soil water and temperature regimes in drip and sprinkler irrigation, and implications to soybean emergence. Agric. Water Manage. 43: 15-28.

Wu L., X. Guo, and A. Harivandi. 2001. Salt tolerance and salt accumulation of landscape plants irrigated by sprinkler and drip irrigation systems. J. Plant Nutr. 24(9): 1473-1490. DOI: 10.1081/PLN-100106996

Direct Root-zone Delivery to Enhance Deficit Irrigation Application

Pete W. Jacoby[1] and Xiaochi Ma[2]

[1]Professor, and [2]Doctoral Student, Dept. Crop & Soil Sciences, Washington State University

Pullman, WA 99164-6420

Written for presentation at the Irrigation Association Technical Program

Long Beach, California

December 3-6, 2018

Abstract: *Direct root-zone (DRZ) is a sub-surface micro-irrigation application that delivers water directly into the plant root zone via plastic pipe placed vertically into the soil to depths of 2-feet (60 cm) or more. Field trials were initiated during 2015 in commercial vineyards to determine: 1) ability of DRZ to achieve grape production at or near commercial production goals with reduced irrigation amounts; 2) determine ideal depth at which to deliver the water; and 3) determine advantage of applying water in interrupted pulses versus continuous application. In 2016, we began to analyze grapes for quality attributes. After four years, we have concluded that: 1) DRZ can achieve grape production at or near full production levels while using only 60 percent the amount of water normally applied as surface drip; 2) grapes receiving reduced amounts of water via DRZ had progressively higher sugar content (% Brix), reduced levels of acidity, and higher total anthocyanin at reduced rates of water delivery; 3) no differences in yield were attributed to either a specific of water delivery depth or to use of pulse delivery.*

Keywords: Water conservation, Subsurface micro-irrigation, Deficit irrigation, Wine grape quality

Introduction

The Columbia Basin of Washington State is one of three areas within the United States capable of producing more than 300 different crops. This ability stems from irrigation from river systems that traverse through the major growing regions and provide irrigation to lands that would otherwise support native desert vegetation typical of the intermountain west. Underground aquifers are limited and not capable of supporting the current level of agricultural production derived from surface water. Additionally, ground water quality is often lower than the surface water, but varies by locale. Most importantly, most irrigation water has been adjudicated and water is demanded by many competing interests, including fisheries, energy production, and domestic use.

Against this backdrop scenario has emerged a profitable crop with an 8 percent annual growth in acreage, winegrapes, which has placed Washington State in second place only to California and quickly achieving an enviable recognition for high quality premium wines. The only factor currently threatening continued growth of this phenomenal industry is availability of water https://www.washingtonwine.org. Therefore, use of water must become more efficient, and new, cost-effective methods must become available for use by growers. Alternatively, acreage and water resources now committed to current crops may need to be converted to permit continued growth of winegrape acreage.

Irrigation is typically applied to wine grapes as surface drip which is considered an efficient and effective method. However, in the hot, dry region of Washington State, water is lost to both soil surface evaporation and weeds. Additionally, vine root systems become mostly concentrated in the upper 18-20 inches (0.5 m) of the soil profile under current irrigation strategies (Stevens and Douglas, 1994; Bauerle *et al.*, 2008; Davenport *et al.*, 2008). This condition is compounded by the fact that wine grapes in this region are mostly own-rooted from varieties that typically develop shallow to medium depth root systems (Keller *et al.*, 2012). These growth patterns may be more susceptible to both cold intolerance and drought during infrequent, short-term climatic perturbations such as experienced in 2010 and 2015.

Deficit irrigation is known to enhance a number of compounds related to production of premium red wines and substantial literature has been published on this topic that is summarized in detail by Chaves *et al.* (2010). Likewise, subsurface drip irrigation has emerged as an effective water saving strategy in a number of row crops (Lamm *et al.*, 2012; 2015). Some growers in Washington had attempted to use subsurface micro-irrigation applied through buried lines and many reported that this system failed their expectations because of soil clogging of emitters and chewing damage by burrowing rodents. Therefore, a newer form of subsurface water delivery was developed in our lab during 2013-2014 which we labeled direct root-zone (DRZ) micro-irrigation. In the following paper, we describe the design and application of this system in two phases (proof-of-concept and commercial application).

To investigate opportunities for increasing water use efficiency, we initiated a coordinated, trans-disciplinary research effort in 2014 with three broad objectives: 1) determine the potential and feasibility for use of a new form of sub-surface drip irrigation which might improve water conservation while improving the use of deficit irrigation in the production of high quality premium red wine grapes; 2) gain a better understanding of the potential advantages of developing deeper root systems of vines to obtain moisture from the available soil profile in the rhizosphere; and, 3) evaluate the potential of remote sensing to monitor water stress on the whole vineyard scale when applying direct root-zone micro-irrigation at rates much lower than used to meet commercial wine grape production goals.

Methods and Materials

To guide the approaches and methods used in the proof-of-concept phase, a stakeholder advisory group was established involving active growers and managers representing leading red wine grape producers in Washington. This group was convened quarterly to confer and advise the project Principal Investigator (PI) and team members on critical industry needs and opportunities. This group assisted in identifying collaborators to host research experiments in commercial vineyards, and they provided letters of support for grant proposals submitted to sponsoring agencies.

Research Site Description. Treatments were installed in a commercial block of Cabernet Sauvignon wine grapes located on Kiona Vineyards, Block 2 (46°16'59" N, 119°26'33" W) in the Red Mountain American Viticulture Area (AVA) near Benton City, WA in early 2015. Soil on the experimental site is of the Aridisol order and classified as a Hezel loamy fine sand (Xeric Torriorthents) on a terrace landform with parent materials being eolian sands over silty glacio-fluvial sediments deposited at the end of the most recent ice age. These soils are well-drained, subject to wind erosion, and relatively infertile, containing very low amounts of organic matter. Depth to nearest water table is more than 80 inches (>2 m). Normal annual precipitation is 8.83 inches (224 mm) and occurs mainly as rainfall during the dormant growing season. Summer temperatures average 70°F (21° C) with mid-day temperatures reaching 90°F (32°C) or higher and cooling during the night, typical of a desert climate. These conditions favor the development of high quality red wine grapes under irrigation. The research site was planted to Cabernet Sauvignon (Clone 2) of own-rooted vines on a spacing of 8' (2.5 m) between rows and 6' (1.8 m) between vines. The vineyard was 8 years old at the beginning of the 2015 growing season.

Experimental Design: During the Proof-of-Concept phase of the study (2015-2017), a randomized complete block with two main effect treatments, irrigation rate and depth of delivery sub-surface was used with a split plot super-imposed to compare pulse irrigation and constant delivery. Each treatment plot involved 15 vines (5 vines x 3 rows) with the center-most 3 vines designated for physiological measurements while being buffered by other vines receiving the same treatment. Each of 18 treatment plots was replicated 3 times (810 vines) and yields were compared with 12 plots (180 vines) receiving full commercial rate of irrigation via surface drip. Irrigation scheduling was determined by the vineyard manager according to long-standing guidelines used to meet commercial production goals. In 2017, pulse irrigation was discontinued and some of those plots were repurposed to allow direct comparisons of surface drip and DRZ (subsurface drip) delivery at equal rates on grape yield, vine stress levels, and fruit quality. Since no consistent differences in yield or grape quality were related to DRZ delivery depth during the first two years of the study, all DRZ treatments were delivered at the 2-ft. depth for the new phase of the project. Additionally, rates of water application to the treatments were raised to 80, 60, and 40 percent of full commercial rate. This modification to the original design permitted direct measurements between surface and subsurface (DRZ) and allowed relative water use efficiencies to be determined for grape production (yield) and grape quality (phenolic and other chemical components).

Direct root-zone delivery device. DRZ is designed to deliver water to deep within the root zone vertically through PVC pipes. This approach was used to avoid the two most common problems associated with use of buried driplines – clogging of emitters in contact with soil and chewing damage of buried lines. An additional advantage was the ease of installing within established plantings of perennial crops such as grapes.

To install this system, a 1 inch (25.4 mm) diameter hole was bored vertically to a depth of 1, 2, or 3 foot (30, 60, and 90 cm respectively) about 1.5 foot (45 cm) either side from the base of each vine and beneath the trellis wire and suspended irrigation dripline. A length of PVC tube was inserted into each hole. Each section of PVC pipe was previously cut to length to reach the desired depth while extending above ground for a given distance and then split about 6 inches (15 cm) from the lower end with a band saw to allow sufficient water passage to move into the soil to reduce the backing of water up the tube. A PVC cap, previously drilled to allow passage of a one foot (30.5 cm) length of ¼ inch (6.35 mm) diameter micro-tubing through the hole prior to attaching one end to a barbed connector and inserted into the main horizontal water line, and attaching a pressure compensating drip emitter at the other end of the feeder tube before placing the emitter snuggly into the top of the PVC tube beneath the drilled out cap. Emitters were selected to deliver 0.5 gallon (2 liter) per hour, thus delivering approximately 1.2 gallon (4 liters) per vine per hour.

Installation of all treatments involving 990 vines was completed prior to beginning of the 2015 growing season. Additionally, soil moisture access tubes containing electron capacitance sensors were installed in designated plots to monitor soil water dynamics both temporally and spatially across depths within the top 6.5 feet (2.0 meters). This technique measured soil moisture content (data not shown) continuously when using the SynTek EnviroSCAN https://www.sentek.com.au or by supplemental spot reading when using the SynTek Diviner. Irrigation events were monitored by data transmitted from the EnviroSCAN probes via cellphone transmitters to a base server AgriNET http://www.grovision.com .

Monitoring of plant water stress. Vine water stress was determined by periodically sampling the center three vines of direct root-zone treatments and control vines by measuring mid-day xylem water potential using the pressure bomb method (Scholander *et al.*, 1965). Leaves were selected on the east side of vines (most sun drenched prior to sampling), then covered with a plastic bag inside an aluminum foil exterior envelope and allowed to equilibrate for about an hour prior to sampling. The encased leaf was detached from the vine by severing the petiole with a razor blade, then quickly removed from the bag and inserted into the pressure chamber, while shielding the leaf from direct sunlight, then pressurized with nitrogen gas to the point of water movement from the cut petiole. Pressures were measured in pounds per square inch (psi), then converted to mega-pascals. These data were correlated with multi-spectral digital images obtained through aerial and ground-based platforms during both growing seasons as described and reported by Zuniga *et. al.* (2017; 2018). Visual observations of plant condition were documented with digital photos to document phenological impacts on vines from treatments as described by Keller (2005). Additionally, periodic measurements of gas exchange were taken with a field portable auto-porometer to gauge vine physiological activity.

Determination of grape production and quality. In 2015, grape clusters from designated vines were collected to determine estimates of production, as well as number and size of berries in each treatment and control block. Sampling was executed during two stages of maturity, early-*veraison* and mid-*veraison*. At the 2015 harvest date (September 26), all grapes were harvested at end of growing season from replicated rows of each treatment and weighed by individual vine. In 2016, cluster sampling was not repeated, but all vines in the treatment plots and a sample of commercially irrigated vines were harvested on September 24 and weighed. Samples of grapes from specific treatments were collected two weeks prior to harvest and submitted to a private commercial laboratory for analytical determination of a number of characteristics associated with quality red wines. Samples of clusters from each treatment were also taken at the date of harvest and submitted to the same commercial lab for juice and phenolics analyses. A similar sampling procedure was used in 2017 and 2018 to directly compare DRZ delivered at the 2-ft. (60 cm) depth with surface drip treatments, both applied as equally reduced rates in replicated plots during the second phase of the study.

Results and Discussion

Proof of Concept Phase: Treatments were successfully implemented during three consecutive growing season from 2015 through 2017. Direct root-zone treatments (DRZ) were effective in delivering water to the designated depths. The electronic controllers worked flawlessly, as did the small mechanical meters, except it was necessity to remove the meters over the winter months and reinstall prior to initiation of irrigation in the spring of the year. We discovered that the meters performed most accurately when placed face-up rather than to the side, owing to an internal design factor.

Comparison of pulse and constant irrigation delivery methods revealed no differences their relative contribution to either plant water stress or grape yield and these comparisons were terminated after 2016. Likewise, we found no consistent influences for delivery of water at any particular depth below the soil surface and no interaction was found between depth and amount of water applied.

Table 1. Seasonal irrigation delivery and water use efficiency based on grape production during 2015 and 2016 comparing commercial surface drip irrigation with season-long deficit irrigation imposed by direct root-zone micro-irrigation delivered subsurface at rates of 60, 30, or 15% the rate of surface drip irrigation.

	Commercial (100 %)	Irrigation Treatments 60 %	Irrigation Treatments 30%	Irrigation Treatments 15%
2015 Water Use (acre ft.)	1.25	0.81	0.40	0.20
Grape production (tons/ac.)	4.54	4.08	3.40	3.18
Production Efficiency (tons/ac. ft. applied)	3.63	5.04	8.50	15.90
Relative Efficiency	1.0	1.39	2.34	4.38
2016 Water Use (acre ft.)	1.27	0.84	0.43	0.23
Grape production (tons/ac.)	8.16	5.02	4.19	3.52
Production Efficiency (tons/ac. ft. applied)	6.43	5.98	9.74	15.30
Relative Efficiency	1.0	0.93	1.51	2.38
2017 Water Use (acre ft.)	1.24	0.74	0.37	0.19
Grape production (tons/ac.)	4.64	4.43	4.06	3.52
Production Efficiency (tons/ac. ft. applied)	3.74	5.99	10.97	18.53
Relative Efficiency	1.0	1.60	2.93	4.95

Water use efficiency, determined as amount of fruit produced per unit of water applied, increased progressively with reduced rates of irrigation in 2015. This trend was repeated in 2016 and 2017, but was not as pronounced, largely owing to a much higher rate of grape production that occurred from the commercial plots in 2016 than in 2015 and high soil water content during the first half of 2017.

One of the objectives for the Proof-of-Concept phase was to ascertain the greatest degree of water conservation that could be achieved while maintaining health and productivity of the vine. The 2015 growing season was determined to be the hottest and driest on record for the area. Fruit production at the commercial irrigation rate and applied by surface drip averaged 10 pounds (4.5 kg) per vine, while DRZ irrigation applied 1-3 feet subsurface at reduced rates of ca. 60, 30, and 15% of full commercial rate produced an average of 9.0, 7.6, and 7.2 pounds per vine, respectively. By contrast, the 2016 growing season was one of the most productive on record with commercial production averaging almost 18 pounds per vine (8.16 tons/acre). DRZ treatments produced an average of 11.1, 9.2, and 7.8 pounds per vine (5.0, 4.2, and 3.5 tons per acre, respectively). There was wider disparity in fruit production between the commercial treatment plots and the DRZ plots than occurred during the previous growing season (Table 1) which we attribute to carryover effects on fruit set from the 2015 growing season and the levels of stress imposed from the reduced rates of water delivery.

Plant water stress, as measured by obtaining xylem pressure potentials among the treatment vines, showed obvious differences among the irrigation delivery rates when measured at 3 dates during the growing season (Table 2). Plant water stress increased proportionately with decreasing irrigation rate and progression of the growing season. Similar measurements were taken at only one date during 2015 and also showed progressively more stress with decreasing water application rates of DRZ irrigation (data not shown). Stress measurements continued on a more frequent basis during the subsequent years of study and consisted of both on plant water status measurement by xylem pressure method (Scholander, 1967) and remote sensing techniques (Zuniga *et. al.*, 2017; 2018).

Table 2. Plant water stress as determined by leaf stem xylem potential during 2016 growing season contrasting commercial surface drip irrigation with season-long deficit irrigation imposed by direct root-zone micro-irrigation delivered subsurface from 1-3' depths at rates of 60, 30, or 15% the rate of surface drip irrigation.

	Irrigation Treatments			
	Surface Drip (DI)	----------DRZ----------		
	(100 %)	(60 %)	(30%)	(15%)
2016	Xylem Pressure Potential (-MPa)			
June 3	-0.529	-0.593	-0.641	-0.781
July 7	-0.635	-0.825	-0.925	-1.188
August 10	-0.869	-1.177	-1.522	-1.593

In 2015, cluster samples from the DRZ treatments showed that cluster weights were slightly lower, but grapes were more numerous, yet smaller in size than in the clusters from vines receiving the higher irrigation rate. These findings suggested that the grapes from the lower irrigation rates might have greater potential to produce higher quality red wine, owing to higher concentration of anthocyanins, tannins and sugars. In 2016, similar effects were noted for grapes receiving the DRZ treatments. Replicated cluster samples were obtained from the commercial and DRZ treatment plots and submitted to a private, commercial analytical lab for determination of a dozen components and ratios. Data summarized in Table 3 illustrates four of these components. Acidity became progressively reduced below the 60% irrigation rate, while sugars (Brix), tannins, and anthocyanins all trended higher with decreasing rate of irrigation. These results are in line with the findings of Casassa *et al.* (2015) who noted that efforts to derive benefits in grape quality and water savings through greatly reduced irrigation levels should recognize the potential for yield reductions and/or physiological impacts on vines. Results from our study provides evidence that use of efficient irrigation application such as DRZ could both sustain vines and produce grapes during drought conditions while yielding grapes with potential to produce premium quality red wines in the hands of skilled viticulturists and enologists.

Table 3. Comparison of selected chemical components influencing red wine quality. Analyses of Cabernet Sauvignon grapes grown under full and reduced rates of irrigation during 2016. Reduced irrigation rates were applied via direct root-zone micro-irrigation (DRZ) delivered 2 feet (61 cm) subsurface.

	Surface drip (DI)		DRZ	
	Control (100 %)	High (60 %)	Moderate (30%)	Low (15%)
2016				
pH	3.41	3.36	3.48	3.55
Brix (degrees)	25.5	27.1	27.6	28.6
Tannins (mg/L)	403	594	600	741
Anthocyanin (mg/L)	1015	1242	1298	1480
2017				
pH	3.38	3.55	3.50	3.47
Brix	25.1	25.3	26.5	25.0
Tannins	730	676	839	784
Anthocyanin	1185	1167	1447	1373

Treatment plots were harvested each year a few days prior to commercial harvest. Each vine was harvested and fruit was weighed on site. Yields within a given irrigation amount were not consistently influenced by depth of the water delivery within the soil profile. No significant differences were attributed to whether or not the water delivery was interrupted into hourly pulses or applied continuously at the selected rate. However, yields were impacted by the total amount of water applied. These rates were applied at the pre-selected rates throughout the entire growing season from start to finish. Under this regime, individual grape size was found to be reduced consistent with the declining amounts of water delivered. Approximately 6 clusters per treatment replication were selected for subsequent juice and phenolic panel analyses by a commercial analytical laboratory. These analyses revealed that pH and titratable acidity increased in proportion to amount of water applied, i.e. more acidic. Total anthocyanin, Brix, and tannins increased in proportion to increasing irrigation deficit, in response to greater plant stress.

Data obtained during the 2015 and 2016 growing seasons revealed that vines tended to show increasing levels of stress with lowering of the rates of water as anticipated; however, rates below 60 percent of commercial delivery resulted in less than 80 percent of commercial production and often led to berry shrivel at the lowest rate of delivery, regardless of depth of water delivery. Additionally, canopy cover at the low level of delivery was reduced to a level that was deemed less than acceptable.

Therefore, prior to the initiation of the 2017 growing season DRZ irrigation rates were adjusted to higher rates of irrigation on selected plots representing approximately 80, 60, and 40 percent the rate of surface drip irrigation used to achieve commercial production goals. In 2017, levels of soil moisture were exceedingly high from high amounts of winter snow moisture and cool, moist weather during spring that deferred the initiation of irrigation until July (Figure 1).

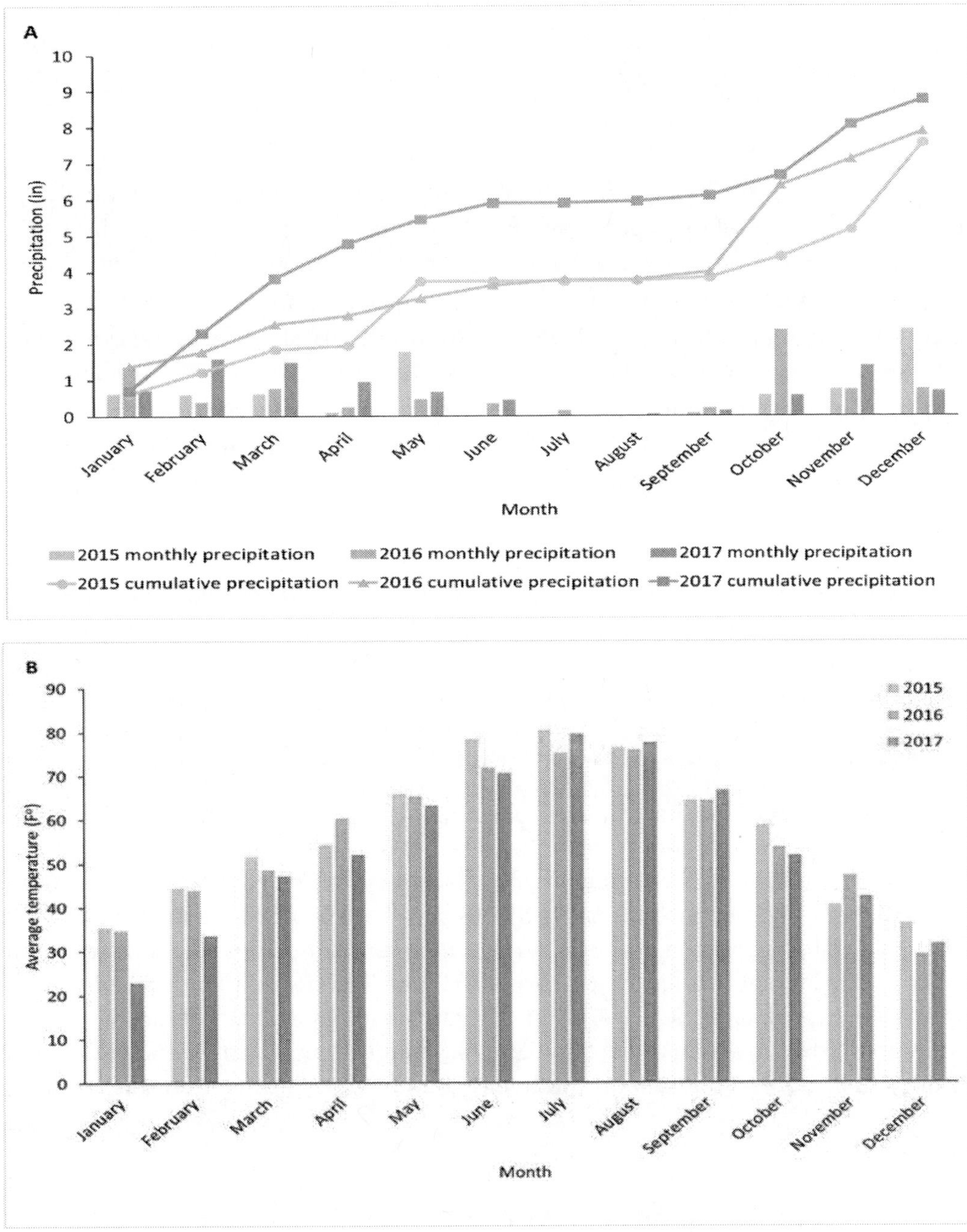

Figure 1. Average monthly and accumulated precipitation (A) and monthly temperature (B) during 2015, 2016, and 2017 at study site near Benton City, WA.

Grape yields from DRZ treatments during 2017 and 2018 proved to be largely statistically insignificant and generally in the same ranges of yield as grapes receiving the commercial rate of irrigation. Vines receiving DRZ subsurface delivery generally produced higher yields than vines receiving the same rate of water delivered as surface drip. Vines receiving DRZ delivery, including those reduced to 40 percent of full surface drip, also consistently produced higher yields than the vines which received full rates of surface drip irrigation. Grape quality factors such as Brix and anthocyanin increased progressively with decreased levels of water (data not shown).

Conclusions

A new form of subsurface micro-irrigation was developed to achieve direct root-zone (DRZ) subsurface delivery of drip irrigation to wine grapes in Washington State. During the concept phase of the project, vines were maintained at production levels with much reduced amounts of water than required by vines managed to meet production goals. In 2015, DZR water delivery reduced to 60, 30, and 15% of commercial irrigation rate produced ca. 90, 75, and 70% of the commercial grape production weight, respectively. Second year production rates were generally lower relative to the commercial grape yield, suggesting a carry-over effect from stress imposed during the previous year. No obvious advantages were found for using pulse irrigation delivery over continuous application sets. No consistent patterns were observed to favor a specific depth among 1, 2, or 3 ft. subsurface delivery points. Rates of deficit irrigation using DRZ delivery were raised to 80, 60, and 40 percent of commercial rate during 2017 and 2018. Comparison of DRZ delivered at the 2-ft. depth with surface drip irrigation consistently showed higher yields at each reduced rate and grape quality was improved over grapes produced with equal rates of surface drip irrigation. DRZ demonstrated the ability to achieve water savings up to 60 percent of the rate applied by surface irrigation to achieve commercial production goals while producing higher yields and quality of grapes than achieved with standard surface drip irrigation.

Acknowledgements

This research was supported in part by the following: 1) Northwest Center for Small Fruit Research, Agreement No. 59-5358-4-010 and grant # 2072-21000-047-16; 2) Washington State Department of Agriculture – Specialty Crop Block Grant Program – Agreement K1768; Washington Grape and Wine Research Committee – Agreement Nos. # 3019-3818 and # 3019-6818. The authors also express sincere thanks to Scott Williams, general manager – Kiona Winery and Vineyards, Benton City, WA for providing access to land and resources for this project. We also thank Rick Hamman of Hogue Ranches, Bill Riley of Ste. Michael Wine Estates, Julia Kock of Klipsun Vineyards, and Jim Holmes of Ciel du Cheval Vineyard for serving as advisors to this research project and others. Appreciation is also extended to Drs. Markus Keller and R. Troy Peters for subject matter consultation related to various aspects of this research.

References

Bauerle, T.L., J.H. Richards, D.R. Smart and D.M. Eissenstat. 2008. Importance of internal hydraulic redistribution for prolonging the lifespan of roots in dry soil. *Plant, Cell & Environ.* 31: 177-186.

Casassa, L.F., M. Keller and J.F. Harbertson. 2015. Regulated deficit irrigation alters anthocyanins, tannins and sensory properties of Cabernet Sauvignon grapes and wine. *Molecules* 20: 7820-7844.

Chaves, M.M., O. Zarrouk, R. Franciso, J.M. Costa, T. Santos, A.P. Regalado, M.L. Rodriques and C.M. Lopes. 2010. Grapevine under deficit irrigation: hints from physiological and molecular data. *Annals Bot.* 105: 661-676.

Davenport, J.R., R.G. Stevens and K.M. Whitley. 2008. Spatial and temporal distribution of soil moisture in drip-irrigated vineyards. *HortSci.* 43: 229-235.

Keller, M., L.J. Mills and J.F. Harbertson. 2012. Rootstock effects on deficit-irrigated winegrapes in a dry climate: vigor, yield formation, and fruit ripening. *Am. J. Enol. Vitic.* 63: 29-39.

Lamm, F.R., J.P. Bordovsky, L.J. Schwanki, G.L. Grabow, J. Enciso-Medina, R.T. Peters, P.D. Colaissi, T.P. Trooien and D.O. Porter. 2012. Subsurface drip irrigation: status of the technology in 2010. *Trans. ASABE* 55: 483-491.

Lamm, F.R., K.C. Stone, M.D. Dukes, T.A. Howell, J.W. Robbins and B.Q. Mecham. 2015. Emerging technologies for sustainable irrigation. ASABE/IA Symposium. *Trans. ASABE* 59: 155-161.

Scholander, P., E. Bradstreet, E. Hemminsen, and H. Hammel. 1965. Sap pressure in vascular plants: negative hydrostatic pressure can be measured in plants. *Science* 148: 339-346.

Stevens, R.M. and T. Douglas. 1994. Distribution of grapevine roots and salt under drip and full-ground cover micro-jet irrigation systems. *Irrig. Sci.* 12:181-186.

Zuniga, C.E., A. P. Rathnayake, M. Chakraborty, A. P. Rathnayake, S. Sankaran, P.W. Jacoby, and L.R. Khot. 2018. Applicability of time-of-flight-based ground and multispectral aerial imaging for grapevine canopy vigour monitoring under direct root-zone deficit irrigation. *Int'l. J. Remote Sensing.* DOI: 10.1080/01431161.2018.1500047.

Zuniga, C.E., L.R. Khot, S. Sankaran, and P.W. Jacoby. 2017. High resolution multispectral and thermal remote sensing based water stress assessment in grapevines to evaluate subsurface irrigation technique effects. *Remote Sensing* 9(9):961-976; http://www.mdpi.com/2072-4292/9/9/961/htm DOI: 10.3390/rs9090961. [(ISSN 2072-4292)].

Sub-surface drip fertigation for precision management of cotton

Mark Dougherty (Corresponding Author)
Biosystems Engineering Dept.
218 Corley Bldg.
Auburn University, AL 36849-5417
Phone: 334-844-8939
Fax: 334-844-3530
doughmp@auburn.edu

Anis Bouselmi
Regional Extension Agent I, Irrigation Specialist
Henry County
101 N. Doswell St.
Abbeville, AL 36310
azb0103@auburn.edu

Larry M. Curtis
Extension Irrigation Engineer, Biosystems Engineering Dept. (Retired)
200 Corley Bldg.
Auburn University, AL 36849-5417
curtilm@auburn.edu

C.H. Burmester
Extension Agronomist, Tennessee Valley Research & Extension Center (Retired)
P.O. Box 159
Belle Mina, AL 35615
burmech@auburn.edu

Hugh D. Harkins
Assoc. Director, Tennessee Valley Research & Extension Center
P.O. Box 159
Belle Mina, AL 35615
harkihd@auburn.edu

B.E. Norris
Director, Tennessee Valley Research & Extension Center
P.O. Box 159
Belle Mina, AL 35615
norribe@auburn.edu

Brad Durham
Advisor III, Tennessee Valley Research & Extension Center
P.O. Box 159
Belle Mina, AL 35615
durhabm@auburn.edu

Abstract

A subsurface drip fertigation study in north Alabama evaluated four fertigation timing scenarios over seven years. Replicated treatments included 40-inch rows with drip tape between every other row of cotton. All treatments received 135# of N and K_2O, 20# of S, and 1.0# of B. Phosphorus was surface-applied to maintain high P soil levels. Treatments were: (1) Non-fertigated control - 60#N pre-plant surface, remainder side-dressed, (2) Pre-plant 20#N surface, remainder drip to bloom+25 days, (3) 100% fertigated (bloom + 25 days) - drip 20#N at planting, (4) 100% fertigated (bloom + 40 days) - drip 20#N at planting, (5) Pre-plant 40#N surface, remainder drip to bloom (50 days). Water was not a limiting factor with adequate irrigation provided to all treatments. Treatments (2) and (5), the two fertigation treatments with the highest amount of pre-plant surface N, had significantly higher yields in four out of seven years. These two treatments out-yielded the non-fertigated control (1) indicating the effectiveness of pre-plant surface N in conjunction with drip fertigation in cotton.

Keywords. Drip irrigation, fertigation, cotton production

Introduction

Subsurface drip irrigation (SDI) systems can have a long economic life when properly designed, installed, and managed. The profitability of drip irrigation is impacted by design life of the system, crop and climate, placement of laterals, and method by which irrigation and fertilization is scheduled and applied. As energy and corresponding pumping costs continue to increase and longer SDI operational life is documented, lower pressure SDI systems have become an alternative to higher pressure center pivot irrigation systems. In many fields in Alabama, SDI is becoming a viable option for producers as they become more familiar with installation and operational technology. SDI fertigation through chemical injection of nutrients is an option that can reduce field travel and input costs.

In an SDI system, improved profitability and reduced environmental contamination is possible because of the ability to manage small applications of water and N fertilizer as needed by the crop (Camp et al., 1997). Application of fertilizer nutrients through irrigation systems (fertigation) has been found to increase seed cotton yield, water use efficiency, and nutrient uptake by researchers in Syria (Janat and Somi 2001a, 2001b; Janat, 2004), Texas (Enciso-Medina et al., 2007), and India (Thind et al., 2008). Irrigation systems permit multiple small dose fertilizer injections at different intervals, reducing the risk of leaching compared to fertilizers applied in a single application. Notwithstanding, Hunt et al. (1998) near Florence, South Carolina found that N fertigation using a single drip-application produced the highest seed cotton yield compared with five split drip-applications. Similar results were reported by Hou et al. (2007) for N applied at the beginning of the irrigation cycle rather than in more frequent, smaller doses throughout the irrigation cycle. Alternatively, Bauer et al. (1997) determined that N application method (single versus five split drip-applications) through SDI had no significant effect on cotton yield. The present study follows closely on these studies to evaluate the effectiveness of varied fertigation timing treatments on cotton yield in northern Alabama.

Methods

This research compares four SDI fertigation management treatments with one conventional side-dress treatment for nitrogen, potassium and phosphorus delivery. The study uses an automated nutrient delivery system to provide in-season fertilizer injection as prescribed by four SDI fertigation treatments. The response to all treatments (four sub-surface fertigation and one conventional surface fertilization) is quantified by measuring seed cotton yield. The focus of the seven-year study is to evaluate the timeliness, precision, and effectiveness of several fertigation treatments. A pressure-compensating SDI tape product was installed at approximately 12-inch depth using a tractor auto-guidance system to ensure all tape runs were precisely located parallel and 80 inches apart between every other row of 40-inch cotton. The resulting experimental plots consist of four replications of five eight-row treatments. An automated tracking (auto-guidance) system was also used to accurately plant cotton rows each season in relation to the SDI tape. All subsequent field operations were completed using an auto-guidance systems to ensure proper equipment traverse across each plot, each year.

Experiment Design

The SDI fertigation study was initiated in 2006 at the Alabama Agricultural Experiment Station's Tennessee Valley Research and Extension Center (TVREC) in Belle Mina, AL (86 ° 52' 30" W, 34 ° 41', 00" N) on a Decatur silt loam (fine, kaolinitic, thermic, Rhodic Paleudult). The treatment design is a randomized complete block (r=4) with five fertilizer application methods (one conventional surface-fertilization vs. four fertigation treatments). The dimension of each harvested treatment plot is 80 inches x 345 feet making up the middle two rows of an eight-row treatment plot. At the outboard edge of each treatment one irrigated buffer or "guard" row is provided to minimize interference between fertilizer/fertigation treatments. Irrigation was provided equally to all treatments ensuring that water was not a limiting factor in this study.

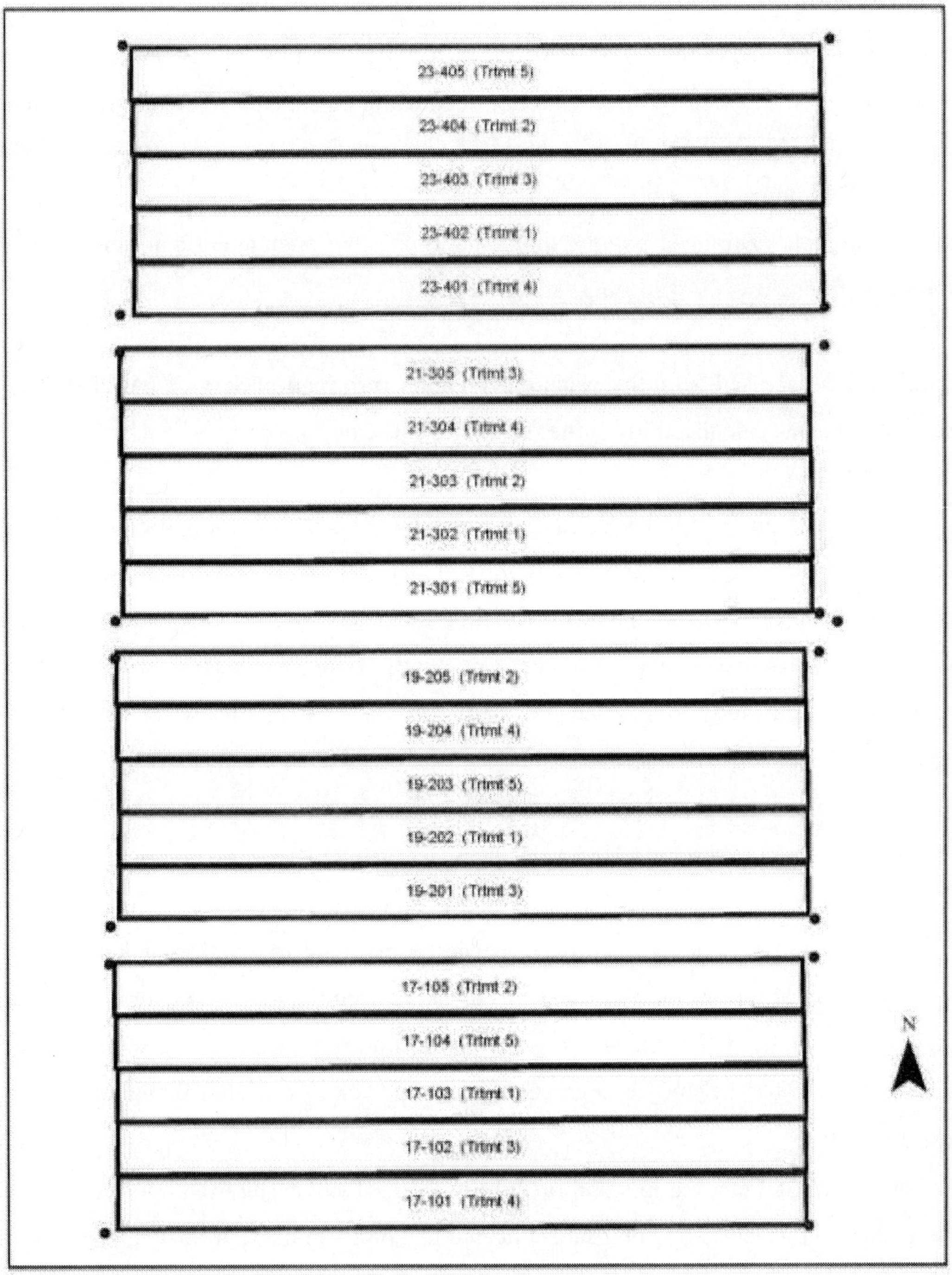

Fig. 1. Five randomized treatments (r=4) making up SDI fertigation study, 2006-2012, Belle Mina, AL. Each treatment plot is made up of eight rows of cotton 360 feet long in 40-inch rows with two irrigated guard (buffer) rows between each treatment plot. The middle two rows of each treatment plot (80"x345') were harvested for yield response.

Irrigation and scheduling

The irrigation system was installed in fall of 2005 before cotton planting in spring 2006. Subsurface drip tape (0.874" ID) was installed between 10 and 14 inches deep using a heat-treated shank with installation tube on the backside (Andros Engineering, Paso Robles, CA) mounted on a standard toolbar behind a 130 hp agricultural tractor. The toolbar assembly was equipped with an integrated reel carrier and platform for handling the self-feeding reels. During the study, the irrigation system operated at a nominal pressure of 10 psi with individual emitter flows (24" spacing) at 0.24 gph.

Daily pan evaporation was accessed from the Alabama Weather Information Service (AWIS). The daily irrigation requirement, ET, was calculated using the following equation;

$$ET = 0.90 \times PAN \times CC,$$

where evapotranspiration, ET (in day^{-1}) was calculated from 90% pan evaporation, PAN (in day^{-1}) adjusted for fractional canopy cover, CC. Canopy cover was determined weekly by measuring the open canopy distance (in) between rows with a tape measure. Fractional canopy cover was calculated using the following formula; (row width - open canopy distance)/row width. Canopy closure measurements were a critical component used in this study to estimate the daily amount of irrigation water as a function of calculated daily pan evaporation.

Crop

Cotton was planted in the second or third week of April each year using a 4-row planter with a 40-inch row spacing and a seeding rate of 4-5 seeds ft^{-1}. Plots were end-trimmed by 3 ft to eliminate edge effects just before picking. The two center rows of each plot were harvested with a 2-row cotton picker and weighed using a boll buggy equipped with electronic load cells to measure accumulated seed cotton yield per plot. Average post-harvest turnout from the gin was used to determine lint yield for subsequent economic analysis. Average lint turnout for the study as a percentage of seed cotton was approximately 41%.

Fertilization treatments

Approximately 7,500 feet of SDI tape, supply and flush lines, control valves, air-vacuum release, and fertilizer injection equipment was installed on four replicated test plots. Four positive displacement liquid fertilizer injectors (Neptune, USA) were installed to supply four replications of five nutrient timing treatments including four fertigated treatments and one non-fertigated control (Table 1). Treatment 1 served as the irrigated control providing conventional surface side-dressing of nutrients. Treatments 3 and 4 provided "spoonfed" fertigation scenarios with 100% of nutrient mix provided through SDI injection. Treatments 2 and 5 provided a combination of pre-plant surface-applied nutrients along with in-season fertigation. All treatments received 135 pounds per acre of nitrogen and potassium (K_2O), 20 pounds per acre of sulfur, and 1.0 pound per acre of boron. Phosphorus fertilizer was surface-applied to maintain P at high soil test levels. Drip fertilizer was 8-0-8-1.2S-0.06B made using 32% liquid N, potassium thiosulfate, fertilizer grade KCL, solubor, and water. Nurse tank and injection pumps were mounted on a concrete pad located in the middle of treatment plots (Fig. 2). Sand filtration was provided ahead of nutrient injection. Drip tape was flushed twice yearly, at beginning and end of each season using a separate positive displacement pump with injection nurse tank for water/acid/water flushing.

Table 1. Treatment description, fertigation management trials, 2006-20127.

Treatment	Description
1. Control - drip irrigated, but all fertilizers are surface applied.	*Preplant* - N and K @ 60 pounds per acre. *Post-Plant* N (75lb/A) sidedressed at early square.
2. Timing 1 – with surface preplant	*Preplant* - 20 pounds of N and K (surface). *Drip* 40 pounds N,K –square to bloom (25 days) *Drip* 75 pounds N,K – bloom to 25 days
3. Drip timing 1 – no preplant	*Planting Drip* - 20 pounds N,K *Drip* 40 pounds N,K –square to bloom (25 days) *Drip* 75 pounds N,K – bloom to 25 days
4. Drip timing 2 – no preplant "spoon-fed"	*Planting Drip* - 20 pounds N,K *Drip* 40 pounds N,K square to bloom (25 days) *Drip* 75 pounds N,K – bloom to 40 days
5. Timing 2 – with surface preplant	*Preplant* - 40 pounds of N and K (surface). *Drip* 95 pounds N,K –square through bloom (50 days)

All treatments received 135 pounds per acre of nitrogen and potassium (K_2O), 20 pounds per acre of sulfur, and 1.0 pound per acre of boron. Phosphorus fertilizer was surface-applied to maintain P at high soil test levels. Drip fertilizer was 8-0-8-1.2S-0.06B made using 32% liquid N, potassium thiosulfate, fertilizer grade KCL, solubor, and water.

Fig. 2. Fertigation mixing tank, chemical injection pumps, and injection header for five replicated SDI fertilizer treatments and two irrigated buffer (guard) rows, Belle Mina, AL

Results and discussion

Observed yields indicated significant differences between treatment means during five out of seven years of this study. Table 2 indicates that treatments 2 and 5, the two fertigation treatments with pre-plant side-dressed N had significantly higher yields in 2008, 2010, 2011, and 2012. In 2006, treatments 2 and 5 were also the highest yielding treatments, but only treatment 2 was significantly higher than the other treatments. During 2007 and 2009, there were no statistical differences observed between treatments.

Seed cotton yields for the first two years of the study, 2006 and 2007 occurred during two of the driest consecutive years on record at TVREC and still yielded an average 3.0 and 2.9 bales of cotton, respectively. In 2007 and 2008, although the non-fertigated control (treatment 1) was the highest yielding treatment it was not significantly different from the two highest fertigated treatments (treatments 2 and 5). Though not significant, the three highest yielding treatments in 2007 and 2008 received at least 20 pounds of surface-applied, preplant nitrogen and potassium (K_2O). When averaged across the seven years of this study, the three highest ranking treatments (1, 2, and 5) verify the conclusions of previous researchers (Hunt et al., 1998; Hou et al., 2007) regarding the importance of early N applications versus continuous or "spoon-fed" applications throughout the growing season.

Table 2. Summary of mean seed cotton yield by fertilizer treatment and year, lb/A.

trt	2006	2007	2008	2009	2010	2011	2012	ave.
1	3167^{c}	3636nd	4649^{a}	2071nd	4092^{b}	2329^{c}	$3618^{a,b}$	**3366**
2	3788^{a}	3328nd	$4226^{a,b}$	2370nd	$4338^{a,b}$	3142^{a}	3996^{a}	**3598**
3	3536^{b}	3164nd	4038^{b}	2395nd	4136^{b}	2726^{b}	3519^{b}	**3359**
4	3437^{b}	3266nd	4008^{b}	2389nd	4226^{b}	$2510^{b,c}$	3232^{b}	**3295**
5	3614^{b}	3333nd	4596^{a}	2282nd	4626^{a}	2787^{ab}	3989^{a}	**3604**
ave.	**3508**	**3345**	**4303**	**2301**	**4284**	**2699**	**3671**	**3445**
Rain†, in	6.63	6.06	11.27	12.39	8.34	12.69	9.24	**11.49**
Temp‡, F	82	85	77	77	82	80	78	

[a] superscript denotes significant difference between treatment means within an individual year (α=0.05).
[nd] denotes no significant difference between treatment means within an individual year (α=0.05).
† Total June through August rainfall. Average represents 83-year total June-August rainfall.
‡ Average June through August temperature.
1. Surface applied N-P-K with drip irrigation (control). 2. Preplant 20# N-K surface with 2 N-K drip timings. 3. 20# N-K drip at planting with 2 N-K drip timings (to 25 days after bloom). 4. 20# N-K at planting with 2 N-K drip timings (to 40 days after bloom). 5. Preplant 40# N-K surface with 1 N-K drip timing (square through bloom).

The higher response of the non-fertigated control treatment in the second year of the study, 2007, was thought due to beneficial downward movement of surface-applied fertilizer early in the season as a result of 4.55" of rain between May and July. A comparable number of storm events in 2006 delivered only 2.87" of rain over the same May through July period. As a result, in spite of nearly equal and unusually low total seasonal rainfall in the 2006 and 2007 growing seasons (6.6" and 6.1", respectively), higher early season rainfall in 2007 may have facilitated delivery of surface-applied nutrients, increasing yield. Several large convectional storms later in the 2007 season may have advantageously moved surface-applied nutrients into the soil horizon, while concurrently leaching fustigated nutrients farther out of reach of roots. In both 2006 and 2007, plant tissue nutrients (not shown) were generally higher in the highest yielding treatments, with plant tissue boron, manganese, and sodium generally lower in the surface-applied control treatment (treatment 1). Similarly, during the 2008 growing season 2.57 inches of rain fell on one day June, potentially moving surface-applied nutrients downward and highlighting the importance of rainfall or surface irrigation as a relevant factor.

Summary

A subsurface drip irrigation study was installed at the Tennessee Valley Research and Extension Center (TVREC), in Belle Mina, Alabama in 2005 to evaluate four precision fertigation management (i.e., fertilizer timing) scenarios. Approximately 7,500 feet of SDI tape and four positive displacement liquid fertilizer injectors were installed on five nutrient timing treatments with four replications in RCB design. The resulting twenty treatment plots were made up of eight, 345-foot rows of cotton on 40-inch row spacing, with drip tape between every other row of cotton. All treatments received 135 pounds per acre of nitrogen and potassium (K_2O), 20 pounds per acre of sulfur, and 1.0 pound per acre of boron. Phosphorus fertilizer was surface-applied to maintain P at high soil test levels. Treatments were as follows: (1) Non-fertigated control - 60#N pre-plant surface, remainder side-dressed at early square, (2) Pre-plant 20#N (surface), remainder drip at square and bloom + 25 days, (3) 100% fertigated (bloom + 25 days) - drip 20#N at planting, remainder drip at square and bloom, (4) 100% fertigated (bloom + 40 days) - drip 20#N at planting, remainder drip at square and bloom, (5) Pre-plant 40#N (surface), remainder drip at square through bloom (50 days). Drip fertilizer was 8-0-8-1.2S-0.06B made using 32% liquid N, potassium thiosulfate, fertilizer CL, solubor, and water. Although growing season rainfall varied considerably across the seven-year study, water was not a limiting nutrient as irrigation was consistently applied to all treatments using an adjusted pan evaporation method. Treatment yield response across all seven years of this study revealed significantly higher yields in treatments (2) and (5) in four out of seven years, the two fertigation treatments receiving the highest amount of pre-plant nitrogen (20# and 40#, respectively). These two fertigated treatments out-yielded, on average, the non-fertigated control (1) which had 60# of pre-plant nitrogen surface applied. Results indicate the effectiveness of drip fertigation in conjunction with pre-plant surface application of nitrogen to boost cotton yields in the Tennessee Valley.

References

Bauer, P. J., P. G. Hunt, and C. R. Camp. 1997. In-season evaluation of subsurface drip and nitrogen-application for supplying nitrogen and water to cotton. *J. Cotton Sci.* 1:29-37.

Camp, C.R., P.J. Bauer, and P.G. Hunt. 1997. Subsurface drip irrigation lateral spacing and management for cotton in the southeastern coastal plain. *Trans. ASAE.* 40: 993-999.

Enciso-Medina, J., P.D. Colaizzi, W.L. Multer, C.R. Stichler. 2007. Cotton response to phosphorus fertigation using subsurface drip irrigation. *Applied Eng. Agric. 23: 299-304.*

Hou, Z., P. Li, B. Li, J. Gong, and Y.Wang. *2007.* Effects of fertigation scheme on N uptake and N use efficiency in cotton. *Plant Soil 290:115-126.*

Hunt, P.G., P. J. Bauer, C.R. Camp, and T.A. Matheny. 1998. Nitrogen accumulation in cotton grown continuously or in rotation with peanut using subsurface microirrigation and GOSSYM/COMAX management. *Crop Sci.* 38:410-415.

Janat, M. *2004.* Assessment of nitrogen content, uptake, partitioning, and recovery by cotton crop grown under surface irrigation and drip fertigation by using isotopic technique. *Commun. Soil Sci. Plant Anal. 35:2515-2535.*

Janat, M., and G. Somi. *2001a.* Performance of cotton crop grown under surface irrigation and drip fertigation. I. Seed cotton yield, dry matter production, and lint properties. *Commun. Soil Sci. Plant Anal. 32: 3045-3061.*

Janat, M., and G. Somi. *2001b.* Performance of cotton crop grown under surface irrigation and drip fertigation. II. Field water-use efficiency and dry matter distribution. *Commun. Soil Sci. Plant Anal. 32:3063-3076.*

Thind, H.S., M.S. Aujla, and G.S. Buttar. *2008.* Response of cotton to various levels of nitrogen and water applied to normal and paired sown cotton under drip irrigation in relation to check-basin. *Agric. Water Manage. 95: 25-34.*

ACKNOWLEDGEMENT: Grants from Alabama Cotton Commission and The Tennessee Valley Authority along with technical and material contributions of drip tape products from private industry for this research are appreciated. In addition, the Auburn University College of Agriculture and Alabama Agricultural Experiment Station provided funding and land and equipment resources for this research as part of Hatch program multi-state research. Comments from peer reviewers are acknowledged.

Soil Moisture and Nutrient Dynamics in the Root Zone of Collard Greens in Different Organic Amendment Types and Rates

Ripendra Awal, Almoutaz Elhassan, Ali Fares, Selamawit Woldesenbet, Ram Ray, Haimanote Bayabil, Peter Ampim, Eric Risch, Richard Griffin, and Hamideh Habibi
College of Agriculture and Human Sciences, Prairie View A&M University, Prairie View, TX 77446

Abstract. *Understanding how organic amendment affects the soil-water-plant-atmosphere continuum has become imperative as the use of organic amendments in conventional and organic agricultural production is increasing. A study was conducted at the Prairie View A&M University Research Farm to investigate the effect of organic amendment types (Chicken manure, Cow manure, and Milorganite) and application rates (0, 168, 336, 672 kg total N ha-1) replicated three times on soil moisture and nutrient dynamics within and below the root zone of collard greens. Soil moisture sensors were installed at all 36 plots to monitor soil moisture within and below the root zone. Suction cups were used to collect soil water samples from 15 and 45 cm depths. Soil water samples were collected six times during the growing season at each plot and analyzed for different nutrients (N, P, K, Ca, Mg, Na, Fe, Mn, Zn, and Cu). The preliminary results showed that both organic amendment types and rates affect the soil moisture and nutrient dynamics. The Chicken manure treatments seemed to have the higher water contents consistently for a major portion of the growing season. However, the water contents of the Cow manure and Milorganite treatments were very comparable and relatively lower than that of the Chicken manure treatments. Most of the total Nitrogen concentration increment occurred after irrigation and rainfall. The Cow manure treatments had the lowest value of total Nitrogen concentration in soil solutions collected within and below the root zone in all amendment rates.*

Keywords. soil moisture, nutrients, collard greens, organic amendments

Introduction

In order to meet the increasing demand for food currently and in the future, it is important to optimize crop yields by minimizing inputs, mainly water, and nutrient application. Many approaches have been developed to improve water management (Fereres and Goldhamer 1990) approaches such as real-time soil water content monitoring sensors (Hanson et al. 2000a) or plant water status monitoring sensors (Goldhamer and Fereres 2001; Intrigliolo and Castel, 2004; Hedley and Yule, 2009).

The role of soil moisture in the unsaturated zone plays a key role in many agricultural and environmental studies (Wang and Qu, 2009; Walker, 1999). Therefore, it is important to accurately monitor and estimate variations of soil moisture within and below the root zone to optimize irrigation scheduling that will minimize crop water stress and excess water leaching below the root zone that can have adverse environmental effects (Fares and Alva, 2000). Real-time soil moisture monitoring used for irrigation scheduling helps conserve water and nutrients (Vellidis et al., 2008). The soil moisture data is used to quantify the water content of root zone that needs to determine when and how much to irrigate to replenish the water content in the root zone to its field capacity level and prevent nutrient leaching due to over irrigation. Good irrigation management is vital for optimum crop production and minimum nutrient losses that might contaminate our water resources (Jones, 2004).

Cow manure, Poultry manure, and Milorganite are sources of Nitrogen (N) as well as other macro- and micro-nutrients (Reiter et al., 2014). As such, organic farmers use manure to achieve adequate crop yield (Lim, 2015). However, improper application of manure may affect the environment through nitrate leaching (Paramasivam et al., 2009; Fares et al., 2008). Once leached below the root zone, nitrate (NO_3–N) is no longer available for plant uptake and may eventually end up in groundwater (Chinkuyu and Kanwar, 2001; Ahmad et al., 2009). The mobility of nutrients in the vadose zone is a function of manure rate and time of application. Leaching of nutrients also depends on the soil water inputs and uptakes. Water flow and nutrient movement through the vadose zone are affected by the soil physical and hydrological properties; it is equally known that organic amendments impact soil physical properties (Fares et al., 2008). The main objectives of this study are: i) to establish a site-specific irrigation scheduling for collard greens under Southeast Texas conditions, and ii) quantify the effect of organic amendment types and rates on the water and nutrients dynamics under this crop. In order to achieve these objectives, first we need to understand the dynamics of soil water content in the root zone of the collard greens in order to establish site-specific irrigation scheduling, and then evaluate nutrients fates and concentrations within and below the root zone.

Materials and Methods

Study Site

The research was conducted at the University Research Farm of the College of Agriculture and Human Sciences, Prairie View A&M University.

Experimental Design

Three organic amendments: (i) Cow manure, (ii) Chicken Manure, and (iii) Milorganite and four application rates (non-amendment control [0 kg N/ha], half recommended rate [168 kg N/ha], recommended rate [336 kg N/ha], and double the recommended rate [672 kg N/ha]) were applied on 36 raised bed plots of size (300 cm x 150 cm). Figure 1 shows the 36-plot experimental layout. The organic amendments applied to each plot and then mixed into the top 15 cm of the soil using hoes and rakes. Soil moisture sensors, suction cups, and collars for CO_2 monitor were installed on each plot before they were seeded. Plots were seeded with collard greens in October 2017. A three line-drip irrigation system was installed at each bed. A complete weather station was also installed at the site. The experimental site was hand weeded throughout the growing season.

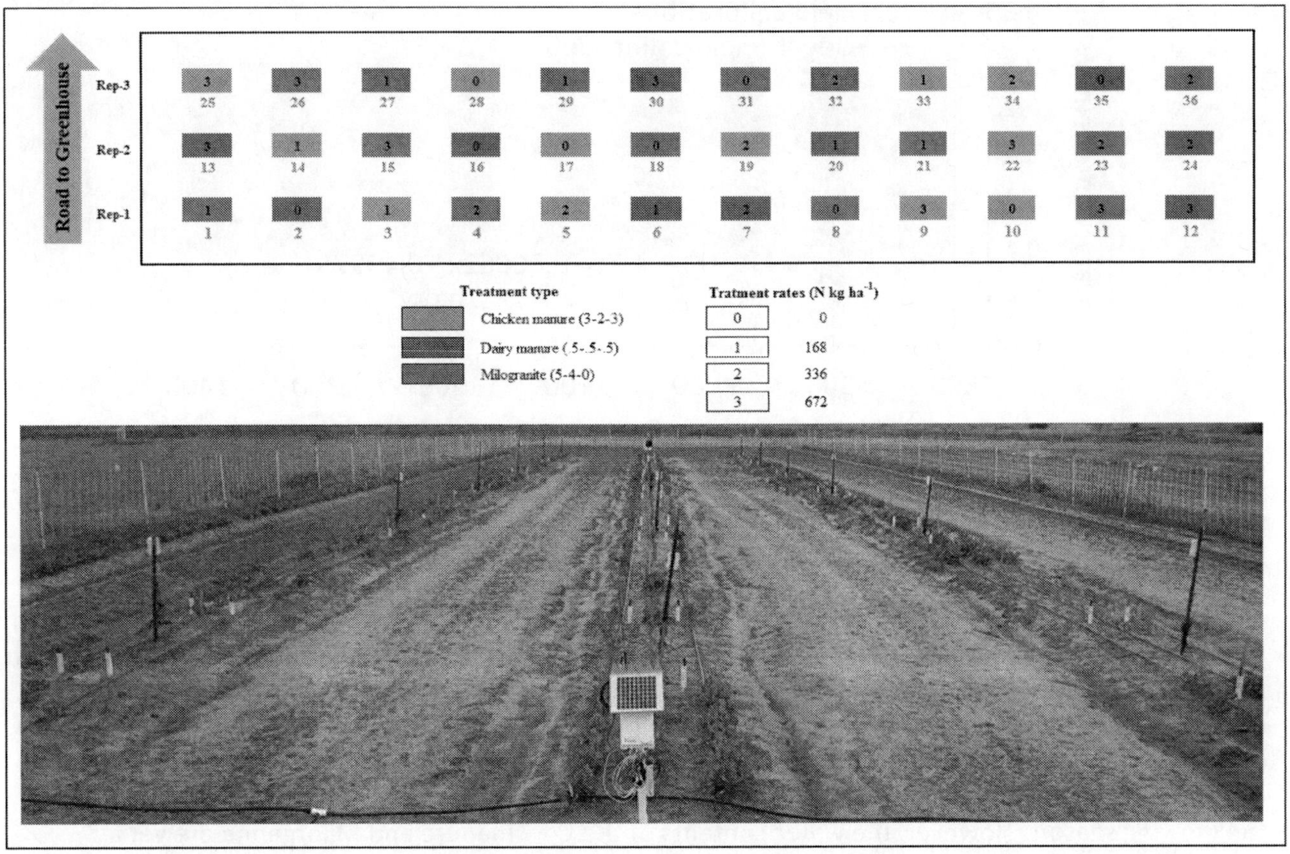

Figure 1. Experimental layout of plots.

Data Collection

Two 10HS soil moisture sensors (METER Group, Inc., Pullman, WA, USA) were installed at each plot to monitor soil moisture in the top 5 cm and below the root zone (at 30 cm). The soil solution suction cups were installed to sample the soil solution at 15 and 45 cm depths. Soil solutions were collected six times during the growing season at each location and analyzed for different nutrients e.g., Total Nitrogen (N), Phosphorus (P), Potassium (K), Calcium (Ca), Magnesium (Mg), Sodium (Na), Iron (Fe), Manganese (Mn), Zinc (Zn), and Copper (Cu).

Preliminary Results

Soil Moisture Dynamics

We calibrated the 10HS soil moisture sensors in non-amendment control plots at the same field. The standard calibration equation overestimated the water content (Figure 2). The findings of Fares et al.

(2016) highlighted the need to account for the effect of soil organic matter content to improve the accuracy and precision of the 10HS sensor. However, in this study, we used the same field calibration equation for all plots ignoring amendment types and rates.

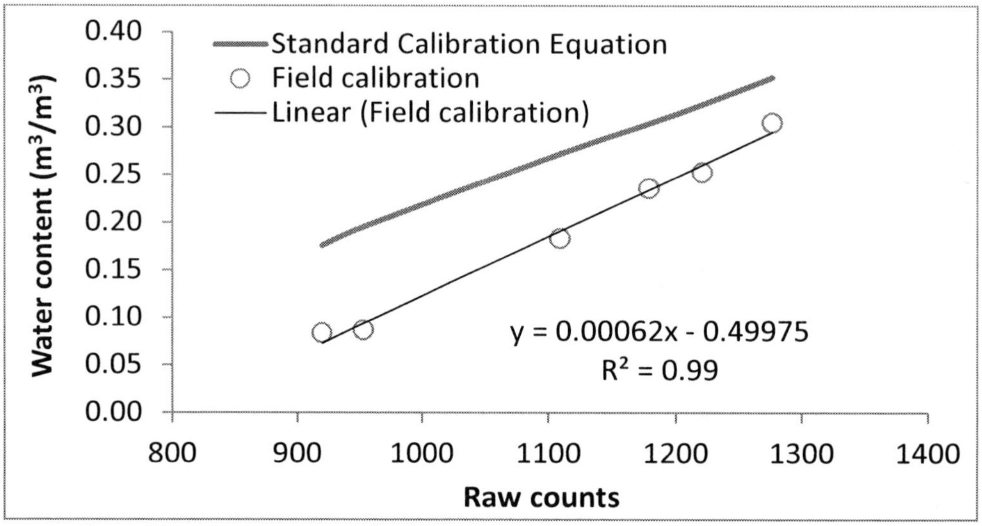

Figure 2. Site-specific calibration equation for the10HS at University Farm of Prairie View A&M University.

Figure 3 shows the temporal variation of soil moisture within and below the root zone for all amendment types (Chicken manure, Cow manure, and Milorganite) and treatment rate 3 (672 kg N/ha). The soil water content within the root zone was very dynamic across the growing season; it went as high as 0.35 $cm^3\ cm^{-3}$ and as low as 0.12 $cm^3\ cm^{-3}$. Increases of the water content within the root zone are in clear response to rainfall and irrigation events; however, the staircase shape of the water content curve, especially during the rain and irrigation free days is in clear response mainly plant water uptake. The Chicken Manure treatments seem to consistently have the higher water contents for a major portion of the growing season. However, the water contents of the Cow manure and Milorganite are very comparable and relatively lower than that of the chicken manure.

The water contents below the root zone were almost consistently higher than those within the root zone; in addition, they were very close to each other; they increased from about 0.29 $cm^3\ cm^{-3}$ in the beginning of the growing season to 0.33 $cm^3\ cm^{-3}$ right before the end of the experiment. The lack of strong dynamics is an indicator that not a substantial amount of crop roots, if any, were able to explore that region of the soil. It is important to note that this soil is duplex soil with the top 30 cm as sandy loam and the part below that has relatively high clay content. Differences in soils properties and lack of plant roots could be the main reason for this slow water content dynamics below the root zone.

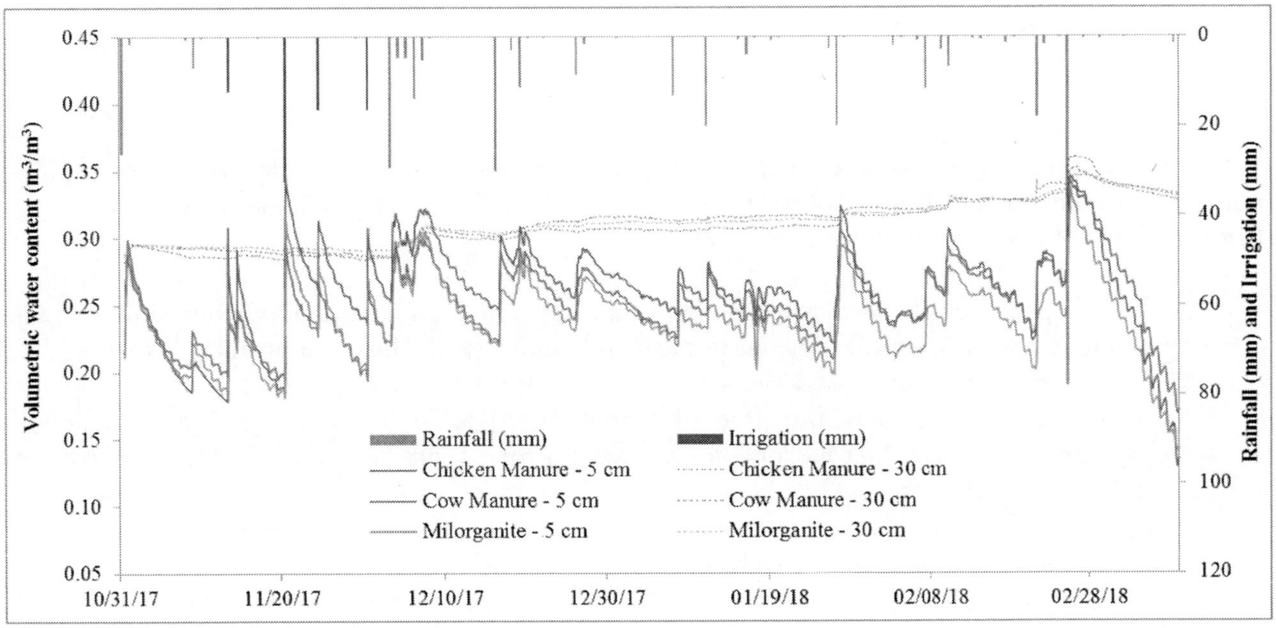

Figure 3. Temporal variation of soil moisture within and below the root zone (Treatment rate 3).

Figure 4 shows the temporal variation of soil moisture within and below the root zone for all treatment rates of Cow manure, from October 2017 to February 2018. The temporal variation of moisture content within the root zone varies in different treatment rates. We also carried out a similar analysis for Chicken and Milorganite amendment rates.

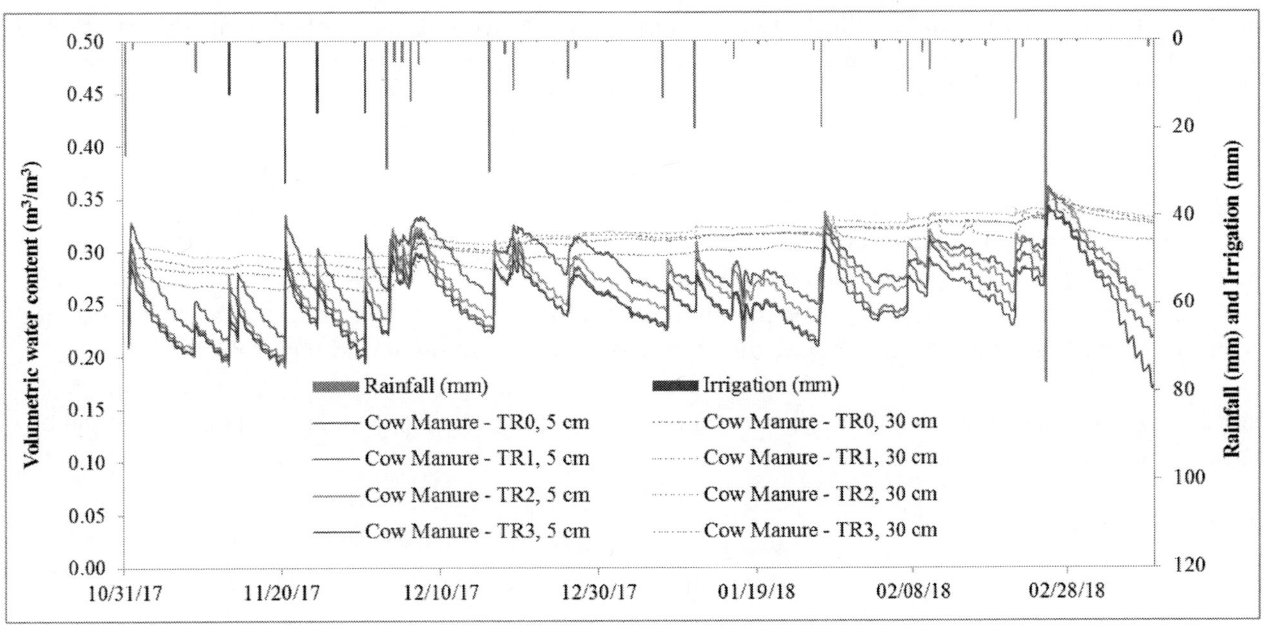

Figure 4. Temporal variation of soil moisture within and below the root zone (Cow manure, treatment rates 0, 1, 2, and 3).

Nutrient Dynamics

Total Nitrogen

Six batches of samples were collected. Each batch contains 72 samples. The samples were analyzed for nutrients including total N, P, K, Ca, Mg, Na, Fe, Mn, Zn, and Cu. The preliminary analysis of total Nitrogen concentration rate within and bellow root zone after applying different amendment types, and rates are as follows:

- Figure 5a shows the existence of Nitrogen in the control plots (non-amendment) which is due to the long period of non-agricultural activities in the experimental field. The concentration of total Nitrogen in the control plots gradually decreased during the samples collection period within and below the root zone that can be the effect of rainfall and irrigation. In addition, a slight reduction of Nitrogen concentration in the root zone and below root zone might be due to the growth of collard greens in the control plots.
- The concentration of total Nitrogen had a significant increase in below root zone, after 32 days in the plots of half recommended rate of Milorganite (Figure 5b). Same significant increase in Milorganite had occurred for recommended (Figure 5c) and double recommended rates (Figure 5d).
- The chicken manure had the highest concentration in the root zone and after 32 days in the below root zone among all amendments (Figure 5c).
- In the double recommended rate plots (Figure 5d), the concentration of N in all amendments, remain almost constant after 10 days, within and bellow root zone.
- Most of the total Nitrogen concentration increment occurred after irrigation and rainfall (Figures 5b, 5c, and 5d).
- The dairy manure had the lowest value of concentration in all cases (Figure 5)
- Figure 6a shows that the total Nitrogen concentration rate for both control and half-recommended plots of chicken manure within and bellow root zone are almost the same. In addition, the total Nitrogen concentration resulted from the recommended rate was almost remained constant after 30 days in the root zone (Figure 6a).
- The total N concentration from dairy manure in all rates is about the same after 32 days within the root zone. The highest concentration below the root zone had occurred from the recommended rate of dairy manure (Figure 6b).
- After 10 days, the recommended and double recommended rate of Milorganite had approximately the same and highest values of concentration within the root zone. The double recommended rate had the highest concentration below the root zone after the same period (Figure 6c).

We conducted a similar analysis for other nutrients (P, K, Ca, Mg, Na, Fe, Mn, Zn, and Cu) which will be reported in the near future.

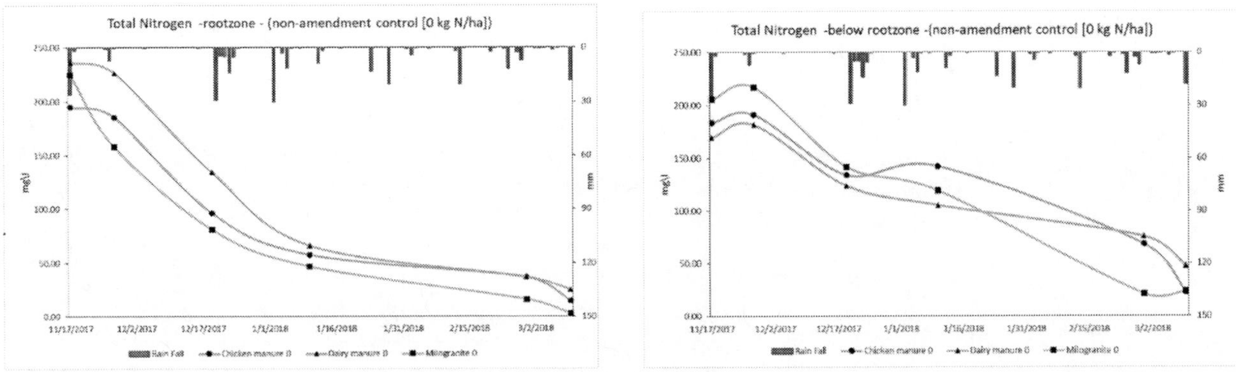

(a) Control plots

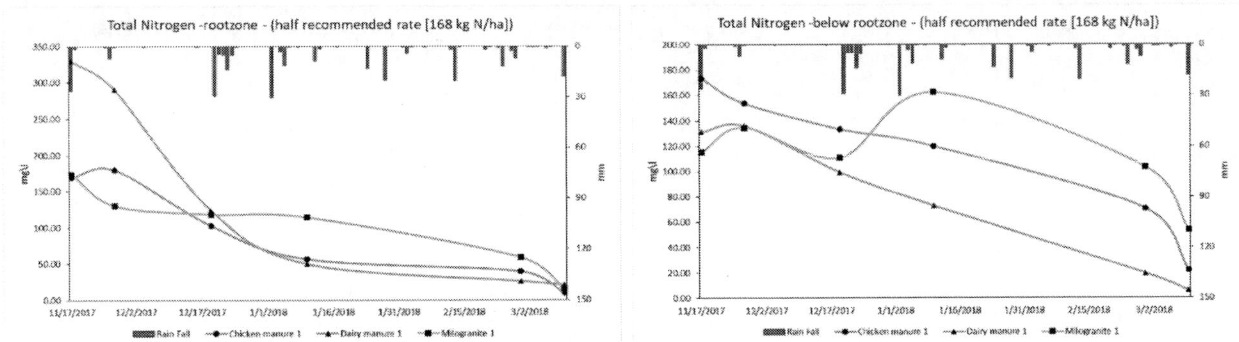

(b) Half recommended rate [168 kg N/ha]

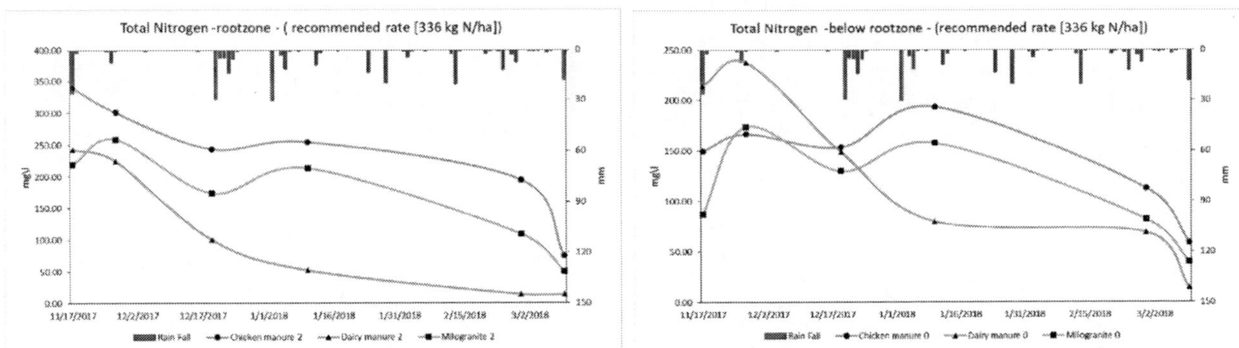

(c) Recommended rate [336 kg N/ha]

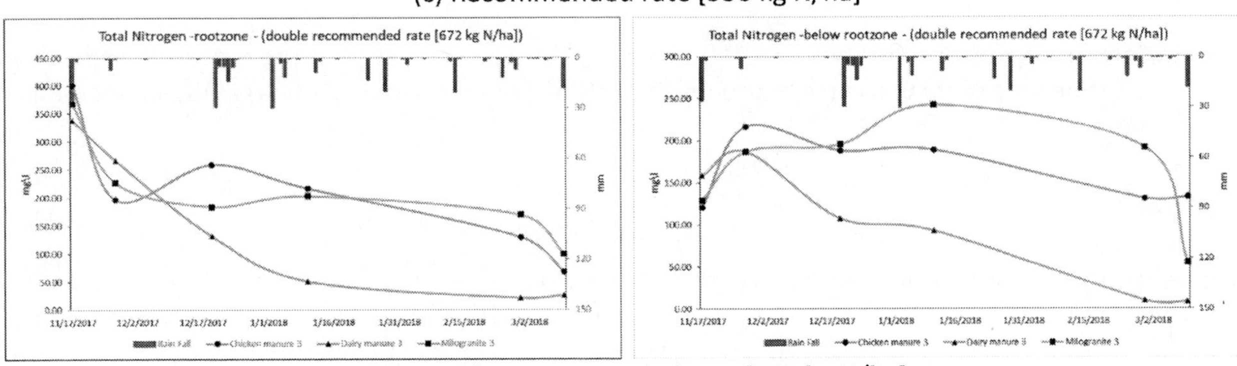

(d) Double recommended rate [672 kg N/ha]

Figure 5. Total Nitrogen concentration (mg/l) from 11/17/2017 to 3/10/2018 at the root zone and below the root zone of (a) Control, (b) Half recommended rate [168 kg N/ha], (c) Recommended rate [336 kg N/ha], and (d) double recommended rate [672 kg N/ha] application plots.

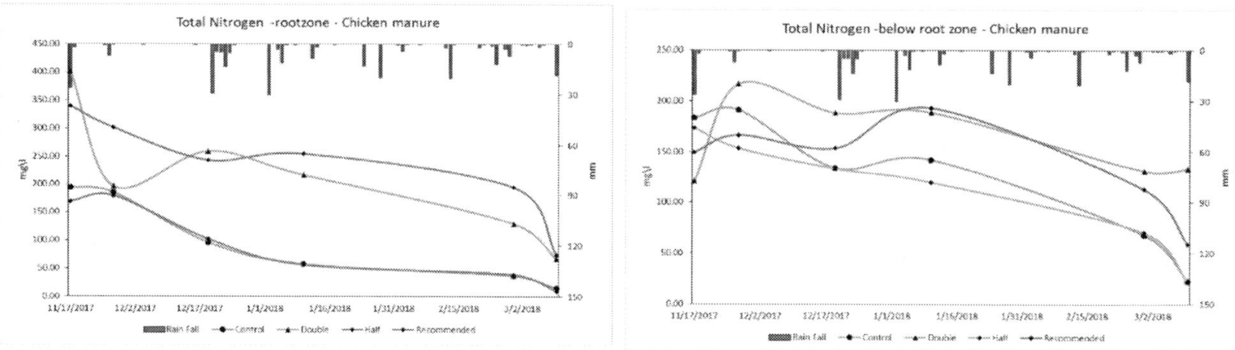

(a) Chicken manure amendment plots

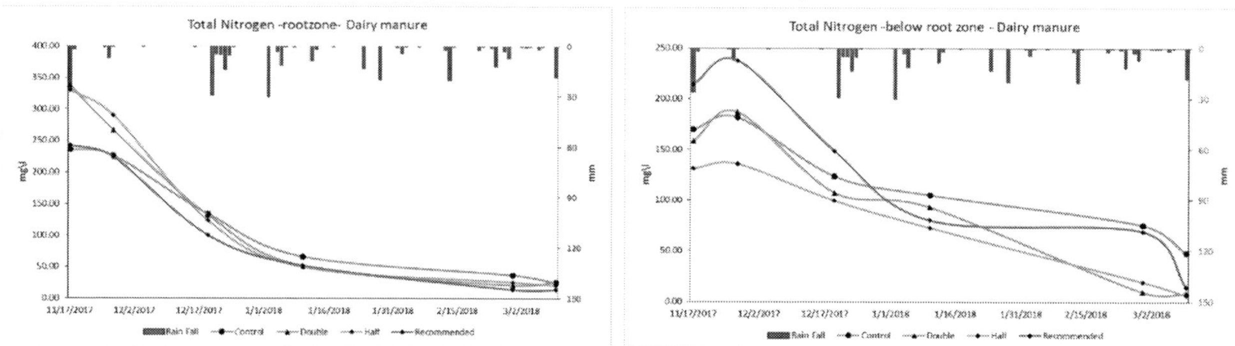

(b) Dairy manure amendment plots

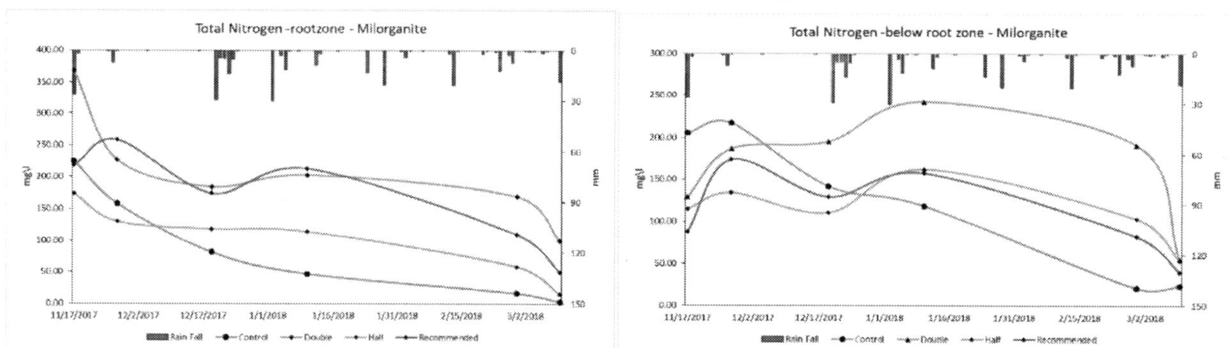

(c) Milorganite amendment plots

Figure 6. Total Nitrogen concentration (mg/l) from 11/17/2017 to 3/10/2018 at the root zone and below the root zone of (a) Chicken manure, (b) Dairy (Cow) manure, and (c) Milorganite amendment plots.

Summary

We monitored soil moisture and nutrient dynamics in the root zone and below the root zone of collard greens grown in different organic amendment types and rates. The preliminary results showed that both organic amendment types and rates affect the soil moisture and nutrient dynamics. The Chicken manure treatments seemed to have the higher water contents consistently for a major portion of the growing season. However, the water contents of the Cow manure and Milorganite treatments were very comparable and relatively lower than that of the Chicken manure treatments. Most of the total Nitrogen concentration increment occurred after irrigation and rainfall. The Cow manure treatments had the lowest value of total Nitrogen concentration in soil solutions collected within and below the root zone in all amendment rates.

Acknowledgments

This work was supported by the USDA National Institute of Food and Agriculture, Evans Allen project. The authors wish to thank all students and researchers for their assistance in conducting the experiment and analyzing soil water samples for nutrient analysis.

Reference

Ahmad, A.A., A. Fares, F. Abbas, and J.L. Deenik. 2009. Nutrient concentrations within and below root zones from applied chicken manure in selected Hawaiian soils. J. Environ. Sci. Heal. B 44(8): 828-843.

Chinkuyu, A.J. and R.S. Kanwar. 2001. Predicting Soil Nitrate-Nitrogen Losses from Incorporated Poultry manure using the GLEAMS Model. Trans. ASAE 44(6): 1643.

Fares, A., F. Abbas, A. Ahmad, J.L. Deenik, and M. Safeeq. 2008. Response of selected soil physical and hydrologic properties to manure amendment rates, levels, and types. Soil Sci. 173(8): 522-533.

Fares, A., and A.K. Alva. 2000. Evaluation of capacitance probes for optimal irrigation of citrus through soil moisture monitoring in an entisol profile. Irrig. Sci. 19(2):57-64.

Fares, A., R. Awal, and H. Bayabil 2016. Soil Water Content Sensor Response to Organic Matter Content under Laboratory Conditions, Sensors 16(8), 1239; doi:10.3390/s16081239.

Fereres, E., and D.A. Goldhamer. 1990. Deciduous fruit and nut trees. In: Stewart BA, Nielsen DR (eds) Irrigation of agricultural crops. ASA, Madison, Wis., pp. 987–1017

Goldhamer, D.A., E. Fereres, M. Mata, J. Girona, M. Cohen. 1999. Sensitivity of continuous and discrete plant and soil water status monitoring in peach trees subjected to deficit irrigation. J. Am. Soc. Hortic. Sci. 1244:437–444

Hedley, C.B., and I.J. Yule. 2009. A method for spatial prediction of daily soil water status for precise irrigation scheduling. Agr. Water Manage. 96(12): 1737-1745.

Intrigliolo, D.S., and J.R. Castel. 2004. Continuous measurement of plant and soil water status for irrigation scheduling in plum. Irrig. Sci. 23(2): 93-102.

Jones, H.G. 2004. Irrigation scheduling: advantages and pitfalls of plant-based methods. J. Exp. Bot. 55(407): 2427-2436.

Lim, S.L., T.Y. Wu, P.N. Lim, and K.P.Y. Shak. 2015. The use of vermicompost in organic farming: overview, effects on soil and economics. J. Sci. Food Agr. 95(6): 1143-1156.

Paramasivam, S., K.A. Richards, A.K. Alva, A.M. Richards, K.S. Sajwan, K. Jayaraman, A. Heanacho, and J. Afolabi. 2009. Evaluation of poultry litter amendment to agricultural soils: Leaching losses and partitioning of trace elements in collard greens. Water Air Soil Poll. 202(1-4): 229-243.

Reiter, M.S., T.C. Daniel, N.A. Slaton, and R.J. Norman. 2014. Nitrogen Availability from granulated fortified poultry litter fertilizers. Soil Sci. Soc. Am. J. 78(3): 861-867.

Vellidis, G., M. Tucker, C. Perry, C. Kvien, and C. Bednarz. 2008. A real-time wireless smart sensor array for scheduling irrigation. Comput. Electron. Ag. 61(1): 44-50.

Walker, J. 1999. Estimating Soil Moisture Profile Dynamics from Near-Surface Soil Moisture Measurements and Standard Meteorological Data. Ph.D. dissertation, The University of Newcastle, Australia

Wang, L., and J.J. Qu. 2009. Satellite remote sensing applications for surface soil moisture monitoring: A review. Front. Earth Sci. China. 3(2): 237-247.

POTENTIAL FOR INTENSIFICATION OF MAIZE PRODUCTION WITH SDI

Freddie R. Lamm
Research Irrigation Engineer
Northwest Research-Extension Center
Kansas State University
Colby, Kansas
Voice: 785-462-6281
Fax: 785-462-2315
flamm@ksu.edu

SUMMARY

In 2017, a study was conducted at the KSU Northwest Research-Extension Center in Colby, Kansas to evaluate the potential for intensification of field corn production with subsurface drip irrigation. Experimental factors were three irrigation levels (115, 100, or 85% of ET-rain replacement), two high-yielding corn hybrids (Pioneer 1151 or Pioneer 1197) and three elevated plant densities (42,000, 38,000, or 34,000 plants/acre). Corn yields were not affected by irrigation level and significantly greater crop water productivity (CWP) was obtained by the lower 85% ET-rain irrigation treatment. There were significant differences between hybrids with Pioneer 1197 averaging nearly 9% more grain yield, primarily due to a greater number of kernels/ear. Increasing plant density from 34,000 to 38,000 or 42,000 plants/acres increased grain yield by 4% (10 bushels/acre). Seasonal profile soil water was relatively stable across irrigation treatments and plant densities further indicating that the 85% ET-Rain irrigation treatment was sufficient for this corn production intensification study. These results are part of a mult-year study. Unfortunately, the 2018 crop was lost due to wind and hail damage. More results are necessary before firm conclusions should be drawn.

INTRODUCTION

Crop production intensification is a key factor for addressing one of the greatest challenges of this century: feeding 9.5 billion people by the year 2050. Inherent in this challenge are the limitations of arable land as well as a shortage of fresh water sources. Ecologically, crop intensification can protect marginal lands from further development and save water resources. Intensification on a smaller land area also has potential to reduce crop production and crop protection inputs. These include seed, fertilizer, herbicides, pesticides, crop scouting, crop insurance, harvesting costs and any other input cost that has a fixed cost per land area basis.

Crop water productivity (CWP, also known as water use efficiency, WUE) is defined as the yield divided by the total water use:

$$CWP = \frac{Crop\,Yield}{Total\,Wate\,Use} \qquad \text{Eq 1.}$$

Thus, it can be easily recognized that either the numerator can be increased, or the denominator can be decreased to increase CWP. Often strategies to increase CWP concentrate on the denominator such as using deficit irrigation to reduce water withdrawals. Implicit with these strategies is the desire to not greatly reduce farm profitability by negatively impacting crop yields to a large extent. However, a traditional definition of deficit irrigation is a level of irrigation anticipated to reduce crop evapotranspiration (ETc) to less than the full potential amount. Since crop yield and ETc are typically linearly related, a reduction in ETc often means a reduction in crop yield. There are limitations to using the denominator to increase CWP, since the overall reason irrigation is practiced is to increase farm profitability. Actually, it can be shown that appropriate levels of irrigation can actually increase CWP (Figure 1).

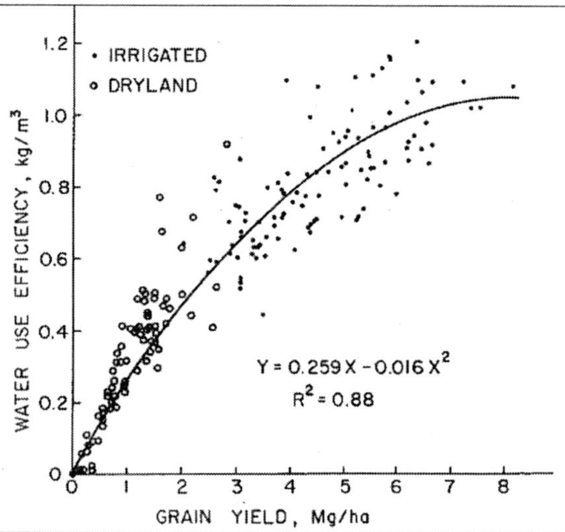

Figure 1. Wise use of irrigation can increase CWP, here expressed as WUE. (Data after Evett et al. 2014, using data from Musick et al., 1994).

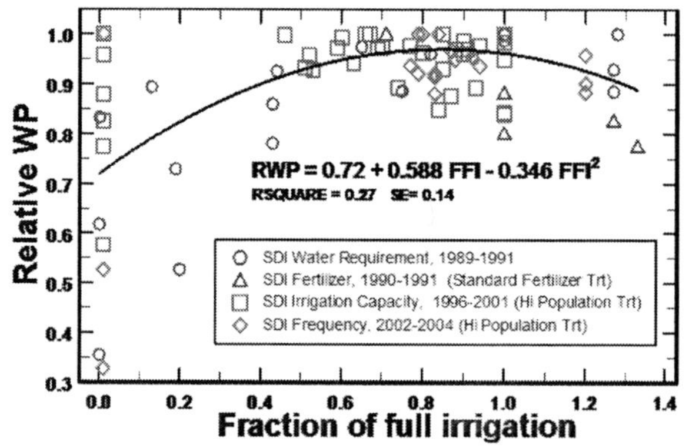

Figure 2. Crop water productivity maximized at about 80% of full irrigation in four different SDI studies at Colby, Kansas from 1989 through 2004. Graph summarized by Lamm and Rogers, 2014.

Subsurface drip irrigation (SDI) has great potential to optimize crop production at a greater level while efficiently using water. Earlier studies have indicated CWP could be maximized with SDI at about 80 percent of full irrigation for corn and still result in high crop yields (Figure 2 and Lamm and Rogers, 2014). Although the corn yields in these earlier studies were high (200-250 bushels/acre), with additional intensification greater yields are anticipated (consistently greater than 280 bushels/acre). These greater yields are not unrealistic, as SDI corn yields as high as 304 bushels/acre were obtained in a research study in Kansas in 1998 (Figure 3). Consistently increasing yields above 280 bushels/acre will require optimal fertilization. SDI allows for timely in-season fertigation (application through the subsurface driplines to the center of the root zone).

Plant density of modern corn hybrids can be increased to reasonably high levels without plant barrenness due to advances in genetics. In SDI studies at Colby, Kansas there appeared to be little yield penalty with greater plant density even when irrigation and precipitation was greatly limited (Figure 3). It is thought that even small amounts of water when applied daily with SDI can help alleviate some of the water stresses that would occur with other types of more infrequent irrigation (e.g., surface or sprinkler irrigation).

Plant hybrids also play a major role in high yielding systems. Pioneer 1151 has been shown to have a large kernel mass (≈380 mg) that is surprisingly stable across a wide range of irrigation regimes (Lamm, 2016). Pioneer 1197 is also a great hybrid which has been documented as a hybrid-of-choice in many high yield contests conducted by producers (Jeschke, 2017).

It is already known that intensification can have positive results. Comparing crop yields for the period pre- and post-introduction of commercial fertilizers is a prime example. Further investigations of increasing CWP with crop intensification are warranted. These efforts will concentrate on optimizing additional inputs: the focus of such efforts will be to increase the numerator of the CWP equation (crop yield).

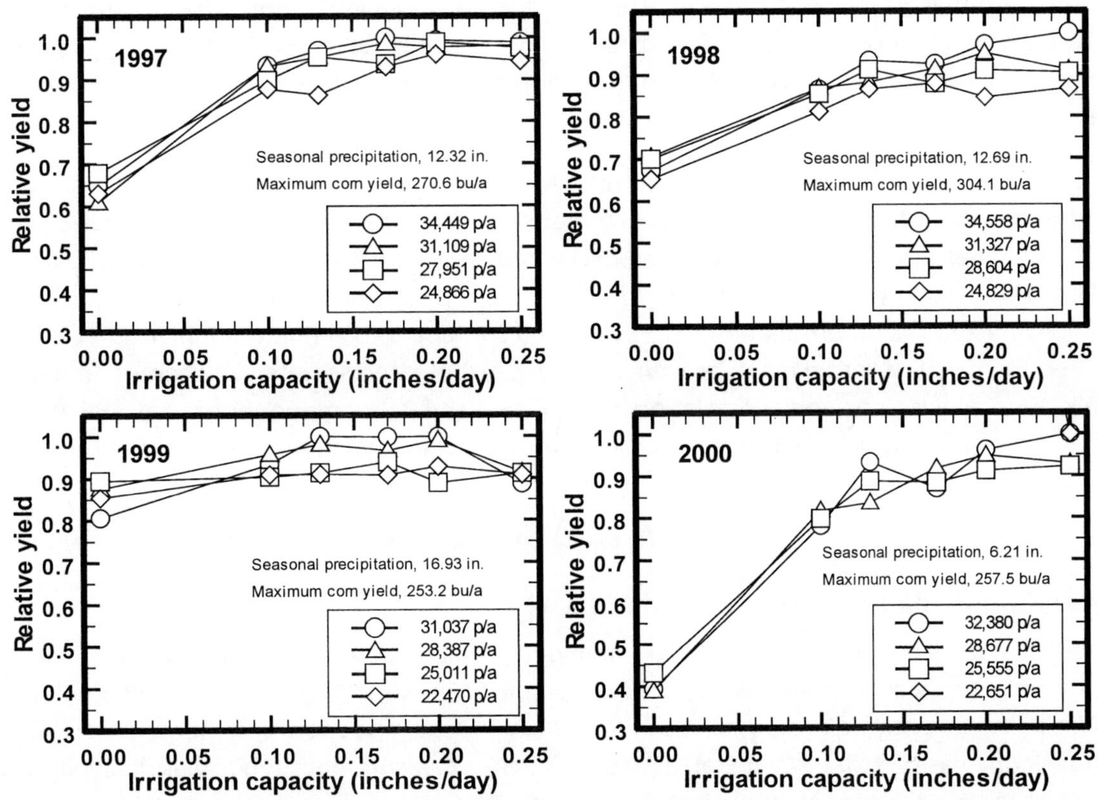

Figure 3. Maximum SDI corn grain yields ranging from 253.2 to 304.1 bushels/acre with modest irrigation capacities and in-season precipitation ranging from 6.21 to 16.93 inches at Colby, Kansas. Average in-season precipitation is approximately 12.3 inches. ***It can be noted that the greatest yield in most years was at the greatest plant density and that the maximum yield (304 bushels/acre) in 1998 still appears to be increasing.***

BRIEF DESCRIPTION OF PROCEDURES

A study was initiated in 2017 at the KSU Northwest Research-Extension Center at Colby Kansas to evaluate intensification of corn production with SDI. The study was also conducted in 2018 but the crop area was abandoned due to heavily damage from wind and hail on June 30. This report will only discuss the 2017 results. The site has a deep silt loam soil and a semi-arid climate. Experimental factors were limited to three irrigation levels, (115, 100, and 85% of ET-Rain), three elevated plant seeding densities (34,000, 38,000 and 42,000 seeds/acre) and two high yielding corn hybrids (Pioneer 1151 and 1197). Fertilization was not included as a study variable in this study due to space limitations at the site, but was fixed at an advanced level with in-season applications of nitrogen, phosphorus and potassium as well as micronutrients. This fertilization scheme was selected from one of the more promising schemes currently being evaluated with SDI at the Center (results are unpublished at this time).

The SDI system was installed in 2016 at a depth of 16 inches with a dripline spacing of 5 ft. The emitter spacing for this system is 12 inches and the emitter flowrate is 0.15 gallons/hour. The SDI application rate is 0.05 inches/hour. Since this system is subsurface, not wetting the soil surface, the irrigation frequency does not affect evaporative losses, so the SDI frequency was more frequent daily applications as governed by the irrigation regimes.

Soil water was measured in the complete root zone (0-8 ft) with a neutron probe periodically throughout the season to help quantify periods of water stress and to determine crop water use. Weather data was measured using an automated KSU weather station approximately 1500 ft. from the study site. Corn grain yield was determined by hand harvesting a representative sample at physiological maturity which enabled the determination of all corn yield components, (grain yield, plant density, ears/plant, kernels/ear, and kernel mass) as well as the important intermediate yield component, kernels/area. Data analyses included correlation of corn yield and yield components, seasonal water use, irrigation as affected by irrigation regime, corn hybrid and plant density.

RESULTS AND DISCUSSION

Weather Conditions and Irrigation Requirements

The 2017 season started out very wet (Figure 4) but ended with cropping season precipitation just slightly above normal. Calculated well-watered corn evapotranspiration (ET) was just slightly below normal. Planting was slightly delayed by wet conditions until May 9 and the crop emerged approximately 5 days late on May 20. Installation of soil water measurement tubes was delayed by wet weather and the initial soil water readings were on June 6. Relatively mild summer weather with timely rainfall and an extended grain-filling period resulted in an exceptionally good corn growing season. Irrigation amounts were 12.00, 14.50 and 16.75 inches for the 85, 100 and 115% ET - Rain irrigation treatments, respectively.

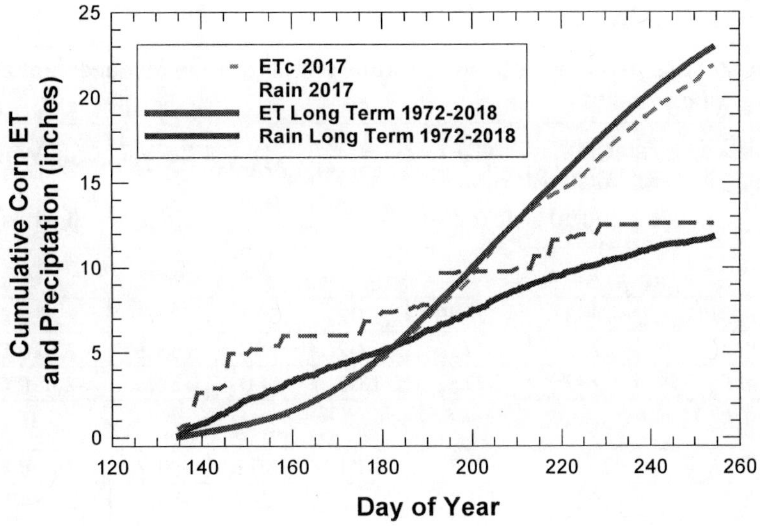

Figure 4. Cumulative calculated well-watered corn evapotranspiration (upper lines) and measured precipitation (lower lines) at Colby, Kansas for May 15 through September 11, 2017 as compared to the long term (1972—2018) average values.

Corn Yields, Yield Components, Water Use, and CWP

Corn yields were excellent in 2017 with the mean of 4 replications ranging from 269 to 314 bushels/acre for the various treatment combinations.

Yields were not affected by irrigation level (Table 1), which corresponded to earlier findings that SDI levels approximating 75 to 80 percent of full irrigation will maximize yields (Lamm and Rogers, 2014 and also illustrated in Figure 2). Irrigation increased crop water use, but this only reflects the higher irrigation amounts which were likely reflected in increased deep percolation. This is further emphasized by the statistically significantly greater crop water productivity (CWP) at the irrigation level designed to match 85 percent of ETc minus precipitation. The fact that greater irrigation levels were not necessary with crop intensification is important in that the denominator of CWP was not increased to result in greater corn yields.

There also was a strong hybrid effect on yield: Pioneer 1197 exceeded Pioneer 1151 by 24 bushels/acre, emphasizing that hybrid selection remains an important factor in intensively managed corn. This yield increase for Pioneer 1197 was caused primarily by greater number of kernels per ear. Pioneer 1197 also had higher crop water productivity (CWP) than Pioneer 1151, but crop water use was slightly greater with Pioneer 1197.

A plant density of 38,000 or 42,000 plants/acre resulted in significantly greater yield than 34,000 plants/acre, but crop water use was not affected at approximately 27.26 inches. Although the lower plant density had greater number of kernels per ear, this value was not able to compensate for the lower plant density. This reflects a growing understanding that maximizing irrigated corn yields often requires maximizing the intermediate yield component of kernels/area (i.e., plant density multiplied by ears per plant multiplied by kernels per ear).

Table 1. Corn yield and water use parameters in an SDI study with intensive management at the KSU-NWREC, Colby, Kansas in 2017.							
Main Effect	**Grain yield (bu/a)**	**Plant Density (p/acre)**	**Ears /Plant**	**Kernels /Ear**	**Kernel Mass (mg)**	**Crop Water Use (inches)**	**Crop Water Productivity (lb/a-in)**
Effect of Irrigation Level							
Irr 1, 115% ETc *(16.75 inches)*	293	37679	1.02	587	33.3	**29.19 A**	**563 C**
Irr 2, 100% ETc *(14.50 inches)*	292	37716	1.02	586	33.3	**27.10 B**	**605 B**
Irr 3, 85% ETc *(12.00 inches)*	289	37752	1.01	580	33.6	**25.50 C**	**638 A**
Effect of Hybrid							
Hybrid 1, Pioneer 1151	**280 B**	37873	1.01	**556 B**	33.7	**26.68 B**	**590 B**
Hybrid 2, Pioneer 1197	**304 A**	37558	1.02	**612 A**	33.1	**27.84 A**	**614 A**
Effect of Plant Density							
Plant Density 1, 42K p/a	**296 A**	**41600 A**	0.99	**552 C**	33.0	27.35	607
Plant Density 2, 38K p/a	**295 A**	**37788 B**	1.02	**587 B**	33.3	27.30	608
Plant Density 3, 34K p/a	**285 B**	**33759 C**	1.03	**614 A**	34.0	27.14	591
Data for a main effect within a column followed by different letters are significantly different at P=0.05 level.							

Seasonal Available Soil Water Change as Affected by Irrigation Level and Plant Density

As an additional illustration of the adequacy of the 85% ET-Rain irrigation treatment, a time series of available soil water (ASW) for the three irrigation treatments as affected by the highest and lowest plant densities was graphed (Figure 5). There was less than 0.47, 0.96, and 1.13 inches differences in the ASW values for a given profile depth, 2, 4, or 8 ft, respectively, on a given date as affected by the three irrigation levels or the highest and lowest plant density. Seasonal ASW for the three profile depths, 2, 4 and 8 ft all had narrow ranges, 3.73 to 4.84 inches/2 ft, 7.07 to 9.05 inches/4 ft, and 12.99 to 16.27 inches/8 ft, respectively. The small differences and the narrow range of values indicate that there was very little water stress experienced by the corn plants. The larger range of ASW values for the 8 ft depth suggests some of the applied water was moving downward for the higher irrigation treatments.

CONCLUSIONS

In this first year (2017) of a multi-year study, corn grain yield was unaffected by irrigation levels ranging from 85 to 115% of ET-Rain replacement. Significantly greater CWP was obtained by the lower 85% ET-rain treatment.

Selections of corn hybrid resulted in significant grain yield differences which emphasize the importance of this factor in intensification studies. Although both hybrids (Pioneer 1151 and Pioneer 1197) yielded exceptionally well, Pioneer 1197 averaged 9% greater grain yield, primarily due to a greater number of kernels/ear.

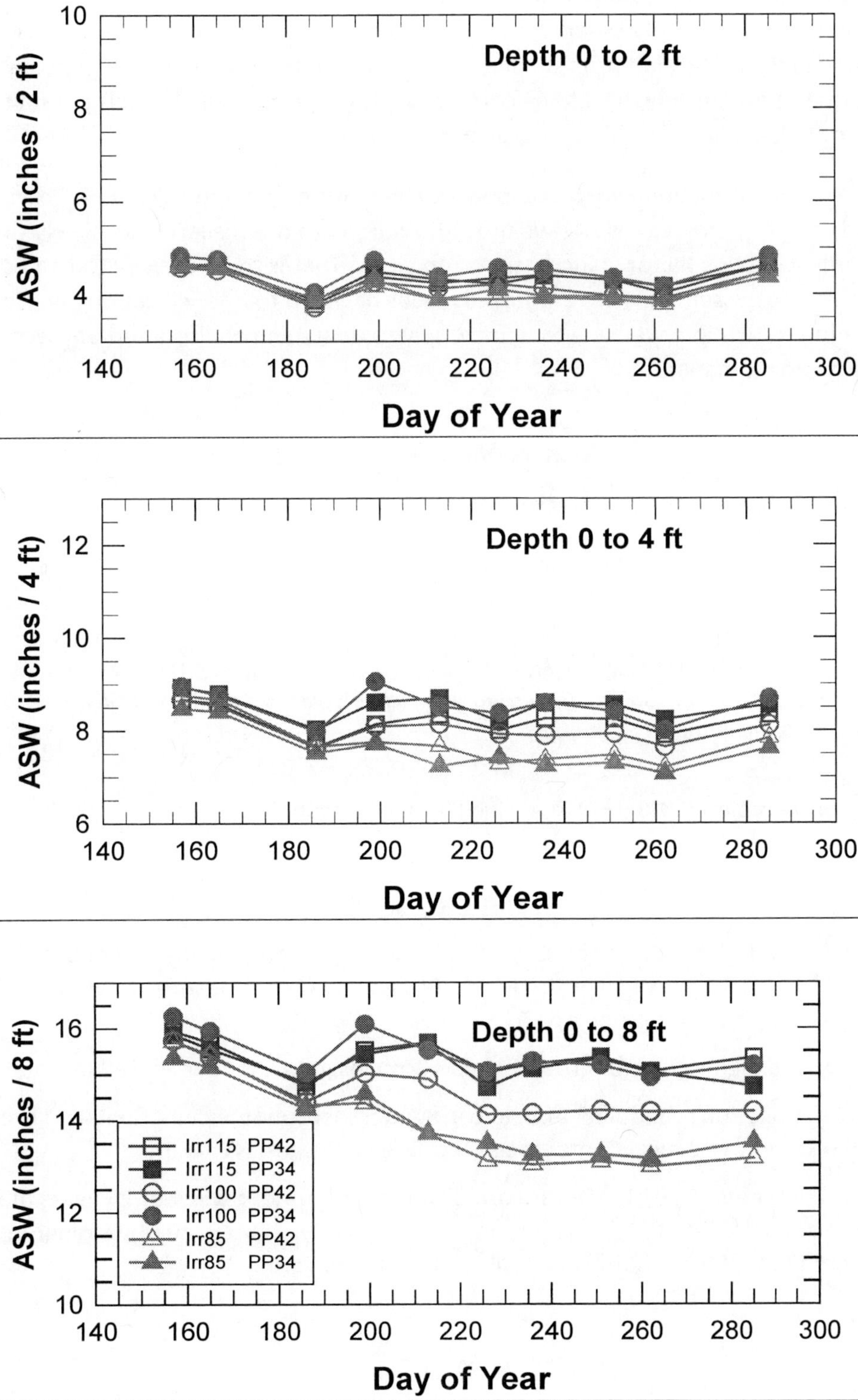

Figure 5. Available soil water for various soil profile depths for the three irrigation levels (115, 100, or 85% of ET –Rain) as affected by the highest and lowest corn plant densities (42,000 and 34,000 plants/acre).

Increasing plant density from 34,000 to 38,000 or 42,000 plants/acre also significantly increased grain yield. Although there were slight statistically nonsignificant reductions in kernel mass and statistically less kernels/ear for the greater plant densities , the 34,000 plant/acre plant density could not compensate enough in grain yield for its smaller number of plants. This reflects a growing understanding that to optimize high yielding corn production systems, the kernels per unit land area must be maximized (i.e., plant density x ears/plant x kernels/ear).

Intensification of corn production with SDI appears to be a promising approach to improving the use of our limited land and water resources. As we move forward with the research, it is likely that some other inputs will become a limiting factor in increasing crop yield. That is to be anticipated and can be addressed at that time. Essentially, all farming advances have led to intensification. As we move towards further intensification, we have to work wisely to do it in an environmentally, ecologically, and economically acceptable manner.

ACKNOWLEDGEMENTS

This paper is part of SDI technology transfer effort beginning in 2009 involving Kansas State University, Texas A&M University and the USDA-ARS and is funded by the Ogallala Aquifer Program. To follow other activities of this educational effort, point your web browser to http://www.ksre.ksu.edu/sdi/.

This paper also is being presented as part of a special session of USDA-NIFA multi-state research project W-3128, ***Scaling Microirrigation Technologies to Address the Global Water Challenge***, which the author is a member.

Ongoing funding for the project (2018⟶) is provided by the Foundation for Food and Agricultural Research (FFAR).

REFERENCES

Evett, S. R., P. D. Colaizzi, S. A. O'Shaughnessy, F. R. Lamm, T. J. Trout, and W. L. Kranz. 2014. The future of irrigation on the U.S. Great Plains. In: Proc. 26th annual Central Plains Irrigation Conference, Feb. 25-26, 2014, Burlington, Colorado. pp. 2-25.

Jeschke, M. 2017. Managing corn for greater yield. Crop Insights, Pioneer Hibred International. 6 pp.

Lamm, F. R. 2016. Irrigation scheduling of field corn under institutional constraints. In: Proc. 2016 Irrigation Association Technical Conference, Dec. 6-8, Las Vegas, NV. 8 pp.

Lamm, F. R. and D. H. Rogers. 2014. SDI for corn production -A brief review of 25 years of KSU research. In: Proc. 26th annual Central Plains Irrigation Conference, Feb. 25-26, 2014, Burlington, Colorado. Available from CPIA, 760 N. Thompson, Colby, Kansas. pp. 55-66.

High-density grapefruit production in open hydroponics system

Rhuanito S. Ferrarezi, Kayla A. Thomason, Mark A. Ritenour, and Alan L. Wright

University of Florida, Institute of Food and Agricultural Sciences, Indian River Research and Education Center, Fort Pierce, FL 34945. Corresponding author: rferrarezi@ufl.edu

Abstract

Precise irrigation and fertigation management provide a less-limiting environment to roots while minimizing over irrigation and leaching of nutrients. This concept can improve tree growth in the presence of HLB and help optimize water and nutrient use. Higher tree density can increase fruit yield per area under high HLB pressure. This study evaluated the efficiency of open hydroponics on 'Ray Ruby' grapefruit production under different irrigation systems and tree density. We tested a combination of rootstocks (Sour orange and US897), tree spacing [standard and high density staggered (HDS)], fertilization (dry granular and fertigation), and irrigation systems (drip and microjet), arranged on five treatments: RR/SO_STD_dry_MS) SO + standard spacing + dry granular fertilizer + micro jet, RR/SO_HDS_fert_DD) SO + HDS + fertigation + drip, RR/897_HDS_fert_MS) US897 + HDS + fertigation + microjet, RR/897_HDS_fert_DD) 'US897' + HDS + fertigation + drip, and RR/SO_HDS_fert_MS) SO + HDS + fertigation + microjet. Foliar nutrient, insecticide and fungicide were sprayed using standard practices. We scouted for psyllids, leaf minors and other citrus pests monthly. HLB incidence reached 100% after five years of planting. Trunk diameter and canopy volume increased over time, and were higher on RR/SO_STD_dry_MS compared to other treatments. Total number of fruit and fruit yield were 226% and 183% higher in 2016 compared to 2015. RR/SO_STD_dry_MS yielded 7,309 kg/ha in 2017 compared to an average of 22,153 kg/ha for other treatments. Soluble solid contents, acidity, and ratio were not significant (p>0.05). Total solids per hectare was always low in RR/SO_STD_dry_MS. High density staggered (HDS) planting resulted in higher fruit yield, irrespective of rootstock and irrigation system, representing an important advance to the grapefruit production system. However, labor cost and effect on plant growth over time still need to be determined for commercial recommendation.

Keywords

High density, plant performance, fruit yield, fruit quality.

Introduction

Citrus (*Citrus* spp.) is Florida's most important agricultural commodity. The state produces citrus for different markets: round oranges (*C. sinensis*) for juice; navels, mandarins (*C. reticulata*), grapefruit (*C. paradisi*) and lemons (*C. limon*) for the fresh fruit industry; and lemons for extracting peel oil for processing. However, citrus production in Florida has declined drastically since 2005 due to citrus greening or Huanglongbing (HLB), caused by *Candidatus* Liberibacter asiaticus (*C*Las), in addition to canker (*Xanthomonas axonopodis*), urban development, and recent hurricanes (Morgan et al., 2009; Kadyampakeni et al., 2016).

The decline in citrus productivity has resulted in an aggregate loss of $ 3 billion in citrus growers' revenue from 2006 through 2014 (Hodges et al., 2014). The Indian River Citrus District maintains approximately 13.5% of Florida's total citrus acreage inventory since 2000/01 (excluding the crop years 2004/05 due to hurricane damage) with drastic reduction in recent years. The number of grapefruit trees planted has increased since 2000/01; conversely, total bearing acreage and total bearing trees in the state have both declined by 38% in the last decade (USDA, 2017). The total number of boxes of grapefruit produced in Florida has decreased from 40.9 million in 2003/04 to 10.8 million in 2015/16 (USDA, 2017). Excluding season 2004/05, Florida's average grapefruit yield is quickly decreasing over time – from 497 boxes/acre in 2000/01 to only 288 boxes/acre in 2016/17 (USDA, 2017).

Once citrus is affected by HLB, it becomes gradually less productive. HLB alters the plants' physiology (reducing photosynthesis and xylem flux) and morphology (e.g. phloem translocation, bloated and corky veins, root length and density) affecting nutrient uptake, translocation, and utilization (Kadyampakeni et al., 2014).

Nutrients are vital in disease control as nutrients influence plant resistance and pathogen growth (Handique et al., 2012). Huanglongbing inhibits root growth, reduces nutrient uptake, promotes leaf and fruit drop, results in deformed fruit with unpleasant flavor, and results in whole tree decline that is often lethal (Handique et al., 2012). Fibrous root length and density are not consistent in HLB-affected trees and are affected by distance from the tree trunk (lateral and vertical), type of rootstock, and age of the tree (Kadyampakeni et al., 2014). Moreover, HLB induced reduction in root density reduces water and nutrient uptake (Graham et al., 2013). The decline in fibrous density is manifested as weak canopy volume, poor fruit quality, and yield losses.

Precise irrigation and fertigation management provide a non-limiting environment to roots while minimizing over irrigation and leaching of nutrients. This concept can improve tree growth in the presence of HLB and help optimize water and nutrient use. Higher tree density can increase fruit yield per area under high HLB pressure.

This study evaluated the efficiency of open hydroponics on grapefruit cultivated under different tree spacing (standard and high-density), fertilizer sources (dry granular and water-soluble fertilizer) and irrigation methods (drip and micro jet).

Material and Methods

The study was conducted at the UF/IFAS Indian River Research and Education Center in Fort Pierce, FL (lat. 27°26'01.8" N, long. 80°26'49.80" W and elevation 10 m). 'Ray Ruby' grapefruit trees were planted in Sept/2013 (total of 2,769 trees in 3.23 hectares).

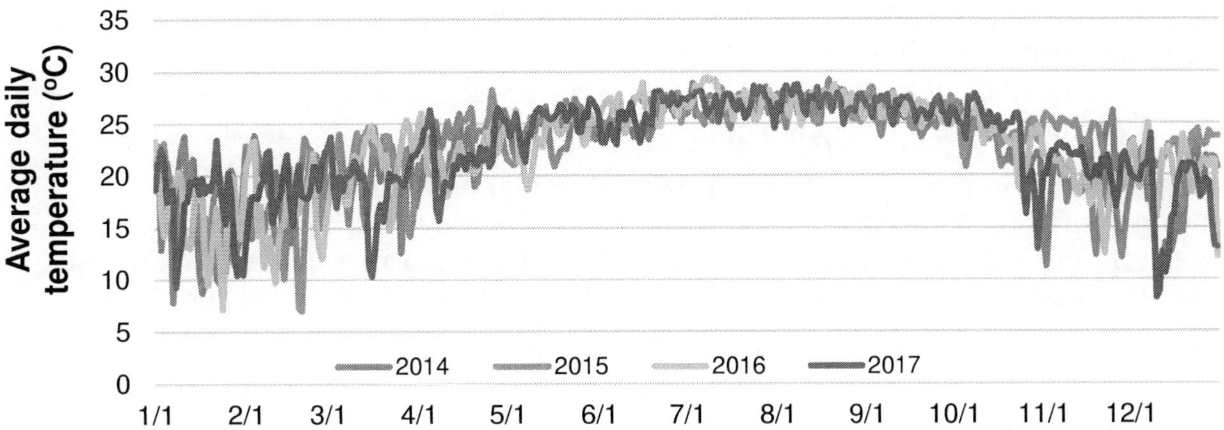

Fig. 2. Average daily temperature during the 2014-2017 experiment. Fort Pierce, FL.

We tested a combination of rootstocks {'Sour orange' [*C. aurantium*] and 'US897' [Cleopatra (*C. reticulata*) × Flying Dragon (*Poncirus trifoliata*)]}, tree spacing [standard and high-density staggered planting (HDS)], fertilization (dry granular and water-soluble fertilizer), and irrigation systems (drip and micro jet), arranged on five treatments:

- RR/SO_STD_dry_MS) 'Ray Ruby' on Sour Orange + standard spacing (3.8 × 7 m, 358 trees/ha) + 16N-2.2P-8.3K dry granular fertilizer applied in-ground + micro jet (one emitter per tree; blue microsprinklers - 40.5 LPH @ 20 psi)
- RR/SO_HDS_fert_DD) 'Ray Ruby' on Sour Orange + HDS [(2.74 × 1.5 × 0.9 m) × 6.1 m, 953 trees/ha)] + 15N-2.6P-22.4K applied by fertigation + drip irrigation (four emitters per tree, installed on double rows; blue dripper - 3.8 LPH)
- RR/897_HDS_fert_MS) 'Ray Ruby' on US-897 + HDS + 15N-2.6P-22.4K applied by fertigation + micro jet (same as above)
- RR/897_HDS_fert_DD) 'Ray Ruby' on US-897 + HDS + 15N-2.6P-22.4K applied by fertigation + drip (same as above)
- RR/SO_HDS_fert_MS) 'Ray Ruby' on Sour Orange + HDS + 15N-2.6P-22.4K applied by fertigation + micro jet (same as above)

The experimental design was a completely randomized with five replications.

We measured HLB incidence, tree size, leaf macro and micronutrient concentrations, total number of fruit and fruit yield, fruit quality parameters (soluble solid contents, acidity, and ratio) and calculated total solids per hectare. Huanglongbing diagnosis of mature leaves for *C*Las titer and activity was measured annually by using the quantitative polymerase chain reaction (q-PCR) (Li et al. 2006). A total of 4-6 mature leaves was collected from summer flush in each of all four cardinal sections per tree, from which the petiole/midribs was used for *C*Las detection.

Tree size was assessed every year by measuring trunk diameter (~8 cm above the bud union), tree height to top of canopy (not including height of vigorous shoots that extend significantly past the top of the canopy), and canopy diameter (in parallel and perpendicular to the tree row). Canopy volume was calculated using the formula: [(diameter parallel to row × diameter perpendicular to row) × height] ÷ 4.

Leaf and soil nutrient concentrations was determined annually in August (on spring flush). Approximately 20-30 mature leaves were collected in different parts of the tree. Leaf tissue samples were dried for 72 h at 65 °C and analyzed by Inductively Coupled Plasma Atomic Emission Spectroscopy (ICP-AES) (Obreza and Morgan, 2008).

Total number of fruit and fruit diameter were determined by harvesting all tree fruit and passing them through an optical sorter (Autoline, Reedley, CA) mounted on a trailer. Measurements were converted into number of fruit per carton within a commercial category. Total fruit yield was determined by direct weighing all the fruit per tree.

Random samples of 20 fruit from each experimental unit were collected for fruit quality analysis on each year. The fruit samples were weighed, and fruit diameter at the equator was measured with a digital caliper. The fruit were weighed and juiced using a press juicer; then, juice was weighed, and expressed as a percentage of the total fruit weight. Total soluble solids content was determined with a digital refractometer (HI96801; Hanna Instruments, Woonsocket, RI) using a few drops of juice. The total acidity was determined by titration of 5 ml of fruit juice with 0.1 N sodium hydroxide (NaOH) to pH 8.1. Total solids per hectare was calculated as: (% juice in fruit ÷ 100) × (soluble solids content ÷ 100) × Yield (kg/ha).

Foliar nutrient, insecticide and fungicide were sprayed using standard practices. We scouted for psyllids, leaf minors and other citrus pests monthly.

Data were analyzed by normality (Proc univariate), analysis of variance (ANOVA) (Proc GLM), and Tukey's multiple comparisons test (Proc lsmeans) using SAS (v. 9.4; SAS Institute, Cary, NC). Probability values ≤ 0.05 were considered statistically significant.

Results and Discussion

HLB incidence increased over time, reaching 100% after 5 years of planting (Fig. 2). The disease progression warrants the rapid HLB spreading in the state of Florida.

Trunk diameter and canopy volume increased over time, and were higher on RR/SO_STD_dry_MS compared to other treatments (p<0.001, Fig. 3). Such response was expected, since trees planted on lower densities receive more solar radiation, water and nutrients, not competing with surrounding trees as the high density plantings.

Leaf macro and micronutrient concentrations were influenced by treatment and sampling date (data not shown).

Total number of fruit and fruit yield were 226% and 183% higher in 2016 compared to 2015. RR/SO_STD_dry_MS yielded 7,309 kg/ha in 2017 compared to an average of 22,153 kg/ha for other treatments. In 2017, data was compromised by Hurricane Irma, which caused 50%-70% fruit drop (visual observation) (Fig. 4).

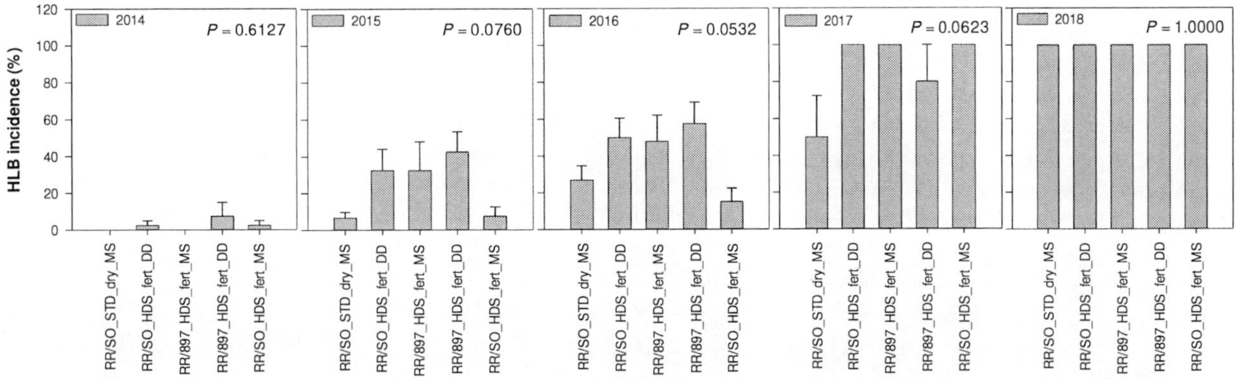

Fig. 2. Huanglongbing (HLB) incidence of 'Ray Ruby' grapefruit trees. Treatments: RR/SO_STD_dry_MS) Sour Orange (SO) + standard spacing [358 trees/ha] + dry granular fertilizer + micro jet, RR/SO_HDS_fert_DD) SO + high density staggered (HDS) [(953 trees/ha)] + fertigation + drip, RR/897_HDS_fert_MS) US-897 + HDS + fertigation + micro jet, RR/897_HDS_fert_DD) US-897 + HDS + fertigation + drip, and RR/SO_HDS_fert_MS) SO + HDS + fertigation + micro jet.

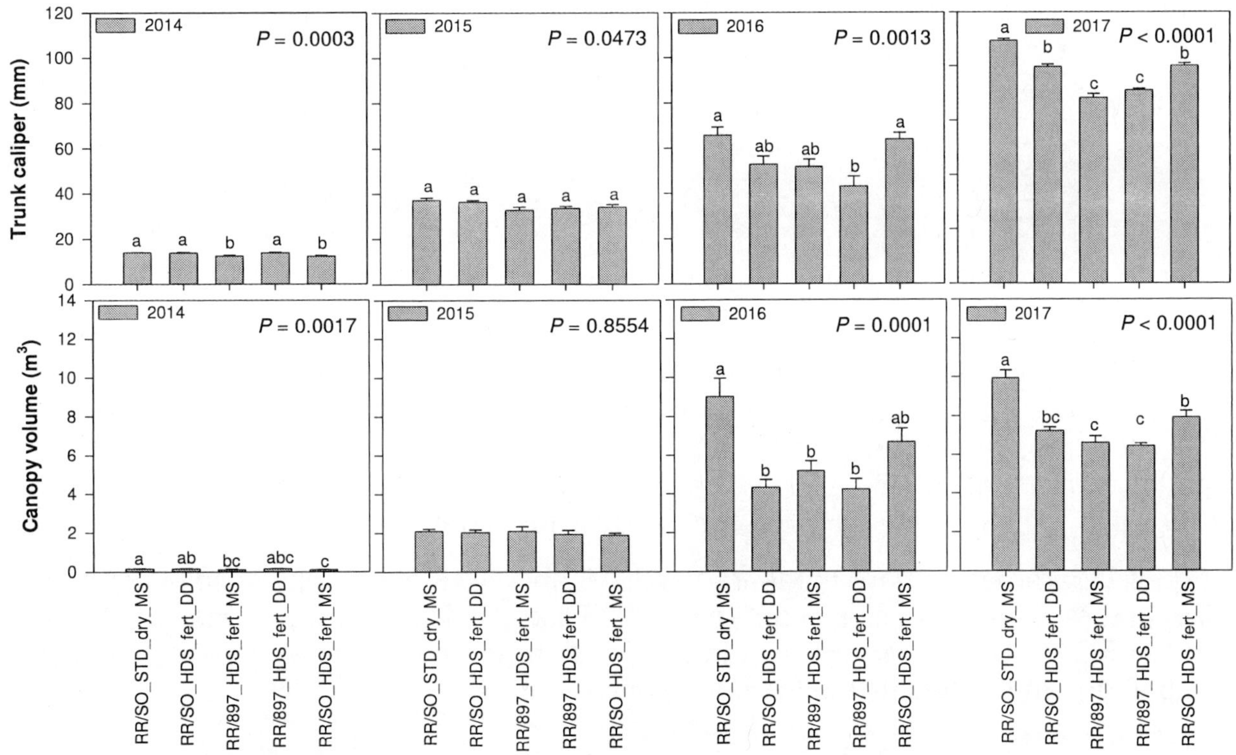

Fig. 3. Trunk caliper and canopy volume of 'Ray Ruby' grapefruit trees. Treatments: RR/SO_STD_dry_MS) Sour Orange (SO) + standard spacing [358 trees/ha] + dry granular fertilizer + micro jet, RR/SO_HDS_fert_DD) SO + high density staggered (HDS) [(953 trees/ha)] + fertigation + drip, RR/897_HDS_fert_MS) US-897 + HDS + fertigation + micro jet, RR/897_HDS_fert_DD) US-897 + HDS + fertigation + drip, and RR/SO_HDS_fert_MS) SO + HDS + fertigation + micro jet.

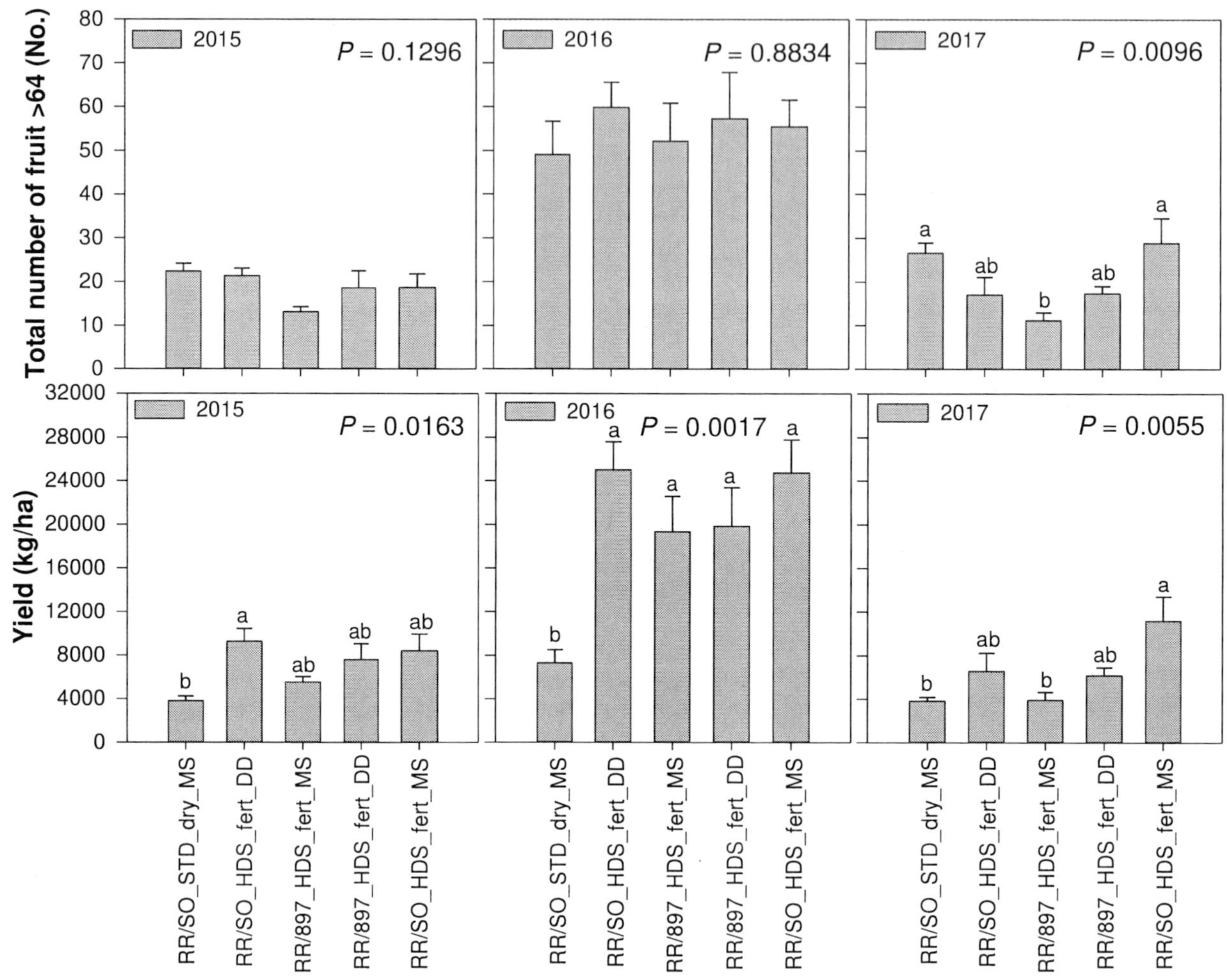

Fig. 4. Total number of fruit and fruit yield of 'Ray Ruby' grapefruit trees. Treatments: RR/SO_STD_dry_MS) Sour Orange (SO) + standard spacing [358 trees/ha] + dry granular fertilizer + micro jet, RR/SO_HDS_fert_DD) SO + high density staggered (HDS) [(953 trees/ha)] + fertigation + drip, RR/897_HDS_fert_MS) US-897 + HDS + fertigation + micro jet, RR/897_HDS_fert_DD) US-897 + HDS + fertigation + drip, and RR/SO_HDS_fert_MS) SO + HDS + fertigation + micro jet.

Fruit quality parameters were measured in 2016 and 2017 only. Soluble solid contents, acidity, and ratio were not significant (p>0.05). Total solids per hectare was constantly low in RR/SO_STD_dry_MS (Fig. 5). In 2017, all parameters were considerably lower than 2016 due to the negative effects of Hurricane Irma on fruit quality.

Conclusions

High density staggered (HDS) planting resulted in higher fruit yield, irrespective of rootstock and irrigation system, representing an important advance to the grapefruit production system. However, labor cost and effect on plant growth over time still need to be determined for commercial recommendation.

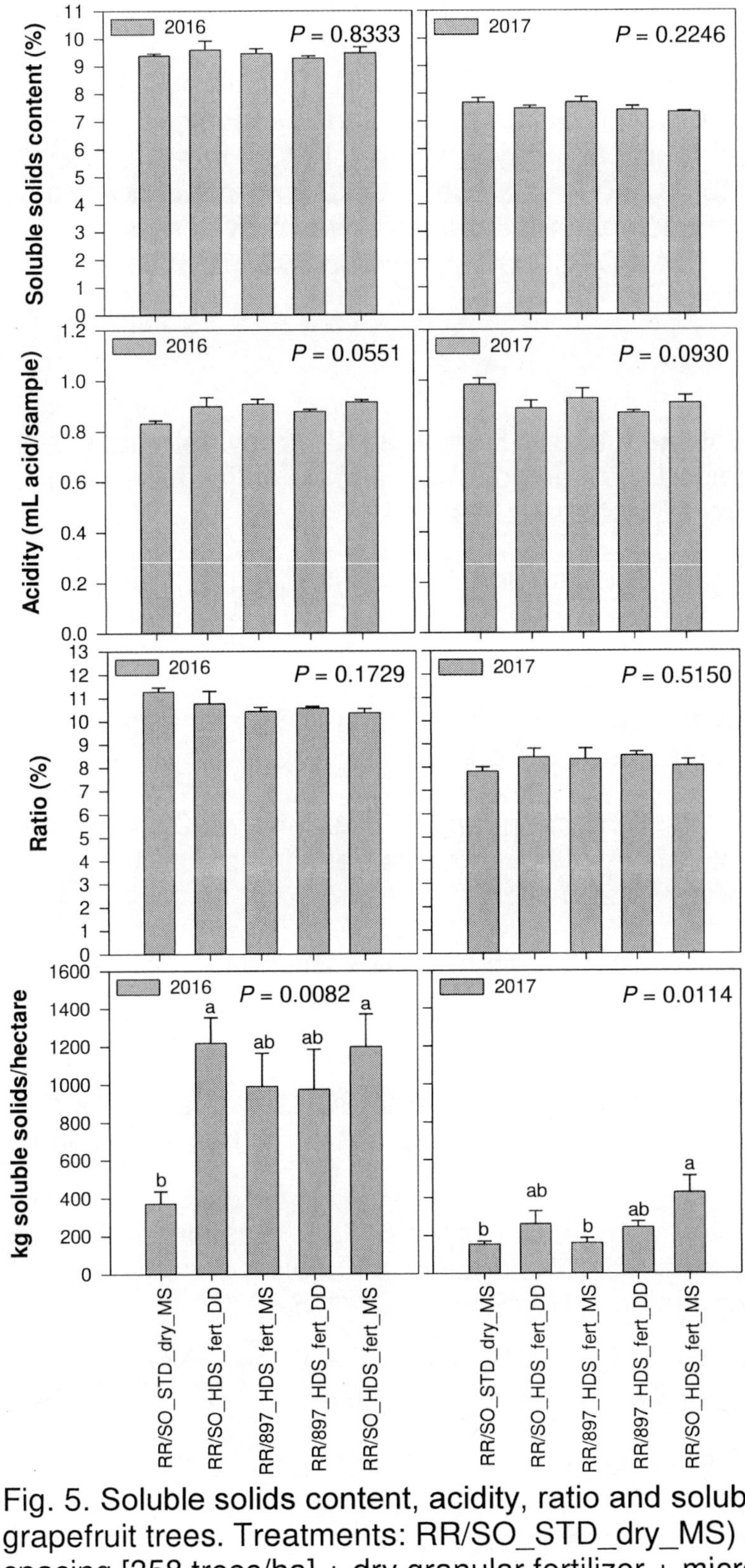

Fig. 5. Soluble solids content, acidity, ratio and soluble solids per hectare of ‘Ray Ruby’ grapefruit trees. Treatments: RR/SO_STD_dry_MS) Sour Orange (SO) + standard spacing [358 trees/ha] + dry granular fertilizer + micro jet, RR/SO_HDS_fert_DD) SO + high density staggered (HDS) [(953 trees/ha)] + fertigation + drip, RR/897_HDS_fert_MS) US-897 + HDS + fertigation + micro jet, RR/897_HDS_fert_DD) US-897 + HDS + fertigation + drip, and RR/SO_HDS_fert_MS) SO + HDS + fertigation + micro jet.

Acknowledgements

We thank R. Cave, R. Burton, C. King, R. Lesmes, D. Phuyal, H. Tom James, T. Meadows, J. Stephens and D. Davis for technical assistance; and KeyPlex micronutrients, Valent, Tessenderlo Group, UPI, Bayer, Hunter, FMC. Harrell's, ICL Specialty Fertilizers, Plant Food Systems and Arysta LifeScience for product donation toFerrarezi's Citrus Horticulture Lab. Funding for this research was provided by UF/IFAS, UF/IFAS Citrus Initiative, USDA-FDACS-SCBG (award #00092195).

References

Graham, J.H., E.G. Johnson, T.R. Gottwald, and M.S. Irey. 2013. Presymptomatic fibrous root decline in citrus trees caused by huanglongbing and potential interaction with *Phytophthora* spp. *Journal of Plant Disease* 97:1195–1199.

Handique, U., R. Ebel, and K.T. Morgan. 2012. Influence of soil-applied fertilizer on greening development in new growth flushes of sweet orange. *Proc. Fla. State. Hort. Soc.* 125:112:9–14.

Hodges, A.W., M. Rahmani, T.J. Stevens, and T.H. Spreen. 2014. Economic impacts of the Florida citrus industry in 2012-13. Gainesville, Florida.

Kadyampakeni, D.M., K.T. Morgan, A.W. Schumann, and P. Nkedi-Kizza. 2014. Effect of irrigation pattern and timing on root density of young citrus trees infected with huanglongbing disease. *HortTechnology* 24:209–221.

Kadyampakeni, D.M., K.T. Morgan, and A.W. Schumann. 2016. Biomass, nutrient accumulation and tree size relationships for drip- and microsprinkler-irrigated orange trees. *Journal of Plant Nutrition* 39:5, 589–599.

Li, W., J.S. Hartung, and L. Levy. 2006. Quantitative real-time PCR for detection and identification of Candidatus *Liberibacter* species associated with citrus huanglongbing. *Journal of Microbiological Methods* 66(1) 104-115.

Morgan, K.T., T.A. Wheaton, S.C. William, and L.R. Parsons. 2009. Response of young and maturing citrus trees grown on a sandy soil to irrigation scheduling, nitrogen fertilizer rate, and nitrogen application method. *HortScience* 44:145–150.

Obreza, T.A. and K.T. Morgan. (eds.) 2008. Nutrition of Florida Citrus Trees. Univ. Florida, Inst. Food Agr. Sci., Electronic Data Info. Source, SL 253.

U.S. Department of Agriculture. 2017. Florida Citrus Statistics 2015-2016. 31 May 2017. Available at https://www.nass.usda.gov/Statistics_by_State/Florida/Publications/Citrus/Citrus_Statistics/2015-16/fcs1516.pdf

Presented at 2018 Irrigation Show & Education Conference, Dez 3-7, 2018
Long Beach Convention Center, Long Beach, California, USA

DRIPPERS PERFORMANCE APPLYING TREATED SWINE EFFLUENT

Gustavo H. S. Vieira. Full Professor. PhD.
IFES *campus* Santa Teresa, Rod. ES 080, km 93, São João de Petrópolis, Santa Teresa-ES, Brazil. 29.660-000. e-mail: ghsvieira@gmail.com

Carlos M. M. Poloni. Agronomy undergraduate student.

IFES *campus* Santa Teresa – Rod. ES 080, km 93, São João de Petrópolis, Santa Teresa-ES, Brazil. 29.660-000.

André D. B. Garcia. Agronomy undergraduate student.
IFES *campus* Santa Teresa – Rod. ES 080, km 93, São João de Petrópolis, Santa Teresa-ES, Brazil. 29.660-000.

Paola A. V. Lo Monaco. Full Professor, PhD.

IFES *campus* Santa Teresa – Rod. ES 080, km 93, São João de Petrópolis, Santa Teresa-ES, Brazil. 29.660-000.

Ismail R. Haddade. Professor, PhD.

IFES *campus* Santa Teresa – Rod. ES 080, km 93, São João de Petrópolis, Santa Teresa-ES, Brazil. 29.660-000.

Eduardo F. Oza. Agronomy undergraduate student.
IFES *campus* Santa Teresa – Rod. ES 080, km 93, São João de Petrópolis, Santa Teresa-ES, Brazil. 29.660-000.

Millena M. dos Santos. Agronomy undergraduate student.
IFES *campus* Santa Teresa – Rod. ES 080, km 93, São João de Petrópolis, Santa Teresa-ES, Brazil. 29.660-000.

Abstract: *This study aimed to evaluate the performance of drippers used in fertigation by applying swine wastewater after treatment in anaerobic bamboo filters at the Federal Institute of Espírito Santo, Santa Teresa Campus, Brazil. The effluent from the institution 's swine sector was stored in fiberglass reservoirs with a capacity of 100 L and conducted by gravity for anaerobic bamboo filters with a capacity of 5.8 L each one. After filtration, the effluent was distributed to the drip hoses supported on a workbench, using two different types of drippers (G1 and G2). We evaluated the Distribution Uniformity Coefficient (DU), the Flow Rate Variation Coefficient (Qvc) and the Relative Flow Rate of the emitters (Qr), which did not show changes, even after 300 h of use. Thus, we conclude that swine wastewater treated by anaerobic* *bamboo filters, can be conducted in drip irrigation systems, using both the self-compensating and the non-compensating models, without compromising the distribution uniformity.*

Keywords: uniformity, wastewater, drip irrigation, clogging.

INTRODUCTION

The pork production is one of the most important agricultural activities in the Brazilian territory. According to data from the Brazilian Association of Animal Proteins (ABPA), Brazil, in 2016, ranked as the fourth largest producer of pork in the world, with a production of 3,731 thousand tons, behind China, the European Union and the United States (ABPA, 2017).

Due to the great importance of this activity, aspects such as swine management, production chain management and manure management must always be observed, otherwise they can cause enormous damage to producers, consumers of the product and the environment.

Among the aspects to be observed is the management of manure, since the average production of liquid pig manure is 14 L day^{-1} per animal (Konzen, 2006). Considering that some 21 million animals were slaughtered in the first half of 2017 (IBGE, 2017), there was at least a daily production of approximately 294 million liters of swine wastewater (SWW).

When improperly disposed of, pork wastewater can cause many environmental problems, ranging from soil leaching and contamination, to the pollution of water resources. Faced with this problem, several studies are being developed with the purpose of reducing the impacts caused by this activity.

Among the studies and alternatives, the utilization of swine wastewater (SWW) as an organic fertilizer for several crops is worth mentioning (Antonelli et al., 2014; Costa et al., 2017; Lo Monaco et al., 2017; Menezes et al., 2017). However, in spite of the several papers indicating the use of SWW, few are those that relate their treatment for use in fertigation, with drip application, as well as its effects on these equipments.

Due to the risk of drippers clogging when using wastewater, it is necessary to perform a treatment before these waters are conducted in localized irrigation systems.

Among the simple solutions, proposals for the treatment of wastewater rich in organic material, as is the case of those coming from smaller swine farms, we highlight the anaerobic filter of upward flow. According to Tonetti et al. (2011) anaerobic filters are a low-cost option in both the construction and operational aspects, removing approximately 70% of the organic matter and producing a small amount of sludge.

One of the impediments to the adoption of real-scale anaerobic filters refers to the cost of the filler as the carrier medium, usually the gravel, which may cost the same value as the filter construction (Tonettti et al., 2011). In this sense, alternative materials to the gravel become essential, especially if they are low cost, accessible and allow high efficiency in the treatment of wastewater. Due to the high availability of bamboo in the Centro Serrana region of Espírito Santo state, Brazil, this study aimed to evaluate the performance of different models of drippers by applying swine wastewater treated in anaerobic filters containing bamboo rings as support material.

MATERIAL AND METHODS

The study was carried out at the Instituto Federal do Espírito Santo, Campus Santa Teresa, Espírito Santo state, Brazil (19° 48 'S, 40° 40' W), at 130 meters of altitude above sea level. The climate of the region, according to the classification of Köppen, is classified as Aw, characterizing as megathermic (tropical humid), with dry winter and maximum rains in the summer.

The SWW used in the experiment came from the medium-sized animal sector of the campus. The effluent was stored in a 500 L capacity asbestos tank, installed in a structure 1 m above the filters, causing them to be fed by gravity, as shown in Figure 1A.

The equipment used for effluent treatment consisted of four upflow anaerobic filters units, constructed with 100 mm diameter PVC pipe segments, with a capacity of 5.8 L each, as can be seen in Figure 1B. As filter material we used circular segments of bamboo with average dimensions of 2 cm in height and 3 cm in diameter.

Figure 1. Effluent storage, distribution and filtration structure (A) and top view of the filter showing the filter material (B).

The time of arrest or permanence of SWW in the filters was in accordance with the recommendation of NBR 7229 (1993), which suggests a period of at least 12 hours to complete its treatment completely, without compromising the effluent quality.

For the wastewater characterization, before and after filtration, the following characteristics were analyzed: total solids (TS) and suspended solids (SS) by gravimetric method, turbidity (T) by nephelometric method, total nitrogen (N_T) and total phosphorus (P_T) by the semi micro process Kjeldahl, biochemical oxygen demand (BOD) by the Winkler process and electrical conductivity (EC) by means of digital conductivity meter. All analyzes were carried out at the Laboratory of Water Quality and Solid Waste at the Santa Teresa campus, following the methods described by Matos (2015).

The effluent was submitted to the analysis to evaluate the pollutant removal efficiency of the bamboo filter. To calculate the removal efficiency, Equation 1 was used, from the concentration of the affluent and the effluent at the time of samples collection.

$$Ef = \frac{(C_{af} - C_{ef})}{C_{af}} 100 \qquad (1)$$

Where:

E_f: removal efficiency, %;

C_{af}: affluent concentration;

C_{ef}: effluent concentration.

After the treatment, the effluent was conducted to a tests bench, in which there were two different models of drippers installed for evaluation, which can be observed in Figure 2.

Figure 2. Tests bench for application of treatments and emitters evaluation.

The test bench was assembled using four segments of dripping tubes with 25 units per segment. Each set of two tubes was composed of a dripper model, totaling 50 evaluated drippers of each type, being: (G1) D5000 Flow Regulated self-compensating Drip Line with nominal flow of 2.1 L h-1 and (G2) Naandanjain no and rated flow rate of 1,6 L h-1.The tests bench was assembled using four segments of dripping tubes with 25 units per segment. Each set of two tubes was composed of a dripper model, totaling 50 evaluated drippers of each type, being: (G1) D5000 Flow Regulated self-compensating Drip Line with nominal flow rate of 2.1 L h^{-1} and (G2) Naandanjain non-compensating with a nominal flow rate of 1,6 L h^{-1}.

The bench structure consisted of a system pressurized by a motor pump of 0.5 hp, a plastic tank of 500 L and a disc filtering system of 120 mesh. The system was in accordance with the manufacturer's specifications and pressure gauges were installed at all the dripping tubes, guaranteeing the pressure system uniformity.

Cada agrupamento de gotejadores passou por avaliações em função do tempo de uso, sendo 0; 48; 120; 200 e 300 horas, aplicando o efluente do filtro anaeróbio de forma contínua, completando-se o nível do reservatório a cada avaliação. Foram avaliados o coeficiente de uniformidade distribuição (CUD), o coeficiente variação de vazão (CVQ) e a vazão relativa dos emissores (Qr), conforme as equações 2, 3 e 4, respectivamente.

Each group of drippers was evaluated according to the time of use, being 0; 48; 120; 200 and 300 hours, applying the effluent of the anaerobic filter continuously, completing the level of the reservoir with each evaluation. The emission uniformity (DU), the flow rate variation coefficient (FVC) and the relative flow rate of the emitters (Qr) were evaluated, according to equations 2, 3 and 4, respectively.

$$EU = \frac{q_{25\%}}{\bar{q}} 100 \quad (2)$$

Where:

EU = emission uniformity, %;

$q_{25\%}$ = average discharge rate of the lowest one-fourth data emitter readings, L h^{-1};

$\bar{q}$ = average discharge rate of all the emitters checked, L h^{-1}.

$$FVC = \frac{S}{q_m} 100 \quad (3)$$

Where:

FVC = flow rate variation coefficient, %;

S = standard deviation os samples, L h^{-1};

q_m = average of measured flow rates, L h^{-1}.

$$Q_r = \frac{Qa}{Qi} 100 \quad (4)$$

Where:

Qr = relative flow rate, %;

Qa = real flow rate, L h^{-1};

Qi = initial flow rate, L h^{-1}.

The interpretation of the emission uniformity (EU) values was based on the methodology proposed by Merriam and Keller (1978), presented in Table 1.

Table 1. Classification of performance values of the emission uniformity.

Evaluated parameters	**Classification**
90% a 100%	Excellent
80% a 90%	Good
70% a 80%	Regular
Lower than 70%	Bad

Adapted from Merriam and Keller (1978).

RESULTS AND DISCUSSION

The physical-chemical attributes of the effluent obtained in the swine wastewater, before and after treatment with bamboo filters, as well as the removal efficiency can be observed in Table 2.

Table 2. Affluent concentration (Caf), removal efficiency (%) and effluent concentration (Cef) values for total solids (TS), suspended solids (SS), turbidity (Tb), total phosphorus (Pt), total nitrogen (Nt), biochemical oxygen demand (BOD), electrical conductivity (EC) and pH

Parameter	Caf	Removal efficiency (%)	Cef
TS (mg L^{-1})	2,931.32	51.08	1,434.00
SS (mg L^{-1})	1,115.17	88.75	125.46
Tb (NTU)	299.6	79.40	61.72
Pt (mg L^{-1})	44.21	70.57	13.01
Nt (mg L^{-1})	6,724.00	24.33	5,088.05
BOD (mg L^{-1})	282.16	23.42	216.08
EC (dS m^{-1})	4.31	15.62	3.64
pH	8.24	-	7.20

We can see that, after the filtration of SWW, there was a significant reduction in the observed parameters, mainly in relation to suspended solids and water turbidity.

According to Ayers and Westcot (1985), water with concentrations greater than 100 mg L^{-1} of SS and 2,000 mg L^{-1} of TS, the chance of drippers clogging becomes severe. After the filtration, the quality of water used for irrigation was classified as a severe risk of clogging for SS (125.46 mg L^{-1}) and moderate risk for TS (1,434.00 mg L-1), because it is within the range between 450 and 2,000 mg L^{-1}.

Although classified as a severe risk, SS concentration was slightly above that recommended by Ayers and Westcot (1985), indicating that the treatments with the bamboo filters showed good results in the removal efficiency of the evaluated parameters.

According to Almeida (2010), care must be taken when using irrigation water with suspended particles, which can cause drip obstructions due to small holes (0.75-1.40 mm) and special geometric shapes (spiral, labyrinth, etc.) through which the water flows.

Figure 3 shows the emission uniformity (EU) in the drippers models self-compensating (G1) and non-compensating (G2), as a function of the time of application of wastewater treated with anaerobic filter, having bamboo rings as support material.

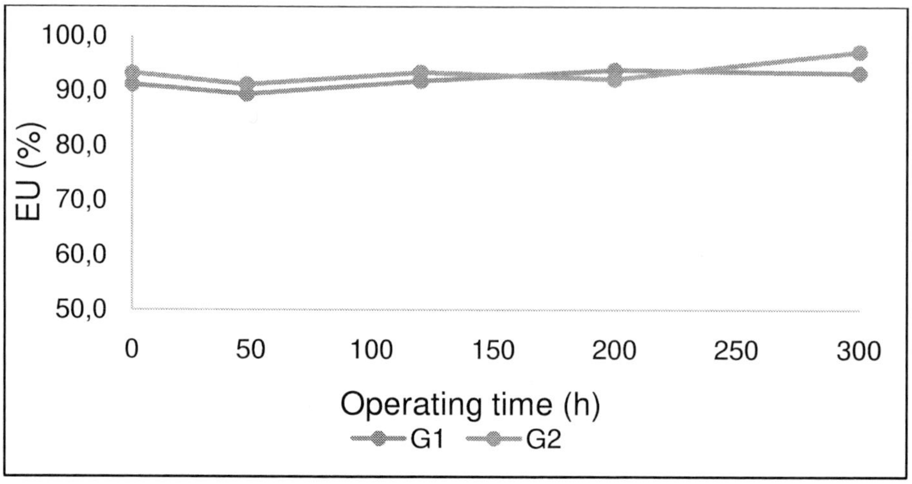

Figure 3. Emission uniformity in the drippers self-compensating (G1) and non-compensating (G2) models, depending on the time of application of wastewater treated with anaerobic filter.

According to the classification proposed by Merriam and Keller (1978) it can be affirmed that the emitters performance is considered excellent throughout the operating time, since the EU average was above 90% for both models. In a study by Batista et al. (2008), the application of swine effluents by drippers, after 160 hours, there was a high reduction in the emission uniformity due to the growth of biofilm internally in the emitters and consequent clogging. The occurrence of biofilms inside the drippers and lateral lines due to the interaction of bacterial mucilages with organic and inorganic particles is the main cause of obstruction in localized irrigation systems that operate with wastewater (Dazhuang et al., 2009; Oliver, et al., 2014).

The maintenance of high levels of emission uniformity was possible due to the low concentrations of particles in the water, indicated by the low turbidity value (61.72 NTU), including, for this parameter, the recommendations of Resolution #430 (BRASIL, 2011), which complements and amends CONAMA Resolution 35,705, which establishes that turbidity of up to 100 NTU for class II waters is suitable for use in irrigation of vegetables, fruit trees and parks, gardens, sports fields and leisure, with which the public can come into direct contact.

Os coeficientes de variação da vazão nos dois tipos de gotejadores em função do tempo de operação são apresentados na Figura 4.

The flow rate variation coefficients for the two types of drippers as a function of the operating time are presented in Figure 4.

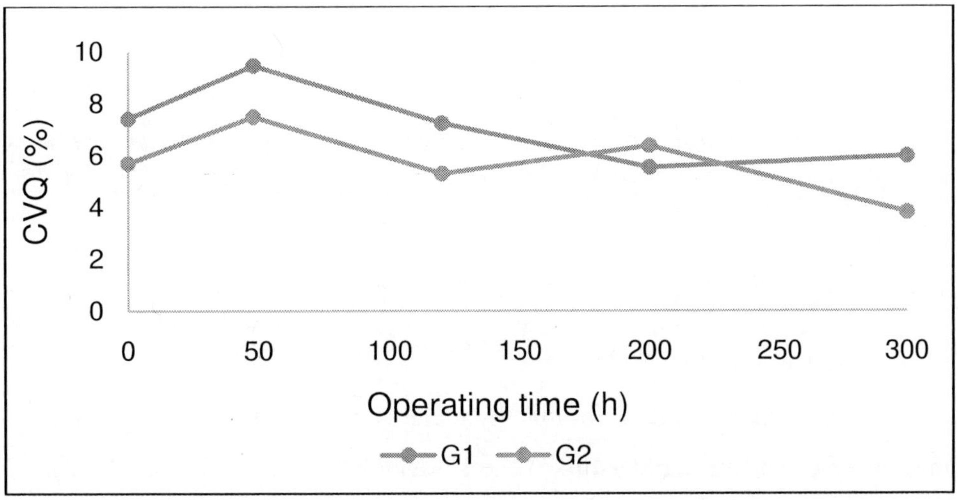

Figure 4. Flow rate variation coefficients for the two types of drippers as a function of the operating time.

In the same way as observed for the EU, there was little flow rate variation during the evaluation period, for both models of drippers, indicating that the SWW treated with anaerobic bamboo filters can be used for fertigation in irrigation systems without compromising the water distribution uniformity.

Table 3 presents the values of flow rate relative to the initial as a function of the operating times of the drippers. It is observed that the relative flow rate remained practically the same throughout the evaluation period, indicating that the dripper models evaluated can be used to apply filter-treated swine wastewater.

Table 3. Relative flor rate of the emitters.

Dripper	Initial flow rate (L h^{-1})	Relative flow rate - Qr (%) Operating time (h)			
		48	120	200	300
G1	2.51	94	93	92	91
G2	2,12	104	102	102	104

The results obtained in this study differ from those obtained by Batista et al. (2011) who, when using treated domestic sewage in maturation ponds, observed that there was a reduction in the flow rate of the drippers of the studied irrigation systems, attributing this result to the higher concentration of suspended solids in the evaluated treatments.

CONCLUSIONS

Swine wastewater, treated by anaerobic bamboo filters, can be appplied by drip irrigation systems, using both the self-compensating and the non-compensating models, without compromising the uniformity of distribution.

ACKNOWLEDGMENT

To Ifes *campus* Santa Teresa for the financial, operational and infrastructure available for the experiment and for providing scholarship to the students. To Prodif-Ifes for the financial support in the trip to present this paper.

REFERENCES

Almeida, O. A. 2010. Water quality for irrigation [electronic resource]. Embrapa Mandioca e Fruticultura, Cruz das Almas, BA, Brazil. Pp. 227. Also at https://www.alice.cnptia.embrapa.br/alice/bitstream/doc/875385/1/livroqualidadeagua.pdf

Antonelli, J., C.A. Lindino, R.F. Santos, S. Souza, W.C. Nadaletti, P. Cremonez, F. Gurgacz. 2014. Linseed response to treatment with swine wastewater as biofertilizer. Journal of Food, Agriculture & Environment. 12(3-4):229-231.

Associação Brasileira de Proteína Animal (ABPA). 2017. 2017 annual report São Paulo, SP, Brazil. Also at http://abpa br.com.br/setores/avicultura/publicacoes/relatorios-anuais/2017

Ayers, R.S.; Westcot, D.W. 1985. Water Quality for Agriculture. Food and Agriculture Organization of the United Nations (FAO): 29. Also at http://www.fao.org/docrep/003/t0234e/t0234e00.HTM

Batista, R.O.; Soares, A.A.; Souza, J.A.R.; Batista, R.O. 2008. Empirical models of localized sanitary sewage treatment. Engenharia na Agricultura. 16(3):369-377.

Batista, R.O., Soares, A.A., Noreira, D.A., Feitosa, A.P., Bezerra, J.M. 2011. Influence of different qualities of domestic sewage on the flow rate of drippers. Revista Caatinga. 24(3):128-134.

Brasil. National Council of Environment. 2011. CONAMA resolution #430, May, 13, 2011. It deals with the conditions and standards for the discharge of effluents, complements and change the Resolution #357, of March, 17, 2005, CONAMA. Diário Oficial da União , Brasília, DF, Brasil, May 16, 2011. Section 1, 92, pp. 89.

Costa, O.M.J., PandolfI, M.A.C., Rodrigues, G.A. 2017. Corn production through no-tillage fertigated with swine manure. Simtec. 3(1): pp. 9.

Dazhuang, Y.; Zhihui, B.; Rowan, M.; Likun, G.; Shumei, R.; Peiling, Y. 2009. Biofilm structure and its influence on clogging in drip irrigation emitters distributing reclaimed wastewater. Journal of Environmental Sciences. 21(6):834-841.

IBGE - Brazilian Institute of Geograply and statistics . IBGE indicators - Livestock Production Statistics March 2017. Also at https://agenciadenoticias.ibge.gov.br/agencia-noticias/2013-agencia-de-noticias/releases/16600-abate-de-suinos-cresce-1-3-e-tem-melhor-2-trimestre-da-serie-historica.html

Konzen, E.A. 2006. Environmental and economic viability of swine manure. Sete Lagoas: Embrapa Milho e Sorgo, Sete lagoas, MG, Brazil. pp.25.

Matos, A.T. 2015. Manual of analysis of solid waste and wastewater. 1th ed. UFV, Viçosa, MG, Brazil. pp.149.

Menezes, J.F.S., Silva, M.P., Neto, I.M., Benites, V.M., Berti, C. L. F. 2017. Productivity of sugarcane and residual nutrient contents in the soil after liquid slurry applications. Agrarian. 10(35):42-51.

Merriam, J.L.; Keller,J. 1978. Farm irrigation system evaluation: a guide for management. Logan: Utah State University. pp.271.

Lo Monaco, P.A.V.L., Soela, D.M., Haddade, I.R., Oza, E.F., Santos, M.M. 2017. Technical note: Treatment of swine wastewater in flooded systems constructed with Heliconia psittacorum and Hedychium coronarium. Engenharia na Agricultura. 25(6):561-568.

NBR 7229 Brazilian Standard. 1993. Design, construction and operation of septic tanks. ABNT, São Paulo, SP, Brazil. pp.

Oliver, M.M. H.; Hewa, G.A.; Pezzaniti, D. 2014. Bio-fouling of subsurface type drip emitters applying reclaimed water under medium soil thermal variation. Agricultural Water Management. 133(1):12-23.

Tonetti, A.L., Coraucci Filho, B., Guimarães, J.R., Cruz, L.M.D.O., Nakamura, M.S. 2011. Evaluation of the start and operation of anaerobic filters having bamboo as filling material. Engenharia Sanitaria e Ambiental. 16(1):11-16.

Agronomic and Economic Assessment of Aerating Drip Irrigation Water

Jim Lauria, Vice President—Sales and Marketing
Mazzei Injector Company, 500 Rooster Drive, Bakersfield, CA 93307

Dave Goorahoo, PhD, Associate Professor Vegetable Production - Plant Science Department
Soil Scientist - Center for Irrigation Technology
California State University, Fresno, 2415 E. San Ramon Ave. MS AS 72, Fresno, CA. 93740

Abstract. *Injecting air into the crop root zone has been a goal of farmers since the dawn of agriculture, initially driving advances in tillage—and, ironically, no-till farming—and now fostering new advances in subsurface drip irrigation (SDI) practices. In this paper, we review the potential agronomic and economic benefits associated with the implementation of irrigation system involving aeration of the water delivered to the root zone via SDI. The importance of oxygen is acknowledged for many of the chemical reactions that take place in the root zone, such as nitrification, that result in the conversion of nutrients into plant-available forms. In contrast, a shortage of oxygen leads to denitrification and the gaseous loss of N to the atmosphere in the form of nitrous oxide, nitric oxide and, ultimately, free nitrogen. While many approaches have been adopted for delivering oxygen via the irrigation water, the system with the venturi injection technology appears to be most efficient. Findings from on-going applied research, combined with assessment of the economic benefits from commercial growers adopting the venturi technology indicate that there is also a biological shift in the soil microbial population due to aerated drip irrigation water. Future steps should explore the impact of these microbial shifts on the gaseous emissions and nitrate/nitrite/ammonia/ammonium ratios within the soil under aerated and non-aerated regimes. In addition to the environmental effects of aerating drip irrigation water, further documentation of the impacts of AirJection® on crop yield, quality, health, and profitability will also be important in stimulating adoption of the practice.*

Note: AirJection® is a registered trademark of Mazzei Injector Company

Keywords: Sub-surface drip irrigation, greenhouse gases, soil health, AirJection, nitrous oxide, venturi injection, aerating drip irrigation water.

Introduction

Injecting air into the crop root zone has been a goal of farmers since the dawn of agriculture, initially driving advances in tillage—and, ironically, no-till farming—and now fostering new advances in subsurface drip irrigation (SDI) practices (Ayars et al., 2015). Aerating subsurface drip irrigation water has been demonstrated to improve crop yield, quality, and profitability (Bhattari et al., 2004, 2005 &

2006; Goorahoo et al., 2001 & 2008). More recent research by Goorahoo et al. (2016) indicated that soil treated with aerated drip irrigation water can cause a beneficial shift in microbial populations, suggesting the possibility of significant improvements in nutrient use efficiency and nutrient availability, as well as reductions in groundwater pollution and the release of greenhouse gases from cropland.

Healthy soils teem with life, much of which depends upon aerobic conditions—earthworms, insects and beneficial microbes all thrive in the presence of air (Wolf, 1999). Growing recognition of the role of biology in the transformation and exchange of nutrients in the root zone helps underscore the importance of a healthy aerobic microbial community in the soil.

Oxygen is also critical for many of the chemical reactions that take place in the root zone, such as nitrification, that result in the conversion of nutrients into plant-available forms. In contrast, a shortage of oxygen leads to denitrification and the gaseous loss of N to the atmosphere in the form of nitrous oxide, nitric oxide and, ultimately, free nitrogen (US EPA, 2018).

Soils that lack adequate subsurface air—heavy clays, compacted soils and saline soils, for instance, or water-saturated soils that have "drowned out"—illustrate the importance of air in healthy plant development.

Growing Importance

In 1941, William D. Durell wrote in *Plant Physiology*, "It has long been recognized among plant physiologists that the air relations of roots have an extremely important bearing on both the vegetative and reproductive phases of plant growth." A study of suitable oxygen carriers, which could be applied as fertilizer and which would release oxygen slowly to the soil during the growing season, may be worthwhile," he added.

Aerating the soil may be even more worthwhile than ever in light of the challenges that today's farmers face. A global soil health crisis rages that includes dramatic reductions in soil organic matter, loss of structure from excessive tillage, salination, and acidification in many areas. Water challenges range from scarcity to the pollution of surface water and ground water with excess nutrients. And air pollution that includes nitrous oxide, nitric oxide and ammonia from agricultural sources puts farmers under significant pressure from their bankers, regulators, environmental activists, and neighbors (Park et al., 2012).

A recent study by Almaraz et al. (2018) of the University of California, Davis estimated that fertilized agricultural soils in California account for 20 to 32 percent of the state's NOx emissions. The U.S. EPA Inventory of U.S. Greenhouse Gas Sources and Sinks: 1990-2016 (2018) estimates agricultural soil management (US-EPA, 2018) causes 77 percent of the country's NOx emissions. The agricultural community is contesting these assertions, but the findings have focused the ire of environmentalists on farmers in California and nationwide and added weight to the argument that reducing the production of NOx caused by farming would be a significant step forward in improving air quality.

Mechanical Attempts at Aeration

The desire to aerate agricultural soils goes back to the era of the very first plows, pulled by ancient oxen to rip farmland, create porosity, and introduce air at least into the seed zone. The advent of moldboard plows in the 19th century turned over several inches of soil at a time, opening heavy northern European and North American soils to seeding—and to the atmosphere.

In recent decades, many farmers have sought to reduce the direct costs of tillage, including diesel, machinery, and labor, by reducing tillage passes or using lighter or more advanced, lower-disturbance equipment. There is also an increasing recognition of the externalities caused by tillage, such as the loss of soil organic matter, increased erosion, the development of impermeable plow pans beneath the level of tillage, and the destruction of natural soil structure—many of which, ironically, reduce the amount of air in many soils, at least over the long term.

Changes in irrigation practices are creating changes in soil management. According to Ayars, et. al. (2015), low-volume irrigation accounted for nearly 40 percent of California's irrigated acreage in 2010, the result of a steady rise since the early 1990s. Within that low-volume category, permanent subsurface irrigation accounts for a significant share of acreage, particularly in certain crops including tomatoes and tree crops.

The growing adoption of permanent subsurface irrigation lines and bedding practices that permit reduced-tillage management are contributing to more precise crop and soil management. (Yuan et al., 2018). They also create demand and opportunity for aeration via drip irrigation water.

Complex Aeration Systems

Subsurface drip lines provide a welcome and ideally positioned delivery system for air into the root zone, and several innovators have explored systems to exploit that pathway. One approach reads as if it were lifted straight from W.D. Durell's wish-list—the injection of hydrogen peroxide into the soil. Hydrogen peroxide is an extremely powerful oxidant. Mixed with water, it forms hydroxyl (OH) radicals that are highly reactive.

However, as a tool for injecting oxygen into the soil, hydrogen peroxide is somewhat worrisome. A key concern is the storage and handling of a highly reactive compound. On a biological level, hydrogen peroxide's role as a potent disinfectant in water treatment hints at the devastation it may cause to soil biota in the treated soil—microbes that play vital roles in the conversion and transport of nutrients to crop roots—and to roots themselves. And, because hydroxyl radicals are so reactive that they are very short-lived, Durell's ideal of a slow-release oxidizer is not realized by hydrogen peroxide.

Another logical approach was the employment of compressors to pump air into irrigation water. However, forcing air has proved to be an expensive approach due to energy, and equipment and maintenance costs required. Inadequate mixing of the forced air with the irrigation water makes it difficult to control the amount of air delivered and to keep it in the root zone.

Super microbubble generating systems, which create air bubbles less than three microns in size, have been demonstrated to be more effective than compressed air in improving photosynthesis in treated crops. However, the equipment and energy costs are high, so treated areas must be small, limiting the technology's effectiveness in field crops.

Venturi Injection Technology

The most efficient and effective aeration technology is the use of venturi injectors, which apply the Bernoulli principle to incorporate air into a moving stream of water (Goorahoo et al., 2001). The injector is a precisely engineered instrument in which the flow of irrigation water is constricted in a conical chamber, then allowed to expand again. The change in available space creates a corresponding change in pressure within the stream; as it expands into the outlet chamber, the flow creates a low-pressure zone that draws air into the stream through a suction inlet and entrains or dissolves it into the water. Because the injection and mixing is driven by the force of the water passing through the injector, the system requires no additional energy and operates with minimal head loss.

Turbulence created within the system allows well-designed venturi injectors to be highly efficient at dissolving oxygen into the stream. A study of venturi injectors in aerating drip irrigation water by Maestre-Valero and Martinez-Alvarez (2010) in Agricultural Water Management, demonstrated that venturi nozzles brought dissolved oxygen levels to 80.6 percent saturation—close to the saturation point. "The installation of the venturi air injector in the pipeline supplied a great quantity of extra dissolved oxygen and air bubbles to the water," they wrote.

Venturi injectors have been studied extensively in the 200 years since their development. In recent decades, they have been employed in systems ranging from wastewater treatment to fuel injection, and their efficacy in aerating, oxygenating or ozonating water is thoroughly proven. In 1996, Mazzei Injector Company of Bakersfield, California, began experimenting with the use of venturi injectors to aerate subsurface drip irrigation water. In 2000, the company brought its early AirJection® system to the Center for Irrigation Technology (CIT) at California State University- Fresno (Fresno State) for academic trials and validation. That year saw a pilot study on peppers. In 2001, Mazzei was awarded its first AirJection patent, and the company and university began field trials. Since 2003, the company and university researchers have tested the AirJection system in honeydew, cantaloupe, tomato, bell pepper, sweet corn, broccoli, and strawberry fields in California. Additional testing of the concept has been conducted by university researchers in Australia, Spain, Egypt, Italy, Japan, and China. Large-scale commercialization of AirJection began in 2005 and continues today.

Significant Benefits

Research by Fresno State and other academic institutions has demonstrated a wide range of benefits of aeration of subsurface drip irrigation water. In cooperation with the Fresno State research team, one major San Joaquin Valley farm compared 1,500 acres outfitted with an AirJection system against rows with conventional buried drip tape for eight years. They recorded a 23-percent average increase in yield in cantaloupes, as well as yield increases in honeydew, sweet corn, and peppers. The yield boost in cantaloupes ranged from 12 to 34 percent over the study period.

In other trials, tomato yields increased by 21 percent in normal soils and 38 percent in saline soil as a result of aerated subsurface irrigation water. Watermelon yields have been demonstrated to nearly double, while soluble solids increased by 4 percent. Studies in broccoli recorded increases in root mass, root size, and canopy among plants treated with aerated drip irrigation water. Aeration resulted in root dry weight increases of 12 percent in broccoli and 16 percent in sweet corn compared to crops irrigated with non-aerated water.

Profitability is also very positively impacted by aeration. In the California cantaloupe study, applying aerated drip water on all their cantaloupe acres would have penciled out to a total of 1.3 million more boxes and added net return of more than $3.7 million over the 8-year study period.

Significantly, the added yield required no additional irrigation water, increasing water use efficiency (WUE) by 23 percent on the aerated acres. In an era in which it is vital to create "more crop per drop" and more output per kilowatt of energy, such efficiencies will be increasingly valued. The beneficial effects of drip irrigation water aeration are likely to be observed in heavy or poorly structured saline soils, where crop stress, reduced macropore space, and low-oxygen conditions make the effects of introducing air to the root zone even more dramatic.

Biological Shifts

Recent replicated blind trials by Goorahoo and Josue Somano Monroy of Fresno State and Adrian Unc and Crystal McCall of Memorial University of Newfoundland indicate that aerating subsurface drip irrigation water can have significant impacts on the diversity of the soil microbial community (Goorahoo et al., 2016). The team used a PowerSoil extraction kit and PCR DNA analysis to assess the presence and proportion of genes in soil samples, discerning among genes from nifH, amoA and Archea—which are active in ammonia oxidation—and genes involved in nitrite or nitrate reductase processes, including narG, napA, nirK, nirS, and nosZ. The study explored the ratios among Archaea and the various bacteria in the samples.

Maintaining a balance of microbes that favor nitrification over denitrification is extremely important to creating an environment for good nutrient use efficiency (NUE), as well as for minimizing air pollution and the emission of greenhouse gases and smog precursors. Soils with adequate air, proper pH, and adequate moisture provide an environment conducive to increased nitrification and ammonification. In healthy soils, microbes Nitrosomonas and other microbes convert ammonia to nitrite, then Nitrobacter and other biota transform the nitrite into nitrate, which plants can use. The process requires two steps and microorganisms from up to four genera to complete. A shortage of oxygen increases denitrification, in which Pseudomonas, Bacillus, and Thiobacillus microbes reduce nitrate into nitrous oxides, nitric oxide, and ultimately into gaseous N2, all of which can volatilize into the atmosphere, depleting the soil of valuable nitrogen. While N2 is benign, nitrous and nitric oxides are harmful to the environment.

In the Goorahoo et al. (2016) study, the distribution of the tested genes within the microbial populations was very distinct among the aerated and non-aerated treatments. AirJection had a clear selective impact on the distribution of the tested genes among the population, and it may thus be hypothesized that total diversity also changed. Moreover, bacteria dominance in the samples treated with AirJection indicates a likely shift to more luxurious growth conditions.

The nature of the shift is extremely important. The microbial populations in the aerated treatments favored beneficial bacteria that convert ammonia and ammonium into plant-available nitrate, while the non-aerated treatments leaned toward higher proportions of reducing bacteria. That suggests that aeration can result in lower NOx production and increased availability of nitrate in the root zone, enhancing nutrient use efficiency while also reducing the release of NOx, NO and N_2 to the atmosphere.

Next Steps

There are two main avenues for future studies of aerating subsurface drip irrigation water. Deeper exploration of the makeup of microbial populations under aerated and non-aerated regimens is warranted, as is detailed investigation of the gaseous emissions and ratios among nitrate, nitrite, ammonia and ammonium within the soil under both types of treatment.

In addition to the environmental effects of aerating drip irrigation water, further documentation of the impacts of AirJection on crop yield, quality, health, and profitability will also be important in stimulating adoption of the practice.

Crop consultants and the irrigation industry can be extremely helpful in establishing demonstration and research plots for AirJection systems. With the growth in adoption of aeration technology and increased documentation of its effects on gaseous emissions and water pollution potential, it should be possible to attract the interest and support of agencies such as: the California Air Resources Board (CARB), the Greenhouse Gas Reduction Fund (SWEEP), California Healthy Soils Initiative, and the US Department of Agriculture's Environmental Quality Improvement Program (EQIP) to provide support for a technology that delivers important benefits to farmers while providing increasingly vital environmental services to society.

References

Almaraz, M., Bai, E., Wang, C., Trousdell, J., Conley, S., Faloona, I., & Houlton, B. Z. 2018. Agriculture is a major source of NOx pollution in California. Science advances, 4(1), eaao3477.

Ayars, J.E., A. Fulton and B. Taylor. 2015. Subsurface drip irrigation in California—Here to stay? Agricultural Water Management. 157: 39-47.

Bhattarai S.P., S. Huber, and D.J. Midmore. 2004. Aerated subsurface irrigation water gives growth and yield benefits to zucchini, vegetable soybean and cotton in heavy clay soils. Ann. Appl. Biol. 144: 285-298.

Bhattarai S.P., L. Pendergast, D. J. Midmore. 2006 Root aeration improves yield and water use efficiency of tomato in heavy clay and saline soils. Scientia Horticulturae. 108: 278-288.

Bhattarai S.P., N. Su, and D.J. Midmore. 2005. Oxygation unlocks yield potentials of crops in oxygen limited soil environments. Advances in Agronomy. 88: 313-377.

Durell, W.D. 1941. The effect of aeration on growth of the tomato in nutrient solution. Plant Physiology. 16(2): 327-341.

Goorahoo D., A. Unc, C. McCall, J. Samano-Monroy, F. Cassel Sharma, T. Thao, G. Seepersad. 2016. AirJection irrigation mitigates denitrification and leaching. Proceedings of the Irrigation Association Conference. 6 December 2016.

Goorahoo D., G. Carstensen and A. Mazzei. 2001. A pilot study on the impact of air injected into water delivered through subsurface drip irrigation tape on the growth and yield of bell peppers. California Agricultural Technology Institute (CATI) Publication # 010201.

Goorahoo D., D. Adhikari, D. Zoldoske, F. Cassel S., A. Mazzei, and R. Fanucchi. 2008. Potential for AirJection® Irrigation in strawberry production. pp 152-155 In: Takeda, F., D.T.Handley, and E.B. Poling (ed.). Proc. 2007 N. American Strawberry Symposium. North American Strawberry Growers Association, Kemptville, ON Canada.

Maestre-Valero J. and V. Martínez-Alvarez. 2010. Effects of drip irrigation systems on the recovery of dissolved oxygen on hypoxic water. Agricultural Water Management. 97. 1806-1812.

Park S., P. Croteau, K.A. Boering, D.M. Etheridge, D. Ferretti, P.J. Fraser, K-R Kim, P.B. Krummel, R.L. Langenfelds, T.D. van Ommen, L.P. Steele and C.M. Trudinger. 2012. Trends and seasonal cycles in the isotopic composition of nitrous oxide since 1940. Nature Geoscience. Vol. 5. April 2012.

US- EPA. 2018. Inventory of U.S. greenhouse gas sources and sinks 1990-2016. EPA 430-R-18-003. U.S. Environmental Protection Agency. 2018 Downloaded via https://www.epa.gov/ghgemissions/inventory-us-greenhouse-gas-emissions-and-sinks.

Wolf B. 1999. The Fertile Triangle. Chapter 1 in "The Interrelationship of Air, Water, and Nutrients in Maximizing Soil Productivity". Haworth Press, Inc. New York, (1999).

L. Yuan, N. Wenquan, W. Jingwei, X. Jian, Z. Mingzhi, L. Kangyong. 2016. Review on advances of AirJection irrigation. International Journal of Agricultural and Biological Engineering. Vol. 9, No. 2. 1-10.

Soil Moisture Sensor-Based Irrigation Scheduling to Optimize Water Use Efficiency in Vegetables

Pappu Kumar Yadav
Center for Irrigation Technology, California State University, Fresno
* Currently at the Department of Biological & Agricultural Engineering, Texas A&M University, College Station, TX
Florence Cassel Sharma
Center for Irrigation Technology, California State University, Fresno
Department of Plant Science, California State University, Fresno
Touyee Thao
Center for Irrigation Technology, California State University, Fresno (past)
Department of Environmental Science, University of California, Merced
Dave Goorahoo
Center for Irrigation Technology, California State University, Fresno
Department of Plant Science, California State University, Fresno

Abstract

Advancement in the precision of soil moisture sensors has led to the adoption of these devices in the design, monitoring and control of irrigation systems in commercial agriculture, urban landscapes, and research. The World's soil sensor market has been estimated to grow at a rate of 16.2% between 2015 and 2020 reaching sales of 206.2 million U.S. dollars. While this rapid growth of the sensor market caters to the need of precision farming and effective irrigation, there still exists a need for researchers to determine effective depths for scheduling irrigation based on crop growth stage, soil texture, and infiltration rate. The overall goal of our ongoing study is to optimize water use efficiency (WUE) and irrigation scheduling in vegetable cropping systems. In this presentation, we focus on the technology and operating principles of the applications based on measurements of crop evapotranspiration (ET) or soil water content. For the ET-based approach, an irrigation-scheduling program was developed to: 1) poll daily ET data from a local state weather station using the California Irrigation Management Information System (CIMIS) web Application Programming Interface (API) over radio and internet links, and 2) calculate daily irrigation applications. For the sensor-based approach, six capacitance-soil moisture devices installed at three locations and two depths (6'' and 12'') were used with a datalogger and a 24VAC solenoid valve. Irrigation scheduling was programmed using upper (field capacity) and lower thresholds (30% maximum allowable depletion) of soil available water. Irrigation applications were triggered when the average soil moisture values reached the lower threshold and ended after field capacity was attained. In addition, some fields were irrigated based on visual crop and soil observations and manually operating irrigation valves. Results indicate that both ET- and soil sensor-based technology can improve water use efficiency and help growers optimize their irrigation scheduling.

Keywords: soil moisture sensor, water use efficiency, optimizing, irrigation scheduling, vegetable crops

Background

Efficient management of water resources is of utmost importance for the sustainability of irrigated agriculture, particularly in areas of diminished water supplies like California. One approach to improve irrigation efficiency is to implement scheduling practices that include the use of sensor technology and/or accurate measurements of crop water requirements based on replenishing evapotranspiration (ET). In

California, water conservation has become a top priority for vegetable producers as they thrive to adopt management practices that optimize irrigation and water use efficiency. One approach to improve irrigation efficiency is to implement irrigation scheduling practices to accurately estimate crop water requirements, and subsequently determine how much water to apply and when to irrigate. Implementation and the accuracy of these schedules could be improved with the use of soil sensor technology.

Soil moisture sensors (SMS) are good source for monitoring spatial variation of soil moisture and hence can be an effective tool for precisely managing water application to various crops. These sensors allow site specific crop management which is the most crucial part of precision agriculture (Badewa et al., 2018). Advancement in sensor technology have made them more accurate and precise in estimating soil water content. Development of technologies like internet of things (IoT) and data loggers have made these sensors to remotely monitor in-situ field dryness or wetness. SMS can be integrated with pump and valve controllers to automatically trigger irrigation based on crop water demand estimated indirectly from soil water content.

Every SMS can be categorized in to two basic types: volumetric water content (VWC) type and soil water potential type. VWC type measures percent volume of water in unit volume of soil while soil water potential type measures soil water potential. These sensors attached with controllers can be set with upper and lower threshold values for triggering and shutting off irrigation events. The upper threshold value is generally set at field capacity (FC) and lower threshold value at management allowable depletion (MAD) of soil. Field capacity is defined as the water content when free drainage ceases, and is equivalent the soil water content at a soil water potential is 0.33 bars and 0.1 bars for clay-textured and sandy soils, respectively (Enciso et al., 2007). MAD is defined as the soil water content below which the plants should not be allowed to stress, and is generally calculated as a percentage of the available water (AW). This value depends on crop types, growth stage and soil type (Enciso et al., 2007).

Besides using SMS, growers also use evapotranspiration (ET_o) data from weather stations and convert those values into equivalent crop evapotranspiration (ET_c) using crop coefficient (K_c) values with growing stage. Many growers still prefer to irrigate manually by visually observing dryness or wetness of soil and the crops.

The overall goal of our ongoing project is to evaluate the effectiveness of all these methods in terms of water use efficiency (WUE) for vegetables and other cropping systems typical of the San Joaquin Valley (SJV), California, USA. The specific objective of this presentation is to review the technology and operating principles of the applications based on measurements of ET_c and oil water content for estimating WUE of various surface drip-irrigated crops.

Materials and Methods

To evaluate these three irrigation scheduling options mentioned above, we conducted field studies on various crops (lettuce, tomato, sorghum, corn) irrigated with the following scheduling approaches (**Figures 1 to 10**): 1) soil moisture sensor (SMS), 2) ET_c, and 3) time-based with visual soil/plant inspection (control).The test site was located at the University Farm at California State University, Fresno. Soil texture at the test site was classified as a Sandy Loam.

For the sensor-based approach, **referred to as SMS**, six capacitance (Decagon 10HS) soil moisture sensors installed at three locations and at two depths (6'' and 12'') were used with a datalogger and a 24VAC solenoid valve. Irrigation scheduling was programmed using upper (field capacity) and lower thresholds (30% maximum allowable depletion) of soil available water. Irrigation applications were triggered when the average soil moisture values reached the lower threshold and ended after field capacity was attained. A Campbell Scientific CR1000 datalogger was used along with NL115 ethernet interface, RF401 radio modem, 900 MHz radio antenna and internet connectivity with 24 VAC solenoid valves in each irrigation treatment lines. Another CR1000 with RF401 radio modem and NL115 ethernet interface was used to give remote access to field data using internet. A PC with LoggerNet software and internet connectivity was used to monitor, control and download field data remotely.

For the climate based approach, **referred to as ET**, an irrigation-scheduling program was developed to: 1) poll daily ET data from a local state weather station using the California Irrigation Management Information System (CIMIS) web Application Programming Interface (API) over radio and internet links, and 2) calculate daily irrigation applications. Data from CIMIS station # 188 (Madera II) was used to poll daily ETo data by one of the CR1000 loggers connected to internet and then transfer this to field logger with irrigation control system. A CRBasic program was developed to calculate daily run time for ET based irrigation treatments.

For the "Control", **referred to as "Manual"**, plots were irrigated based on visual crop and soil observations and manually operating irrigation valves. Typically, the operator would turn on the water in the morning and irrigate until the crop appeared turgid and/or when water began ponding on the soil surface. This was considered to be the practice adopted by grower based on the availability of water and the scheduling approach that took into consideration other crops plant on the farm.

Figure -1 Block Diagram of Overall System Design for ET and Soil Sensor Based Irrigation (*Left*), and photos of Lettuce plot (*Middle)* and Tomato plot (*Right*).

Fig. 2. Flowmeters with Valves (Tomato & Lettuce)

Fig. 3. Controller with Telemetry (Tomato & Lettuce)

Fig. 4. Soil Moisture Monitoring Station (Corn/Sorghum)

Fig. 5. Flowmeter Monitoring Station (Corn/Sorghum)

Fig. 6. Irrigation Controller (Corn/Sorghum)

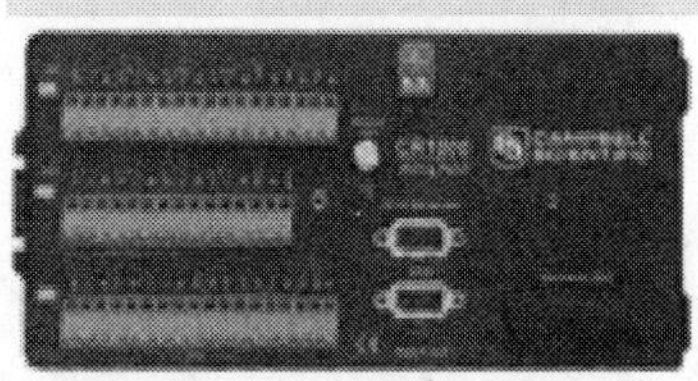

Fig. 7 CR1000 Datalogger

- Can be used to record analog and digital sensors.
- Can be programmed using CRBasic programming language for automation and control.
- Supports different communication protocols like SDI-12, RS232, PakBus, Modbus, DNP3, TCP/IP, FTP.

Fig. 8. Meter 10 HS Sensor

- Measures soil volumetric water content.
- Type: Capacitance (frequency domain).
- Range: 0-50% VWC.
- Measurement Time:10 ms.
- Output Type: 300-1250 mV.

Fig. 9. Seametrics MJR-200-1G Flowmeter

- Measures flow of liquid at higher rates.
- Sensor: Reed Switch.
- Output :Square Pulse.
- Flow Range: 0.44-52 GPH.
- No Power Required.
- Max Psi : 150 psi.
- K Factor: 1 Pulse/Gallon.

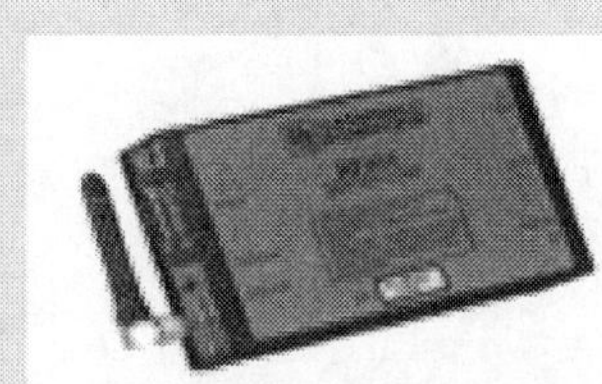

Fig. 10. RF 401 Radio

- Used for transmission and reception of data signals via radio frequency.
- Frequency range: 910-918 MHz.
- Designed to be used in PakBus network.
- Distance Range: up to 10 miles.

Results and Discussion

In Figure-1, FM1, FM2, FM3 and FM4 represent the flowmeters used in each irrigation line for tomato project. In the lettuce project, there were only three flowmeters as there were only three irrigation treatments. Flowmeters used for lettuce were analog type and hence no data from them could be recorder by datalogger. Therefore, all the readings were taken manually. However, for the tomatoes, Seametrics MJR-100-1P flowmeters were used that could send output pulses equivalent to amount of water flowing through irrigation lines.

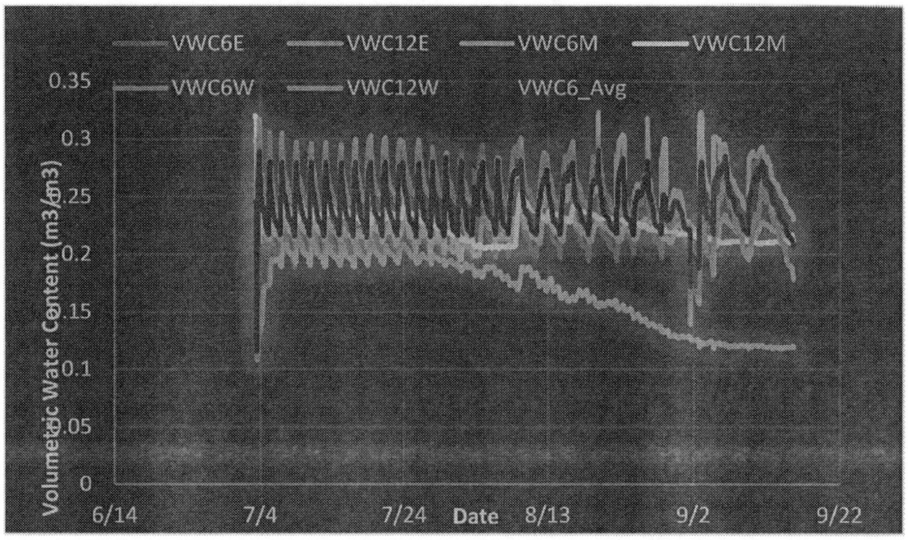

Figure-11 Soil Moisture Profile for SMS-Based Irrigation in Tomato.

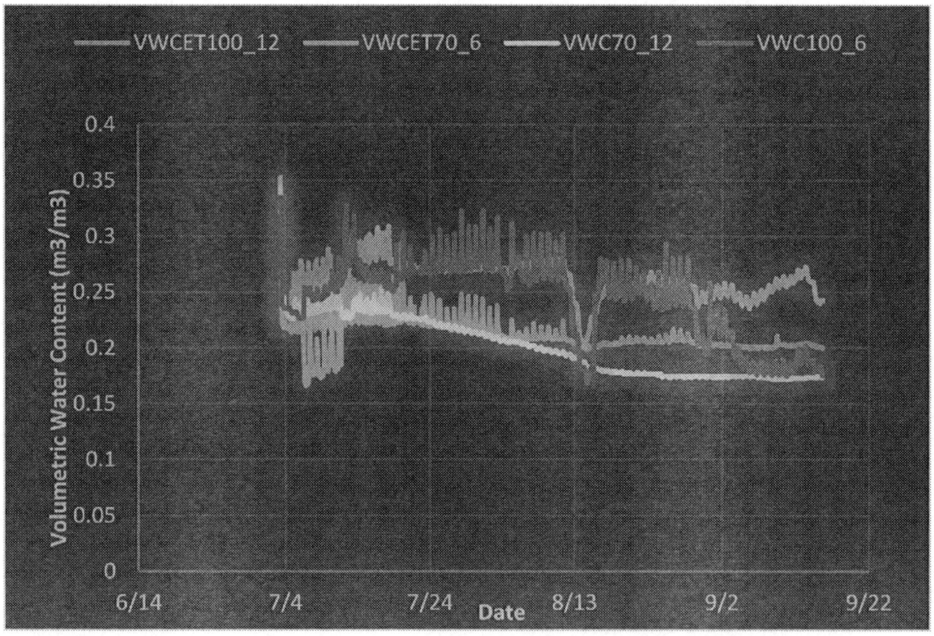

Figure-12 Soil Moisture Profile for ET-Based Irrigation in Tomato.

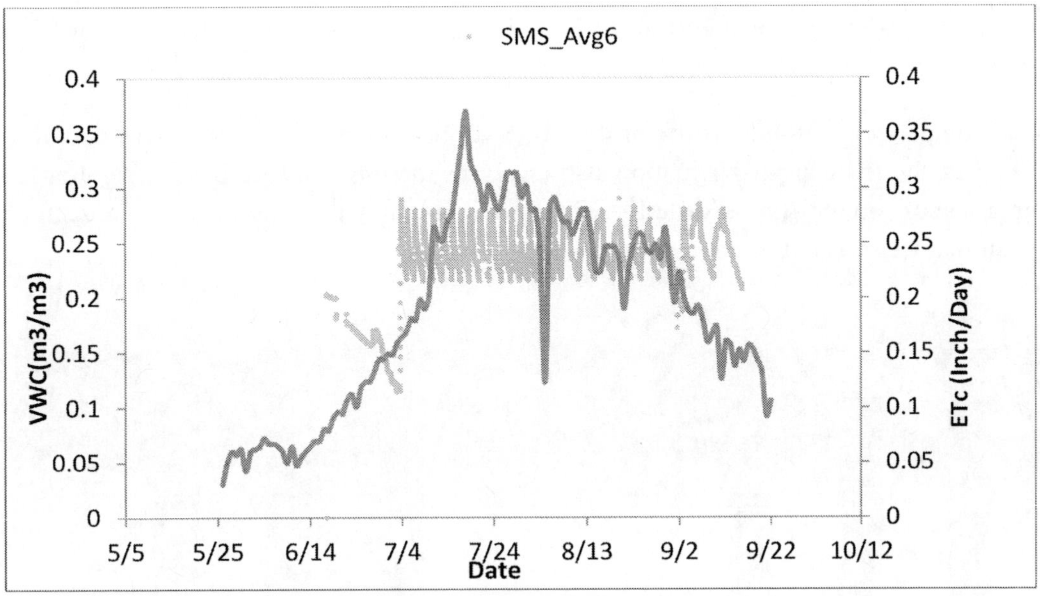

Figure-13: Soil Moisture vs. Daily ET_c (SMS-Based)

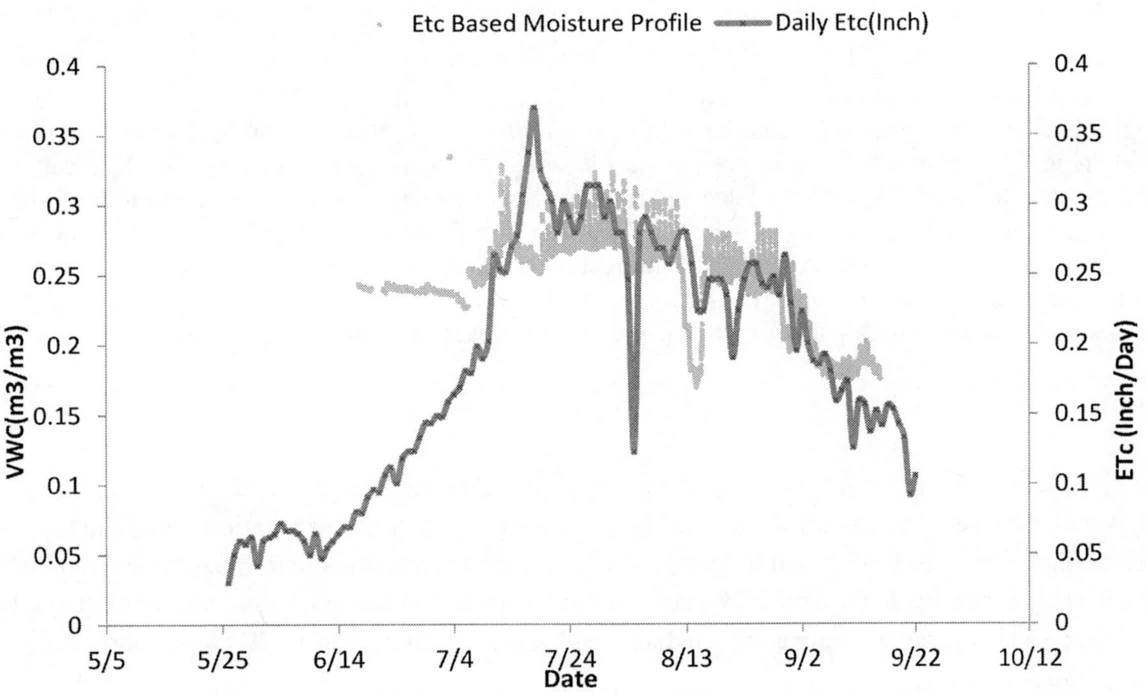

Figure-14: Soil Moisture vs. Daily ET_c (ET-Based)

Figures 11 and 12 show soil moisture sensor profiles for SMS- based and ET based irrigations. Both show that irrigation was triggered whenever average moisture value for three six-inch depth sensors reached the set lower threshold value of 21% VWC (30% less than field capacity of 31% VWC) and was shut off whenever this value reached 31% VWC. These values were locally determined in lab by bringing soil samples and following manufacturer's recommended process. In Figure12, around date 8/13 there can be seen a bigger dip in soil moisture profile. This was because web server was down and hence, no ET data could be polled during a week time which resulted in no irrigation for the entire week time for ET based irrigation zone.

Figure-13 shows comparison of soil moisture profile for SMS based irrigation with that of daily ET. With increasing ET values, the frequency of irrigation also increases for soil moisture-based irrigation. Figure 14 shows comparison of soil moisture profile for ET based irrigation with daily ET. It ca be seen that both of them follow similar trend as expected.

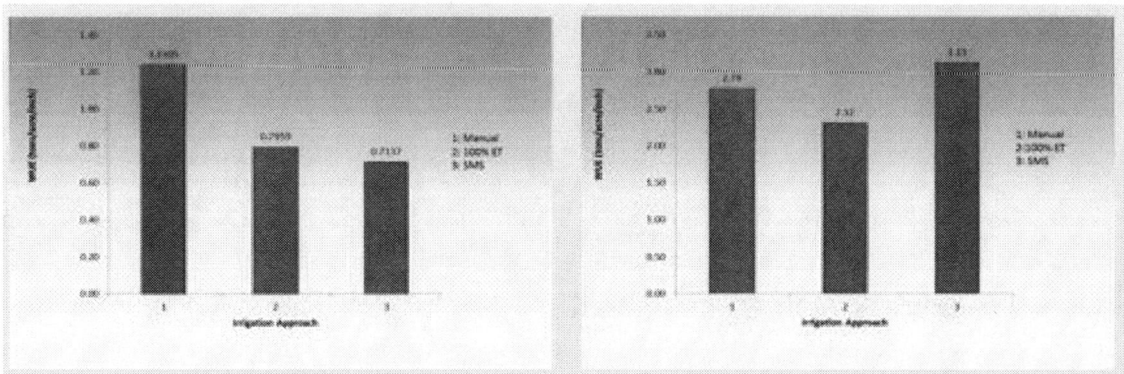

Fig. 15: WUE under various irrigation scheduling approaches for Tomato (left) and Lettuce (right)

The WUE, calculated as yield per unit of water applied was lowest under the SMS-based approach for tomato but found to be the highest for lettuce with values of 0.714 tons/acre/inch and 3.13 tons/acre/inch, respectively (Fig. 15). These differences were in part due to: 1) sensor depths which remain constant during the growing seasons, 2) number of sensors used to trigger irrigation events, and 3) timing of the growing seasons. In lettuce, the depth of water applied was significantly lower under the SMS approach (5.74 in) compared to the ET method (10.38 in), indicating that in-situ measurement of soil water content is critical to optimize irrigation efficiency. The lower WUE values obtained for the ET-based approach are most likely attributed to high water depth applied despite elevated yields.

Conclusion

The most significant conclusion from these preliminary findings is that both ET- and soil sensor-based technology can improve water use efficiency and help growers optimize their irrigation scheduling. However, it appears that for SMS based irrigation, moisture sensors responsible for triggering irrigation need to be buried at varying depth with growing season of crops for optimized water use and better yield. Additional studies are currently being conducted addressing the differences in WUE values observed in this phase of the study.

References

Badewa, E., Unc, A., Cheema, M., Kavanagh, V., & Galagedara, L. (2018). Soil Moisture Mapping Using Multi-Frequency and Multi-Coil Electromagnetic Induction Sensors on Managed Podzols. *Agronomy*, *8*(10), 224.

Enciso, J., Jifon, J., & Wiedenfeld, B. (2007). Subsurface drip irrigation of onions: Effects of drip tape emitter spacing on yield and quality. *Agricultural Water Management*, *92*(3), 126-130.

SDI Flush Line Design and Precision Flush Management

Dale H Allred, PE, MS
AES International, PLLC, 1203 E 700 N, American Fork, UT 84003, dale5790@gmail.com

Abstract. *A methodology is presented for designing SDI sub-units with Flush Manifolds using UDU (Ultimate Distribution Uniformity) and the CHLR (Cost Head-Loss Ratio). Using CHLR/UDU design protocols solves for the unknowns, qExtra, qHelped, the qNeutral location along the flush manifold, and the qZero array of distal lateral dead flow locations. Solving for these unknowns not only leads to optimal sub-unit design, it forms the basis for operational and flushing solutions in the field. Implementing CHLR/UDU design protocols for sub-unit SDI irrigation design and leveraging finesse to precision manage sub-unit flushing can dramatically reduce system cost, improve the efficacy of sub-unit flushing compared to traditional solutions, and more correctly assess DU in terms of defined risk.*

Keywords. SDI, Flush Manifolds, DU, Drip Irrigation Design, Submain Design, Manifold Design, Database Drip Irrigation Design Application, Drip Irrigation, Irrigation System Risk Assessment.

Introduction

The first time I ever saw a flush manifold was in 1984. It was of my own design and making. I had gone to Mexico in 1982 and was involved in the very first line source drip irrigation system ever installed in Central Mexico. By 1985 I had designed and installed thousands of acres of drip irrigation in vegetable crops in Central Mexico from Colima to San Luis Potosi to Sinaloa and beyond. These were surface drip tape systems or minimally buried tape systems with the lateral ends such that each dripline could be opened and individually flushed. Within a year of designing and installing that first flush manifold system in 1984, individual line flushing gave way to flush manifolds as the preference for nearly all my growers.

In those days we didn't have the computing power nor the understanding nor indeed the need to optimize flush manifold design. Tape systems were seasonal and if the flushing was not done optimally, it didn't matter so much provided it were done to some extent. The tape would be replaced for the coming season anyway, giving a fresh beginning. Flush line design ended up being really token hydraulic design with many assumptions because the hydraulics were dynamic and complicated. To hopefully insure that each line was properly flushed and avoid dead zones I always called for multiple flush manifolds having flush valves on the ends of each (2, 3, or 4 per each source manifold or submain depending on width). Though not optimally designed, overall flushing throughout the season was greatly improved using these flush manifolds.

Times have changed. Permanent subsurface drip irrigation (SDI) systems require operational viability for ten years or more. It is here that flush manifold design and precision management is critical to system longevity.

Traditional Flush Manifold Design

In 1996, Dr. Charles Burt, a classmate on mine at Utah State University in the mid 1970's, published an article titled "Sizing of Header and Flushing Manifolds for Row Crop Drip" in the May/June edition of Irrigation Journal. This was a call to designers to consider the implications in both header and flush manifold design to achieve tape flow-through velocities of at least 1 fps for tape flushing. Further, there was a call for an increased pressure at the field valve to bring more water and pressure to bear on the subunit to achieve proper drip line flushing. Figures 1 and 2 from that paper are shown here as Figure 1.

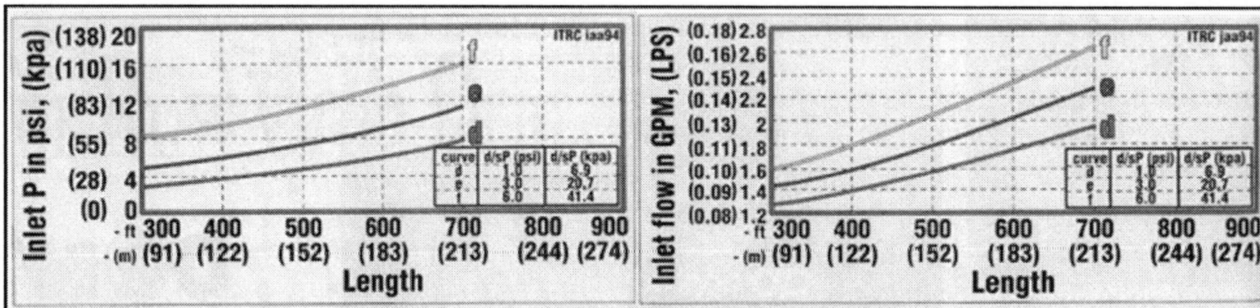

Figure 1. Inlet pressure vs. length for 0.625-inch ID drip tape. Q = 0.22 GPM/100 feet at eight psi (excluding c.v. consideration). Various downstream pressures and and fixed flushing flow of one GPM, (0.063 LPS) x = 0.5, zero slope.

Figure 3. Inlet flow vs. length for 0.625-inch ID drip tape. Q = 0.22 GPM/100 feet at eight psi (excluding c.v. consideration). Various downstream pressures and fixed flushing flow of one GPM (0.063 LPS), x = 0.5, zero slope.

Figure 1. Figures 1 & 3 from Charles Burt, "Sizing of Header and Flushing Manifolds for Row Crop Drip"

As can be seen in these figures, required flush flows can easily reach double the operating flows and the corresponding pressures to achieve those flows can easily triple or quadruple depending on the emitter exponent. Logical conclusions would be bigger manifolds, adjustable pressure regulators, and thicker walled tape to withstand high flush mode pressures. Because of the "tremendous variability of design" as noted by Dr. Burt and due to the shear complexity of analysis of the interactive hydraulics of an irrigation subunit in flushing mode, it is difficult to give a single or even multiple rules for header and flush manifold sizing.

Dr. Burt concludes, "Hopefully this article will encourage designers to plan for larger flushing lines as well as larger valves and adjustable pressure regulators at the inlet to the blocks." These qualitative recommendations have evolved into traditional flush manifold design criteria.

In a day where SDI is proliferating in many crops and extending deep into nearly every clime, it is time for a new more quantitative look at how we design and manage SDI systems, particularly sub-unit flushing.

The Flush and Source Manifold System

The dynamic complexity of sub-unit hydraulics is modeled to find strategic elements that define system operation. AES International, PLLC has developed a database software application to solve for the unknowns, qExtra, qHelped, the qNeutral location along the flush manifold, and the qZero array of distal lateral dead flow locations as shown in Figure 2. Solving for these unknowns not only leads to optimal sub-unit design, it forms the basis for operational and flushing solutions in the field.

In taking a more digital approach to sub-unit design wherein every emission point within the sub-unit is evaluated, two new logical design parameters became apparent. These duo sub-unit design parameters are UDU (Ultimate Distribution Uniformity) and CHLR (the Cost Head-Loss Ratio). CHLR/UDU sub-unit design more logically elucidates design results and more properly defines risk compared to traditional design. CHRL/UDU sub-unit design delivers the most economical field solution for a given risk for both operational and flushing modes.

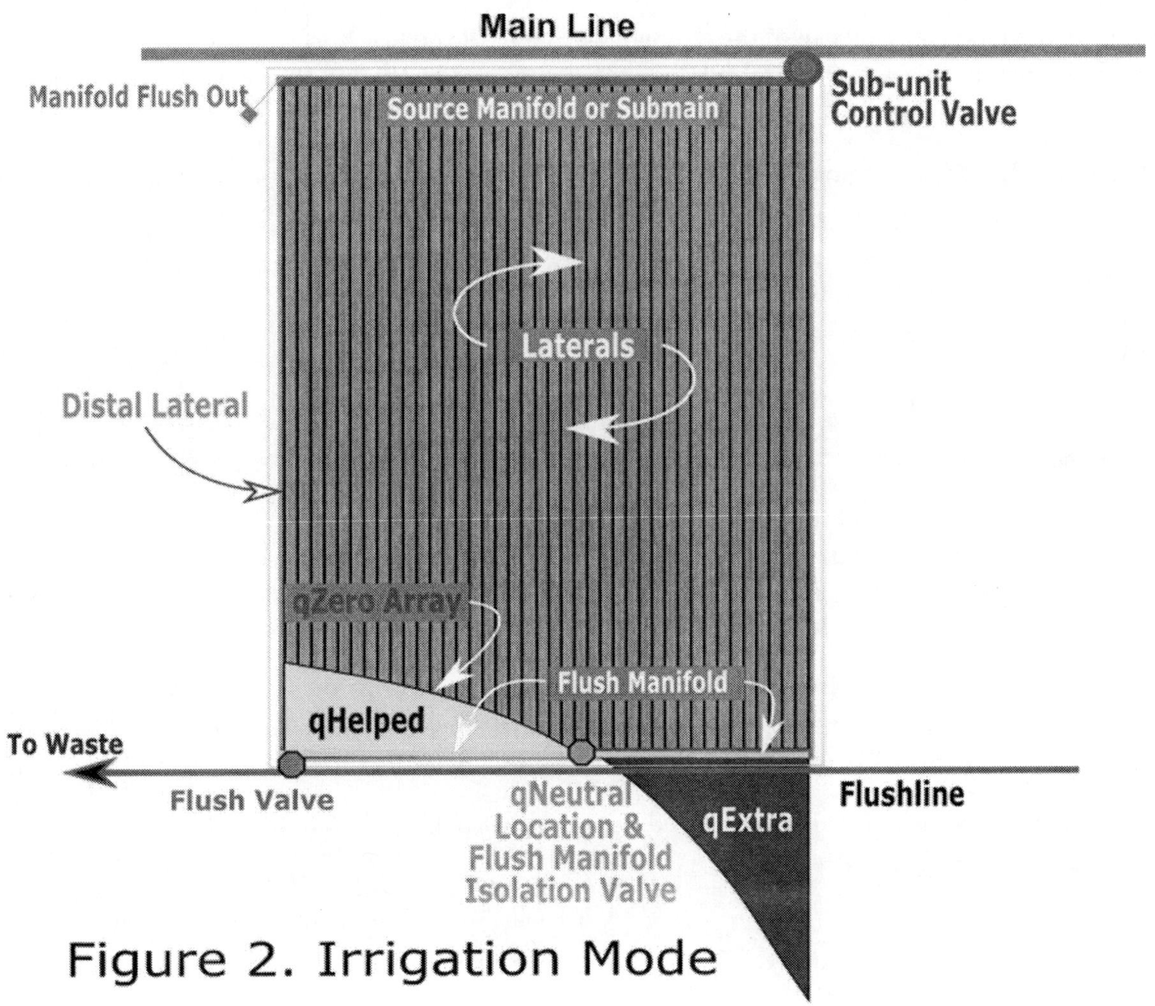

Figure 2. Irrigation Mode

UDU – Ultimate Distribution Uniformity or Ultimate DU

In 1986, standards for structural design changed from ASD (Allowable Stress Design) to USD (Ultimate Strength Design). USD gives the ultimate strength capacity of a structure at no risk. USD is a direct function of a structural configuration with design loads at zero risk. In final design, risk factors are evaluated and applied to define maximum safe loads for the structure. UDU (ultimate DU) in drip irrigation sub-unit design is the equivalent of USD for structural design.

UDU does not include emitter variability risk. Also, as with USD, UDU is a direct function of sub-unit configuration and entrance pressure along with the pressure vs discharge relationship of the emitter to be used. UDU is obtained by evaluating and tabulating the discharge of each emission point in the sub-unit for a given sub-unit configuration and entrance pressure. The sub-unit submain or manifold piping configuration is determined by the value of the CHLR.

For a given value of the CHLR, the corresponding UDU is the ratio of the average of the low ¼ of the sub-unit emission point discharges to the overall average emission point discharge. UDU is a stable measure which is completely reliable to the extent design inputs are reliable. This is the logical design measure to which risk factors can be applied.

This follows the long-standing use of the average discharge of the low ¼ as the practical value for minimum discharge. This was the methodology recommended by the old U.S. Soil Conservation Service (now the NRCS) for field evaluation of irrigation systems.

Jack Keller & David Karmeli also utilized the concept of the average discharge of the low ¼ in their epic development of the expression for the drip irrigation sub-unit design EU (design emission uniformity). EU was the precursor to DU. As part of that development, they came up with is an emitter variability risk factor called the AMDR (Adjusted Manufacturers' Discharge Ratio)

$$\text{AMDR} = \left(1.0 - \frac{1.27}{\sqrt{e}}\, v\right) \qquad \text{Equation 1.}$$

in which

- e is 1.0 or the number of emission points per plant
- σ is the standard deviation for the emission device at the average emission pressure
- qa is the emission flowrate for the emission devise at the average emission pressure
- *v* is the manufacturers' coefficient of variation which is (σ/qa)

A discharge of 1.27σ below qa is precisely the average discharge of the low ¼ based on the standard curve. The AMDR is the ratio of the average discharge of the low ¼ to the overall average discharge at the average emission pressure for an emission device with a manufacturers' coefficient of variation of *v*.

Emitter variability risk as defined by Keller and Karmeli, AMDR, is applied to UDU to get a more logical and precise system design efficiency, DU.

$$\text{DU} = \text{UDU x AMDR} \qquad \text{Equation 2.}$$

in which

- DU is Distribution Uniformity
- UDU is Ultimate Distribution Uniformity
- AMDR is the Adjusted Manufacturers' Discharge Ratio

All other risk factors such as mapping and elevation irregularities, ET miscalculations, potential water source failures, and cyclical extended heat waves are handled within the number of extra daily operational hours build into the system.

CHLR – Cost to Head-Loss Ratio

The Cost to Head-Loss Ratio (CHLR) is made up of a specific measure for the Numerator and a different specific measure for the Denominator as follows:

CHLR Numerator: The CHLR numerator is the cost of increasing the size of a pipe segment to the next larger pipe size versus continuing with the pipe size of the previous pipe segment.

CHLR Denominator: The CHLR denominator is the decrease in the pipe segment head loss from changing to the next larger segment pipe size versus continuing with the previous segment pipe size.

Each pipe segment between laterals making up a source manifold or submain has a different flowrate. The maximum flowrate occurs immediately downstream of the sub-unit control valve. The flowrate diminishes as water leaves, feeding the laterals along the manifold. The minimum flowrate occurs in the last pipe segment feeding the distal lateral.

CHLR/UDU design protocol is an iterative process starting with the distal pipe segment diameter and an initial CHLR target value. A unique UDU sub-unit value will result from a given CHLR target value. The design protocol begins at that distal pipe segment of the manifold with an initial diameter carrying only the distal lateral flowrate. Due to this low distal pipe segment flowrate, the CHRL value will be many multiples above the target value at the start.

Moving along pipe segments from the distal pipe segment toward the control valve, the manifold flowrate increases while the CHLR numerator remains constant. A constant CHLR numerator combined with an increasing CHLR denominator due to escalating manifold flowrates lowers the CHLR toward the target value with each passing pipe segment. When the CHLR target is reached the process calls for a pipe size increase. After the pipe diameter increase the CHLR numerator moves to a new constant value. The CHLR denominator also changes. These changes again result in the CHLR value spiking upward many multiples above the target value for that first increased diameter pipe segment. The process is repeated through all the pipe segments until the control valve is reached.

The optimal economic design solution is where a chosen CHLR delivers the desired UDU. The first design pass results in an initial UDU which is unique to the initial CHLR and sub-unit configuration. If UDU is at the proper level for the project circumstances, design is complete. If UDU is too low, then the CHLR target is increased and the design protocol is repeated. If the UDU is too high, then the CHLR target is reduced and again, the design protocol is repeated.

Sub-unit Flushing Dynamics

The flushing manifold serves a dual function. In flush mode it carries flush water out away from the sub-unit to waste. In irrigation mode, it carries lateral flow-through water from the laterals closest to the sub-unit control valve to the bottom reaches of the laterals furthest from the control valve.

As can be seen in Figure 2, there is a qNeutral location along the flush manifold in which sub-unit hydraulics are such that lateral flow-through water is neutral. All laterals upstream from the neutral location are constantly being flushed during normal irrigation operations. This natural flushing increases to a maximum for the lateral starting closest to the control valve. Conversely, there is a dead flow point for each lateral, defined by the qZero array, from the distal lateral all the way to the qNeutral location along the flush manifold.

For effective sub-unit flushing when opening a valve from the flush manifold to atmosphere, sub-unit hydraulics must be such that flushing occurs on all laterals. The desirable lateral flushing flowrate should be at a maximum on the distal lateral decreasing to a minimum at the qNeutral location.

Contrary to common thought and practice, the worst thing to do for sub-unit flushing hydraulics is to increase the inlet pressure to the sub-unit. That action sends more flow-through water down the rows nearest the sub-unit control valve. It also pushes higher flows through the manifold pipe sections closest to the control valve, burning up manifold pressure needed at the distal end of the manifold. This action exacerbates the problem. Maintaining the control valve outlet pressure the same in flush mode as in irrigation mode, or even reducing the pressure, would provide better flushing hydraulics.

A low-pressure swing check valve is installed in the line between the Flush Manifold and the Flushline near the distal end. See Figure 3. This allows for multiple sub-units to be flushed using a single Flushline and a single Flush Valve hydraulic closing/opening line. A hydraulically operated valve which is either opened or close is placed in the flush manifold precisely at the qNueutral location. This is the Flush Manifold Isolation Valve shown in Figure 3. A second hydraulically operated valve is placed right after the check valve. This is the Flush Valve shown in Figure 3. With the Flush Valve opened, flush water can flow from the flush manifold into the Flushline and then on to atmosphere. Both the Flush Manifold Isolation Valve and the Flush Valve can be operated remotely via a ½" Sch 40 PVC hydraulic line.

Sub-unit Flushing Procedures

The two hydraulically operated valves can be opened and closed multiple times to create localized surge flushing to take further advantage of momentary elevated local pressure differentials across the sub-unit set up by using CHLR/UDU design protocols.

1. Close the Flush Manifold Isolation Valve. See Figure 3, 1st Flushing Mode
   a. The source manifold pressure will initially increase down the line toward the distal end. This will cause a surge in lateral flow-through just downstream of the Isolation Valve giving a good flushing to the laterals in that section.

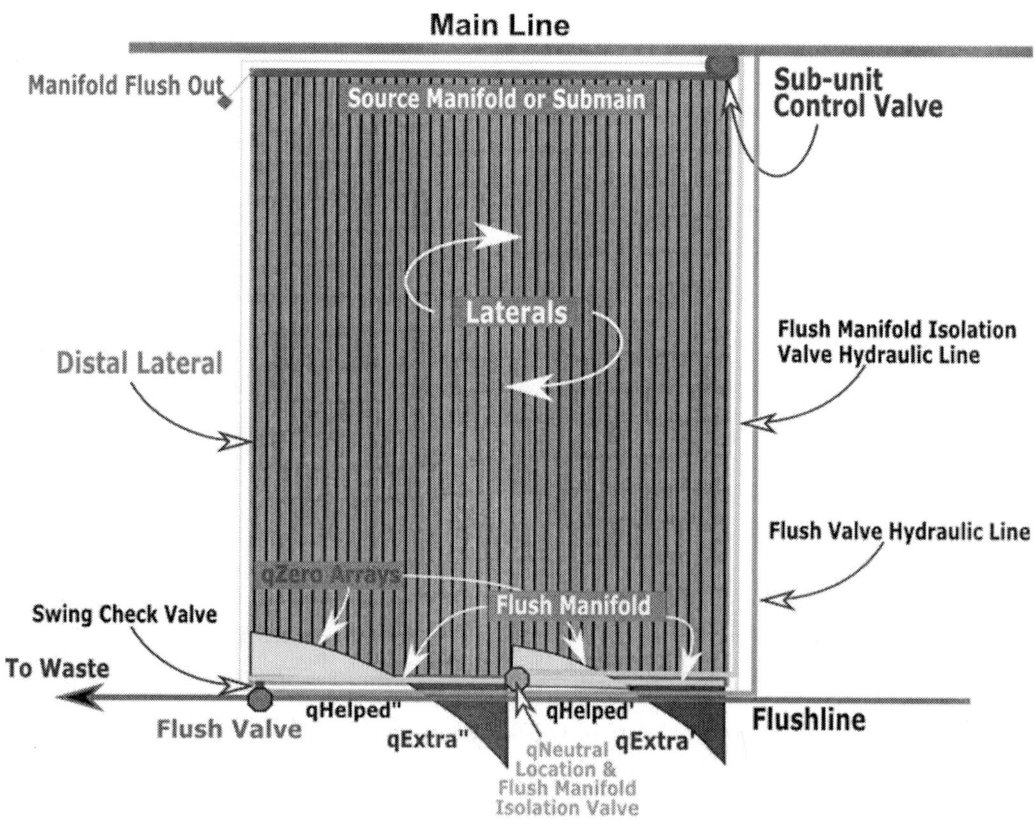

Figure 3. 1st Flushing Mode

b. New dual gExtra and qHelped equilibrium hydraulics will emerge.
c. Leave this configuration until after the system comes to equilibrium.

2. Open the Flush Valve. See Figure 4.
    a. This will reverse the previous hydraulics causing an initial surge of flow-through flushing through the distal laterals, which will decrease to a minimum just downstream of the Isolation Valve where before Flush Valve opening the highest lateral flow-through occurred. This creates localized cycle flushing.
    b. The sub-unit hydraulics will gradually come to equilibrium and approach the steady state flush designed for the manifold/flushing manifold sub-unit configuration together with the flush line configuration from the distal end flush valve to where the Flushline flow outlets to atmospheric pressure.
    c. In this configuration a pressure differential will reach a maximum across the Flush Manifold Isolation Valve.

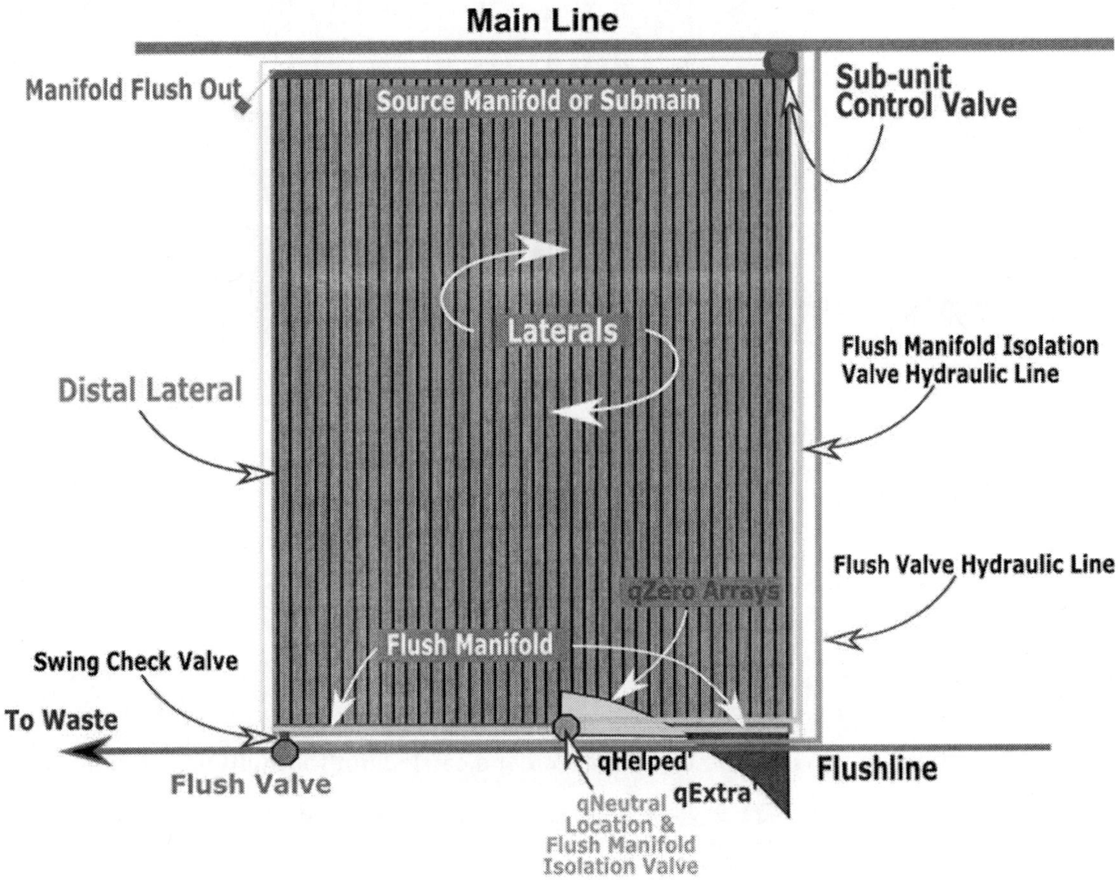

Figure 4. 2nd Flushing Mode

3. Open the Flush Manifold Isolation Valve. See Figure 5.

a. This will cause an initial surge of flow-through flushing from the laterals just upstream from the isolation valve in response to the elevated pressure differential.
b. This flushes the irrigation mode qExtra section laterals that have the least lateral flow-through during normal irrigation operation.

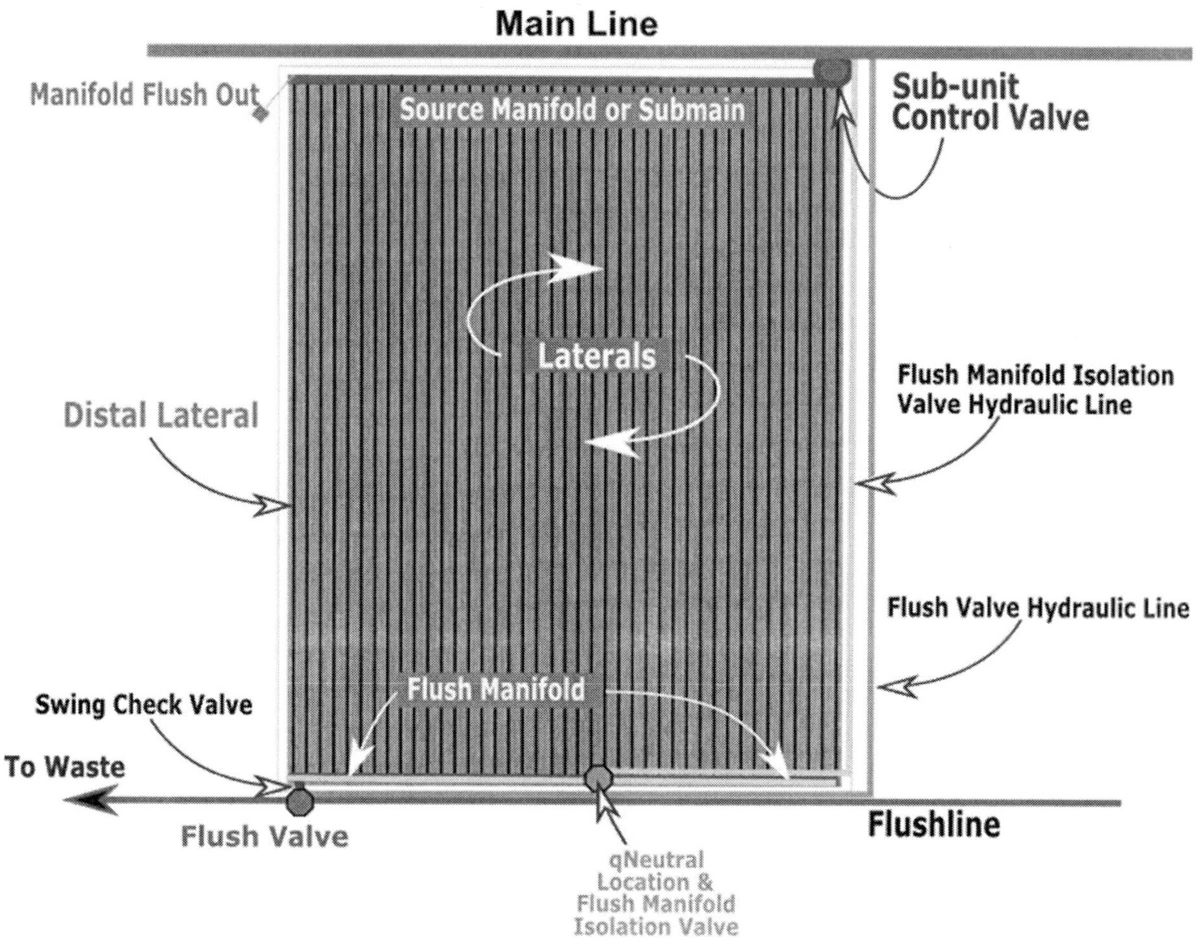

Figure 5. 3rd Flushing Mode

4. Close the Flush Valve to return system to irrigation mode.
5. The flushing system can also be configured and carried out for multiple sub-units.

Conclusion

Implementing CHLR/UDU design protocols for sub-unit SDI irrigation design and leveraging system operation finesse enabled by the design protocols can dramatically reduce system cost, improve the efficacy of sub-unit flushing compared to traditional solutions, and more correctly assess DU in terms of defined risk.

References

Burt, CM. 1996. Sizing of Header and Flushing Manifolds for Row Crop Drip. Irrigation Journal. May/June p 24-27.

Keller, J and Karmeli, D. 1975. Trickle Irrigation Design. Chapter 4 in
Selection and Efficiency of Emitters, Rain Bird Sprinkler Manufacturing Corporation

Development of a Decision Support System, DOMIS, for Designing Micro Irrigation Systems

T. B. S. Rajput*, Neelam Patel** and Deepak Dinkar***
*Emeritus Scientist, ** Principal Scientist and ***Research Associate
Water Technology Centre, Indian Agricultural Research Institte, New Delhi, INDIA

Abstract

A decision support system namely DOMIS (Design of Micro Irrigation Systems) was developed for standardizing the designs of Drip, Sprinkler and Micro Sprinkler irrigation systems. The decision support system was developed using standard formulae with richness of knowledge and expert opinions. Flexibility and simplicity of use make the DSS DOMIS a superior tool for designing micro irrigation systems. The DSS provides knowledge, expert opinion and necessary data on crops, soil, water and climate in the form of default options at different interactive stages of the software. The DSS also allows the user to make appropriate changes in the parameters provided by the DSS, if the user so desires. DSS-DOMIS suggests most optimal layout plan for main, sub-main and lateral pipes with appropriate sizes of different components including pipes, pumping system, filters and fertilizer application systems. The web based system has been hosted on the net with domain name http://domis.iari.res.in. Besides the design of micro irrigation systems, the DSS provides complete list of concerned State and district officials, approved micro irrigation system suppliers and the general information about the district, different Government schemes and their implementing agencies all over the country. The decision support system will be useful for industry, farmers, students, researchers and policy makers. The DSS is available for use or free on line and the author will appreciate the feedback and suggestions from the users for its further improvement.

Keywords:
Micro Irrigation System Design, Decision Support System, Micro irrigation software

Introduction

India has limited land and water resources with only 2.3 % land and 4% of water but with a population of over17% of the World. Immediately after independence in 1947, Government of India started investing rather heavily in increasing its water resources for irrigation from a mere 22.6 Mha to the present level of over 100 Mha out of a total potential of 140 Mha. Introduction of high yielding varieties of cereals combined with developed water resources and of course with the use of chemical fertilizers brought about a green revolution in the country. Government of India also initiated several National and State level organizations for developing package of practices appropriate for local agro climatic and socio economic situations. All these strategies resulted in quantum jump in food grains production from a mere 52 MT in 1952 to 295 MT in the year 2017. This quantum jump in production for food sufficiency adversely affected the water resources availability, particularly ground water. To increase the production further to meet the requirement of growing population there is an urgent need to conserve the depleting water resources through policy, peoples' participation, capacity building, technological advancement and technology transfer pertaining to savings of each and every drop of water.

Agriculture is the major consumer of water (about 80 %) with very poor utilization. Initially very big dams were created for storing of water and then linked with canal water distribution network. Timely completion of big irrigation projects was all the time challenging and having several litigations. Irrigation experts then started realizing that instead of creating additional large dams/ irrigation projects which have long gestation periods and are no longer economically viable due to cost escalation (many fold) and low water-use-efficiency (as low as 38%) such projects have. This happened because of greater emphasis only on the development of water resources and little/ no emphasis on its efficient utilization. Since surface and underground water resources were made available to the farmers free of charge, these resources were exploited mercilessly. This has been reported by the experts that water would be the single most critical limiting factor for India's economical development requiring continuous promotion of micro irrigation for reducing water consumption in agriculture.

In the year 1985, the Government realized the urgent need to promote water saving technologies in agriculture to sustain against the increasing water demands of progressively developing urbanisation, industrialization and other sectors of the economy requiring water. In India land holdings are very small (more than 80% lad holdings are less than 1 ha) and the farmers are resource poor. Micro irrigation is the most efficient method of irrigation but is expensive too. It was understood that small and marginal farmers in India will not be able to benefit from microirrigation technology owing to its high cost and only a few rich farmers having large holdings only will be able to use the technology. Government after analysing the situation decided to offset the high cost of micro irrigation equipment through a subsidy scheme for small and marginal farmers (with land holdings less than 5 ha only). To address the problem of very poor and very small farmers provision of additional subsidy was kept in the scheme. The scheme also addressed the issue of regional inequity in agricultural development by keeping different levels of subsidies.

With the financial support schemes and other enabling initiatives of the Government including, funding of research for developing package of practices for growing different crops under different agro-climatic conditions under micro irrigated environment, the adoption of micro irrigation in the country started increasing. The Government started putting stringent conditions on the system suppliers to manufacture locally to be eligible to participate in the scheme partly funded by the Government. Government also established a National Committee in the Bureau of Indian Standards for developing and certifying irrigation equipments manufactured in India for their desired quality.

The author has had the privilege of being associated with the scheme since its inception in 1985 and has lead the largest Precision Farming Development Centre located at Water Technology Centre at Indian Agricultural Research Institute, New Delhi for over two decades; He has been involved in different Guidelines Formation Committees and subsidy committees of the Government. He has had close interaction with industry. He has been a regular member for over two decades and continues to be a regular member of the Bureau of Indian Standards Committee for formulation of Indian Standards for Irrigation equipments. The author has been a researcher and has published more than two hundred research articles, ten books and twelve bulletins on micro irrigation. His contributions have earned him the top national recognitions and awards.

The current Government has launched a very ambitious irrigation scheme in the country on June 2016 namely, PMKSY (Prime Minister Krishi *Sinchai Yojna* i.e. Prime Minister's Agriculture Irrigation Scheme). Under the scheme incomplete 99 irrigation projects lingering for long period will be taken up on priority for time bound completion which will increase estimated irrigation potential by 2 Mha. Under the scheme micro irrigation has been given another thrust as under the scheme it has been made mandatory (1) to bring at least 10% canal command areas under micro irrigation particularly in the tail end reaches of canal commands to address the problem of lesser water availability at the tail end of the canal owing to excess withdrawals on head reaches/ inadequate canal supplies. (2) to use harvested rain water under all watershed projects through micro irrigation only and (3) include at least 25% non horticultural crops too under micro irrigation scheme. It can be concluded that these Government policies not only have brought India to number one position in micro irrigated area coverage but will also attain a height which will not be easy for any other country to surpass in near future.

All these developments resulted in adoption of micro irrigation systems in very large number of fields (mainly small fields) across the country for different crops under different agro-ecological situations. Standardization of design of micro irrigation systems and the quality control on the type of systems became critical parameters for successful adoption of micro irrigation technologies. It was almost impossible to train all the persons involved in designing the micro irigation systems in the country therefore a need was felt to develop a technological tool for designing micro irrigation systems efficiently for different crops under different agro climatic conditions.

Design of microirrigation systems requires the data of crop, soil, climate and the hardware of the system. All the data may not be readily available to multitudes of the persons engaged in micro irrigation design at field level across the country. Therefore a Decision Support system equipped with complete data set of all the parameters needed for designing efficient micro irrigation systems was considered as the need of the hour for the success of micro irrigation scheme in the country.

Development of DSS

A DSS is a computerized system for helping in any decision-making process, which integrates multidisciplinary databases, modelling tools and multi criteria analysis methodologies that are useful to analyse and rank a set of alternatives. Supporting a decision means helping decision makers to generate alternatives, rank them and make choices, which is particularly useful for design. Design of micro irrigation systems is an appropriate field for application of DSS as it requires in depth knowledge, a large number of data relating to soil, crop, climate water resource and its quality. Since the user may or may not have ready access to all these data his designs may be suboptimal and inefficient and uneconomical too. Therefore a DSS namely, DOMIS was developed for designing efficient micro irrigation systems even when the basic data required on soil, crop, climate and water source is not readily available with the user.

DOMIS- Design of Micro Irrigation Systems:

DOMIS was developed to design different micro irrigation systems including drip, sprinkler and micro sprinkler systems. The DSS provides help and guidance to the users about how to design appropriate micro irrigation systems for efficient water utilization under any given agro-climatic conditions. DOMIS with its huge data bank enables the user to design micro irrigation systems efficiently even without having the required data on soil, crop, climate and water source. The DOMIS is a web based DSS and can be accessed from any internet enabled device including desktop, laptop, mobiles etc. In the DSS the open source programming languages PHP and DHTML are used for front end design but for back end data connectivity MySQL is used.

DSS architecture

The DSS DOMIS has been developed in a simple architecture with five sequential building blocks as follows (Table 1).

Table 1. Building blocks of DOMIS architecture

SN	Name of block	Functions performed
1	Presentation block	web based interface, tools and applets are embedded
2	Data block	different data cache on crops, soil, climate and system hardware data are arranged
3	Information block	the option for the user to change / modify different data are provided
4	Algorithm block	All the logics and computations are done under this block
5	Final/ Report block	The fifth and final block presents the report of system design and the cost estimates.

Computational procedure in DOMIS

The following procedure of computation is followed in the DOMIS in its algorithm/ logic block. At each stage the user is provided with a default option with an option for the user to chose and use his input, if he so desires.

i. Division of the field into blocks (of desired dimensions, based on the field size and the selected system)
ii. Determination of appropriate system layout plan (for network of main, sub main and lateral pipes, based on the location of source and to minimize system cost)
iii. Estimation of water requirements (based on reference crop evapo-transpiration and dynamic crop coefficient and canopy factors)
iv. Determination of suitable water application rate (based on soil type, crop and water source)
v. Determination of sizes of different components of system network (sizes of main, sub main and lateral pipes on the basis of crop water requirements and by optimally balancing the cost and frictional losses within the pipes in limit)
vi. Estimation of size & type of accessories (filters & fertigation equipment, based on water quality and fertigation needs)
vii. Determination of total head requirement (suction, delivery heads and friction losses)
viii. Determination of size of motor pumping unit (based on flow requirements and the total head requirement)
ix. Estimation of the cost of the designed system (Based on the costs of different components of the designed system)

Data embedded in DOMIS and its source

The data embedded in the DSS were collected mainly from different Government authorized sources including historical daily agro-climatic data i.e. potential evapotranspiration (ET0 mm/day), soil properties, ground water table situation, wind velocity information of all the 718 districts of 29 States and 7 UT's in India. The crop database includes all crop parameters of different crops grown across the country. The crop parameters include effective rooting depths, crop coefficients (Kc) and crop spacing. Many crops data were taken from open source FAO publications Number 24 and 56 namely, Crop water needs. The DOMIS provides the user with stored data as default option but with flexibility to the user to modify the default data as per best of their knowledge.

Operation and use of DOMIS

The web based DSS DOMIS has been hosted on the internet with the domain name as http://domis.iari.res.in. In the DSS an interactive Graphical User Interface (GUI) and responsive home page was designed (Fig 1). Besides helping in designing different micro

Fig. 1. Opening screen of DSS DOMIS

irrigation systems, DOMIS also provides general information about Government schemes, different micro irrigation systems, details about districts, implementing agencies in different States of India, approved system suppliers in different States and other general information about micro irrigation.

The DSS presents a little introduction, its scope and capabilities in in its opening screen (Fig 1). Interactively the user is guided in stepwise data inputting either by accepting default option or feeding his data. For example, on the crops data screen, the user is asked to identify a crop from the available list and its photograph and all other parameters, crop evapotranspiration, crop factor, canopy factor and plant to plant and row to row spacing are presented by default (Fig 2.). These may be accepted by the user as such or he can make appropriate changes as per his knowledge. Similarly the user is presented with the concerned appropriate data on different parameters dealt on different screen for the benefit and help of the user of the DSS. Using the input data and the standard design procedures the DSS helps the user in finalizing the layout plan for main, sub main and lateral pipes, dripper discharge and spacing, size and types of filters, fertigation unit and the motor and pumping units for the selected micro irrigation system. It also provides the estimated cost of the designed micro irrigation system based on the prices of different micro irrigation system components at the price index of 2017 (Fig. 3).

Fig. 2. Crop and its evaporation data sheet

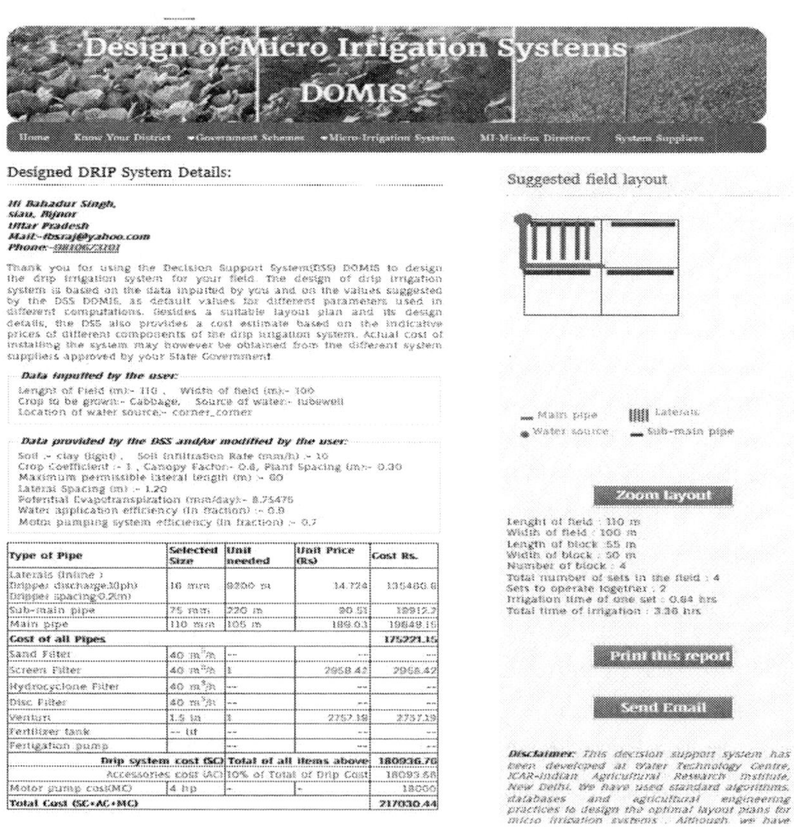

Fig. 3. Design specifications and estimated costs of the different components of the designed drip irrigation system

Conclusion:

The developed decision support system for Design of Micro Irrigation Systems (DOMIS) will provide all the information related to design of system as well as cost estimation for drip, sprinkler, micro sprinkler. It will evaluate the economic feasibility with details cost calculation of the suggested layout in given agro-climate details. The tool will prove to be very useful in optimizing the designs of micro irrigation systems designs under different soil, crop and climatic conditions by multitudes of the personnel involved across the country. The web based system was hosted on the internet with domain name http://domis.iari.res.in on December 15, 2017 and has already touched over 4000 hits which is indicative of its use and acceptance across the country. International Commission on Irrigation and Drainage has also hosted the DSS on their website for use of everybody.

Acknowledgements

The decision support system DOMIS was developed at W,ater Technology Centre, Indian Agricultural Research Centre, New Delhi-110012 under the Precision Farming Development Centre (PFDC) project activities. The financial support from the National Committee on Plasticulture Applications in Horticulture (NCPAH), Ministry of Agriculture & Farmers Welfare, Government of India, is thankfully acknowledged.

Presented at 2018 Irrigation Show & Education Conference, Dez 3-7, 2018
Long Beach Convention Center, Long Beach, California, USA

PERFORMANCE OF IRRIGATED BLACK PEPPER

Gustavo Haddad Souza Vieira. Full Professor. PhD.
IFES *campus* Santa Teresa, Rod. ES 080, km 93, São João de Petrópolis, Santa Teresa-ES, Brazil. 29.660-000. e-mail: ghsvieira@gmail.com

Géssica Monteiro dos Santos. Agronomy undergraduate student.

IFES *campus* Santa Teresa – Rod. ES 080, km 93, São João de Petrópolis, Santa Teresa-ES, Brazil. 29.660-000.

Marcelo Rodrigo Krause. MSc student.
Universidade Federal de Viçosa. Campus Universitário, Viçosa-MG, Brazil. 36570-900.

Milena Monteiro dos Santos. Agronomy undergraduate student.

IFES *campus* Santa Teresa – Rod. ES 080, km 93, São João de Petrópolis, Santa Teresa-ES, Brazil. 29.660-000.

Paola Alfonsa Vieira Lo Monaco. Full Professor, PhD.

IFES *campus* Santa Teresa – Rod. ES 080, km 93, São João de Petrópolis, Santa Teresa-ES, Brazil. 29.660-000.

Abstract: *This study aimed to evaluate irrigation depths in the performance of black pepper (Piper nigrum), in São Mateus-ES, Brazil. We performed the experiment with five treatments and four replications, with depth irrigation replacement being varied according to the crop evapotranspiration (ETc). In the first year we evaluated: number of leaves per plant (NLP), plants height (PH), stem diameter (SD) and leaf temperature (LT). The best results were, respectively, 52.5 leaves per plant; 173.1 cm and 13.5 mm, with 100% replacement of ETc, at 199 days after planting. For leaf temperature, we observed lowest values in the 100% treatment. In the second year we evaluated the number of bunches per plant (NBP) and the productivity of fresh grains (PFG) and dry grains (PDG). For the NBP and PDG variables, the best results occurred in the depths of 75% and 100%, respectively. We recommend 100% of ETC replacements for the first and second year.*

Keywords: water, efficiency, yield, development, leaf temperature, evapotranspiration.

INTRODUCTION

Black pepper (*Piper nigrum* L.) is a climbing plant, belonging to Piperaceae family, of high productivity and one of the most valued spices in the world, with great economic value and also known as Indian pepper or black gold (Lima et al., 2010).

Introduced in Brazil in the seventeenth century in the state of Bahia, and then brought to the States of Paraíba, Maranhão and Pará (Pedeag, 2016), black pepper presented economic importance only in 1993 year, when it was reintroduced in the state of Pará by the Japanese people. Currently, it has a major presence in the municipalities of the North of Espírito Santo state, making it the second largest producer and exporter at the national level, behind only Pará. The plantations in the North of the State are concentrated in the municipalities of São Mateus and Jaguaré, with more than 75% of cultivated area and production (Ramos, 2015).

Lima et al. (2010) state that Brazil is one of the largest producers of black pepper, oscillating between the second and third position in the world market. Due to the good adaptation of the different types of soil, there are several commercially accepted cultivars of black pepper in Brazil, among them Bragantina, also known as Panniyur, which has a denomination of the original cultivar in India (Lemos et al. 2014; Oliveira et al., 2007)

The black pepper is the most used spice by Brazilian cuisine. It is mainly used in industrialized products (salami, sausage, mortadella, ham etc) and for food seasoning. Brazil consumes only 10% in the form of whole grains, the other 90% are used in the form of milled grains, in mixtures with other condiments (Duarte and Albuquerque, 2005).

Grain productivity is indicative of the profit for the producer, being directly related to the profitability of the crop. Therefore, irrigated crops tend to be more profitable, since they can present a longer period of grain filling (flowering until maturation) and green leaf area to perform photosynthesis and remobilize the reserves, providing a greater supply of grain assimilates (Gomes, 2016 and Maehler et al., 2003).

According to Bonomo et al. (2013), irrigation management is a very important technique from an economic and environmental point of view in an irrigated agricultural activity, because through proper irrigation management, one can save water, energy, increase crop productivity and improve product quality. Therefore, irrigation practices are important to ensure production in the regions most susceptible to water deficit. Based on this, it is essential to know the water requirement of the black pepper, cultivated under different irrigation depths, so that it has a good productivity and a better use of the applied water, avoiding excess or water deficit.

The objective of this study was to evaluate the effect of irrigation depths on the initial development and productivity of black pepper cultivar bragantina and to identify the percentage of irrigation that provides the best water use efficiency.

MATERIAL AND METHODS

The cultivation was carried out in a field in the municipality of São Mateus, Espírito Santo state, Brazil, at coordinates 18° 44' S and 40° 06' W, in a altitude of 77 m in relation to sea level, as shown in Figure 1.

Figure 1. Location of the experiment.

The experimental design was a randomized block design (RBD), containing five treatments and four replications. Four useful rows were planted, spaced 2.5 m between rows and 2 m between plants. Six plants were planted for each treatment, two plants were considered as borders and four useful plants, totaling 144 plants in the experiment, being 96 useful plants.

Pepper seedlings were planted in December 2016 and conducted until the beginning of April 2018 for 470 days. The period was divided into two phases, the first of which from planting until 199, called Year 1 (Vegetative stage), and the second, from day 200 to day 470, called Year 2 (Productive stage). The local soil has sandy texture, field capacity of 18%, permanent wilting point of 9% and apparent density of 1.15 g cm^{-3}. The effective root system depth considered was 0.4 m and the water availability factor of the crop (f factor) was 0.4.

For irrigation depths variation, we installed different numbers of emitters per plant, so that all irrigations occurred at the same time. The treatments consisted of different irrigation depths, proportional to crop evapotranspiration (ETc), as follows: T1: 25% of ETc (1 emitter); T2: 50% (2 emitters); T3: 75% (3 emitters); T4: 100% (4 emitters); and T5: 125% (5 emitters). A localized irrigation system microspray type was used, working with pressure of 100 kPa and flow rate of 10 L h^{-1}. Figure 2 shows images of the experimental area, on planting (A and B) and the production (C) stages.

Figure 2. Experimental area at planting time and in the production stage.

After the installation of the irrigation system and the plots demarcation, we assessed the irrigation system, and calculated the CUC (Christiansen Uniformity Coefficient) and the emitters flow rate (Thompson and Ross, 2011). For the evaluation of the irrigation system efficiency, treatment 4 (T4 - 4 emitters) was used as a reference, applying the 100% irrigation depths related to the crop evapotranspiration. In the evaluation, a 1,000 mL beaker and measuring cups were used, evaluating 16 emitters in total, measuring the volume of water collected from each emitter in a time of 20 seconds.

Irrigation management was performed using a spreadsheet (Figure 3), determining the black pepper water demand, using coefficients of adjustment (soil moisture coefficient - K_S and coefficient due to the location of the irrigation - K_L) on the reference evapotranspiration (ETo). The values of K_S were calculated with the logarithmic model (Mantovani et al, 2009). The percentage of wetted area was 40%, resulting in a K_L of 0.63, calculated by the method of Keller and Bliesner (1990). The gross irrigation depths were calculated by means of a water balance, in which the inputs of water were irrigation and rainfall and the exit was crop evapotranspiration (ETc).

	A	B	C	D	E	F	G	H	I	J	K
1	**Irrigation scheduling using the Hargreaves-Samani method for ETo calculation**										
2	Spreadsheet developed by Gustavo Haddad Souza Vieira										
3	ghsvieira@gmail.com			Yellow cells = data entry					Crop	Black Pepper	
4		**Graus**	**Minutos**	**Equipamento**							
5	**Latitude**	18	45	Emitter flowrate		40	L/h		Kc initial	0.6	
6	Hemisphere	S		Emitter spacement		2	m		Kc medium	0.7	
7	altitude (z)	79	m	Lateral lines spacement		2.5	m		Kc final	0.9	
8	**Soil**			CUC		93	%		Plants spacing	2	m
9	Field capacity	18	%		**Net irrigation**	16.6	mm		Rows spacing	2.5	m
10	Permanent wilti	9	%		**Gross irrigation**	17.8	mm		PAW (%)	40	%
11	Bulk density	1.15	g/cm³		**CAD**	41.4	mm		PAS (%)	20	%
12	f factor	0.4			**Aplication rate**	8.0	mm/h		K_L	0.63	
13	Soil root depth	40	cm								
14	**Date**	**Tmax (°C)**	**Tmin (°C)**	**Kc**	**ETc (mm/day)**	**Irrigation time (min)**	**Rainfall (mm)**	**Recommended irrigation time (min)**	**Measured soil moisture (%)**	**Calculated soil moisture (%)**	
15	12/29/16	34.90	24.10	0.6	2.3	21.0		0			
16	12/30/16	34.80	24.30	0.6	2.2	10.0		8		17.8	
17	12/31/16	32.40	23.30	0.6	2.0	10.0		14		17.6	
18	1/1/17	34.40	23.80	0.6	2.2			32		17.1	
19	1/2/17	35.00	23.90	0.6	2.3	65.0		0	14.7	18.0	
20	1/3/17	34.30	24.80	0.6	2.1	142.0		0		18.0	
21	1/4/17	34.10	23.60	0.6	2.2			18		17.5	
22	1/5/17	32.80	24.10	0.6	2.0			34		17.1	

Figura 3. Spreadsheet used to calculate evapotranspiration and water balance.

For the ETo estimation method, the available meteorological elements (maximum and minimum air temperatures) were used, following the model of Hargreaves and Samani (Allen et al., 1998). The temperature data used to perform the experiment were obtained on a maximum and minimum thermometer installed in the farm. The values of Kc used were 0.6 in the initial stage (21 days), 0.7 in the intermediate stage (314 days) and 0.9 in the final stage (136 days).

In the first year, we evaluated the number of leaves per plant (NLP), the plant height (PH), the stem diameter (SD) and the leaf temperature (LT), following the methodology described by Vieira et al. (2014), at 64, 138 and 199 days after planting (DAP). In the second year, we evaluated at 353, 409, 445 and 470 DAP, measuring the number of bunches per plant (NBP), the productivity of fresh grains (PFG) (kg ha^{-1}) and the productivity of dry grains (PDG) (kg ha^{-1}). Leaf temperature values were compared graphically with the maximum temperature recorded on the day of measurement.

To obtain the values of the water demand, the values of the water depths applied in the two cycles of the crop were summed for all treatments. From the data obtained from the productivity of the dry grains, we calculated the water use efficiency for each treatment.

To verify the assumptions for validation of the analysis of variance, the evaluated variables were submitted to the normality test (Shapiro Wilk). As the variables met the assumptions, it was adopted as a procedure the decomposition of the degrees of freedom of the treatments in regression models by the orthogonal polynomials method, being the choice of the model based on the coefficient of determination and the level of significance. For all procedures, the significance of variance analysis was presented in the graphs.

RESULTS AND DISCUSSION

Figure 4 shows the calculated and measured soil moistures, irrigation, rainfall and soil water limits (field capacity, permanent wilting point, and safety soil moisture) during the experimental period for the T4 treatment (100% of ETc). In this treatment, the soil moisture was, in the majority of experimental period, kept above the safety limit, however, in some moments there was a decrease of soil moisture to values below the limit. The measured moisture values served as a reference for adjustments of the values used in the irrigation management at the beginning of the experiment and for validation of the calculations from the third month.

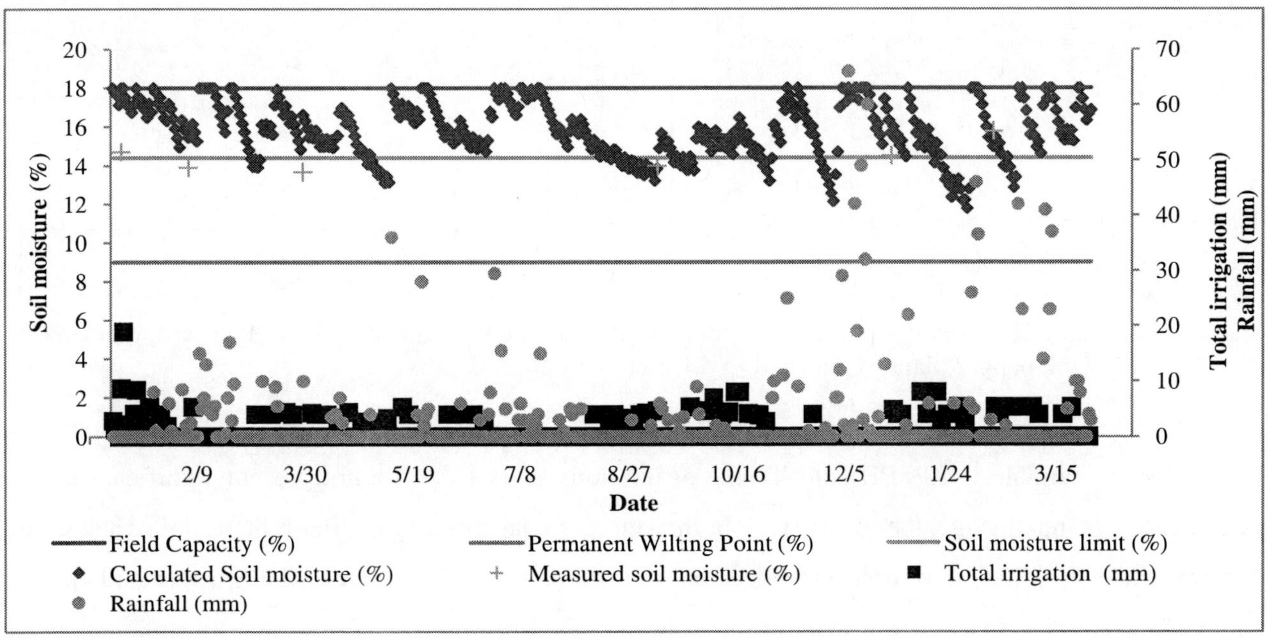

Figure 4. Calculated and measured soil moistures, irrigations performed, rainfall and soil water limits during the experimental period, for T4 treatment (100% ETc).

In the figures 5 (A), (B), (C) and (D), we can see the number of leaves per plant (NLP), the plant height (PH), the stem diameter (SD) and the leaf temperature (LT). According to the results, it can be observed that there was a linear behavior for all the variables NLP, PH, and SD, for the evaluations at 138 and 199 DAP, where the best means were obtained in the water depth referring to 100% of the crop evapotranspiration. For the TF variable, the highest value was obtained in the 25% depth at 64 DAP.

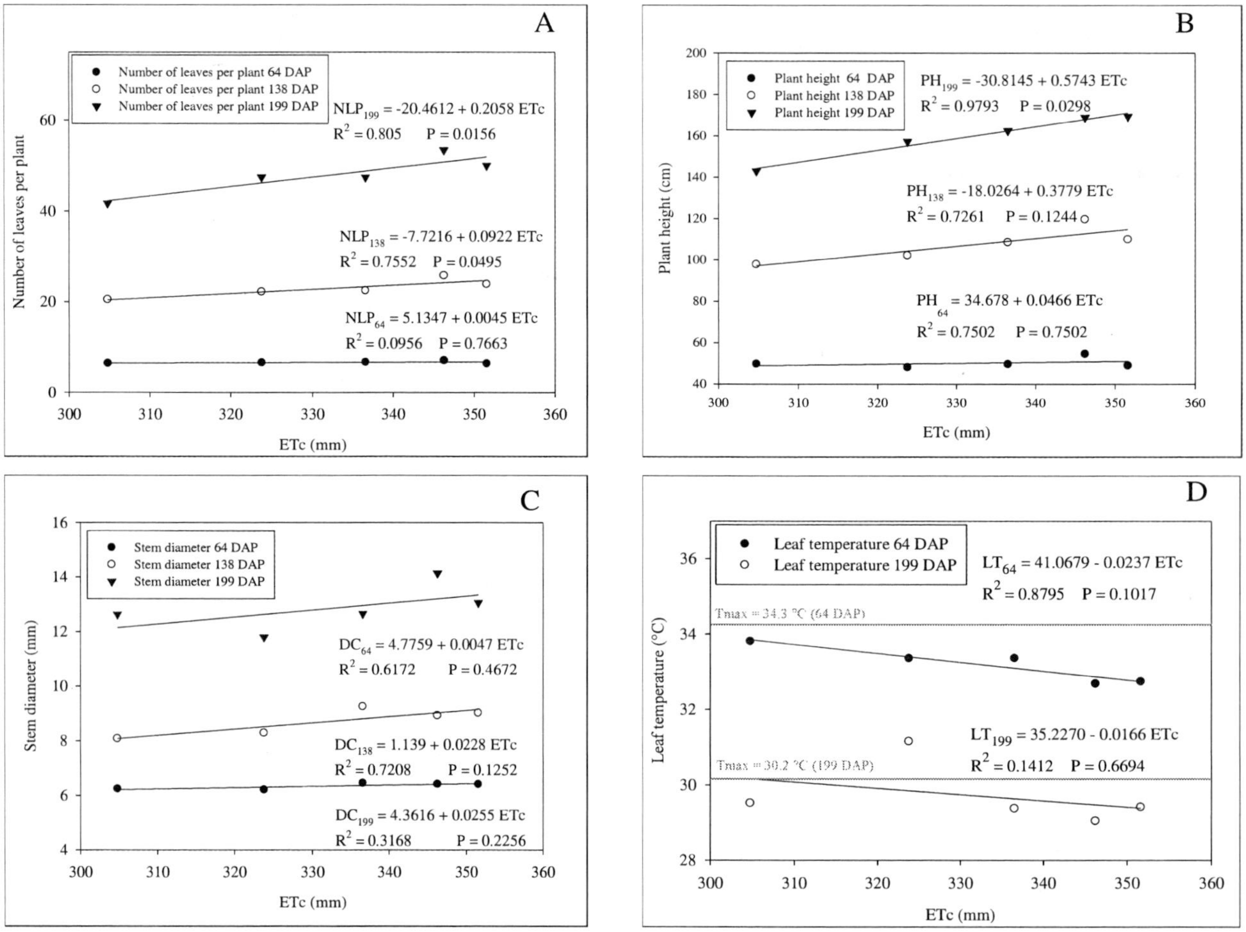

Figure 5: Number of leaves per plant (A), plants height (B), stem diameter (C) and leaf temperature (D) of black pepper plants submitted to different irrigation depths.

For the variables NLP, PH and SD, the evaluations at 64 DAP did not present significant results, because at this initial stage, the plants were in the stage of adaptation to the field, therefore irrigation did not influence its initial development, since the contact between the surface of the roots and the soil provided a surface area for the absorption of water (Lacerda, 2007). Therefore, transplanted plants demand greater protection against water losses as soon as they go into the field, because in this period, the new roots that are being emitted will restore their root-soil connection, which may take a few days until this process occurs totally, causing then a water stress.

According to Gobbo Neto and Lopes (2007), there are physiological factors, such as photosynthesis, stomatal behavior (opening and closing), reserve mobilization, leaf expansion and growth, which are considered critical when they are submitted to water stress. Thus, one of the first processes that are affected in the plant with water deficit is the division and the cellular expansion, where the growth of the leaves and stems diminish considerably before it becomes severe, causing the stomata to close and cause a photosynthesis decreasing (Duarte , 2012).

It was verified that 25 and 52.5 leaves per plant were obtained at 138 DAP and 199 DAP, respectively. A similar result was found by Martins et al. (2006), working with Conilon coffee (*Coffea canephora*), and observed that the increase of the applied depth provided greater fresh matter production of the plants and that the low availability of the same caused a smaller roots development.

At 138 DAP, an average plant height of 116 cm was obtained and at 199 DAP, it was obtained 173 cm. Alves et al. (2000), when working with Conilon coffee (*Coffea canephora*), observed that the mean height values of the plants maintained an upward trend as a function of the applied irrigation depth, which is similar to the results found in this study, where the highest plant height was found in treatments applying 100 to 125% ETc at 199 DAP.

We measured, at 138 DAP, 9.2 mm of SD and at 199 DAP, 13.5 mm. Rodrigues et al. (2009), when working with castor beans, observed that there was an increase in the SD, with greater sensitivity in the initial phase of growth, deducing that plants cultivated without water restriction should be more resistant to tipping due to the sturdier stalks. It was also noticed that, the smaller the depth applied, the smaller the diameter of the stem, a result similar to that found in this study.

It was verified that, for the variable LT at 138 and 199 DAP, there was no adjustment for the different levels of the polynomials ($P > 0.5$). Comparing the leaf temperature values with the maximum air temperature recorded on the days of measurements, a closer approximation is observed in the treatments with the lowest irrigation depths (lower ETc values). These results corroborate with those of Trentin et al. (2011) in a greenhouse study with sugarcane, cultivar RB 86-7515, in which plants maintained under adequate water supply had lower temperatures, when compared to those under conditions of severe water stress and high solar radiation.

According to Dias and Marenco (2007), the increase in leaf temperature also caused a reduction in stomatal conductance, a process in which the stomata open allowing the vapor escaping and the diffusion of the carbon dioxide from the atmosphere to the interior of the leaf, this being the raw material of photosynthesis. Another factor that contributed to the variation in the stomatal and transpiratory behavior of the plant is the variation of the angle of exposure of the leaves to the solar rays, being, therefore, an important mechanism of defense of the plants, that occurs mainly in periods of water stress, decreasing the leaf temperature and, consequently, stomatal transpiration (Oliveira et al., 2005). The opening of the stomata is also related to the water condition of the soil and not only by solar radiation. Apparently, the plant seeks to reduce leaf area to reduce transpiration without abruptly affecting water absorption capacity by roots (Lacerda, 2007). In order to use leaf temperature in irrigation management, as an indicator of crop water conditions, as recommended by Lobo et al. (2004), it is necessary to establish water stress indexes, which determine the momentum and irrigation depth.

The number of bunches per plant (NBP), productivity of fresh grains (PFG) and productivity of dry grains (PDG) are presented in Figures 6 (A), (B) and (C). According to the results, the linear model was chosen due to the higher values of significance for all variables, where the best means were obtained near the depth of 100% ETc.

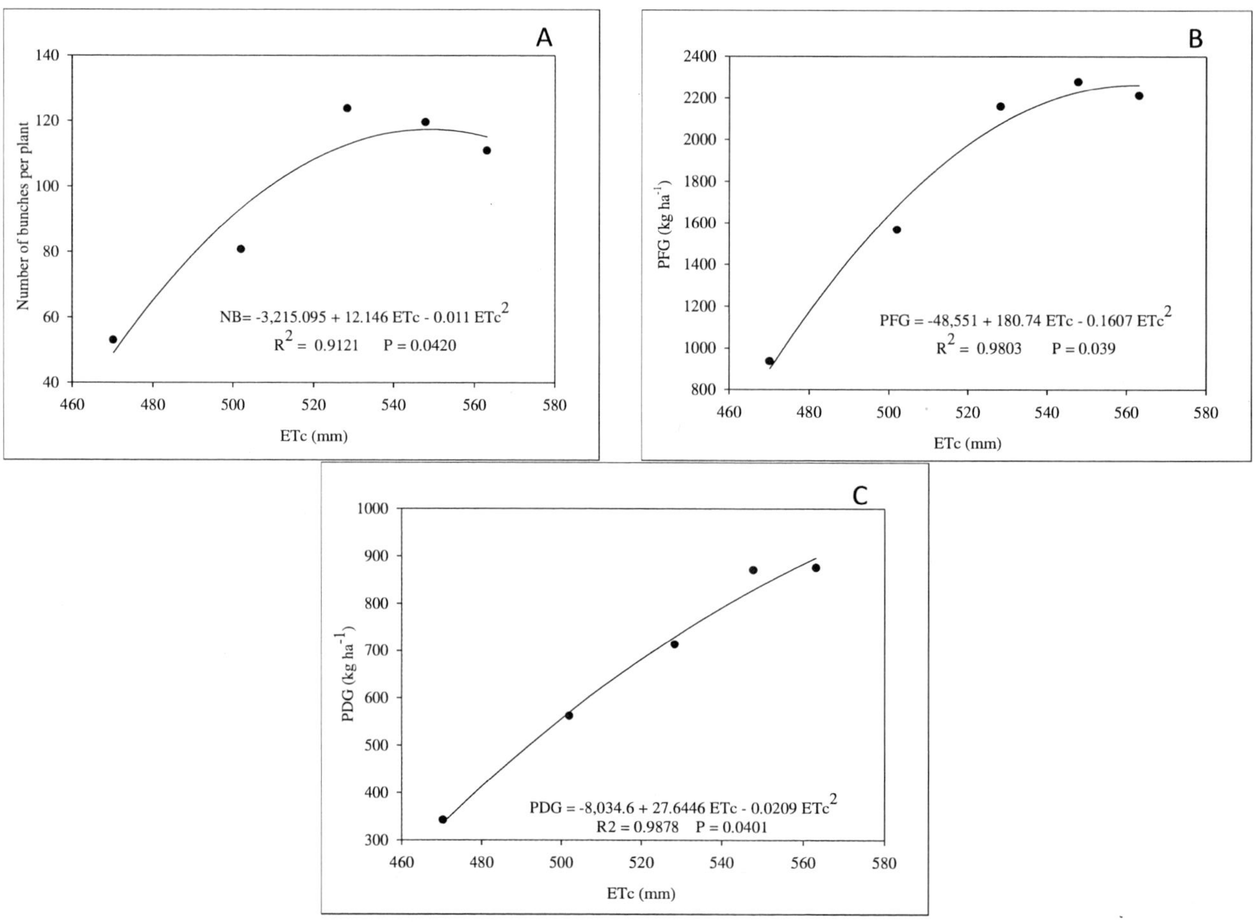

Figure 6. Number of bunches per plant (A), productivity of fresh grains (B) and productivity of dry grains (C) of black pepper plants submitted to different ETc.

The lowest averages were obtained in the depth of 25% ETc, with the productivity increasing proportionally to the increase of the applied depth, up to the 100% ETc. These higher averages occurred due to the water supply in quantities adequate to the needs of the crop, mainly in the pre-flowering stages until the ripening of the grains, when the water stress could affect the translocation of sap in the plant and its productivity.

For the NBP, the reduction of the observed productivity was due to the low applied depths. According to Santos and Carlesso (1998), the limitation of the water availability in the soil during the pre-flowering period affects the development of vegetative structures of the plants, reducing the biomass production capacity by the crop. In this way, the low depths applied in the pepper affected the production of inflorescences or permanence of the same in the plant, occurring the abortion.

Another important factor that interferes directly in the production is the amount of leaf area present in the evaluated plants. According to Martins et al. (2006), the increase of the applied depths provides greater production of fresh matter of the plants, that is, the larger the depth applied, the greater the leaf area, providing greater photosynthesis, that is, greater production of chemical energy. However, when submitted to high temperatures and low water availability, as in the case of treatments with 25% and 50% ETc, they tend to close their stomata, reducing stomatal conductance and consequently reducing the raw material of photosynthesis (Dias and Marenco, 2007).

For the PFG and PDG variables, the highest averages were also obtained on the 100% ETc depth. This is due to the higher water availability for the processes of cell expansion and, consequently, formation of the grain. On the other hand, the lowest rates of accumulated matter is the result of the low water availability for the initial process of grain formation, occurring low cell expansion, which can be limited to small grains or few grains per bunch, which justifies the low productivity in treatments with the smallest depths. Parallel to this process, the production of photosynthesis provides greater amounts of photoassimilates that are used in the formation and development of grains, in which after completing cell expansion, accumulation of dry matter occurs in the developing grains.

Bergamaschi et al. (2004), when working with maize plants, reported that the reduction in the harvest was a consequence of the lack of rainfall that occurred in December and January, when the great majority of the corn plantations of the Espírito Santo State was in the critical period, that is, from the pennon formation to the beginning of grain filling. Similar fact was reported by Bassoi et al. (2011), when evaluating grapevine plants, where they observed that the water deficit between anthesis (opening of flowers) and veraison (beginning of berries ripening) decreases the final size of the berry irreversibly, even if there is a wetting after the veraison. This fact corroborates Santos and Carlesso (1998), who affirm that the grain formation phase is the phase of greater dependence on assimilates. According to the author, this dependence is due to the low reserve that the plant has, and may not complete the grain development, that is, the grain reaches maturity with low dry matter rate or is detached from the plant before of maturation.

Table 1 presents data of crop evapotranspiration, irrigation depths applied during the cycle, crop productivity and water use efficiency, ie the relationship between yield (obtained by PDG) and crop evapotranspiration, for each treatment, in the second year.

Table 2. Treatments and applied depths, water demand and water use efficiency in the second year of crop development in each treatment

Treatment	T1	T2	T3	T4	T5
Perncentual depth (%)	0.25	0.50	0.75	1.00	1.25
Irrigation depth (mm)	47	94	141	188	235
Effective rainfall (mm)	426.5	414.7	397.2	373	348.7
Total water depth (mm)	473,5	508.7	538.2	561	583.7
Productivity(kg ha^{-1})	342.6	562.7	713.8	870.3	875.2
Crop evapotranspiration (mm)	470.2	502.1	528.3	547.8	563.2
Water use efficiency (kg m^{-3})	0.073	0.112	0.135	0.159	0.155

It can be observed that, as the irrigation depths increase and consequent higher crop evapotranspiration, the crop yield increases, reaching values of up to 875.2 kg ha^{-1} for the evapotranspiration of 563.2 mm. This productivity is lower than the average found in commercial crops in Brazil (2,230 kg ha^{-1}); however, it is worth mentioning that the crop cycle was reduced, harvesting only the first bunches. The best efficiency of water use occurred in the 100% replacement of the ETc, with the production of 0.159 kg of dry pepper per 1,000 L of evapotranspirated water.

CONCLUSIONS

In the first year of cultivation of Bragantina pepper, the application of irrigation depths for 100% ETc treatment is recommended to achieve higher plant height, stem diameter and number of leaves per plant, as well as smaller ones leaf temperatures. From the second year of cultivation, the irrigation depth that provided the highest averages for the number of bunches per plant (NBP), productivity of fresh grains (PFG) and productivity of dry grain (PDG) was the irrigation depths of 100% of ETc. The total evapotranspiration that promoted the highest PDG was 563.2 mm, with a yield of 875.2 kg ha^{-1}.

ACKNOWLEDGMENT

To Ifes *campus* Santa Teresa for the financial, operational and infrastructure available for the experiment and for providing scholarship to the students. To Prodif-Ifes for the financial support in the trip to present this paper.

REFERENCES

Allen, R.G., L.S. Pereira, D. Raes, M. Smith, 1998. Crop evapotranspiration: Guidelines for computing crop water requirements. Rome: FAO (56). 300 pp.

Alves, M.E.B.; M.A. Faria, R.J. Guimarães, J.A. Muniz, E.L. Silva. 2000. Coffee plant growth under different irrigation and fertirrigation depths. Revista Brasileira de Engenharia Agrícola e Ambiental. 4(2):219-225.

Bassoi, L.H.; S.O. Goncalves, A.R.L Santos, J.A. Silva, M.A.C. Lima. 2011. Influence of irrigation management on aspects of ecophysiology and production of grapevine cv. Syrah. Irriga. 16:395-402.

Bergamaschi, H.; G.A. Dalmago, J.I. Bergonci, C.A.M. Bianchi, A.G. Müller, F. Comiran, B.M.M. Heckler. 2004. Water distribution in the critical period of corn and grain production. Pesquisa Agropecuária Brasileira. 39:831-839.

Bonomo, D.Z., R. Bonomo, F.L. Partelli, J.M. Souza, M. Magiero, 2013. Vegetative development of the Conilon coffee plant submitted to different irrigation depths. Revista Brasileira de Agricultura Irrigada. 7(2):157-169.

Dias, D. P., R.A. Marenco. 2007. Photosynthesis and photoinhibition in mahogany and acariquara as a function of luminosity and leaf temperature. Pesquisa agropecuária brasileira. 42(3):305-311.

Duarte, M.L.R., F.C. Albuquerque. 2005. Black pepper Production System. Embrapa Eastern Amazon. Brazil.

Duarte, A.L.M. 2012. Effect of water on growth and nutritive value of forage plants. Pesquisa & Tecnologia.9:1-6.

Gobbo Neto, L., N.P. Lopes. 2007. Medicinal plants: factors influencing the content of secondary metabolites. Química nova. 30(2):374-381.

Gomes, D.G. Growth and production of Runner IAC 886 peanut under different irrigation depths. 2016. Doctorate Thesis. Escola Superior de Agricultura "Luiz de Queiroz.

Keller, J., R.D. Bliesner. 1990. Sprinkle and trickle irrigation, pp. 652, Van Nostrand Reinold, New York.

Lacerda, C.L. Part I, 2007. Soil-water-plant relationships in natural and agricultural environments of the Brazilian Northeast. Recife, PE, Brazil.

Lemos, O.D., C.R. Tremacoldi, M.C., Poltronieri. 2014. Good agricultural practices to increase productivity and quality of black pepper in the State of Pará. Embrapa Brasília, DF, Brazil. Also at https://www.infoteca.cnptia.embrapa.br/bitstream/doc/994882/1/CartilhaPimenta.pdf

Lima, J.S.S., R.B., Oliveira, W. Rocha, P.C. Oliveira, W.Z. Quartezani. 2010. Spatial analysis of soil chemical attributes and production of the black pepper (*Piper nigrum* L.). Idesia. 28:31-39.

Lobo, F.A., M.A., Oliva, M. Resende, N.F. Lopes, M. Maestri. 2004. Infrared thermometry to schedule irrigation of common bean. Pesquisa Agropecuária Brasileira. 39(2):113-121.

Maehler, A.R., J.A. Costa, J.L.F.; Pires, L. Rambo. 2003. Grain quality of two soybean cultivars as a function of soil water availability and plant arrangement. Ciência rural. 33(2):13-218.

Mantovani, E.C.; S. Bernardo, L.F. Palaretti. 2009. Irrigation: principles and methods. 3.ed. pp;355. UFV, Viçosa, MG, Brazil.

Martins, C.C., E.F. Reis, C. Busato, J.E.M. Pezzopane. 2006. Initial growth of Conilon coffee (*Coffea canephora* Pierre) under different irrigation depths. Engenharia na Agricultura. 14(3):193-201.

Oliveira A.P., E.U. Alves, J.A. Silva, A.U. Alves, A.N.P. Oliveira, F.A.P. Leonardo, M.F. Moura, I.S. Cruz. 2007. Productivity of black pepper as a function of cattle manure. Horticultura Brasileira. 25:408-410.

Oliveira, A.D., de; E. J. Fernandes, T.J.D. Rodrigues. 2005. Stomatal conductance as an indicator of water stress in beans. Engenharia Agrícola. 25:86-95.

Pedeag. 2016. Strategic plan for the development of Capixaba Agriculture: Black pepper. Also at http://www.incaper.es.gov.br/pedeag/setores08.htm

Ramos, A.L.R. 2015. Black pepper conjuncture of Espírito Santo. Conag: Companhia Nacional de Abastecimento. Also at: http://www.conab.gov.br/OlalaCMS/uploads/arquivos/15_08_26_16_06_06_conjuntura_de_pimenta-do-reino_2015_.pdf

Rodrigues, L.N., A.R. Nery, P.D. Fernandes, N.E.M. Beltrão, H.R. Gheyi. 2009. Growing and production of berries of the castor bean irrigated with domestic wastewater. Revista Brasileira de Engenharia Agrícola e Ambiental. 13(0):825–835.

Santos, R.F., R. Carlesso. 1998. Water deficit and the morphological and physiological processes of plants. Revista Brasileira de Engenharia Agrícola e Ambiental. 2(3):287-294.

Thompson, K.K.; G.A. Ross. 2011. Performance audits. Chapter 14 in Irrigation. 6 ed. pp. 567-612, Irrigation Association, Falls Church, VA.

Trentin, R., S. Zolnier, A. Ribeiro, A.J. Steidle Neto. 2011. Transpiration and leaf temperature of sugarcane unde different values of matric potential. Engenharia Agrícola. 31(6):1085-1095.

Vieira, G.H.S.; E.C. Mantovani, G.C. Sediyama, F.T. Delazari. 2014. Morpho-physiological indicators of water stress on sugarcane as a function of irrigation depths. Bioscience Journal. 30:65-75.

Irrigated Corn Response to Reduced Well Capacity

Alan Schlegel
Kansas State University, Southwest Research-Extension Center, Tribune, KS. schlegel@ksu.edu
Freddie Lamm
Kansas State University, Northwest Research-Extension Center, Colby, KS. flamm@ksu.edu
Isaya Kisekka
University of California-Davis, Dept. of Biological and Agricultural Engineering, Davis, CA
ikisekka@ucdavis.edu.

ABSTRACT

Many of the irrigation systems today in the Central Great Plains no longer have the capacity to apply peak irrigation needs during the summer and must rely on soil water reserves to buffer the crop from water stress. The objective of this study was to determine grain and biomass yield response of corn hybrids to different irrigation capacities. Field studies were conducted at the KSU-NWREC near Colby, KS and KSU-SWREC near Garden City, KS from 2014 to 2017 and at KSU-SWREC near Tribune, KS from 2015 to 2017. The studies were a factorial design of irrigation capacities (ranging from 0.08 to 0.25 inch/day [2 to 6 mm/day]) and corn hybrids (drought tolerant and conventional). Average grain yields increased 7 to 15% when increasing irrigation capacity from about 0.08 to 0.25 inch/day (2 to 6 mm/day). Yield increases were due to an increase in the number of kernels/ear, greater seed mass or both components. Grain yields were not increased by use of a drought tolerant hybrid. At one site, increasing seeding rate from 24,000 to 40,000 kernels/acre (59,000 to 99,000 kernels/ha) increased grain yields an average of 14% with some increase observed at all irrigation capacities.

KEYWORDS: Grower/Farmer, Sprinkler, Deficit irrigation,

INTRODUCTION

Irrigated crop production is a mainstay of agriculture in western Kansas. However, with declining water levels in the Ogallala aquifer, optimal utilization of limited irrigation water is required. The most common crop grown under irrigation in western Kansas is corn. Almost all of the groundwater pumped from the High Plains (Ogallala) Aquifer is used for irrigation. Irrigators are faced with the problem of declining well capacities due to water withdrawals from the Ogallala aquifer for irrigation exceeding mean annual recharge. In addition to limited well capacities, public policy may also impose limits on total amounts of water that can be pumped. For example the 20% reduction in pumped water that is being implemented as part of a Local Enhanced Management Area (LEMA) policy in parts of Groundwater Management District (GMD) 4 and several Water Conservation Areas (WCAs) that have been implemented in GMD 1 and GMD 3. One of the major challenges facing irrigated corn producers in Kansas is how to maintain or increase yields under declining well capacities or limited water supplies.

The major corn seed companies have extensive hybrid development work underway in the western Corn Belt to develop hybrids that are drought tolerant. The overall goal is to develop hybrids that will not incur a yield reduction under ideal conditions, yet stabilize yield under water-stressed conditions. A major secondary trait associated with drought tolerance in corn is the shortening of the anthesis to silk interval (ASI) which has a strong influence on biomass partitioning (Bolaños and Edmeades, 1996). The flowering period of corn often coincides with the period of greatest irrigation demand and consequently a limited irrigation capacity during this critical stage can markedly affect the ASI, biomass partitioning, and grain yield.

Although corn yields have increased greatly during the last 30 years, much of the increase can be attributed to achieving greater final kernel set at higher plant densities (i.e., greater number of kernels/unit area). However, Lobel et al. (2014) suggested that greater plant densities result in greater sensitivity to drought.

Most of the irrigation systems today in the Central Great Plains no longer have the capacity to apply peak irrigation needs during the summer and must rely on soil water reserves to buffer the crop from water stress. Therefore, this study was conducted to evaluate whether management factors such as hybrid selection and seeding rate can be used to increase productivity when well capacity is limited and insufficient to fully meet crop requirements.

MATERIALS AND METHODS

Field studies were conducted at three sites in Kansas. All sites evaluated multiple irrigation capacities and corn hybrids but some treatments differed among sites. At the NWREC near Colby and SWREC near Garden City, whole-plot treatments were sprinkler irrigation capacities of 1 inch every 4, 6, 8, 10, or 12 d (0.25, 0.17, 0.13, 0.10, and 0.08 inch/day [6, 4, 3, 2.5, and 2 mm/day], respectively) in randomized complete block designs. Garden City also had a dryland treatment. Two DeKalb corn hybrids DKC 62-27 DGVT2PRO (DroughtGard) and DKC 62-98 VT2PRO planted at 32,000 seeds/acre (79,000 seeds/ha) were superimposed as split plot treatments. At the SWREC site near Tribune, the whole plot treatments were sprinkler irrigation capacities of 1.5 inch (38 mm) every 1, 2, or 3 weeks (approximates about 1 inch [25 mm] every 4, 8, or 12 days) in a randomized complete block design. Subplots were three corn hybrids (the two Dekalb hybrids along with Pioneer P1151AMX [AquaMax]) at three seeding rates (24,000, 32,000, and 40,000 seeds/acre [59,000, 79,000, and 99,000 seeds/ha]).

Corn was planted in late April to early May under lateral move sprinkler systems at all locations. Irrigations were scheduled as needed according to weather-based water budgets, but limited to the irrigation capacity treatments as indicated in the specific site procedures. Soil water was measured in the complete root zone (0 to 240 cm) periodically throughout the season to help quantify periods of water stress and to determine crop water use. Weather data was measured using automated weather stations that exist on all sites. Corn grain yield was determined by harvesting a representative sample after physiological maturity. Determinations were made of all corn yield components; (grain yield, plant density, ears/plant, kernels/ear, and kernel mass) as well as the important intermediate yield component, kernels/area. Total dry aboveground biomass was also determined through destructive sampling of 5 adjacent plants and drying the samples and weighing the resultant sample.

Crop water use was calculated by summing soil water depletion (soil water at planting less soil water at harvest) plus in-season irrigation and precipitation. In-season irrigations and precipitation amounts at each site-year are shown in Table 1. Crop water productivity (CWP) was calculated by dividing grain yield (lb a^{-1} [kg ha^{-1}]) by crop water use (in [mm]).

RESULTS AND DISCUSSION

Growing conditions were generally favorable during the years of this study. At Colby, growing season precipitation (May through August) was above normal 3 years of the 4 years (Table 1). At Garden City, growing season precipitation was above normal in 3 of the 4 years and, at Tribune, precipitation was above normal every year. However, at Tribune in 2017, hail damage caused some reduction in grain yield.

Colby

Averaged across 4 years, grain yields were increased about 7% (16 bu/a [1,004 kg/ha]) by increasing irrigation capacity from 1 inch (25 mm) every 12 days to 1 inch (25 millimeter) every 4 days (Table 2). This was due mainly to an increase in number of kernels/ear. Biomass was also increased by increasing irrigation capacity. Crop water productivity decreased with increased irrigation capacity. There was little difference (5 bu/a [315 kg/ha] or less) in grain yield between the conventional and drought tolerant hybrids at any irrigation level with an average of 3 bu/a (188 kg/ha) greater yield with the drought tolerant hybrid. The drought tolerant hybrid had greater number of kernels/ear but less seed mass. Biomass production was least with the lowest irrigation capacity with little difference between hybrids.

Garden City

Irrigation at the lowest capacity (1 inch [25 mm] every 12 days) increased yields 66% (79 bu/a [4,955 kg/ha]) compared to non-irrigated (Table 3). This was due to increased seed mass and greater number of kernels/ear. Increasing capacity to 1 inch (25 mm) every 4 days increased yields and additional 15% (30 bu/a [1,882 kg/ha]) due primarily to greater seed mass. Biomass was about 6500 lb/a (407,680 kg/ha) greater with the lowest irrigation capacity compared to dryland and increased an additional 3000 lb/a (188,160 kg/ha) with the highest irrigation capacity. Crop water productivity was greater for all irrigation capacities than for dryland. However, CWP was similar for all levels of irrigation capacity. Hybrid selection had little effect on grain yield when averaged across all irrigation capacities (including dryland) with 4 bu/a (251 kg/ha) less yield with the drought tolerant

Tribune

Increasing irrigation capacity from 0.08 inch/day (2 mm/day) to 0.25 inch/day (6 mm/day) increased average corn grain yields by 11% due primarily to greater kernel mass and a tendency towards greater number of kernels/ear as there were no differences in plant or ear population (Table 2). Increasing seeding rate from 24,000 to 40,000 seeds/acre (59,000 to 99,000 seeds/ha) increased average yields by about 14%. The increase in seeding rate and corresponding increase in ear population more than made up for the decreases in kernel mass and kernels/ear. The total number of kernels/ft^2 (kernels/m^2) was increased with increased seeding rate. The drought tolerant hybrid did not produce greater yield than the conventional hybrid. The conventional hybrid had greater plant and ear populations, and kernel mass than the drought tolerant hybrids but fewer number of kernels/ear. Total biomass production was also greater for the conventional hybrid. Increasing seeding rate and irrigation capacity also increased biomass production. Crop water productivity was increased by increased seeding rate but decreased by increased irrigation capacity. Crop water productivity was greater with the conventional hybrid than the drought tolerant hybrids.

CONCLUSIONS

Increasing irrigation capacity from about 0.08 to 0.25 inch/day (2 to 6 mm/day) increased average grain yields at all sites with the magnitude of the increase ranging from 7 to 15%. The yield increase was due to an increase in the number of kernels/ear and/or greater seed mass. Grain yields were generally similar for the conventional and drought tolerant hybrids. At the site with a population variable, increasing seeding rate from 24,000 to 40,000 kernels/acre (59,000 to 99,000 kernels/ha) increased grain yields an average of 14% with some increase observed at all irrigation capacities. Crop water productivity decreased with increased irrigation capacity at two sites and was similar for all irrigation capacities at the other site.

ACKNOWLEDGEMENTS

This research was supported in part by the Ogallala Aquifer Program, a consortium between USDA Agricultural Research Service, Kansas State University, Texas AgriLife Research, Texas AgriLife Extension Service, Texas Tech University, and West Texas A&M University.

REFERENCES

Bolaños J and G.O. Edmeades GO (1996). The importance of the anthesis-silking interval in breeding for drought tolerance in tropical maize. Field Crops Research 48: 65-80.

Lobell, D. B., M.J. Roberts, W. Schlenker, N. Braun, B. B. Little, R. M. Rejesus, and G. L. Hammer. 2014. Greater sensitivity to drought accompanies maize yield increase in the U.S. Midwest. Science 344: 516-519.

Table 1. Growing season precipitation (May to August) and in-season irrigation amounts at Colby, Garden City, and Tribune from 2014 to 2017.

Year	Colby	Garden City	Tribune
Precipitation		inches (mm)	
2014	13.28 (337)	16.93 (430)	--
2015	15.94 (405	16.37 (416	13.78 (350)
2016	7.61 (193)	12.35 (314)	14.19 (360)
2017	16.05 (408)	7.20 (183)	15.63 (397)
Normal	12.25 (311)	11.41 (290)	10.50 (267)
Irrigation			
2014	7.5 – 13.4 (191-340)	5.0 – 12.0 (127-305)	--
2015	6.7 – 15.4 (170-391)	4.0 – 8.0 (102-203)	6.1 – 15.3 (155-389)
2016	7.7 – 14.4 (196-366)	3.0 – 9.0 (76-229)	6.2 – 12.1(157-307)
2017	7.7 – 14.4 (196-366)	5.0 – 12.0 (127-305)	6.0 – 14.1 (152-358)

Table 2. Average crop parameters as affected by irrigation capacity and corn hybrids, Colby, KS, 2014 - 2017.

Irrigation	Hybrid	Yield	CWP[†]	Plant Pop.	Ear Pop.	1000 seed	Kernels		Biomass
1" per		bu/a (kg/ha)	lb/ac-in (kg/ha-mm)	10^3/acre (10^3/ha)		oz (g)	no/ear	no/foot^2 (no/m^2)	lb/a (kg/ha)
4 days	DKC 62-98	248 (15555)	472 (8.44)	33.5 (82.7)	33.6 (83.0)	12.51 (355)	537	414 (4455)	21737 (24345)
	DKC 62-27 DG	251 (15743)	479 (8.56)	33.1 (81.8)	33.2 (82.0)	10.96 (311)	623	475 (5111)	23311 (26108)
6 days	DKC 62-98	245 (15366)	502 (8.97)	33.2 (82.0)	33.2 (82.0)	12.51 (355)	535	408 (4390)	22748 (25478)
	DKC 62-27 DG	248 (15555)	507 (9.06)	33.3 (82.3)	32.7 (80.8)	10.83 (308)	643	482 (5186)	22671 (25392)
8 days	DKC 62-98	242 (15178)	519 (9.28)	33.5 (82.7)	33.3 (82.3)	12.51 (355)	528	403 (4336)	22071 (24720)
	DKC 62-27 DG	243 (15241)	519 (9.28)	33.4 (82.5)	33.1 (81.8)	11.01 (313)	603	459 (4939)	22642 (25359)
10 days	DKC 62-98	243 (15241)	568 (10.15)	33.4 (82.5)	33.2 (82.0)	12.48 (354)	535	408 (4390)	22028 (24671)
	DKC 62-27 DG	245 (15366)	546 (9.76)	33.1 (81.8)	33.0 (81.5)	10.79 (306)	631	478 (5143)	22070 (24718)
12 days	DKC 62-98	232 (14551)	521 (9.31)	33.4 (82.5)	33.4 (82.5)	12.56 (357)	504	386 (4153)	20513 (22975)
	DKC 62-27 DG	237 (14865)	537 (9.60)	33.1 (81.8)	33.0 (81.5)	10.73 (305)	604	458 (4928)	21779 (24392)
MEANS									
4 days		250 (15680)	475 (8.49)	33.3 (82.3)	33.4 (82.5)	11.74 (333)	580	445 (4788)	22529 (25232)
6 days		246 (15429)	504 (9.01)	33.3 (82.3)	32.9 (81.3)	11.67 (331)	589	446 (4799)	22709 (25434)
8 days		242 (15178)	519 (9.28)	33.4 (82.5)	33.2 (82.0)	11.76 (334)	565	431 (4638)	22357 (25040)
10 days		244 (15304)	557 (9.96)	33.3 (82.3)	33.1 (81.8)	11.64 (331)	583	443 (4767)	22050 (24696)
12 days		234 (14676)	529 (9.46)	33.3 (82.3)	33.2 (82.0)	11.65 (331)	554	422 (4541)	21149 (23687)
	DKC 62-98	242 (15178)	516 (9.22)	33.4 (82.5)	33.3 (82.3)	12.52 (356)	528	404 (4347)	21821 (24440)
	DKC 62-27 DG	245 (15366)	517 (9.24)	33.2 (82.0)	33.0 (81.5)	10.87 (309)	621	470 (5057)	22494 (25193)
2014		238 (14927)	499 (8.92)	32.9 (81.3)	33.0 (81.5)	11.81 (335)	551	418 (4498)	22155 (24814)
2015		249 (15617)	473 (8.45)	32.9 (81.3)	32.9 (81.3)	11.21 (318)	616	465 (5003)	22061 (24708)
2016		238 (14927)	585 (10.46)	32.9 (81.3)	32.5 (80.3)	11.33 (322)	601	448 (4820)	21516 (24098)
2017		249 (15617)	511 (9.13)	34.4 (85.0)	34.3 (84.7)	12.41 (352)	529	416 (4476)	- - -

[†] CWP = crop water productivity.

Table 3. Average crop parameters as affected by irrigation capacity and corn hybrids, Garden City, KS, 2014 - 2017.

Irrigation	Hybrid	Yield	CWP[†]	Plant Pop.	Ear Pop.	1000 seed	Kernels		Biomass
1" per		bu/a (kg/ha)	lb/ac-in (kg/ha-mm)	10^3/acre (10^3/ha)		oz (g)	no/ear	no/foot^2 (no/m^2)	lb/a (kg/ha)
4 days	DKC 62-98	230 (14426)	601 (10.74)	29.8 (73.6)	29.8 (73.6)	16.98 (482)	406	290 (3120)	20888 (23395)
	DKC 62-27 DG	226 (14175)	549 (9.81)	27.8 (68.7)	27.8 (68.7)	14.47 (411)	474	308 (3314)	20802 (23298)
6 days	DKC 62-98	218 (13673)	591 (10.56)	29.0 (71.6)	28.8 (71.1)	17.01 (483)	404	275 (2959)	19707 (22072)
	DKC 62-27 DG	214 (13422)	568 (10.15)	27.8 (68.7)	27.7 (68.4)	14.70 (417)	461	299 (3217)	19882 (22268)
8 days	DKC 62-98	219 (13736)	622 (11.12)	29.9 (73.9)	29.9 (73.9)	17.13 (486)	379	263 (2830)	19642 (21999)
	DKC 62-27 DG	219 (13736)	606 (10.83)	28.0 (69.2)	27.6 (68.2)	14.01 (398)	495	320 (3443)	19707 (22072)
10 days	DKC 62-98	207 (12983)	624 (11.15)	29.9 (73.9)	29.7 (73.4)	15.65 (444)	396	272 (2927)	17852 (19994)
	DKC 62-27 DG	195 (12230)	608 (10.87)	27.0 (66.7)	26.6 (65.7)	13.37 (380)	493	309 (3325)	17242 (19311)
12 days	DKC 62-98	202 (12669)	613 (10.96)	30.6 (75.6)	30.2 (74.6)	15.27 (434)	367	250 (2690)	17798 (19934)
	DKC 62-27 DG	195 (12230)	555 (9.92)	28.0 (69.2)	27.9 (68.9)	12.23 (347)	478	316 (3400)	17767 (19899)
Dryland	DKC 62-98	119 (7464)	427 (7.63)	28.8 (71.1)	27.6 (68.2)	11.26 (320)	332	192 (2066)	11452 (12826)
	DKC 62-27 DG	119 (7464)	446 (7.97)	28.0 (69.2)	27.8 (68.7)	8.99 (255)	405	258 (2776)	10987 (12305)
MEANS									
4 days		228a (14300a)	575a (10.28a)	28.8 (71.1)	28.8 (71.1)	15.72a (446a)	440a	299a (3217a)	20845a (23346a)
6 days		216ab (13548ab)	580a (10.37a)	28.4 (70.1)	28.3 (69.9)	15.86a (450a)	432a	287a (3088a)	19794a (22169a)
8 days		219ab (13736ab)	614a (10.97a)	28.9 (71.4)	28.7 (70.9)	15.57a (442a)	437a	291a (3131a)	19674a (22035a)
10 days		201b (12607b)	616a (11.01a)	28.5 (70.4)	28.2 (69.7)	14.51b (412b)	445a	290a (3120a)	17547b (19653b)
12 days		198b (12419b)	584a (10.44a)	29.3 (72.4)	29.0 (71.6)	13.75b (391b)	422a	283a (3045a)	17783b (19917b)
Dryland		119c (7464c)	437b (7.81b)	28.4 (70.1)	27.7 (68.4)	10.13c (288c)	369b	225b (2421b)	11219c (12565c)
LSD $_{0.05}$		25 (1568)	85 (1.52)	1.6 (4.0)	1.7 (4.2)	0.80 (23)	43	31 (334)	1783 (1997)
	DKC 62-98	199 (12481)	580 (10.37)	29.7a (73.4a)	29.3a (72.4a)	15.55a (442a)	381b	257b (2765b)	17890 (20037)
	DKC 62-27 DG	195 (12230)	555 (9.92)	27.8b (68.7b)	27.6b (68.2b)	12.96b (368b)	468a	302a (3250a)	17731 (19859)
	LSD $_{0.05}$	8 (502)	30 (0.54)	0.8 (2.0)	0.9 (2.2)	0.58 (16)	20	15 (161)	807 (904)
2014		175b (10976b)	635a (11.35a)	29.9a (73.9a)	29.3a (72.4a)	13.16b (374b)	392b	265b (2851b)	15711c (17596c)
2015		184b (11540b)	538b (9.62b)	27.2c (67.2c)	27.1b (66.9b)	- - -	- - -	- - -	17910b (20059b)
2016		217a (13610a)	529b (9.46b)	28.6b (70.6b)	28.2ab (69.7ab)	15.53a (441a)	456a	294a (3163a)	19809a (22186a)
2017		211a (13234a)	- - -	29.2ab (72.1ab)	29.1a (71.9a)	- - -	- - -	- - -	- - -
LSD $_{0.05}$		11 (690)	37 (0.66)	1.2 (3.0)	1.2 (3.0)	0.58 (16)	20	15 (161)	988 (1107)
ANOVA (P>F)									
Irrigation		0.001	0.004	0.803	0.528	0.001	0.016	0.001	0.001
Hybrid		0.284	0.116	0.001	0.001	0.001	0.001	0.001	0.697
Hybrid*Irrigation		0.947	0.739	0.678	0.330	0.908	0.412	0.259	0.993
Year		0.001	0.001	0.001	0.002	0.001	0.001	0.001	0.001

[†] CWP = crop water productivity

Table 4. Average crop parameters as affected by irrigation capacity, corn hybrids, and seeding rate, Tribune, KS, 2015 - 2017.

Irrigation	Hybrid	Seed Rate	Yield	CWP†	Plant Pop.	Ear Pop.	1000 seed	Kernels		Biomass
1" per		10^3/acre (10^3/ha)	bu/a (kg/ha)	lb/ac-in (kg/ha-mm)	10^3/acre (10^3/ha)		oz (g)	no/ear	no/foot^2 (no/m^2)	lb/a (kg/ha)
4 days	DKC 62-98	24 (59)	211 (13234)	414 (7.40)	24.3 (60.0)	24.0 (59.3)	14.01 (398)	566	312 (3357)	17302 (19378)
		32 (79)	233 (14614)	454 (8.11)	32.5 (80.3)	32.0 (79.0)	13.68 (389)	478	351 (3777)	19096 (21388)
		40 (99)	235 (14739)	449 (8.03)	39.7 (98.1)	39.0 (96.3)	13.22 (375)	411	368 (3960)	23180 (25962)
	DKC 62-27 DG	24 (59)	199 (12481)	398 (7.11)	23.3 (57.6)	23.2 (57.3)	11.96 (340)	649	344 (3701)	16994 (19033)
		32 (79)	222 (13924)	426 (7.61)	30.8 (76.1)	30.0 (74.1)	11.42 (324)	587	402 (4326)	19272 (21585)
		40 (99)	229 (14363)	452 (8.08)	36.8 (90.9)	35.3 (87.2)	11.19 (318)	528	425 (4573)	20805 (23302)
	P1151 AMX	24 (59)	191 (11980)	375 (6.70)	21.8 (53.8)	22.7 (56.1)	13.31 (378)	572	297 (3196)	15088 (16899)
		32 (79)	201 (12607)	386 (6.90)	27.7 (68.4)	27.3 (67.4)	12.44 (353)	537	338 (3637)	16363 (18327)
		40 (99)	235 (14739)	456 (8.15)	36.4 (89.9)	35.7 (88.2)	12.12 (344)	493	404 (4347)	19351 (21673)
8 days	DKC 62-98	24 (59)	199 (12481)	460 (8.22)	23.4 (57.8)	23.2 (57.3)	13.82 (392)	560	298 (3206)	16029 (17952)
		32 (79)	220 (13798)	498 (8.90)	32.9 (81.3)	32.1 (79.3)	13.00 (369)	470	347 (3734)	19397 (21725)
		40 (99)	223 (13987)	511 (9.13)	39.3 (97.1)	38.2 (94.4)	12.90 (366)	407	356 (3831)	20203 (22627)
	DKC 62-27 DG	24 (59)	185 (11603)	410 (7.33)	23.0 (56.8)	22.5 (55.6)	11.66 (331)	634	326 (3508)	15184 (17006)
		32 (79)	205 (12858)	449 (8.03)	30.5 (75.3)	30.2 (74.6)	11.07 (314)	553	381 (4100)	16615 (18609)
		40 (99)	206 (12920)	465 (8.31)	35.8 (88.4)	35.1 (86.7)	10.77 (306)	500	397 (4272)	18419 (20629)
	P1151 AMX	24 (59)	180 (11290)	407 (7.27)	21.5 (53.1)	21.6 (53.4)	12.90 (366)	586	290 (3120)	14354 (16076)
		32 (79)	200 (12544)	464 (8.29)	28.6 (70.6)	28.0 (69.2)	12.42 (353)	522	334 (3594)	16910 (18939)
		40 (99)	203 (12732)	458 (8.19)	34.3 (84.7)	33.5 (82.7)	12.07 (343)	483	353 (3798)	17213 (19279)
12 days	DKC 62-98	24 (59)	183 (11478)	449 (8.03)	23.6 (58.3)	23.5 (58.0)	13.44 (382)	517	279 (3002)	15684 (17566)
		32 (79)	190 (11917)	460 (8.22)	31.6 (78.1)	31.5 (77.8)	12.54 (356)	426	309 (3325)	17611 (19724)
		40 (99)	214 (13422)	518 (9.26)	38.7 (95.6)	38.1 (94.1)	12.54 (356)	399	350 (3766)	19109 (21402)
	DKC 62-27 DG	24 (59)	181 (11352)	438 (7.83)	22.9 (56.6)	23.3 (57.6)	11.21 (318)	623	332 (3572)	14568 (16316)
		32 (79)	206 (12920)	481 (8.60)	30.4 (75.1)	30.3 (74.8)	10.72 (304)	567	395 (4250)	16723 (18730)
		40 (99)	195 (12230)	465 (8.31)	38.3 (94.6)	37.8 (93.4)	10.33 (293)	449	390 (4196)	18484 (20702)
	P1151 AMX	24 (59)	180 (11290)	433 (7.74)	22.2 (54.8)	23.1 (57.1)	12.67 (360)	558	293 (3153)	14965 (16761)
		32 (79)	201 (12607)	484 (8.65)	29.1 (71.9)	28.9 (71.4)	11.94 (339)	525	348 (3744)	15512 (17373)
		40 (99)	203 (12732)	477 (8.53)	35.1 (86.7)	34.7 (85.7)	11.82 (336)	452	359 (3863)	16513 (18495)

† CWP = crop water productivity. ** Hail event on 8/18/17 **

Table 4 (cont.). Average crop parameters as affected by irrigation capacity, corn hybrids, and seeding rate, Tribune, 2015 - 2017.

Irrigation	Hybrid	Seed Rate	Yield	CWP[1]	Plant Pop.	Ear Pop.	1000 seed	Kernels		Biomass
1" per		10^3/a (10^3/ha)	bu/a (kg/ha)	lb/ac-in (kg/ha-mm)	10^3/acre (10^3/ha)		oz (g)	no/ear	no/foot^2 (no/m^2)	lb/a (kg/ha)
MEANS										
4 days			217a (13610a)	423b (7.56b)	30.4 (75.1)	29.9 (73.9)	12.59a (358a)	536	360 (3874)	18606a (20839a)
8 days			202b (12669b)	458a (8.19a)	29.9 (73.9)	29.4 (72.6)	12.29ab (349ab)	524	343 (3691)	17147b (19205b)
12 days			195b (12230b)	467a (8.35a)	30.2 (74.6)	30.1 (74.3)	11.91b (338b)	502	339 (3648)	16575b (18564b)
LSD $_{0.05}$			13 (815)	30 (0.54)	1.7 (4.2)	1.7 (4.2)	0.46 (13)	31	24 (258)	1419 (1589)
	DKC 62-98		212a (13297a)	468a (8.37a)	31.8a (78.5a)	31.3a (77.3a)	13.24a (376a)	470c	330b (3551b)	18624a (20859a)
	DKC 62-27 DG		203b (12732 b)	443b (7.92b)	30.2b (74.6b)	29.7b (73.4b)	11.15c (317c)	565a	377a (4057a)	17452b (19546b)
	P1151 AMX		199b 12481b)	438b (7.83b)	28.5c (70.4c)	28.4c (70.1c)	12.41b (352b)	525b	335b (3605b)	16252c (18202c)
	LSD $_{0.05}$		7 (439)	15 (0.27)	0.9 (2.2)	0.9 (2.2)	0.19 (5)	15	12 (129)	734 (822)
		24 (59)	190c (11917c)	420c (7.51c)	22.9c (56.6c)	23.0c (56.8c)	12.78a (363a)	585a	308c (3314c)	15574c (17443c)
		32 (79)	208b (13046b)	456b (8.15b)	30.5b (75.3b)	30.0b (74.1b)	12.14b (345b)	518b	356b (3831b)	17500b (19600b)
		40 (99)	216a (13548a)	472a (8.44a)	37.2a (91.9a)	36.4a (89.9a)	11.88c (337c)	458c	378a (4067a)	19253a (21563a)
		LSD $_{0.05}$	7 (439)	15 (0.27)	0.9 (2.2)	0.9 (2.2)	0.19 (5)	15	12 (129)	734 (822)
		2015	182c (11415c)	438b (7.83b)	27.3b (67.4b)	26.8c (66.2c)	12.66b (360b)	502c	297c (3196c)	14918c (16708c)
		2016	223a (13987a)	444b (7.94b)	31.7a (78.3a)	30.7b (75.8b)	12.97a (368a)	519b	355b (3820b)	19636a (21992a)
		2017	209b (13108b)	467a (8.35a)	31.5a (77.8a)	31.9a (78.8a)	11.17c (317c)	541a	389a (4186a)	17774b (19907b)
		LSD $_{0.05}$	7 (439)	15 (0.27)	0.9 (2.2)	0.9 (2.2)	0.19 (5)	15	12 (129)	734 (822)
ANOVA (P>F)										
Irrigation			0.015	0.028	0.826	0.573	0.030	0.089	0.152	0.031
Seed rate			0.001	0.001	0.001	0.001	0.001	0.001	0.001	0.001
Seed rate*Irrigation			0.605	0.755	0.719	0.750	0.972	0.697	0.563	0.182
Hybrid			0.002	0.001	0.001	0.001	0.001	0.001	0.001	0.001
Hybrid*Irrigation			0.232	0.169	0.770	0.562	0.701	0.641	0.325	0.413
Hybrid*Seed rate			0.600	0.925	0.312	0.099	0.958	0.024	0.452	0.542
Irrigation*Seed rate*Hybrid			0.521	0.206	0.819	0.894	0.890	0.397	0.553	0.766
Year			0.001	0.001	0.001	0.001	0.001	0.001	0.001	0.001

† CWP = crop water productivity * Hail event on 8/18/17

Optimizing Irrigation Scheduling With Limited Water Using the iCrop Decision Support Tool

Isaya Kisekka

Assistant Professor

Departments of LAWR and BAE

UCDAVIS
UNIVERSITY OF CALIFORNIA

From Experiments to Models to Decision Support

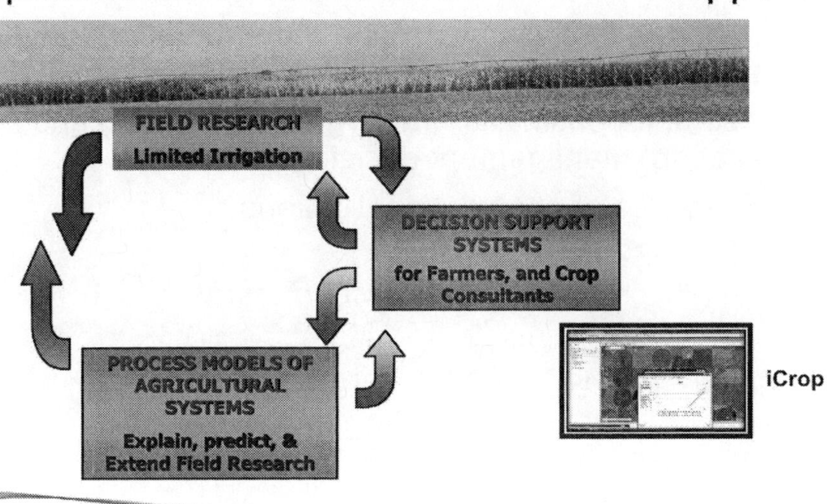

UCDAVIS

DSSAT-CSM CERES-Maize

Calibration

Validation

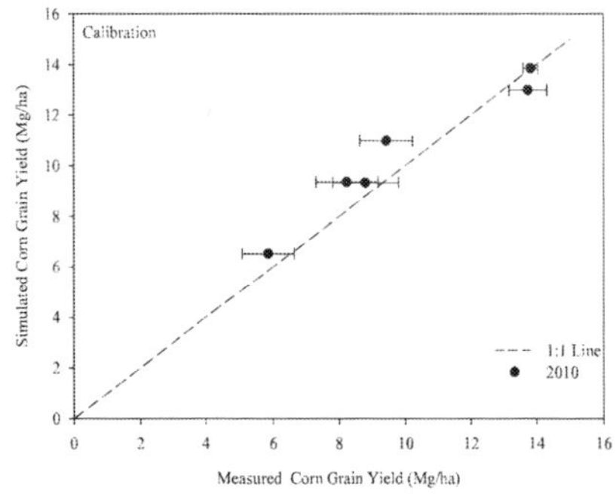

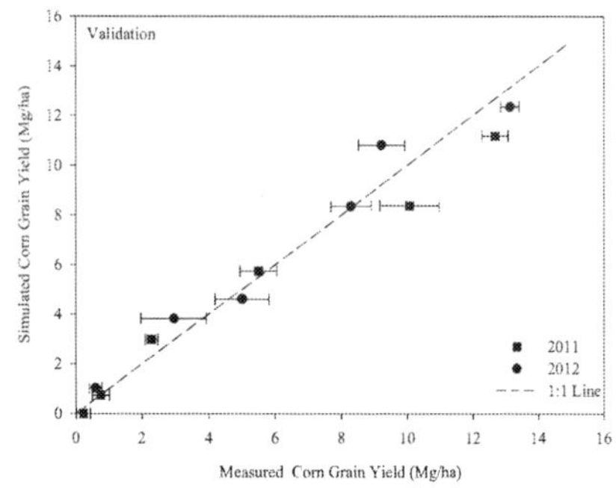

Kisekka et al. 2016

UC DAVIS

iCrop

- **I**ntegrated **C**rop water management model-driven decision support tool.
- Useful for optimizing **strategic** (preseason) and **tactical** (in season) management decisions.
- Examples of potential applications:
  1. Land-water allocation
  2. Hybrid selection and seeding rate
  3. When to initiate irrigation
  4. When to terminate irrigation
  5. Effect of splitting nitrogen applications
  6. etc.

UC DAVIS

iCrop Conceptual framework

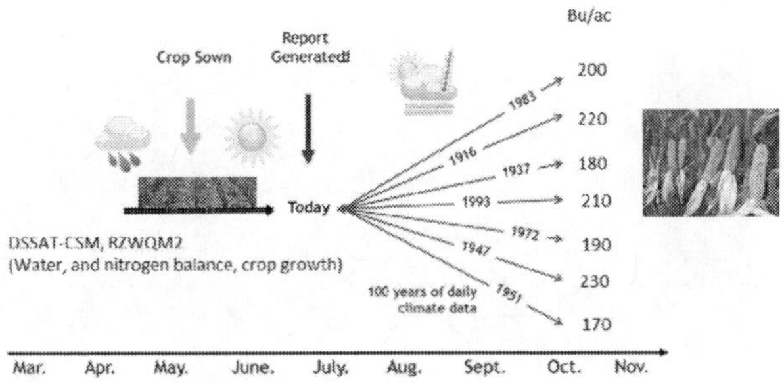

Modified from Yield Prophet: http://www.yieldprophet.com.au/YP/HowItWorks.aspx

UCDAVIS

eRAMS
Home My Account My Groups Resource Center
Map Fields
Select Field: TestK1
Show Field Polygons
Weather Stations
Weather Stations -State Stations (MESONET)
Gridded Data -PRISM
User Data refresh
Click on the search button to display a list of weather stations closest to your crop field(s). After collecting, you can click on each station to see its location and related weather information.
If Kansas Mesonet data until Dec 31st 2016 is missing, it will be filled out with PRISM.
Radius: 20 miles
MESONET
Garden City (3.0 mi)
Soils Summary
Ulysses Silt loam
Richfield Silt loam
Google

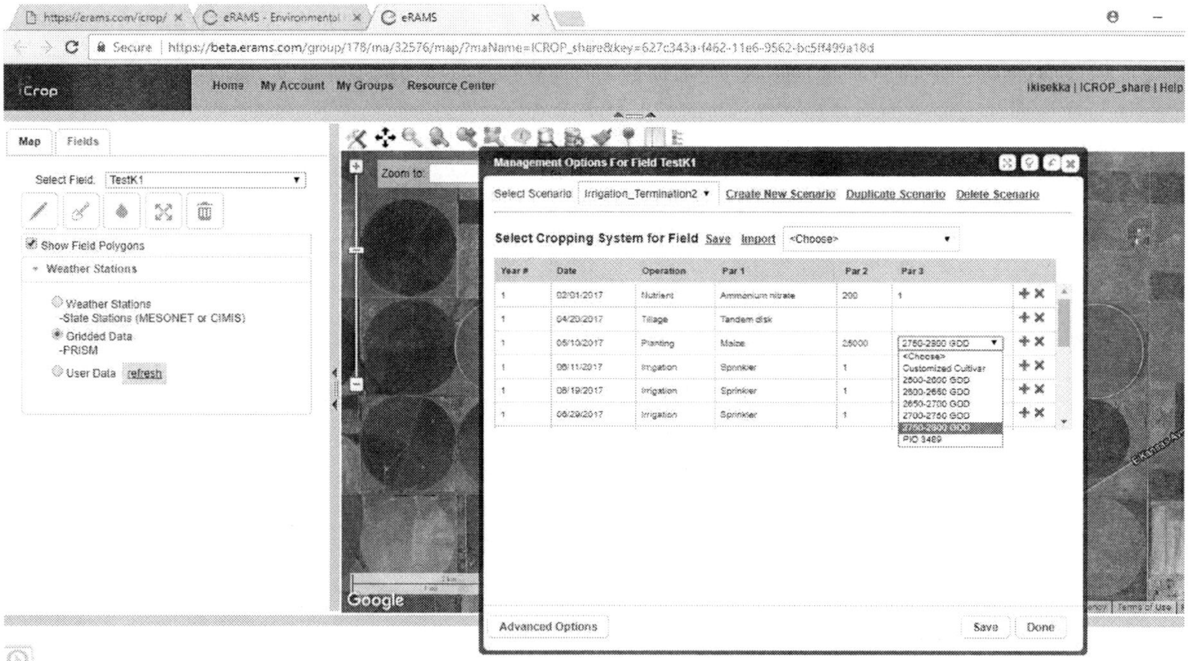
eRAMS
Crop
Home My Account My Groups Resource Center
ikisekka | ICROP_share | Help
Map Fields
Select Field: TestK1
Show Field Polygons
Weather Stations
Weather Stations -State Stations (MESONET or CIMIS)
Gridded Data -PRISM
User Data refresh
Management Options For Field TestK1
Select Scenario Irrigation_Termination2 Create New Scenario Duplicate Scenario Delete Scenario
Select Cropping System for Field Save Import <Choose>
Year # Date Operation Par 1 Par 2 Par 3
1 02/01/2017 Nutrient Ammonium nitrate 200 1
1 04/20/2017 Tillage Tandem disk
1 05/10/2017 Planting Maize 25000 2750-2800 GDD
1 06/11/2017 Irrigation Sprinkler 1
1 06/19/2017 Irrigation Sprinkler 1
1 06/29/2017 Irrigation Sprinkler 1
<Choose>
Customized Cultivar
2500-2600 GDD
2600-2650 GDD
2650-2700 GDD
2700-2750 GDD
2750-2800 GDD
PIO 3489
Advanced Options
Save Done
Google

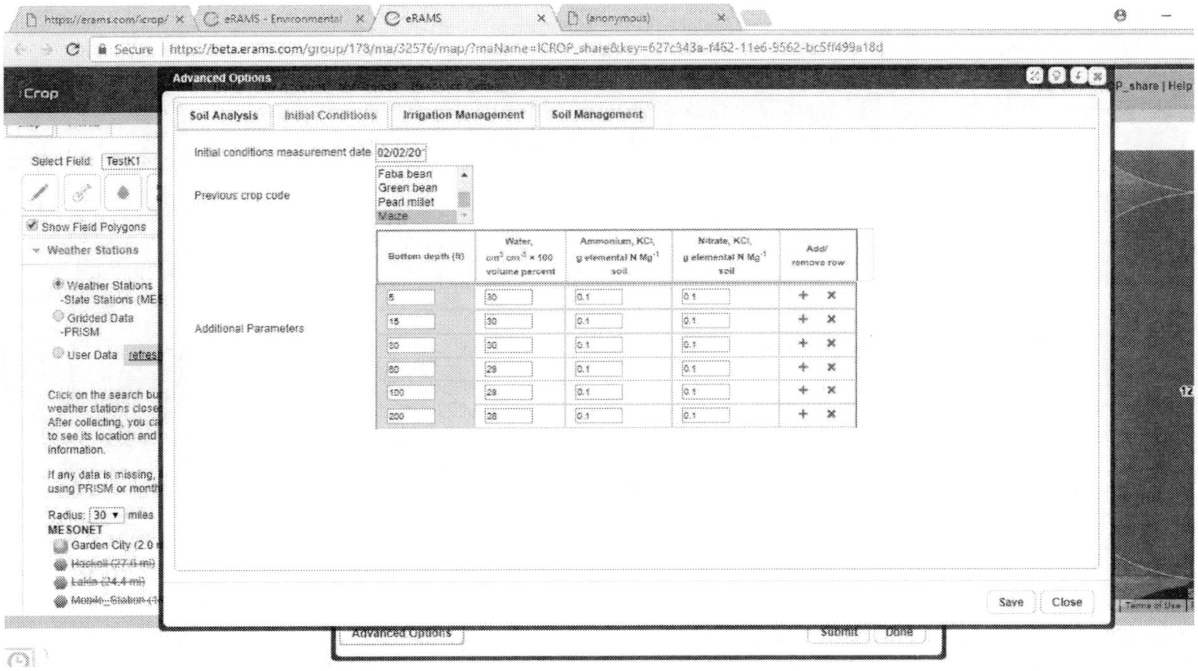
Advanced Options
Soil Analysis
Initial Conditions
Irrigation Management
Soil Management
Initial conditions measurement date 02/02/20
Previous crop code
Faba bean
Green bean
Pearl millet
Maize
Bottom depth (ft)
Add/ remove row
Additional Parameters
Select Field: TestK1
Show Field Polygons
Weather Stations
Gridded Data
-PRISM
User Data
Radius: 30 miles
MESONET
Save
Close
Advanced Options
Submit
Done

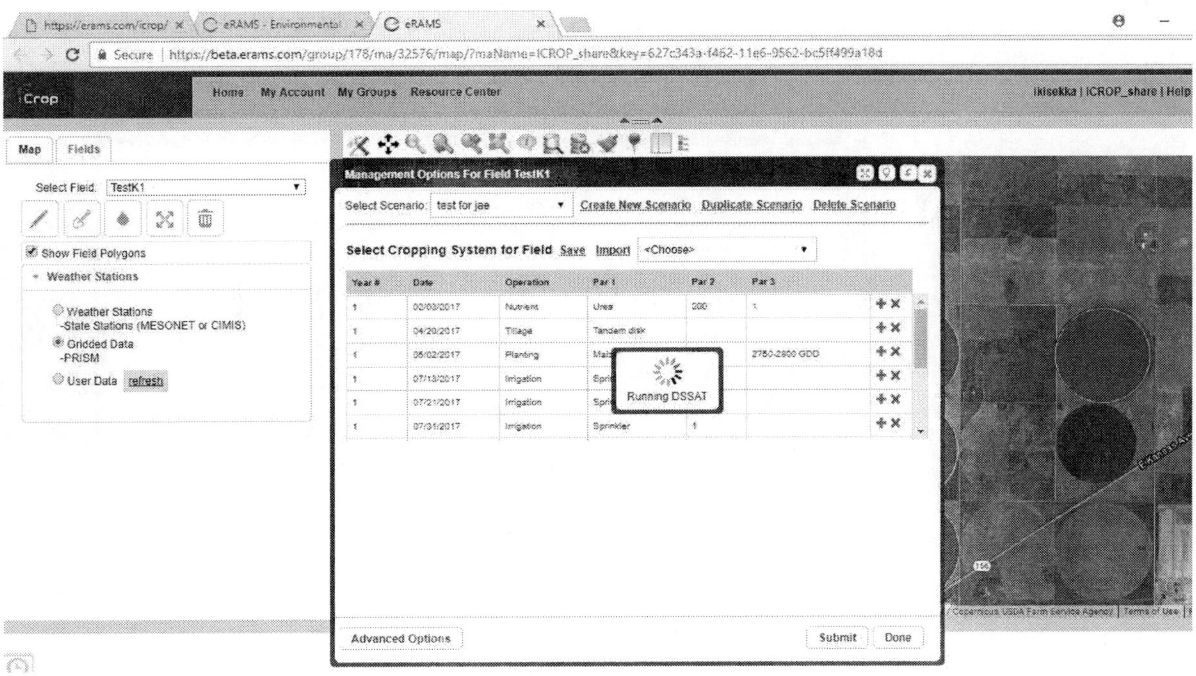
Home My Account My Groups Resource Center
Map Fields
Select Field: TestK1
Show Field Polygons
Weather Stations
Weather Stations
-State Stations (MESONET or CIMIS)
Gridded Data
-PRISM
User Data refresh
Management Options For Field TestK1
Select Scenario: test for jae
Create New Scenario Duplicate Scenario Delete Scenario
Select Cropping System for Field Save Import <Choose>
Year # Date Operation Par 1 Par 2 Par 3
Running DSSAT
Advanced Options
Submit
Done

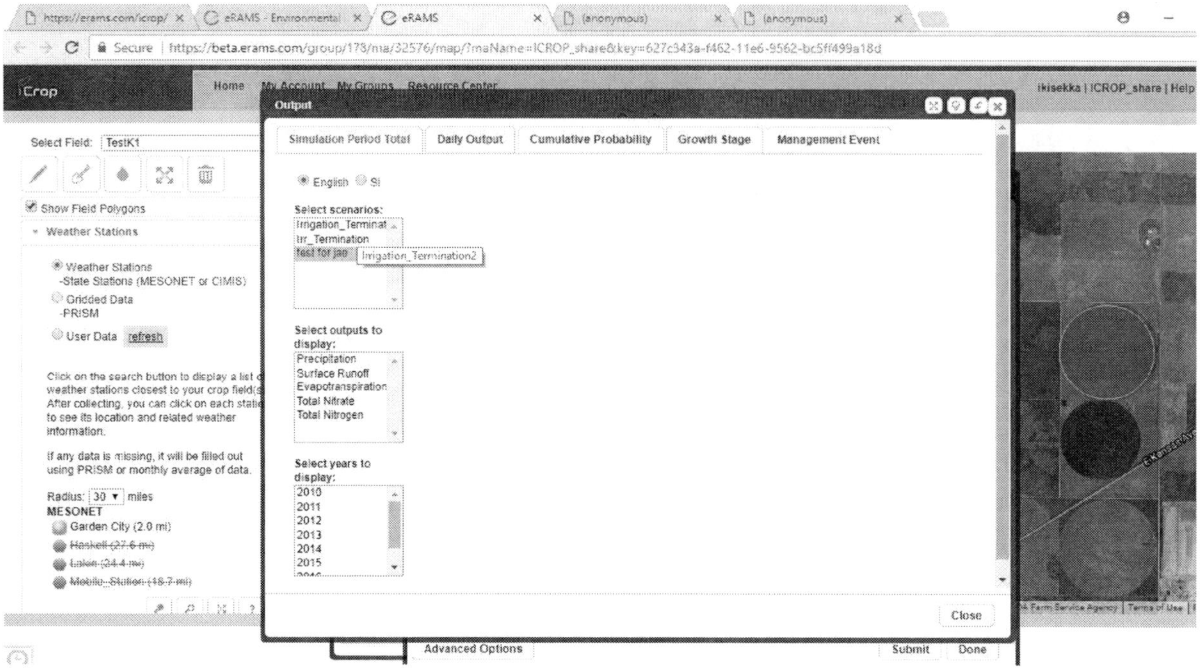

Home My Account My Groups Resource Center
Output
Simulation Period Total
Daily Output
Cumulative Probability
Growth Stage
Management Event
English SI
Select scenarios:
Irrigation_Terminat
Irr_Termination
test for jae
Irrigation_Termination2
Select outputs to display:
Precipitation
Surface Runoff
Evapotranspiration
Total Nitrate
Total Nitrogen
Select years to display:
2010
2011
2012
2013
2014
2015
Close
Select Field: TestK1
Show Field Polygons
Weather Stations
Weather Stations
-State Stations (MESONET or CIMIS)
Gridded Data
-PRISM
User Data refresh
If any data is missing, it will be filled out using PRISM or monthly average of data.
Radius: 30 miles
MESONET
Garden City (2.0 mi)
Advanced Options
Submit
Done

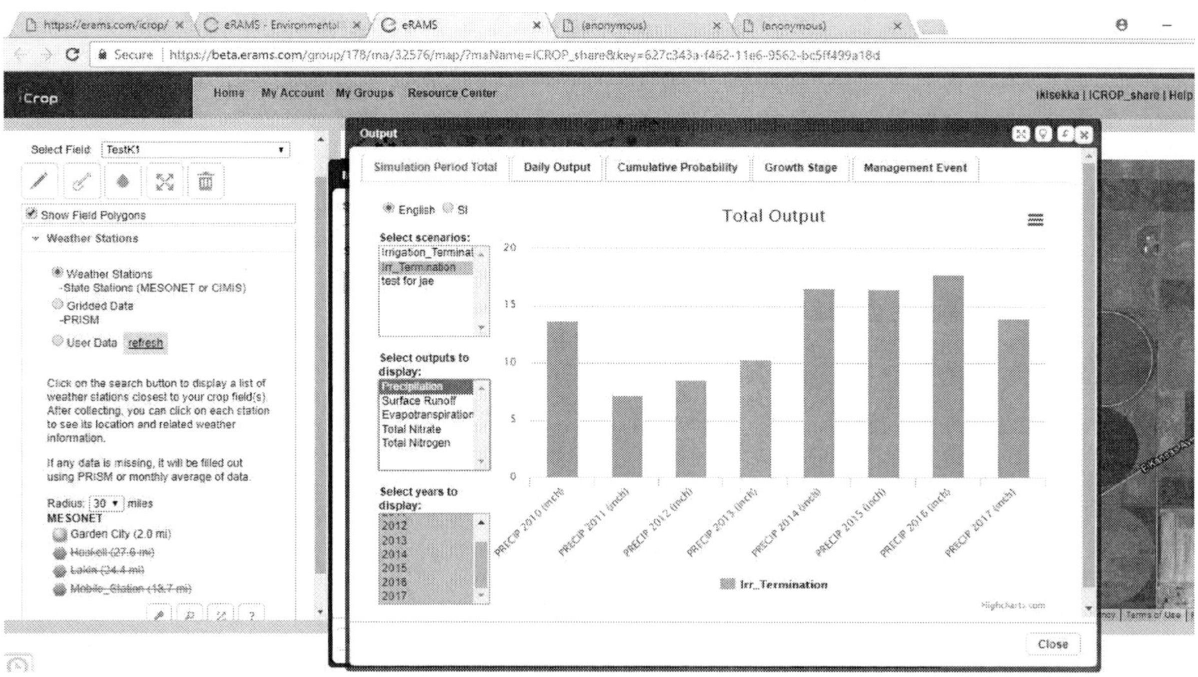

Home My Account My Groups Resource Center
Select Field: TestK1
Show Field Polygons
Weather Stations
Weather Stations
-State Stations (MESONET or CIMIS)
Gridded Data
-PRISM
User Data refresh
Click on the search button to display a list of weather stations closest to your crop field(s). After collecting, you can click on each station to see its location and related weather information.
If any data is missing, it will be filled out using PRISM or monthly average of data.
Radius: 30 miles
MESONET
Garden City (2.0 mi)
Output
Simulation Period Total
Daily Output
Cumulative Probability
Growth Stage
Management Event
English SI
Total Output
Select scenarios:
Irrigation_Terminat
Irr_Termination
test for jae
Select outputs to display:
Precipitation
Surface Runoff
Evapotranspiration
Total Nitrate
Total Nitrogen
Select years to display:
2012
2013
2014
2015
2016
2017
20
15
10
5
0
PRECIP 2010 (inch)
PRECIP 2011 (inch)
PRECIP 2012 (inch)
PRECIP 2013 (inch)
PRECIP 2014 (inch)
PRECIP 2015 (inch)
PRECIP 2016 (inch)
PRECIP 2017 (inch)
Irr_Termination
Close

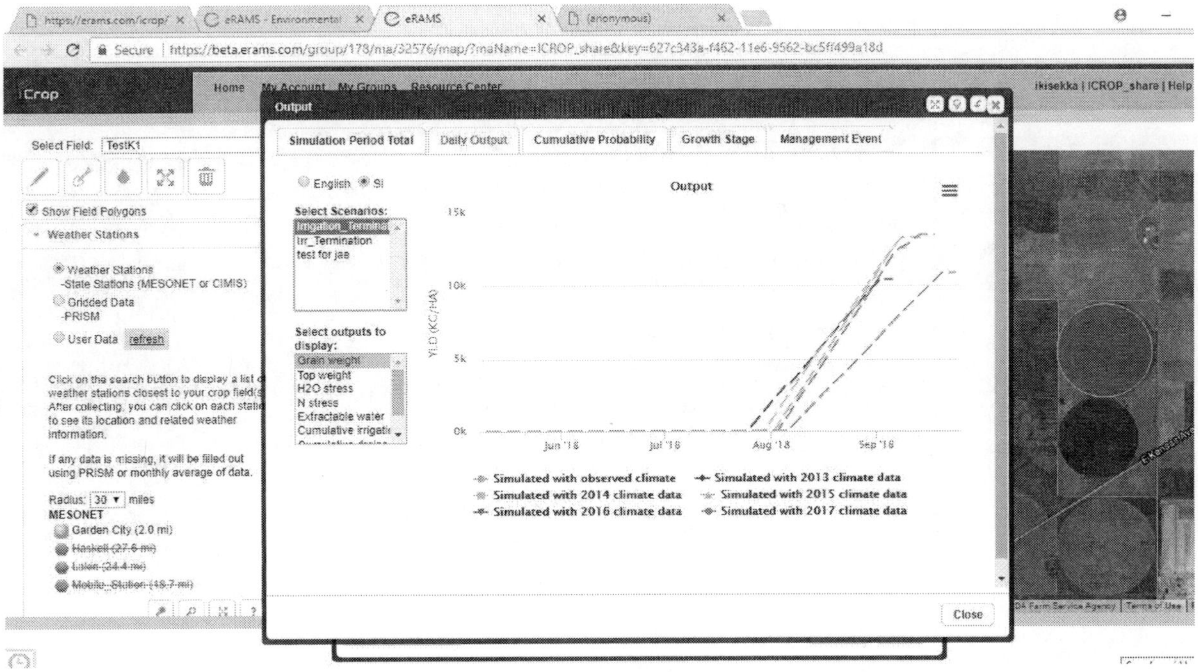
https://beta.erams.com/group/178/ma/32576/map/?maName=ICROP_share&key=627c343a-f462-11e6-9562-bc5ff499a18d
iCrop
Home
ikisekka | ICROP_share | Help
Output
Simulation Period Total
Daily Output
Cumulative Probability
Growth Stage
Management Event
Select Field: TestK1
Show Field Polygons
Weather Stations
Weather Stations
-State Stations (MESONET or CIMIS)
Gridded Data
-PRISM
User Data refresh
English SI
Select Scenarios:
Irrigation_Terminat
Irr_Termination
test for jae
Select outputs to display:
Grain weight
Top weight
H2O stress
N stress
Extractable water
Cumulative irrigatio
Output
YLD (KG/HA)
15k
10k
5k
0k
Jun '18
Jul '18
Aug '18
Sep '18
Simulated with observed climate
Simulated with 2013 climate data
Simulated with 2014 climate data
Simulated with 2015 climate data
Simulated with 2016 climate data
Simulated with 2017 climate data
If any data is missing, it will be filled out using PRISM or monthly average of data.
Radius: 30 miles
MESONET
Garden City (2.0 mi)
Haskell (27.6 mi)
Lakin (24.4 mi)
Mobile_Station (18.7 mi)
Close

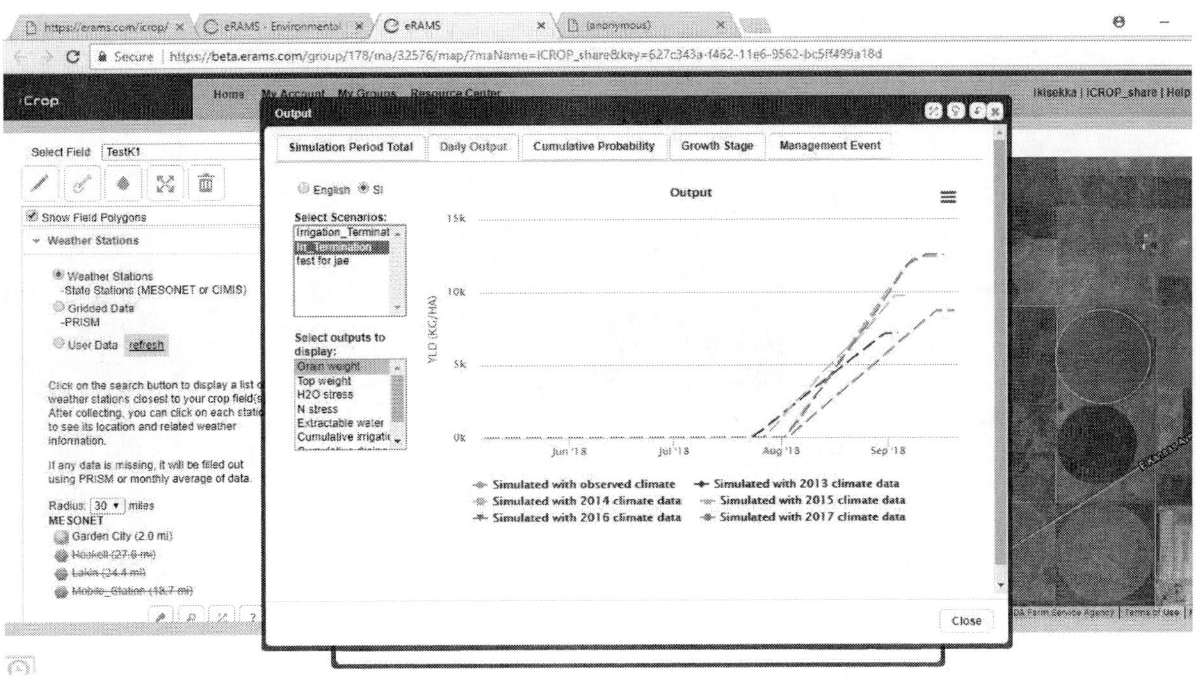
https://beta.erams.com/group/178/ma/32576/map/?maName=ICROP_share&key=627c343a-f462-11e6-9562-bc5ff499a18d
iCrop
Home
ikisekka | ICROP_share | Help
Output
Simulation Period Total
Daily Output
Cumulative Probability
Growth Stage
Management Event
Select Field: TestK1
Show Field Polygons
Weather Stations
Weather Stations
-State Stations (MESONET or CIMIS)
Gridded Data
-PRISM
User Data refresh
English SI
Select Scenarios:
Irrigation_Terminat
Irr_Termination
test for jae
Select outputs to display:
Grain weight
Top weight
H2O stress
N stress
Extractable water
Cumulative irrigatio
Output
YLD (KG/HA)
15k
10k
5k
0k
Jun '18
Jul '18
Aug '18
Sep '18
Simulated with observed climate
Simulated with 2013 climate data
Simulated with 2014 climate data
Simulated with 2015 climate data
Simulated with 2016 climate data
Simulated with 2017 climate data
If any data is missing, it will be filled out using PRISM or monthly average of data.
Radius: 30 miles
MESONET
Garden City (2.0 mi)
Haskell (27.6 mi)
Lakin (24.4 mi)
Mobile_Station (18.7 mi)
Close

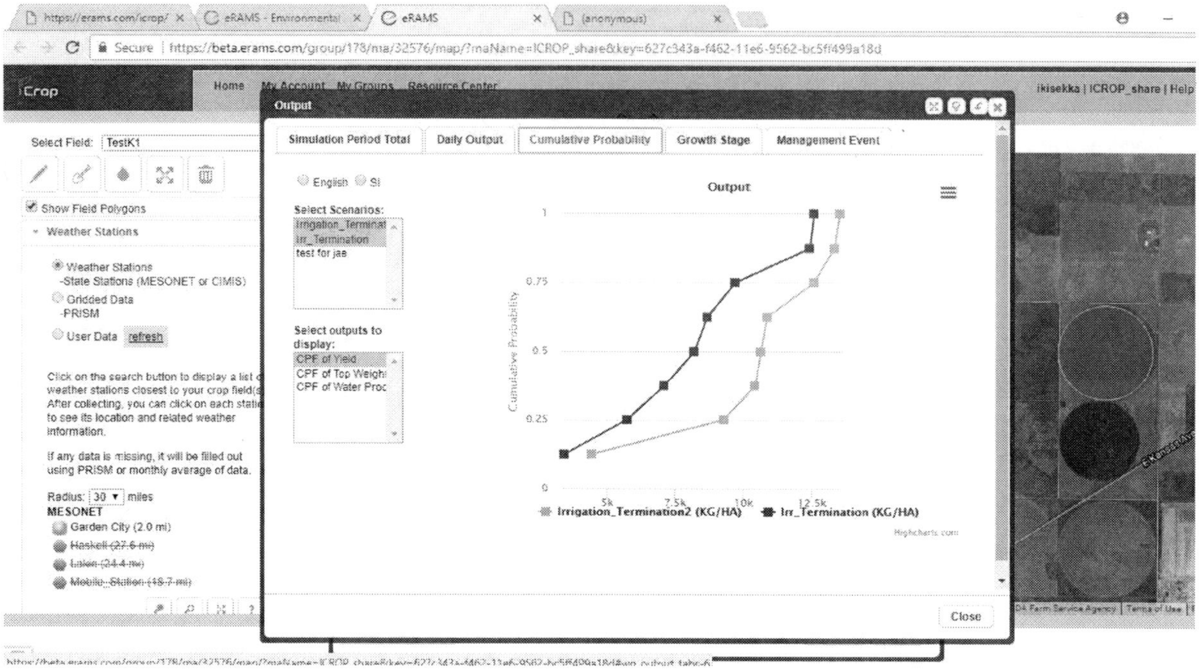

Crop
Home
Select Field: TestK1
Show Field Polygons
Weather Stations
Output
Simulation Period Total
Daily Output
Cumulative Probability
Growth Stage
Management Event
English
SI
Select Scenarios:
Irrigation_Terminat
Irr_Termination
test for jae
Select outputs to display:
CPF of Yield
CPF of Top Weigh
CPF of Water Proc
Output
Cumulative Probability
1
0.75
0.5
0.25
0
5k
7.5k
10k
12.5k
Irrigation_Termination2 (KG/HA)
Irr_Termination (KG/HA)
Highcharts.com
Close
Weather Stations
-State Stations (MESONET or CIMIS)
Gridded Data
-PRISM
User Data refresh
Click on the search button to display a list of weather stations closest to your crop field. After collecting, you can click on each station to see its location and related weather information.
If any data is missing, it will be filled out using PRISM or monthly average of data.
Radius: 30 miles
MESONET
Garden City (2.0 mi)
Haskell (27.6 mi)
Lakin (24.4 mi)
Mobile_Station (18.7 mi)
ikisekka | ICROP_share | Help

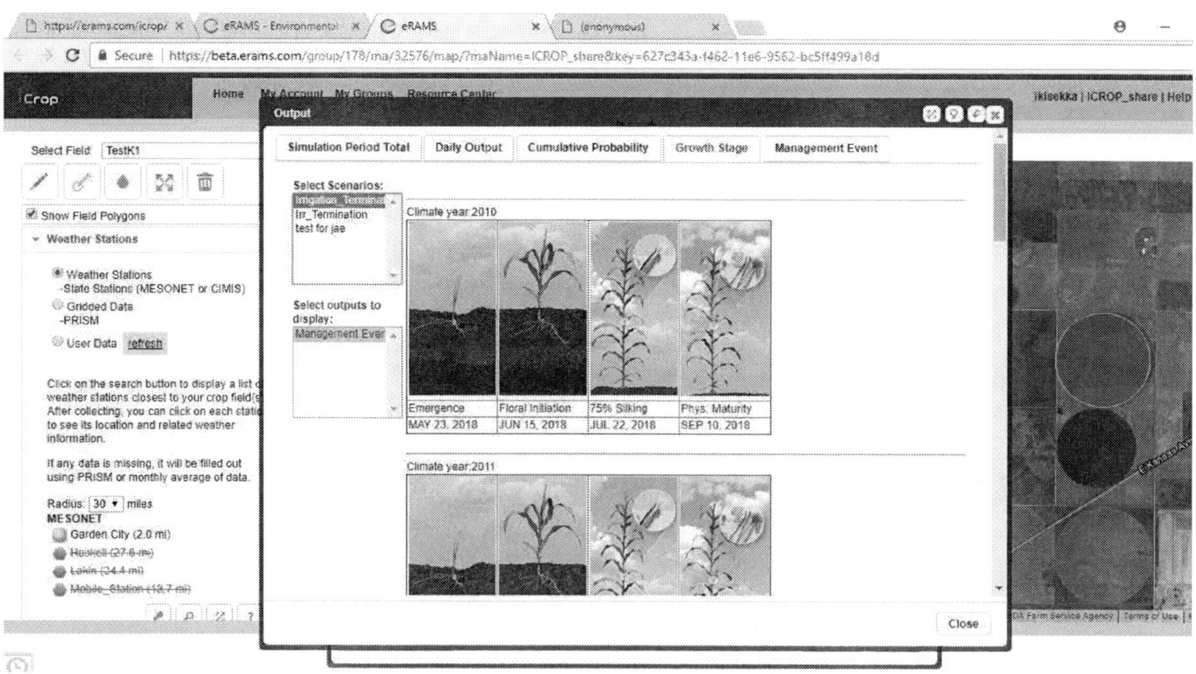

Crop
Home
Select Field: TestK1
Show Field Polygons
Weather Stations
Output
Simulation Period Total
Daily Output
Cumulative Probability
Growth Stage
Management Event
Select Scenarios:
Irrigation_Terminat
Irr_Termination
test for jae
Select outputs to display:
Management Ever
Climate year:2010
Emergence
Floral Initiation
75% Silking
Phys. Maturity
MAY 23, 2018
JUN 15, 2018
JUL 22, 2018
SEP 10, 2018
Climate year:2011
Close
Weather Stations
-State Stations (MESONET or CIMIS)
Gridded Data
-PRISM
User Data refresh
Radius: 30 miles
MESONET
Garden City (2.0 mi)
Haskell (27.6 mi)
Lakin (24.4 mi)
Mobile_Station (18.7 mi)
ikisekka | ICROP_share | Help

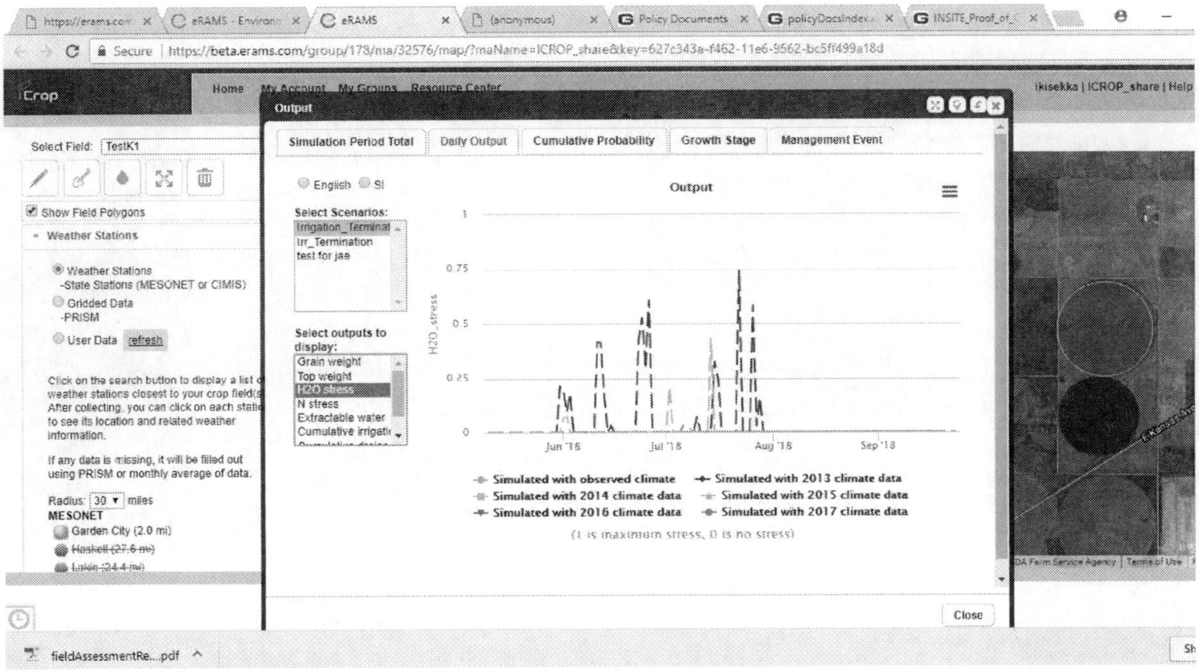
https://beta.erams.com/group/178/ma/32576/map/?maName=ICROP_share&key=627c343a-f462-11e6-9562-bc5ff499a18d
Output
Simulation Period Total
Daily Output
Cumulative Probability
Growth Stage
Management Event
English
SI
Select Scenarios:
Irrigation_Terminat
Irr_Termination
test for jae
Select outputs to display:
Grain weight
Top weight
H2O stress
N stress
Extractable water
Cumulative irrigatio
Output
H2O_stress
Jun '18
Jul '18
Aug '18
Sep '18
Simulated with observed climate
Simulated with 2013 climate data
Simulated with 2014 climate data
Simulated with 2015 climate data
Simulated with 2016 climate data
Simulated with 2017 climate data
(1 is maximum stress, 0 is no stress)
Select Field: TestK1
Show Field Polygons
Weather Stations
Weather Stations
-State Stations (MESONET or CIMIS)
Gridded Data
-PRISM
User Data refresh
If any data is missing, it will be filled out using PRISM or monthly average of data.
Radius: 30 miles
MESONET
Garden City (2.0 mi)
Haskell (27.6 mi)
Close
fieldAssessmentRe....pdf

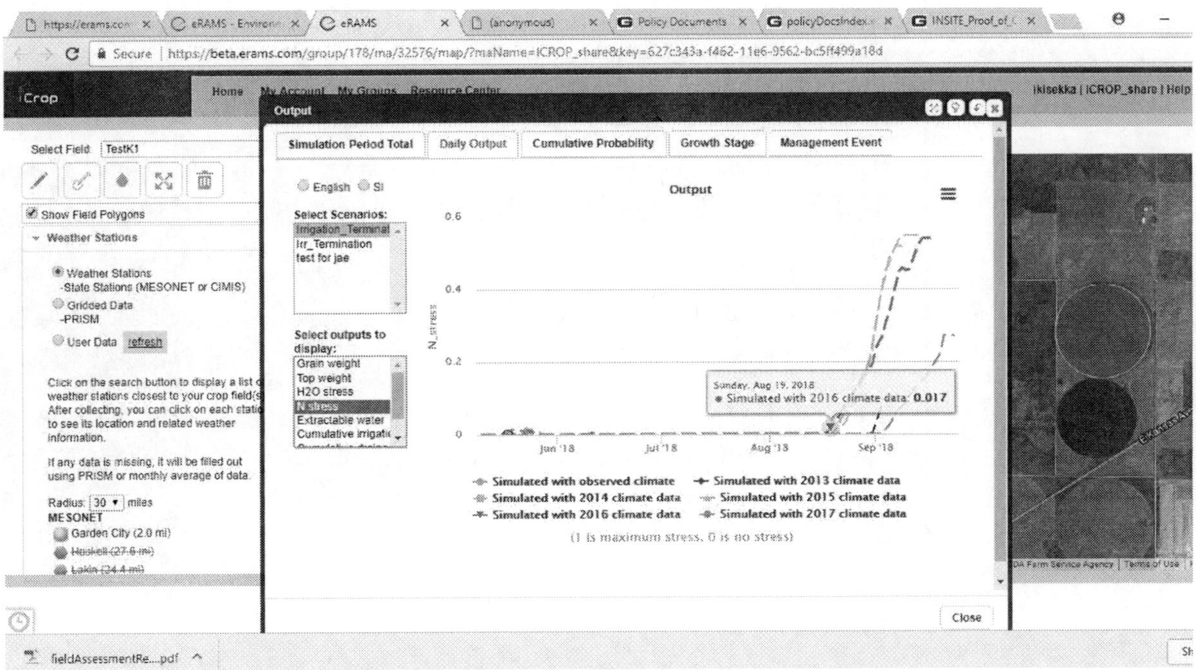
https://beta.erams.com/group/178/ma/32576/map/?maName=ICROP_share&key=627c343a-f462-11e6-9562-bc5ff499a18d
Output
Simulation Period Total
Daily Output
Cumulative Probability
Growth Stage
Management Event
English
SI
Select Scenarios:
Irrigation_Terminat
Irr_Termination
test for jae
Select outputs to display:
Grain weight
Top weight
H2O stress
N stress
Extractable water
Cumulative irrigatio
Output
N_stress
Sunday, Aug 19, 2018
Simulated with 2016 climate data: 0.017
Jun '18
Jul '18
Aug '18
Sep '18
Simulated with observed climate
Simulated with 2013 climate data
Simulated with 2014 climate data
Simulated with 2015 climate data
Simulated with 2016 climate data
Simulated with 2017 climate data
(1 is maximum stress, 0 is no stress)
Select Field: TestK1
Show Field Polygons
Weather Stations
Weather Stations
-State Stations (MESONET or CIMIS)
Gridded Data
-PRISM
User Data refresh
If any data is missing, it will be filled out using PRISM or monthly average of data.
Radius: 30 miles
MESONET
Garden City (2.0 mi)
Haskell (27.6 mi)
Close
fieldAssessmentRe....pdf

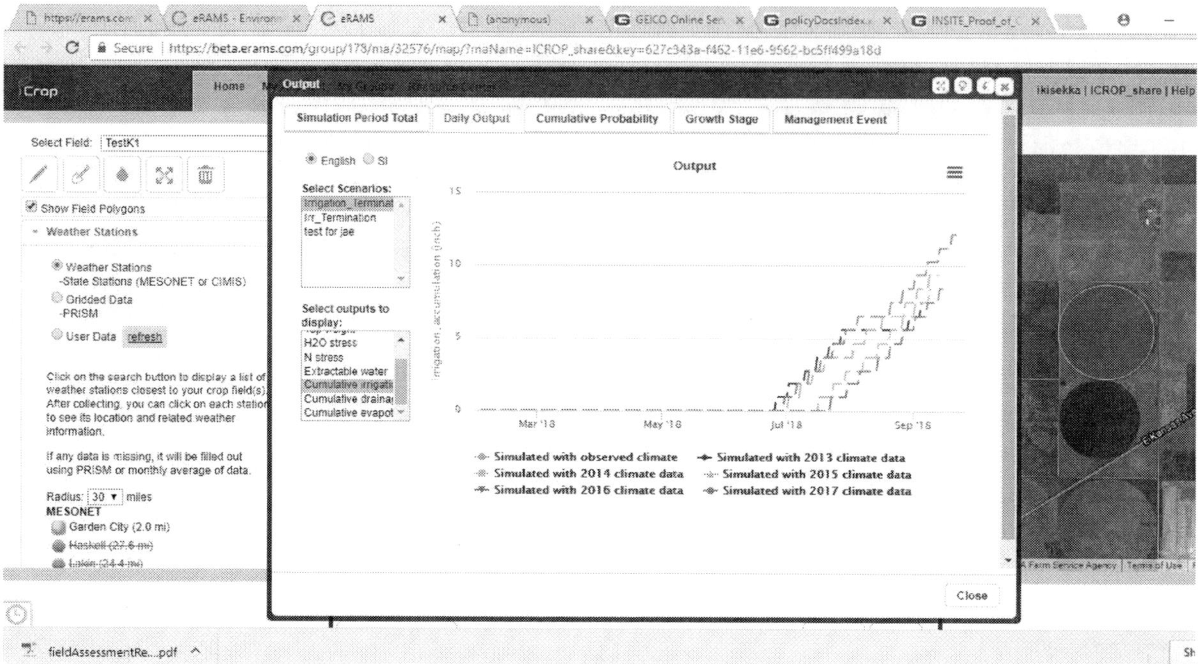
https://beta.erams.com/group/178/ma/32576/map/?maName=ICROP_share&key=627c343a-f462-11e6-9562-bc5ff499a18d
Output
Simulation Period Total
Daily Output
Cumulative Probability
Growth Stage
Management Event
English SI
Select Scenarios:
Irrigation_Terminat
Irr_Termination
test for jae
Select outputs to display:
H2O stress
N stress
Extractable water
Cumulative irrigati
Cumulative draina
Cumulative evapot
Output
Irrigation_accumulation (inch)
15
10
5
0
Mar '18
May '18
Jul '18
Sep '18
Simulated with observed climate
Simulated with 2013 climate data
Simulated with 2014 climate data
Simulated with 2015 climate data
Simulated with 2016 climate data
Simulated with 2017 climate data
Close
Select Field: TestK1
Show Field Polygons
Weather Stations
Weather Stations
-State Stations (MESONET or CIMIS)
Gridded Data
-PRISM
User Data refresh
Click on the search button to display a list of weather stations closest to your crop field(s). After collecting, you can click on each station to see its location and related weather information.
If any data is missing, it will be filled out using PRISM or monthly average of data.
Radius: 30 miles
MESONET
Garden City (2.0 mi)
Haskell (27.6 mi)
Lakin (24.4 mi)
ikisekka | ICROP_share | Help
fieldAssessmentRe....pdf

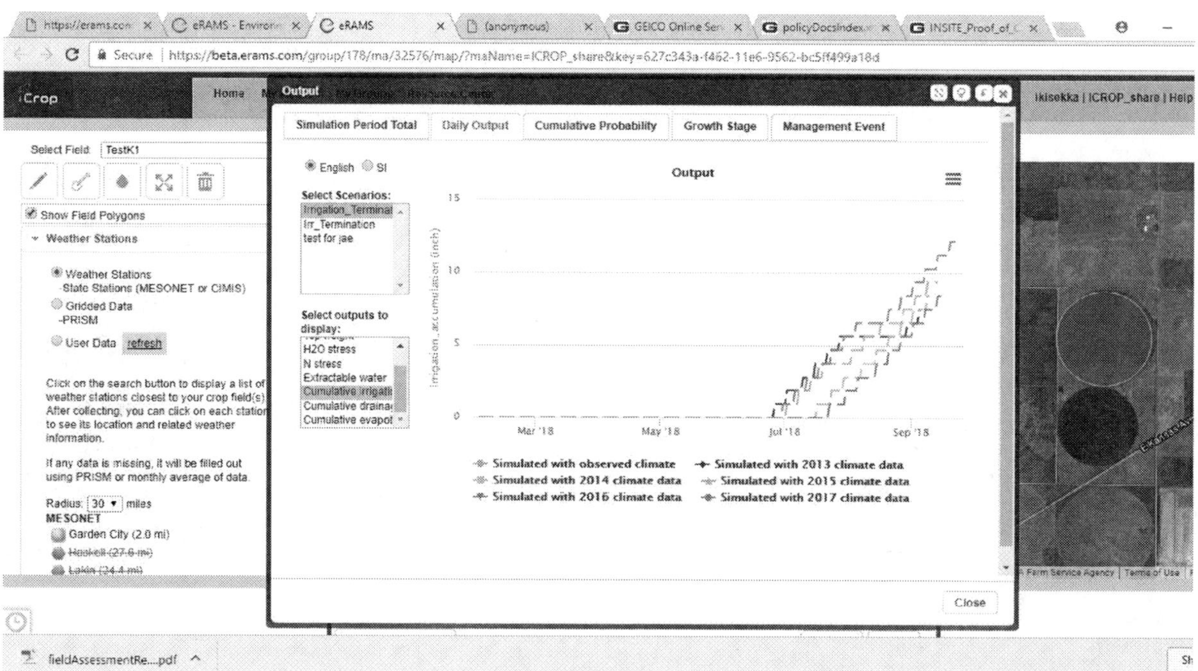
https://beta.erams.com/group/178/ma/32576/map/?maName=ICROP_share&key=627c343a-f462-11e6-9562-bc5ff499a18d
Output
Simulation Period Total
Daily Output
Cumulative Probability
Growth Stage
Management Event
English SI
Select Scenarios:
Irrigation_Terminat
Irr_Termination
test for jae
Select outputs to display:
H2O stress
N stress
Extractable water
Cumulative irrigati
Cumulative draina
Cumulative evapot
Output
Irrigation_accumulation (inch)
15
10
5
0
Mar '18
May '18
Jul '18
Sep '18
Simulated with observed climate
Simulated with 2013 climate data
Simulated with 2014 climate data
Simulated with 2015 climate data
Simulated with 2016 climate data
Simulated with 2017 climate data
Close
Select Field: TestK1
Show Field Polygons
Weather Stations
Weather Stations
State Stations (MESONET or CIMIS)
Gridded Data
-PRISM
User Data refresh
Click on the search button to display a list of weather stations closest to your crop field(s). After collecting, you can click on each station to see its location and related weather information.
If any data is missing, it will be filled out using PRISM or monthly average of data.
Radius: 30 miles
MESONET
Garden City (2.0 mi)
Haskell (27.6 mi)
Lakin (24.4 mi)
ikisekka | ICROP_share | Help
fieldAssessmentRe....pdf

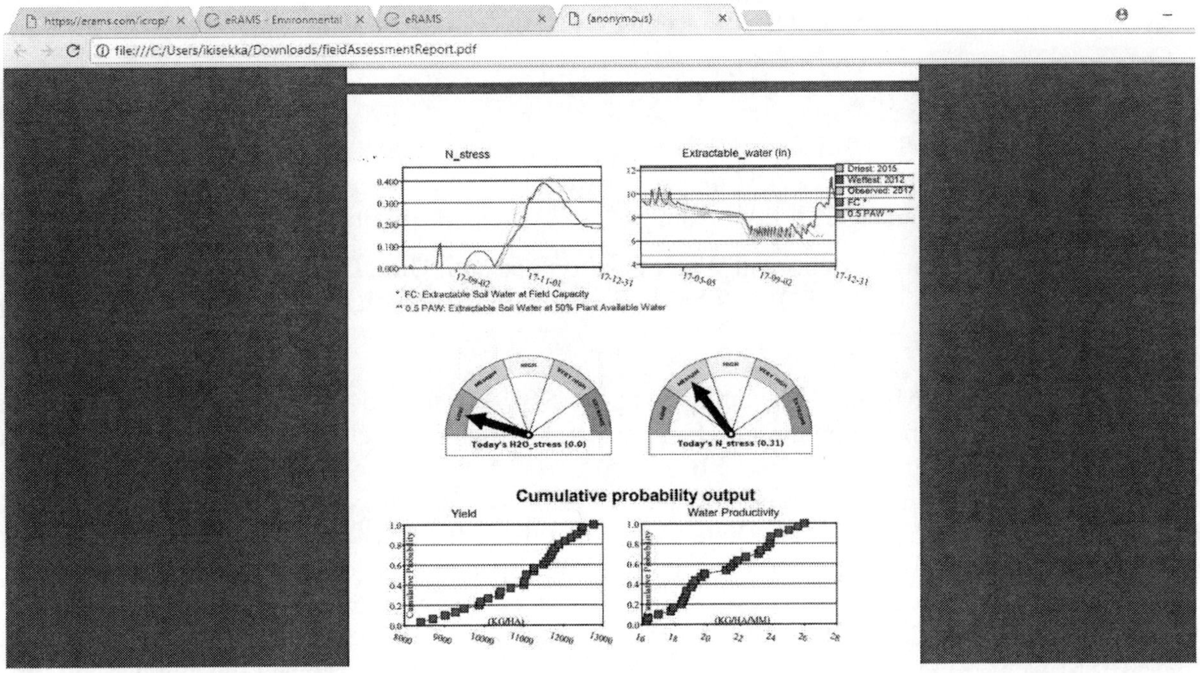
file:///C:/Users/ikisekka/Downloads/fieldAssessmentReport.pdf
N_stress
Extractable_water (in)
Driest: 2015
Wettest: 2012
Observed: 2017
FC *
0.5 PAW **
* FC: Extractable Soil Water at Field Capacity
** 0.5 PAW: Extractable Soil Water at 50% Plant Available Water
Today's H2O_stress (0.0)
Today's N_stress (0.31)
Cumulative probability output
Yield
Water Productivity
(KG/HA)
(KG/HA/MM)

DASHBOARD / RESEARCH FIELD
Collapse Sensors
Reload
Settings
Print
Research Field
WEATHER DATA
64°F
Wind: 31.09 mph
Humidity: 17%
Pressure: 1,025 hPa
FORECAST
Tue Wed Thu Fri Sat
CROP INFORMATION
Crop: Corn
Growth Stage: R6 - maturity
Planting Date: 05/08/2017
Current ET: 0.03
Avg 4 Day ET: 0.03
Estimated GDUs to Maturity: 2800
GDUs
3096
PIVOT
Power: ON
Pressure: 0.0PSI
Direction of travel: FWD
Angle: 50
Vigor
8/2/2017
Pivot History
A. 53428 3RD TOWER DRAGONLINE300
B. 53405 3RD TOWER NOZZLES 300
C. 53403 2ND TOWER DRAGONLINE300
D. 53867 2ND TOWER NOZZLES 300
Soil: Silt Loam
59%
53%
64%
71%
4:22 PM
10/23/2017

Comparison of Sensor-Based Irrigation Scheduling Method and Arkansas Irrigation Scheduler

Ruixiu Sui[1] and Earl Vories[2]

[1]USDA-ARS Sustainable Water Management Research Unit, Stoneville, Mississippi 38776; Phone: (662)686-5382; FAX: (662) 686-5322; email: ruixiu.sui@ars.usda.gov

[2]USDA-ARS Cropping Systems and Water Quality Research Unit, Delta Center, P.O. Box 160, Portageville, MO 63873; Phone: (573) 379-5431; FAX: (573) 379-5875; email: earl.vories@ars.usda.gov

Abstract: Sensor-based irrigation scheduling methods (SBISM) use sensors to measure soil moisture and schedule irrigation events based on the soil-water status. With rapid development of soil moisture sensors, more producers have become interested in SBISM. Arkansas Irrigation Scheduler (AIS) is a weather-based irrigation scheduling tool and has been adopted in the Mid-South for many years. Field studies were conducted for two years in Mississippi Delta to compare the SBISM with the AIS. Soil moisture sensors were installed in multiple locations of a soybean field. Soil water contents of the field were measured across the growth season. Meanwhile, the AIS was installed in a computer. A weather station near the soybean field was employed to obtain all data required by the AIS. Number and time of the irrigation events triggered by the SBISM were compared with those scheduled by the AIS. Results showed the number and time of irrigation events scheduled using the SBISM were often different from those predicted by the AIS, especially during the 2018 growing season. Both the sensor-based irrigation scheduling method and the AIS could be used as tools for irrigation management in the Mid-South region, but extra attention to the effective portion of rainfall or irrigation would be needed in some years.

INTRODUCTION

Irrigated agriculture in the US is a major consumer of freshwater, accounting for 80% of the nation's consumptive water use (Schaible and Aillery, 2015). Irrigation is essential for crop production in arid and semiarid regions. However, in recent years, acreage of irrigated land has increased rapidly the Mid-South region including the Mississippi Delta (MD). MD is one of the major crop production regions in the United States. Though typical annual precipitation is about 130 cm in this area, only about 18% of the precipitation occurs during June to August when crops require a large quantity of water. Furthermore, the precipitation patterns in summer frequently include heavy rainfall events that increase runoff from cropland with only a small amount of rainfall percolated into the soil profile and available for plant use. Uncertainty in the amount and timing of precipitation is one of the most serious risks to crop production in MD. Timely irrigation has been shown to increase yields of corn (Sui et al., 2015; Vories et al., 1993) and cotton (Sui et al., 2017; Vories et al., 2007). Producers in this region have become increasingly reliant on irrigation to ensure adequate yields and reduce production risks. Approximately 90 percent of irrigated cropland in this region relies on the groundwater supply from the Mississippi River Valley Alluvial Aquifer. Excessive withdrawal of the groundwater has resulted in a decline in aquifer levels across the region. Ongoing depletion and stagnant recharging of the aquifer jeopardize the long-term availability of the aquifer and place irrigated agriculture in the region on an unsustainable path. It is necessary to seek improved irrigation technologies to increase water use efficiency for sustainable use of water resources.

Irrigation scheduling is one of the irrigation technologies to determine the time and amount of water to apply. Irrigation scheduling methods include weather-based, soil moisture-based, and plant-based methods. Weather-based methods schedule irrigation based on the estimated amount of water lost by plant evapotranspiration (ET) and the amount of effective rainfall and irrigation water entering into the plant root zone. Soil-based methods measure soil moisture or water potential levels in the plant root zone and water is applied when there is a water shortage for plants. Plant-based methods directly detect plant responses to water stress and irrigation is initialized as plants indicate suffering from water stress.

In soil moisture-based irrigation scheduling, soil moisture content can be directly determined using manual gravimetric soil sampling by weighing and drying the soil sample. The gravimetric method is simple. However, it is time consuming and expensive as frequent measurements are required. Electromagnetic (EM) sensors, such as electrical capacitance and resistance type sensors, and time-domain reflectometer (TDR) devices have been rapidly developed and widely adopted for soil moisture measurement in irrigation scheduling (Dukes and Scholberg, 2004; Evett and Parkin, 2005; Fares and Alva, 2000; Miranda et al., 2005; Seyfried and Murdock, 2001; O'Shaughnessy and Sui, 2018; Sui, 2018; Sui and Baggard, 2015; Vellidis et al., 2008).

The Arkansas Irrigation Scheduler (AIS) was developed in the 1970s and early 1980s under the leadership of Dr. James Ferguson to aid Arkansas farmers in managing irrigation. It has been in use for almost forty years in Arkansas and surrounding states. The AIS uses a water-balance approach to scheduling irrigation. The system balance represents the soil water deficit (SWD), the difference between the soil's existing moisture content, summed over the rooting depth, and the moisture content of the soil at its well-drained upper limit. Deposits to the system include rainfall and irrigation. Withdrawals from the system include crop evapotranspiration (ET_c), runoff, and deep percolation below the root zone. Deep percolation is considered negligible when the SWD > 0 and is therefore not considered in the program. Rooting depth is not used explicitly in the program, but is implicit in the choice of a maximum allowable SWD (Vories et al., 2009).

The objective of this study was to compare the sensor-based irrigation scheduling method (SBISM) with the Arkansas Irrigation Scheduler (AIS) in irrigation scheduling.

MATERIAL AND METHODS

The study was conducted in 2017 and 2018 in a 6.7-ha field at the USDA ARS Research Farm in Stoneville, Mississippi, USA (latitude: 33°26'30.86", longitude: -90°53'26.60"). Silt loam was the predominant soil type in the field. The field was a quadrant of the area under a center pivot irrigation system. Three irrigation treatments were set up in the field, which are variable rate irrigation (VRI), uniform rate irrigation (URI), and rainfed. VRI and URI treatments were assigned in the area under the pivot while the area in the corner of the field was assigned for rainfed treatment. Irrigation scheduling in both VRI and URI treatments was performed using the SBISM. Soybean was planted in the field on April 7 in 2017 (cultivar: HBK4950) and April 20 in 2018 (cultivar: HBK4855). The studies were harvested on September 9 in 2017 and October 4 in 2018. Insects and weeds in the field were controlled with generally recommended procedures in the region throughout the growing seasons.

For irrigation scheduling, GS-1 soil water content sensors (Decagon Devices, Pullman, Washington, USA) were employed to measure soil volumetric water content (VWC) in four locations within the field. In each location, three sensors were installed at three depths (15, 30, 61cm).

To install the sensors, a hole was drilled at the center of the crop row using a soil auger. The sensors were inserted horizontally into the soil at the designated depths. One data logger (EM50R/G, Decagon Devices, Pullman, Washington, USA) was used to collect soil moisture data from the sensors. Data loggers were set up to continuously make one measurement of soil water content every minute and calculate the hourly average of the measurements. Readings of the soil water content from the logger were downloaded wirelessly.

Soil water content measured using the sensors was used for irrigation scheduling. Soil water content measurements at the three depths were interpreted using a weighted average method to reflect the importance of soil water in different depths across the plant root zone. A weight was assigned to each sensor measurement based on the sensor depth as 0.45, 0.35, and 0.2 for the sensor depth of 15, 30, 61cm, respectively. The weighted average of the VWC was used for irrigation scheduling. In this region, it is very common that long-lasting rainfalls occur in early crop growing season to saturate the soil. Soil VWC measured by the sensors in field at 48 hours after soil was saturated was used as the sensor-measured field capacity (FC). VWC at permanent wilting point (PWP) of the soil was determined based on the soil type (Sandall, 2018). Irrigation was triggered when the plant available water (PAW) dropped close to 50%.

PAW is defined as follows:

$$PAW = \frac{(Sensor\_measured\ VWC) - (VWC\ at\ PWP)}{FC - (VWC\ at\ wilt\ point)}$$

In general, at each irrigation event, a 19mm depth of water was applied to the 100% rate zone of VRI treatment and all zones of URI treatment. Water depth applied to the other zones of VRI treatment was scaled down according to the rate assigned. Irrigation water was delivered using a center pivot VRI system.

A weather station located near the soybean field was used to collect weather data including daily ET_o. Irrigation events were triggered by the soil water content measured by the sensors. The amount of irrigation water applied in each scheduled irrigation event was recorded.

A stand-alone version of the Arkansas Irrigation Scheduler (AIS, version 2.3.4) was installed in a computer. According to the operation instructions of the AIS, the data were entered in the AIS program, including planting date, initial soil moisture status, daily ET_o, amounts of rainfall and irrigation. The AIS calculated the daily water deficit with these input data. It was assumed that any rainfall was effective (i.e., entered the soil) until the SWD was replaced and any additional amount ran off the field.

RESULTS AND DISCUSSION

The total precipitation and ET_o were 656mm and 433mm, respectively in the 2017 growing season. The soil water content change across the growing season in 2017 was illustrated in Figure 1. Soil moisture sensors responded well to major precipitations, especially the sensor in the 15cm depth. The sensor-measured soil moisture was used to trigger the irrigation events. Based on the weighted soil VWC, the first irrigation in the amount of 19mm was scheduled on July 21 as the weighted soil VWC dropped to 31%. Another three irrigations were conducted on July 27 in amount of 19mm, August 1 in amount of 22mm, and August 4 in amount of 19mm (Fig. 2) to maintain a low water deficit considering the crops were in the R3-R4 stages, in which the plant is more susceptible to water stress than the other growth stages. Though the weighted soil VWC dropped below the irrigation trigger level around May 20th and June 15th, irrigation was not scheduled for two reasons: the plants were at stages in which they are more tolerant to water stress, and the weather forecast indicated a precipitation event coming soon.

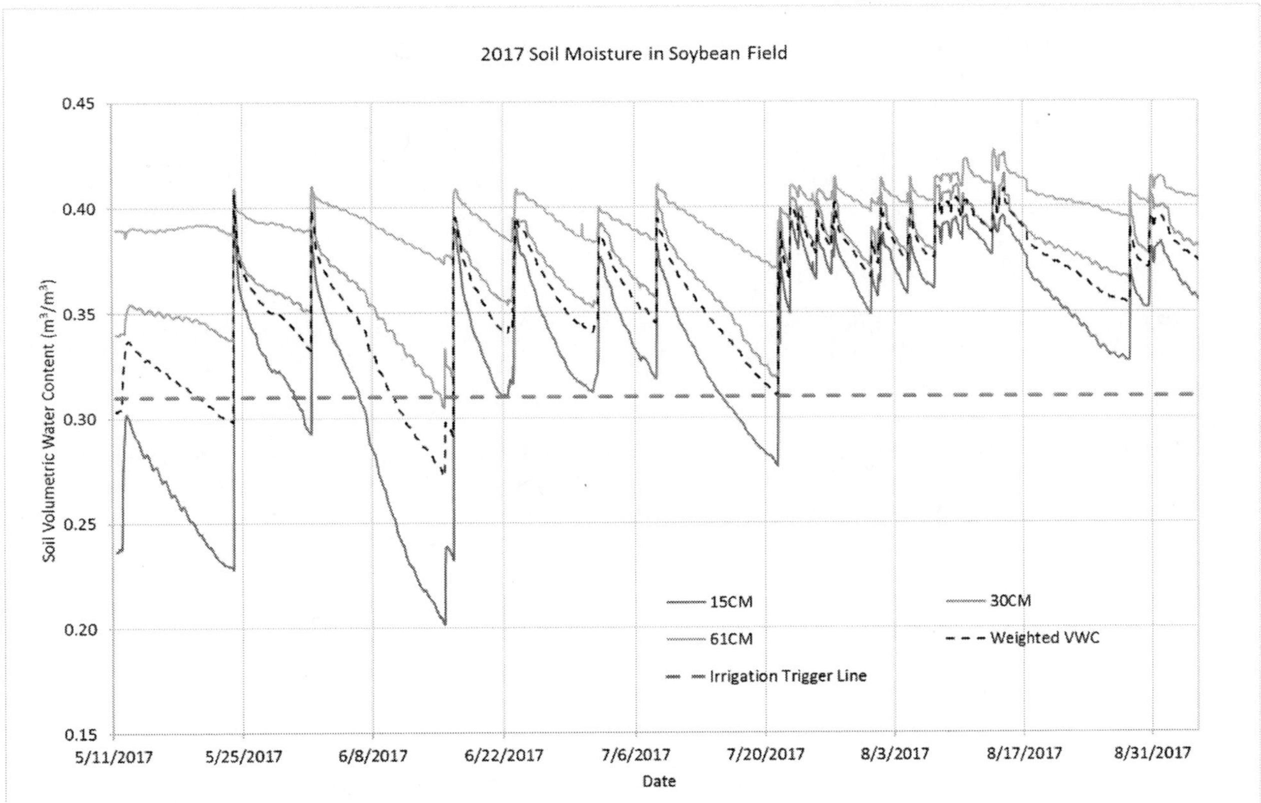

Fig. 1. Soil water content measurements across the 2017 season in soybean field. Irrigation trigger level was set at 31% weighted average soil volumetric content.

The daily inputs to the AIS (amount of ET_o, precipitation, irrigation) and the SWD predicted by the AIS in the 2017 growing season are shown in Figure 2. An allowable SWD of 37 mm was selected to call for an irrigation event, although all irrigations were scheduled by the SBISM. The deficit estimated by the AIS on July 20th was 43mm, which indicated an irrigation was needed (Fig. 2). This prediction matched fairly well with the measurement results of the soil moisture sensors (Fig. 1), although the AIS date would have been 3 days earlier. On June 14th, when the sensor-measured soil moisture suggested to schedule an irrigation (Fig. 1), the AIS deficit prediction was 31mm, which was slightly below the irrigation trigger level of 37mm (Fig. 2). The AIS also indicated a predicted deficit of 32mm on August 27, near the 37mm trigger. However, the sensor-based scheduling method did not suggest the irrigation at that time (Fig. 1). This could be caused by deficit over-prediction by the AIS at the late growth stage of the soybean plants. It was also observed that the sensor-measured soil moisture dropped below the irrigation trigger line on May 20th, but the AIS predicted deficit at that time was 14mm which is considerably lower than the irrigation trigger level of 37mm (Fig. 1). It should be noted that the AIS adjusts the crop coefficient curve based on soybean maturity group (MG); however, actual soybean cultivars tend to vary more than the adjustments can account for. While the cultivar HBK4855 is a MG 4, as a late MG 4 it might fit the MG 5 curve more closely.

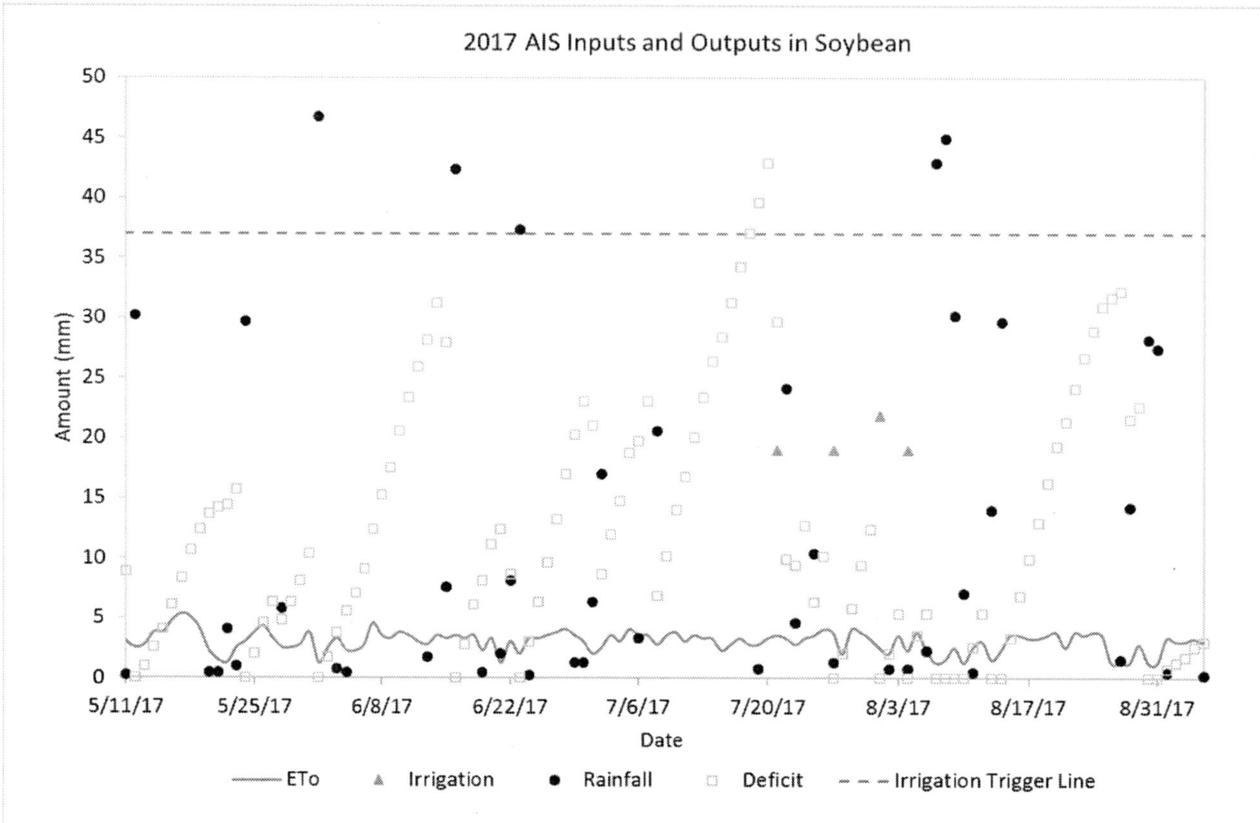

Fig. 2. Inputs to the AIS including ET_o, rainfall, irrigation, and the AIS output of estimated deficit in 2017 soybean growing season. Irrigation trigger level was set at the AIS predicted deficit of 37mm.

The total precipitation and ET_o were 644mm and 438mm, respectively, in 2018. Starting from June 9th, seven irrigations were scheduled based on the soil moisture measured by the sensors in 2018 (Fig. 3). 19mm water was applied in each irrigation event. The 1st irrigation on June 9th and the 2nd on June 12th did not bring the soil moisture content back up to the irrigation trigger level. The rain on June 13th and 19th increased the soil moisture and charged the soil VWC above the irrigation trigger level. The apparent fluctuation of the VWC at 61cm depth on June 4th was caused by the failure of a sensor at that depth in one location. On July 3rd and July 6th, the third and fourth irrigations were triggered as the soil moisture decreased to 31%. The other three irrigation events on July 26, July 27, and August 15 were scheduled based on the sensor-measured soil moisture. Irrigations should have been conducted between July 28 and August 15 according to the sensor-based irrigation scheduling (Fig. 3). However, due to a problem with the center pivot, no irrigation could be made during that period. The apparent lack of response to irrigation observed in the soil moisture sensors suggests that soil compaction and/or crusting prevented much of the applied water from entering the soil. Unfortunately, such a situation is fairly common in the region.

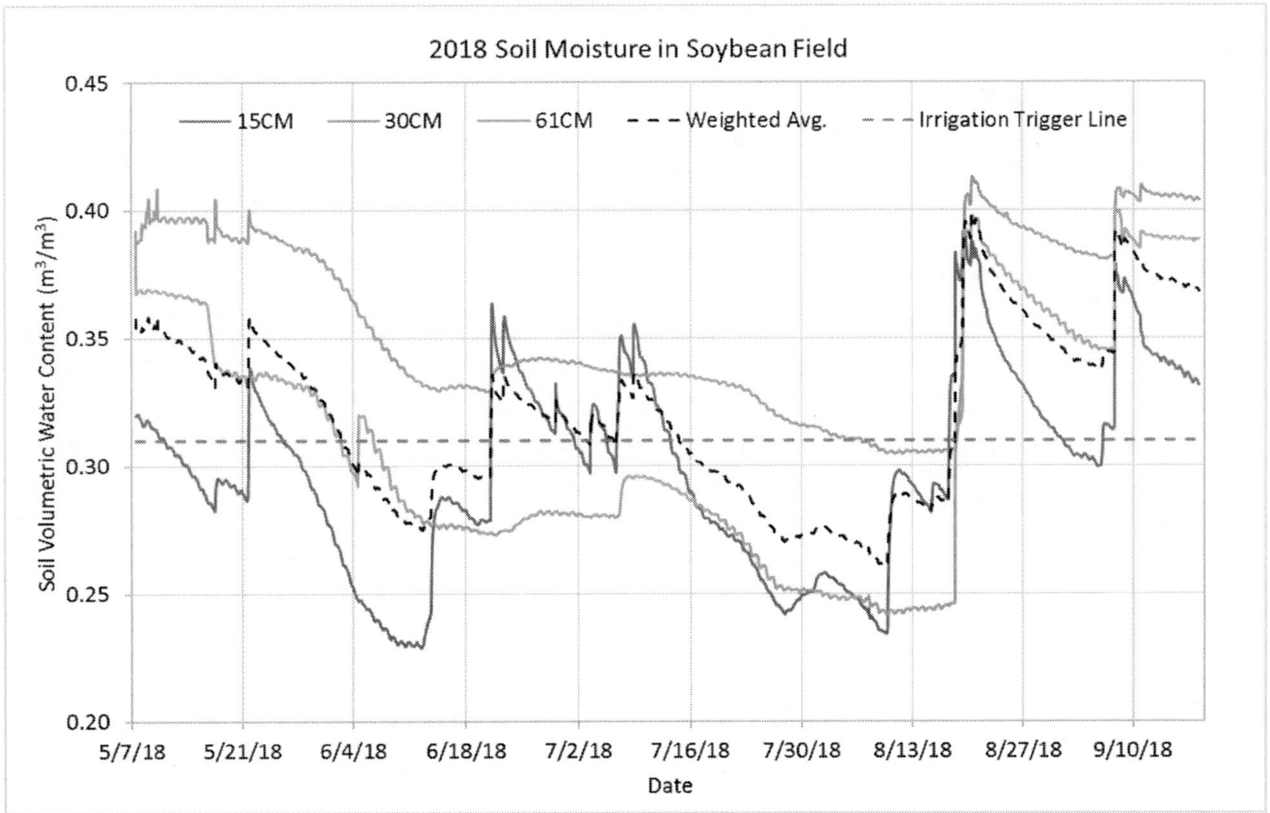

Fig. 3. Soil water content measurements across 2018 season in soybean field. Irrigation trigger level was set at 31% weighted average soil volumetric content.

When the first four irrigations in the 2018 season were scheduled on June 9th, June 12th, July 3rd, and July 6th, the deficits predicted by the AIS were 14mm, 3mm, 27mm, and 22 mm, respectively, which were less than the AIS irrigation trigger level of 37mm (Fig. 4). When the fifth sensor-based irrigation was scheduled on July 26th, the deficit predicted by the AIS was 44mm, which surpassed the AIS irrigation trigger level of 37mm; however, it was 30mm when the sixth irrigation was scheduled on July 27th. When the seventh irrigation was scheduled on August 15th, the estimated deficit was 26mm, which was below the irrigation trigger level. The AIS assumed that all surface water had run off the field within 24h and started accumulating the deficit from zero in the next day, while in reality, such a large rain period often has a longer-lasting effect. The discrepancies between the two methods earlier in the season probably resulted from reduced soil intake suggested by the sensors, while the AIS estimates assumed that the water reached the root zone. This points out a serious shortcoming with water-balance methods like the AIS, knowing how much of a rainfall or irrigation was actually effective and should be used to adjust the deficit.

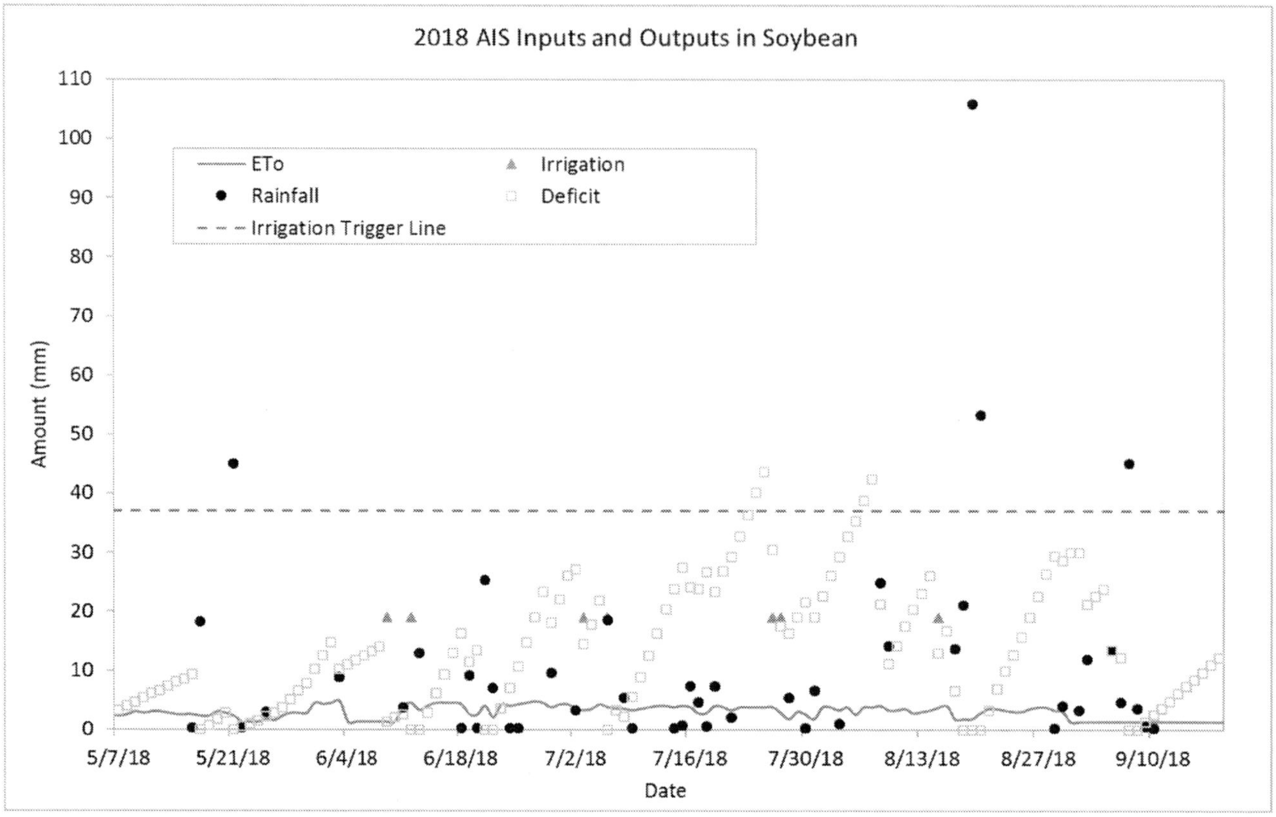

Fig. 4. Inputs to the AIS including ETo, rainfall, irrigation, and the AIS output of deficit output in 2018 soybean growing season. Irrigation trigger level was set at the AIS predicted deficit of 37mm.

CONCLUSION

Field studies were conducted to compare the sensor-based irrigation scheduling method (SBISM) with the AIS (Arkansas Irrigation Scheduler). Two years of tests in soybean crops showed the number and time of irrigation events scheduled by the SBISM differed from those predicted by the AIS, especially in 2018. A couple of mismatching of the irrigation recommendations between these two approaches occurred at the early and late growth season, which might have been caused by the lack of precision in the AIS crop coefficient functions and very high soil moisture content due to large amounts of precipitation in the late growing season. The most pronounced differences were observed in 2018, when the sensors indicated problems with applied water reaching the root zone while the AIS assumed that it did. Both of the SBISM and the AIS could be used as tools for irrigation management in the Mid-South region, but extra care is needed when soil crusting or compaction reduce the effectiveness of irrigation. The wireless soil moisture sensors, which can provide real-time in-situ soil moisture measurement, made the SBISM easy to use. The AIS requires daily weather data from a weather station near the field, which makes it less user friendly. The prediction accuracy of the AIS might be significantly influenced if this requirement could not be met.

Disclaimer

The mention of trade names of commercial products in this article is solely for the purpose of providing specific information and does not imply recommendation or endorsement by the U.S. Department of Agriculture.

REFERENCE

Dukes, M. D. and J. M. Scholberg. 2004. Soil moisture controlled subsurface drip irrigation on sandy soils. Applied Engineering in Agriculture, 21(1):89-101.

Evett, S. R., and G. W. Parkin. 2005. Advances in soil water content sensing: The continuing maturation of technology and theory: *Vadose Zone J.,* 4:986–991, doi:10.2136/vzj2005.0099.

Fares, A. and A. K. Alva. 2000. Evaluation of capacitance probes for optimal irrigation of citrus through soil moisture monitoring in an entisol profile. Irrig. Sci., 19:57-64

Miranda, F. R., R. E. Yoder, J. B. Wilkerson, and L. O. Odhiambo. 2005. An autonomous controller for site-specific management of fixed irrigation systems: *Comp. Elec. Agric.*, 48(3)183-197.

O'Shaughnessy, S.A. and R. Sui. 2018. Chapter 11- Advanced Tools for Irrigation Scheduling. In Advances in Agricultural Machinery and Technologies. Ed. Guangnan Chen. CRC Press, Boca Raton, FL. Pp. 263-288.

Sandall, L. Irrigation Management. http://croptechnology.unl.edu/pages/informationmodule.php?idinformationmodule=1130447123&topicorder=3&maxto=13&minto=1. Accessed online on Oct. 13, 2018.

Sandall, L. Physical Properties of Soil and Soil Water. https://passel.unl.edu/pages/informationmodule.php?idinformationmodule=1130447039&topicorder=10&maxto=10&minto=1. Accessed online on Oct. 13, 2018.

Schaible, G. and M. Aillery. 2015. Irrigation & water use. http://www.ers.usda.gov/topics/farm-practices-management/irrigation-water-use.aspx. Accessed on Oct. 9, 2018.

Seyfried, M. S., and M. D. Murdock. 2004. Measurement of soil water content with a 50-MHz soil dielectric sensor. Soil Sci. Soc. Am. J., 68:394–403.

Sui, R. 2018. Irrigation scheduling using soil moisture sensors. *J. Agric. Sci.* 10(1) 1-11. doi:10.5539/jas.v10n1p1.

Sui, R. and J. Baggard. 2015. Wireless sensor network for monitoring soil moisture and weather conditions. Appl. Engr. Agric. 31(2):193-200. doi:10.13031/aea.31.10694.

Sui, R., R. K. Byler, and C. D. Delhom. 2017. Effect of nitrogen application rate on yield and quality in irrigated and rainfed cotton. *J. Cotton Sci.* 21:113-121.

Sui, R., D. K. Fisher, and K. N. Reddy. 2015. Yield response to variable rate irrigation in corn. J. Agric. Sci. 7 (11):11-18. doi:10.5539/jas.v7n11p11.

Vellidis, G., M. Tucker, C. Perry, C. Kvien, and C. Bednarz. 2008. A real-time wireless smart sensor array for scheduling irrigation. Computers and Electronics in Agriculture, 61:44-50.

Vories, E. D., R. Hogan, P. Tacker, R. Glover, and S. Lancaster. 2007. Estimating the impact of delaying irrigation for Midsouth cotton on clay soil. Trans. ASABE 50(3):929-937.

Vories, E. D., D. J. Pitts, and J. A. Ferguson. 1993. Sprinkler irrigation response of corn on clay. Ark. Farm Res. 42, 6-7.

Vories, E., P. Tacker, and S. Hall. 2009. The Arkansas Irrigation Scheduler. Proceedings of World Environmental and Water Resources Congress. P. 3998-4007.

Effectiveness of Irrigation Water Management Practices for Mid-South Furrow Irrigation

C.G. Henry, Associate Professor and Water Management Engineer, University of Arkansas, Rice Research and Extension Center, Stuttgart, AR 72160, cghenry@uark.edu

J. P. Gaspar, Irrigation Program Associate, University of Arkansas, Rice Research and Extension Center, Stuttgart, AR 72160

A. P. Horton, Irrigation Program Technician, University of Arkansas, Rice Research and Extension Center, Stuttgart, AR 72160

Abstract

Arkansas is the third largest irrigated state and is only 46% sustainable in its groundwater withdraws for irrigation. The objective of this research was to evaluate the efficacy of Computerized Hole Selection (CHS), Surge Irrigation (SURGE), and Soil Moisture Monitoring (SMM). The research consisted of 12 paired furrow irrigated fields where water use was measured with portable propeller flow meters and yield was measured on cooperating farmer fields. Water use and relative yields were compared between the treated field (IWM) and control field (CONTROL). IWM fields that used SURGE, CHS and SMM were found to use 27% less water than the CONTROL field. Fields that used CHS and SMM used 19% less water than CONTROL fields (p=0.015). There were no significant difference in yields (p=0.86) between IWM and CONTROL fields or less than a 1% difference. On-farm pump testing identified a water savings of $511 per site. This study demonstrates that widespread adoption of these practices can significantly reduce the overdraft of the alluvial aquifer in the mid-south and can be done without any yield penalties.

Keywords: Irrigation Water Management, furrow irrigation, surge Irrigation, Soil Moisture Sensors, Computerized Hole Selection

Introduction

Arkansas is the third largest irrigated state (5.0 million acres) in the United States after Nebraska (8.3 million acres) and California (7.5 million acres) (USDA, 2012). The Arkansas Natural Resources Commission (2014) projects that about 70 percent of water demand in Arkansas is supplied using ground water with 80 percent of the water used to irrigate field crops. The largest category of irrigation expense for producers in these states is energy cost associated with pumping of water from surface and subsurface water resources. Approximately 4.3 million of the 4.9 million irrigated acres in Arkansas utilize flood or furrow irrigation (USDA, 2012).

Furrow and flood irrigation (surface irrigation) are the most common form of irrigation methods practiced in Arkansas (Maupin et al., 2014). These forms of surface irrigation are appealing to farmers because of the minimal capital investment and lower energy costs in comparison to pressurized irrigation systems. However, more labor is required than sprinkler irrigation. However, with the advent of lay-flat irrigation pipe, which became available and quickly adopted in Arkansas in the mid-1990's. The labor needs are greatly lessened compared to rigid gated pipe. The challenge with lay-flat pipe is that a hole is punched after the pipe is filled with water and there is no adjustment compared to gated pipe systems that can be adjusted by the irrigator to maintain even furrow advances.

Computerized Hole Selection (CHS) is a tool used for lay-flat polyethylene irrigation pipe products. In lay-flat irrigation pipe, the pipe comes in rolls, the irrigator lays the pipe out and fill it with water. Then when the pipe is full of water the irrigator punches holes in the pipe, typical hole sizes range from ¼ inch to 1 inch. Inefficiencies arise when the holes are too large or too small, resulting in over pressuring the pipe when holes are too small and bursting, or when holes are too large, inadequate volume is delivered to the holes farthest from the inlet. Additionally when fields are not square, longer furrows require higher flowrates than shorter ones. Finally, the pipe crown slope when not level must be accounted for in the hydraulic calculations for each hole to deliver the appropriate volume. Thus using a computer to iterate out the proper hole sizes and changes along the pipe often can result in pipe distribution uniformities of around 90% (low quarter distribution) or higher, even on very irregular and undulating slopes where the pipe is placed.

Surge irrigation (SURGE) is the practice of intermitting applying water to furrow irrigated fields for the purpose of improving down furrow uniformity. Water delivery is alternated between a "right" and "left" side of an irrigation set. Initially the cycles are longer in order to advance the water through the field, then are shortened in order to reduce deep percolation near the top end of the field and deliver more water to the lower end. Infiltration rates are high when soil is dry and reduce as the soil seals from being wetted. Surge irrigation has been shown to reduce deep percolation, increase advance time, decease total water applied and reduce tail water ratios. (Bishop et al., 1981; Izuno et al., 1985; Goldhamer et al., 1987b; Musick et al., 1987; Testezlaf et al., 1987; Israeli, 1988; Eid et al., 1999).

Soil Moisture Monitoring (SMM) is the use of soil moisture sensors and telemetry to monitor the soil water balance. SMM is used for scheduling irrigation events, but also verify the efficacy of the irrigation event in refilling available water capacity. SMM are comprised of soil moisture sensors that may include tensiometers, gypsum blocks, granular matrix potential sensors, heat dissipation sensors, and soil psychrometers (Munoz-Carpena, 2004). One of the most commonly used sensors is the Watermark Model 200SS (Irrometer Company Inc, Riverside, CA) and is a matric potential sensor comprised of two electrodes in a granular matrix. The sensors are attached to schedule 20 polyvinyl chloride pipe and placed in the soil at depths appropriate for representing the managed depth selected by the irrigators. In addition to sensors, SMM are comprised of electronics that measure soil moisture sensor output, provide power to sensors, record, and relay data wirelessly to a server. A cellular modem or other radio communication is used to relay data wirelessly from the SMM to the internet. Sensor data is then displayed to irrigators through web browsers or smartphone devices.

Materials and Methods

Cooperating farmers volunteered with this research by contacting their local Extension Agent during the 2015 crop year. The Agents and farmers located comparable fields that were the same soil texture, slope and productivity. Field were planted to the same variety and had planting dates that were within a few days. Computerized Hole Selection (CHS, SURGE irrigation, and Soil Moisture Monitoring (SMM) was implemented on each Irrigation Water Management Field (IWM). In some fields, sets were divided so that there were two sets utilizing SURGE and one without but did have CHS implemented, so the effect of SURGE relative to CHS and a CONTROL could be assessed. Each CONTROL and IWM field was instrumented with portable propeller mechanical flow meter (McCrometer Corporation, Hemet, CA) to totalize the water use of the field during the season. Farmers managed CONTROL fields according to their conventional practices, punching hole sizes and irrigation timing according to their experience. Generally farmers irrigated their control fields every 7-10 days on a calendar method. County Extension agents communicated with farmers to help time irrigation using the SM. SMM was accomplished using AgSense Aquatrac data collection and telemetry units and Watermark™ soil matric potential sensors.

Sensors were installed at 6, 12, 18 and 30 inch depths. The SMM units provided data to smartphones and through a website, so farmers, agents and the authors could monitor soil moisture trends and status. For each site, an allowable depletion was developed from soil samples taken at the top foot and second foot depths. Soil texture, bulk density, organic matter, field capacity and wilting point were determined through laboratory analysis. This data was used to develop an allowable depletion using the pedotransfer functions published by Saxton and Rawls (2006). This information was used to set a range of irrigation thresholds in the interface for the telemetry unit for each IWM field. The vast majority of sites were a silt loam texture with a few loam or sandy loam soil textures.

P and R Surge valves (P&R Surge, Lubbock, Texas) were used and coupled to the portable propeller flow meters (McCrometer, Hemet, CA). Totalizers were read before the first irrigation, between and after the last irrigation. Each field had a CHS plan developed using either a computer program called PHAUCET (Pipe Hole And Universal Crown Evaluation Tool) or an online web tool provided by Delta Plastics Pipe Planner (Delta Plastics, Little Rock, AR). Agents were trained through in-service trainings on how to implement CHS, read and install meters, surge valve programming (Henry and Krutz, 2017) and soil moisture sensors. Advance time adjustments were only made during the first irrigation. Often the first irrigation takes longer than subsequent irrigations, so additional improvement would be possible if irrigators adjusted advance times after the first irrigation. Irrigation decisions were made by the county Agent and farmer with consultations with the authors when needed. Terminations were done according to UA Extension recommendations for R6.5 and good soil moisture and 50% of the starch line for corn and good soil moisture. For cotton termination was determined by DD60 and 5 Nodes Above White Flower. Good soil moisture was defined as an average reading in the profile were less than a 30% MAD was available or about -60 cb for silt loams and less than -30 cb for sandy loams.

After termination, valves, meters, sensors and telemetry units were removed from the fields. Yields were calculated for each CONTROL, CHS and IWM field in entirety by either certified scale tickets or calibrated yield monitors. In some cases herbicide or other crop damage occurred in part of the field, and yield monitor data was used to adjust yields for direct comparison between the paired fields.

Irrigation Water Use Efficiency (IWUE) was calculated according to Vories et al (2005) as the yield in bushels per acre-inch divided by the irrigation water applied in acre-inches per acre. Data was analyzed with SYSTAT version 13.2 (Systat, San Jose, CA).

For each site, the cost of water was determined to calculate potential energy savings by optimizing the irrigation pumping plant with the CHS Plan. Testing and analysis was conducted using the methodology described by McDougall (2012) for pumps that did not have pump monitors and electric meters energy consumption was conducted as described by Henry and Bingham (2013). For the electric pump sites, the cost of water was determined by measuring the flow from the on-site flowmeter and by measuring the energy consumption by reading the electric meter during the test. The data was done once mid-season. For diesel power units, the fuel use was measured by isolating the engine from the fuel tank and suppling fuel from a 2.5 liter graduated cylinder plumbed to the supply and return lines of the engine. The consumption of fuel or electricity over a several minute time period was measured with a stopwatch. For diesels the tests were conducted over the operating range of the engine as reported by the irrigator. Also the cost of energy in kWh and $/gallon as reported by the irrigator were used to determine the cost of water for each site.

Results

Yield data was converted to relative yield between the treatment (IWM or CHS) to the control yield for comparison between the three crop types, of corn, soybean, and cotton. There were 11 fields (n=11) comprised of 11 soybean, 1 corn and 1 cotton field. These fields had 26 treatments of CHS (4), CONTROL (12) and IWM (10 SURGE AND CHS). All CONTROL fields (100%) applied more water than IWM treatments. The non-parametric Kruskal-Wallis one way Analysis of Variance (ANOVA) on Ranks was used to analyze the data. The Dunns's Method was used for a multiple pairwise comparison to test for differences between water use and yield between the three treatments. A 0.6-1% increase in relative yield was observed. There was no significant difference in relative yields found between the treatments (p=0.864). However, there was a difference in relative water use between CHS and CONTROL (p=0.015) and IWM and CONTROL (p<0.001). No difference between CHS and IWM was found (p=1). Although the water use was 8% less with IWM than CHS. These results show a reduction in water use using IWM of 27% equating to a 2.6 ac-in/ac savings. The analysis was repeated using the parametric ANOVA test and yielded the same results.

The data strongly supports the conclusion that the use of surge, CHS, and SMM can reduce water use with no yield penalty. There was no significant difference between CHS and IWM, suggesting that surge may not have a treatment effect. However, there are only a four data points, and generally the surge sets were twice the size of the CHS sets. In fields that compared CHS and IWM, there were usually three sets, two were put together to make one surge set, while the third was a left over set and only used CHS. In retrospect, the experimental design should had CHS and IWM sets of equal size for a fair comparison. So the insignificance between CHS and IWM, representing the treatment effect of surge irrigation likely does not represent the true treatment effect of surge irrigation.

Table 1. Relative Yield and Relative Water Use of IWM, CHS and CONTROL Paired Fields

	Relative Yield		**Relative Water Use**		**Dunn's Method**
	ratio	% Difference from Control	ratio	% Difference from Control	Relative Water Use Difference from Control
IWM	1.006	**+ 0.6 %**	0.738	**- 27 %**	YES
CHS	1.01	**+ 1 %**	0.811	**- 19 %**	YES
Control	1.00	**0**	1	**0**	
P-value	p=0.864		IWM, p<0.001; CHS, p=0.015		

The Hawthorne effect is described in Roethlisberger (1941) as "human subjects of an experiment change their behavior, simply because they are being studied." As these studies were being implemented the authors and County Extension Agents noted that farmers changed their irrigation scheduling based on the information or request of the Agent to irrigate the IWM field. The field were always nearby, sometimes serviced by the same pump, so it was often convenient to irrigate the two fields at the same time or one right before or after the other. Additionally, as the cooperators were looking at sensor data and learning how to use the information from the Agents managing the IWM fields, the irrigators may have subconsciously used this information to manage the CONTROL field. The larger differences were often found when a different person, such as a father or other hired person managed the CONTROL field and the main decision maker and the agent managed the IWM field. However it is not always possible to facilitate such management, and even if agreed to, the close working relationship often cannot maintain separation of management of the paired fields. Thus, these results may underestimate the potential savings because of the Hawthorne effect.

Cost of Water for Irrigation Pumping Plants Studied

The cost of water for several of the sites was collected. Cost of water in US dollars per acre-inch of water are reported in Tables 2 and 3. Data for sites was collected on motor size, make, speed, and flowrate. The area the pumps serviced was checked and the capacity and cost of water was determined for each pump. Pumping plants for rice-soybean rotations should have at least 10 gpm/ac capacity, only one site did not have adequate capacity. The cost of water for electric pumps varied greatly, from $0.10/ac-in/ac to $1.86/ac-in/ac. Generally the cost of water for electric powered pumps was about half or a third of the diesel powered pumps, but some of the electric pumps cost as much as diesel.

Table 2. Electric Pump Size, Make, RPM, Flowrate, Acres, Capacity and Cost of Water

County	Motor Power (HP)	Motor Make	RPM	Flow Rate	Acres	GPM/Ac	Cost of Water $/Ac-in
Ashley	60	Emerson	1775	1300	101	12.9	0.10
Cross	*	Submersible	1700	600	36.3	16.5	1.38
Cross	75	North American Electric	1780	870	36.3	24.0	1.58
Cross	*	Submersible	1770	850	*	*	0.50
Greene	60	WorldWide Electric	1750	1467	79.8	18.4	0.53
Lee	*	Submersible	1770	667	37.5	17.8	0.53
Lonoke	75	WEG	1770	833	45.9	18.1	1.86
Mississippi	60	Emerson	1775	2708	80	33.9	0.58
St. Francis	50	North American Electric	1770	1083	65	16.7	0.20
White	20	Submersible	1770	300	32.6	9.2	1.49
White	15	Submersible	1770	400	32.6	12.3	1.11
White	15	Submersible	1770	400	32.6	12.3	1.09

For diesel powered pumps, an additional analysis was conducted. For each field, the CHS plan was obtained. The CHS plan was adjusted from the plan as designed by the irrigator to a plan that used the lowest cost of water engine speed but accomplished the irrigation in less than 40 hours. Typical water use for crops in rotation or other fields serviced by the pump were used to estimate potential savings. For example, for one site the CHS plan was designed for 24 hours at 1650 RPM that delivered 2000 gpm.

However after testing it was found that this engine speed resulted in a cost of water of $2.11/ac-in. Reducing the engine speed to 1500 RPM provided 1700 gpm at a cost of $1.68/ac-in. The pump serviced two fields of equal size that were rotated between soybeans and rice. Water use estimates of 13 ac-in/ac for soybeans and 32 ac-in/ac for rice were used to estimate total volume pumped per year. Making the engine speed change increased the irrigation time from 24 hours to 28 hours but resulted in a difference of $0.43/ac-in/ac which worked out to an annual savings of $1,518 in fuel savings for the fields the pump supplied. Similar analysis was performed for each site (Table 3) where a savings could be identified. For one electric site that was traditionally serviced by three submersible wells, a savings of $1,008 was found, by only using two pumps instead of all three.

Table 3. Diesel Powered Irrigation Pumping Planted Tested and Cost of Water and Potential Savings

County	Motor Size (HP)	Motor Make	Motor Model	RPM	Flow Rate	Acres	GPM/Ac	Cost of Water $/Ac-in	Potential Savings ($/ac-in)
Clay	99	John Deere	4045TF285B	1750	2250	157.9	14.2	2.05	0.43
Craighead	105	Case IH	4390T	1250	950	37.7	25.2	1.35	0.00
Crittenden	152	Cummins	6BT5.9-C	1692	1550	65.8	23.6	2.05	0.04
Lawrence/Randolph	66	Deutz	F4L912	*	1200	37.9	31.7	1.76	0.20
Phillips	60	John Deere	4039D-001	900	800	19	42.1	1.45	0.21
Poinsett	127	Isuzu	AI-4JJ1X	1800	1600	93.6	17.1	2.14	0.25

Diesel powered pumping plants had a cost of water between $1.35/ac-in/ac and $2.14/ac-in/ac. All pumps tested, albeit one, found a savings that could be generated by optimizing the cost of water and irrigation set time to engine speed. The Craighead site was experiencing cavitation, and thus as engine speed increased the flow decreased. The average savings from the eight locations was $511.66 with a range of $0 for the pump that was experiencing cavitation to $1,518 for a fairly new installation in Clay County. There did not appear to be any obvious trends, between age of pumps or power units. There seems to be an opportunity to reduce irrigation costs by testing the cost of water of when multiple electric pumps are used and diesel pumps. Given that some of the electric pumps cost almost as much as diesel to operate, measuring the cost of water for electric pumps could identify pumps and motors that have reached the end of their service life.

Table 4. Annual Site Savings

Site	Site Annual Savings
White	$1,008
Clay	$1,518
Phillips	$304
Lonoke	$104
Poinsette	$104
Craighead	$0
Crittenden	$35.36
Arkansas	$1,020
Average	$511.66

Conclusions

Water use and relative yields were compared between the fields that utilized Computerized Hole Selection, SURGE irrigation, and Soil Moisture Monitoring relative to a CONTROL field. Water use from both fields were measured with portable propeller flowmeters and yield were measured with combine yield monitors or scale tickets. IWM fields that used SURGE, CHS and SMM were found to use 27% less water than CONTROL ($p<0.001$). Fields that used CHS and SMM used 19% less water, but there were only a few observations of CHS and set sizes were not equal, so the real treatment effect of surge irrigation likely was not captured in this study. There were no significant difference in yields ($p=0.86$) between IWM and CONTROL fields or less than a 1% difference. On-farm pump testing identified a water savings of $511 per site and showed that measuring the cost of water for a surface irrigation pump and operating at is lowest cost can produce substantial annual energy savings. This study demonstrates that widespread adoption of these practices can significantly reduce the overdraft of the alluvial aquifer in the mid-south with no yield penalty.

References

Arkansas Natural Resources Commission. (2014). Arkansas Water Plan 2014 Update Appendix C Water Availability Report. Available at http://arkansaswaterplan.org/plan/ArkansasWaterPlan/2014AWPWaterPlan/App%20C_Water%20Availability%20Report_Final_11.24.14.pdf (Accessed as of on Feb 25, 2016)

Bishop, A.A.,W.R.Walker,N.L.Allen, and G.J.Poole. 1981. Furrow Advance Rates under Surge Flow Systems. Am. Soc. Civil Engr. J. Irrig. Drain. 107:257-264.

Carpena, R. M. 2004. Field Devices for Monitoring Soil Water Content. Agricultural and Biological Engineering. Publication #BUL343. University of Florida.

Eid, S.M., M.M. Ibrahim, S.A. Gaheen and S.A. Abd El-Hafez. 1999. Evaluation of surge flow irrigation system in caly soil under different land leveling practices. Paper presented at the third Conference on Farm Irrigation and Agroclimatology. Agri Rees. Center. Soil and Water Enviornment Res. Inst., Dokki, Egypt. 25-27 January.

Goldhamer, D.A., M.A. Alemi, and R.C. Phene. 1987. Comparison of surge and continuous flow irrigation in California. In: L.G. James and M.J. English, editors, Irrigation systems for the 21st Centruy. Proc Irrigation and Drain. Div Special Conf. Portland, OR. 28-30 July Am. Soc. Civil Eng., New York. pp 392-408.

Henry, C.G. and L.J. Krutz. 2017. Surge Irrigation Information. Extension Factsheet. University of Arkansas. Accessed on November 1, 2018 at https://www.uaex.edu/environment-nature/water/Surge%20Irrigation%20Factsheet%202017.pdf

Henry, C.G. and B. Stringham. 2013. How to Read Electric Meters. Irrigating Smart Factsheet Series. LSU AgCenter Publication 3241-I

Israeli, I. 1988. Comparison of surge and cablegation for continuous furrow irrigation. Paper presented at the summer meeting of the ASAE, St Joseph, MI, ASABE paper No. 141897470.

Izuno, F.T., T.H. Podmore, and H.R. Duke. 2985. Infiltration under surge irrigation. Trans. ASAE 28:517-521. Doi:10.13031/2013.32289.

Maupin, M. A., Kenny, J. F., Hutson, S. S., Lovelace, J. K., Barber, N. L., & Linsey, K. S. (2014). Estimated use of water in the United States in 2010. United States Geological Survey (No. 1405). Retrieved from http://pubs.usgs.gov/circ/1405/ on 6-20-2016

McDougall, W. 2015. A Pump Monitoring Approach to Irrigation Pumping Plant Performance Testing. Master's Thesis. Biological and Agricultural Engineering, University of Arkansas. Fayetteville, Arkansas.

Musick, J.T., J.D. Walker, A.D. Schneider, and F.B. Pringle. 1987. Seasonal evaluation of surge flow irrigation in corn. Appl. Eng in Agriculture. 3:247-251. Doi:10.13031/2013.26683.

Saxton, K.E. and W. J. Rawls. 2006. Soil Water Characteristic Estimates by Texture and Organic Matter for Hydrologic Solutions. *Soil Science of America Journal*. 70:1569-1578. Doi:10.2136/sssaj2005.0117.

Roethisberger, F.J. 1941. Management and morale. Harvard University Press. Cambridge, MA. doi:10.4159/Harvard.9780674420540

Testezlaf, R.R., L. Elliot and J.E. Reyes, and R. Pujols. 2004. Comparison between surge irrigation and conventional furrow irrigation for covered black tobacco cultivation. Spanish J. Agri. Res. 2:445-458.

U.S. Department of Agriculture, National Agricultural Statistics Service (NASS). (2012). Census of Agriculture, Farm and Ranch Irrigation Survey. Vol.3, Special Studies, Part 1 Available at:http://www.agcensus.usda.gov/Publications/2012/Online_Resources/Farm_and_Ranch_Irrigation_Survey/ (Accessed as of on Feb 25, 2016)

Presented at 2018 Irrigation Show & Education Conference, Dez 3-7, 2018
Long Beach Convention Center, Long Beach, California, USA

SUGARCANE PRODUCTIVITY ESTIMATION USING AGROMETEOROLOGY

Gustavo H. S. Vieira. Full Professor. PhD.
IFES *campus* Santa Teresa, Rod. ES 080, km 93, São João de Petrópolis, Santa Teresa-ES, Brazil. 29.660-000. e-mail: ghsvieira@gmail.com

Demetrius P. Nascimento. Agronomy undergraduate student.

IFES *campus* Santa Teresa – Rod. ES 080, km 93, São João de Petrópolis, Santa Teresa-ES, Brazil. 29.660-000.

Natália G. Silvestre. Agronomy undergraduate student.
IFES *campus* Santa Teresa – Rod. ES 080, km 93, São João de Petrópolis, Santa Teresa-ES, Brazil. 29.660-000.

Paola A. V. Lo Monaco. Full Professor, PhD.

IFES *campus* Santa Teresa – Rod. ES 080, km 93, São João de Petrópolis, Santa Teresa-ES, Brazil. 29.660-000.

Ismail R. Haddade. Professor, PhD.

IFES *campus* Santa Teresa – Rod. ES 080, km 93, São João de Petrópolis, Santa Teresa-ES, Brazil. 29.660-000.

Abstract: *This study aimed to obtain predictions of sugarcane productivity, at different planting times, in São Roque do Canaã, Espírito Santo state, Brazil. The simulations were performed using the AquaCrop software, using local climate (maximum and minimum daily air temperature, precipitation, wind speed and air relative humidity), soil, crop and irrigation data. We performed simulations for planting in the months of March, April, May, October and November and obtained productivity data (t ha^{-1}), net irrigation requirement (mm) and water use efficiency (kg m^{-3}). April showed the best planting time, requiring a lower irrigation depth (895.6 mm), reaching satisfactory productivity (89.5 t ha^{-1}), as well as having the highest water use efficiency (2.6 kg m^{-3}). Thus, we conclude that the Aquacrop software can be used as a tool for decision making regarding the best planting season for sugarcane.*

Keywords: Irrigation, *Saccharum officinarum*, yield, planting time.

INTRODUCTION

São Roque do Canaã is a regional and national highlight in the production of cachaça, and in 2011 there were 21 distilleries in the city, performing the work of production and bottling of cachaça. These factories constitute important generators of employment and income for the municipality.

Therefore, the production of sugarcane is highlighted in the municipal scenario, being the second crop with the largest planting area, with 750 ha (Proater, 2011). In this context, factors such as edaphoclimatic monitoring aimed at improving the management of sugarcane, as well as the choice of its best cultivars, significantly interfere in production.

The availability of water in the soil interferes with plant production and, in places that do not have irrigation systems, this water availability is governed by the distribution of rainfall and water storage in the soil, factors conditioned by retention capacity and drainage (Maule et al. al, 2001). The water requirements for sugar cane range from 1,500 to 2,500 millimeters per cycle, which should be evenly distributed during the period of vegetative development, according to data from the United Nations Food and Agriculture Organization (FAO). However, recent studies have shown that the amount of water required for the crop to reach its maximum potential is around 1,200 to 1,300 millimeters (Marin, 2005 and Vieira et al., 2014).

The climatic classification of Köppen in the municipality of São Roque do Canaã is Aw, with rains distributed unevenly, raining much more in the summer than in the winter, being 1,155 mm annual precipitation in the historical average. In the last three years, this average has not been reached.

Thus, this study aimed to estimate sugarcane yield using the Aquacrop model; predict possible adverse climatic conditions detrimental to the crop, based on historical meteorological data; and determine the best planting times and the most appropriate cultural treatments for sugarcane.

MATERIAL AND METHODS

Foi utilizado o modelo agrometeorológico AquaCrop versão 6.0 (Figura 1), ferramenta importante por incorporar os conhecimentos atuais das respostas fisiológicas das culturas, fazendo uma previsão de suas biomassa e produtividade, em resposta à água disponível no solo. O modelo abrange uma ampla variedade de culturas herbáceas, incluindo forrageiras, hortaliças de folhas, grãos, frutas, raízes e tubérculos. Além disso, estabelece o equilíbrio entre precisão, simplicidade, robustez e facilidade de uso, sendo destinado a usuários práticos, tais como extensionistas, gestores de recursos hídricos, economistas e especialistas em políticas públicas que utilizem modelos simples para planejamento e análise de cenários (HSIAO et al.,2009 in LIMA et al., 2013).

We used the Agro-meteorological model AquaCrop version 6.0 (Figure 1), an important tool to incorporate the current knowledge of the physiological responses of the crops, making a prediction of their biomass and productivity, in response to the water available in the soil. The model covers a wide variety of herbaceous crops, including forage crops, leafy vegetables, grains, fruits, roots and tubers. In addition, it strikes a balance between accuracy, simplicity, robustness and ease of use, and is intended for practical users such as extensionists, water managers, economists and public policy experts who use simple models for scenario planning and analysis (Hsiao et al., 2009 and Lima et al., 2013).

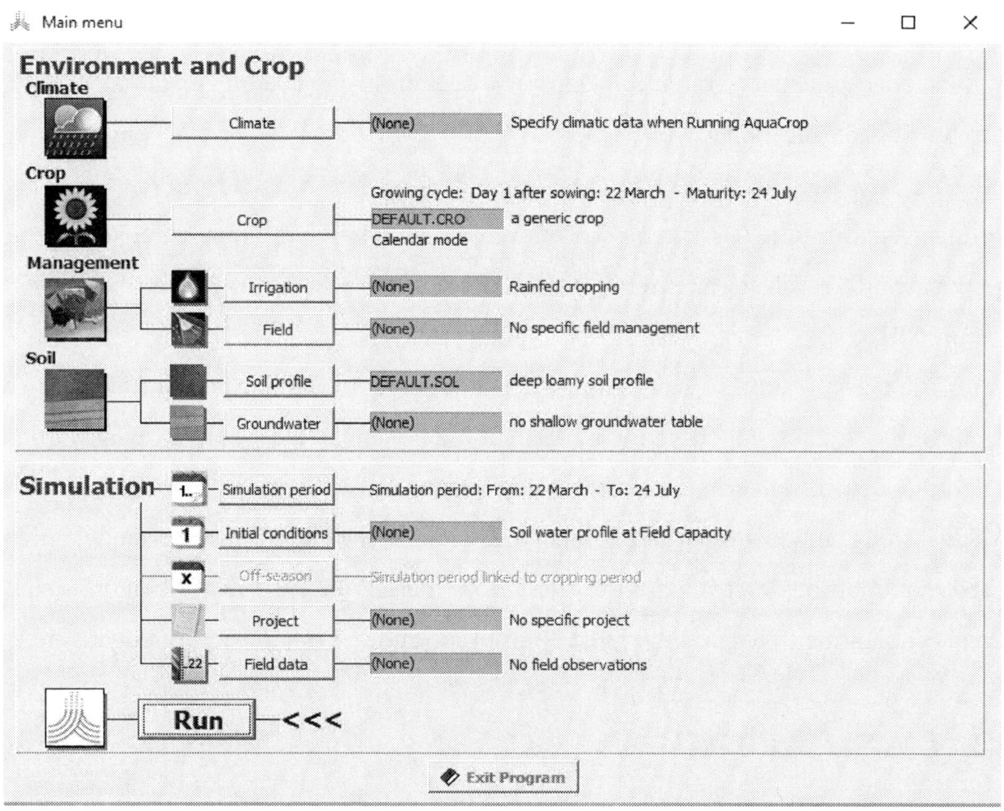

Figura 1 – Initial interface of AquaCrop 6.0.

The structure of the AquaCrop presents the soil, through its water balance, the plant with its growth, development and production processes, and the atmosphere (rainfall, evaporative demand and carbon dioxide concentration). In addition, some aspects of management are presented, with emphasis on irrigation, soil fertility levels that affect crop development, biomass production in relation to water use, in addition to crop adjustments to water stress, resulting in their final income.

For the evaluation of the AquaCrop model, input variables related to climate, soil and crop information were used. In relation to these, the climate variables required to execute the model are: maximum and minimum daily air temperature, precipitation and reference evapotranspiration (ETo). These local weather data were obtained in a meteorological station installed at IFES - Campus Santa Teresa. These variables were saved in a text file, being retrieved by the model through the user interface. The average annual atmospheric CO_2 concentration of AquaCrop is 369, 47 ppm (716.0 mg m^{-3}) recorded at the Mauna Loa Observatory in Hawaii in the year 2000 (Raes et al., 2009) and adjusted by the model, for the year of simulation.

The characteristics of the soil profile (input file), required by AquaCrop, are: profile thickness, texture, volumetric water content at saturation, field capacity, permanent wilting point and saturated hydraulic conductivity (Ksat). The soil data of cultivated areas of the IFES - Campus Santa Teresa were used to carry out the simulations.

The agronomic performance variables evaluated during the sugarcane cycle were: yield (kg ha^{-1}), cycle length (days), adjusted harvest index (%), soil moisture in several depths (%), net irrigation requirement (mm) and water use efficiency in evapotranspirated water productivity (kg m^{-3}).

AquaCrop is represented by a simple linear function, as presented in Equation 1.

$$Y = B\ IC \qquad (1)$$

Where:

Y – crop yield (kg ha^{-1});

B – total biomass (kg ha^{-1});

IC – harvest index (%).

The daily production of biomass (g m^{-2} or t ha^{-1}) above the soil is given by Equation 2.

$$m_i = Ks_b\ WP_i^* \left(\frac{Tr_i}{ETo_i}\right) \qquad (2)$$

Where:

Ks_b - adjustment factor for air temperature as a function of day-degrees (depending on the number of degree-days generated in a day, which can vary between 0 and 1)

WP_i^* - daily biomass productivity in relation to the use of water for the crop (kg ha^{-1});

Tr_i - daily crop transpiration on day i calculated by AquaCro p(mm day-1);

ETo_i – daily reference evapotranspiration (mm dia^{-1}).

The total biomass production (t ha^{-1}) above the soil is represented by Equation 3.

$$B = Ks_b\, WP^* \sum \left(\frac{Tr}{ETo}\right) \tag{3}$$

Where:

Ks_b - adjustment factor for air temperature as a function of day-degrees (depending on the number of degree-days generated in a day, which can vary between 0 and 1)

WP^*- standardized biomass productivity in relation to the use of water for the crop (kg ha^{-1});

Tr- total transpiration of the crop on a day calculated by AquaCrop (mm day^{-1});

ETo – reference evapotranspiration in the cycle (mm $cycle^{-1}$).

In the WP * adjustments for atmospheric CO_2 concentration, synthesized products and soil fertility are described by Raes et al. (2009). The model manual presents in detail the equations used in the calculation of WP and IC adjustments, as well as the subroutines for obtaining evapotranspiration through the water balance.

RESULTS AND DISCUSSION

The evaluations were carried out during the period between March 1, 2016 and March 28, 2017. Five planting simulations were carried out in March, April, May, October and November. The AquaCrop model simulated the crop cycle, assuming that it lasted twelve months, which made it possible to obtain two simulations with complete cycle (planting in March and April). The simulation with planting in May considered only eleven months, and the plantation in October and November considered only six and five months of the cycle, respectively.

Transpiration (Tr) and canopy cover (CC) graphs were generated for each planting month, as shown in Figures 2 A, B, C, D and E,

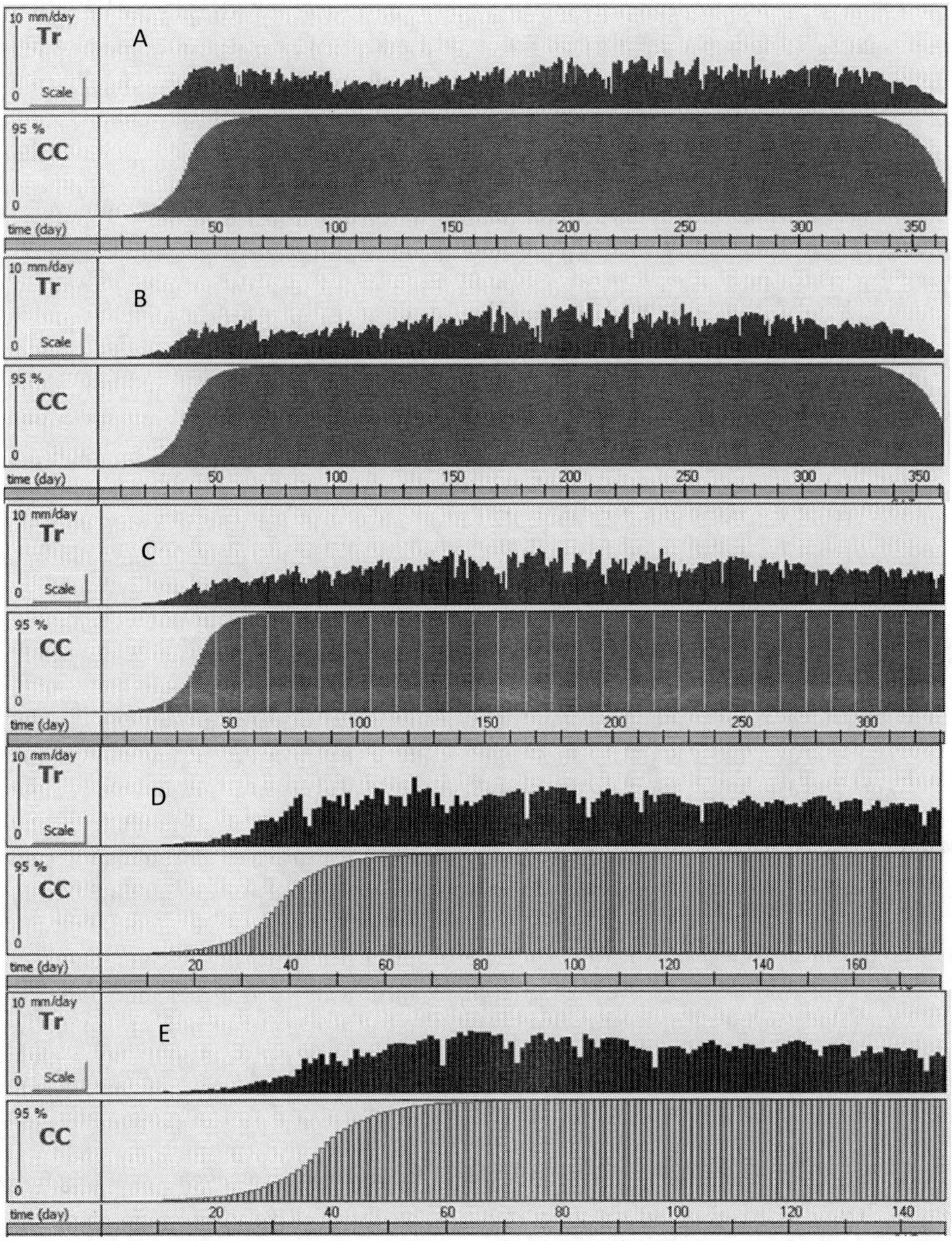

Figure 2 - Crop transpiration (Tr) and canopy coverage (CC) for simulated plantings in the months of March (A), April (B), May (C), October (D) and November (E), respectively.

The graphs that represented the CC showed no differences in their behavior. At the beginning of development, the relative growth rate is constant and the crop is mainly vegetative, characterizing the exponential phase. After the development of the root system and the expansion of the leaves, the plant increases water and nutrient withdrawal from the soil and develops photosynthesis processes, passing to a linear growth phase, with the greatest increase in biomass accumulation. When reaching the definitive size, the plant enters the senescence phase and shoots maturation, with less interception of light energy, decrease in the accumulation of dry matter and translocation of sugars to the storage organs.

In the simulations of Tr, which covered the entire crop cycle, the planting in April had less transpiration, reducing the need for irrigation during the cycle. The plantings carried out in the months of October and November, even though they did not simulate their complete cycle, had greater sweating for the period in the imposed climatic conditions.

Based on the climatic data used, the model calculated ET_0 values and estimated rainfall and irrigation values for each planting period, as shown in Figure 3. For the full-cycle simulations, in April presented the lower evapotranspiration, the higher volume of rainfall and consequently the lower volume of water used in irrigation, compared to the month of March.

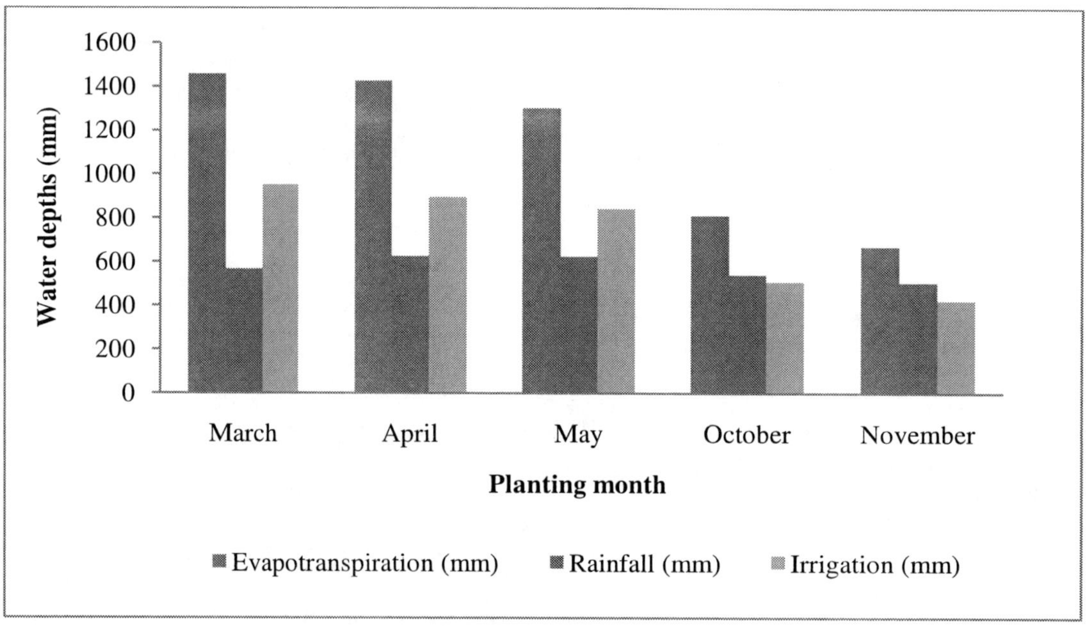

Figure 3. Evapotranspiration, rainfall and irrigation values for the sugarcane cycle according to the planting month.

The simulation in May, despite showing only 11 months of the cycle, showed similar behavior to the months of March and April, besides having rainfall volume higher than March, and consequently lower irrigation and evapotranspiration, but the planting in May can be risky, since sugarcane in the tillering phase (40 to 120 days after planting) and initial growth (120 to 270 days after planting), is sensitive to the water deficit coinciding with the period dry in the planting period. Ramesh and Mahadevaswamy (2000) stated that drought during the formation phase will result in lower yields.

For the simulation of planting in October and November, considering only six and five months of the cycle, rainfall volumes close to the value obtained when considering the complete cycle, in the planting situation in March, were worth highlighting. The fact happened because the simulation started in months considered rainy, when in the normal environmental conditions for the region.

The simulated values for fresh and dry mass (Figure 4) in March and April were similar, with an insignificant difference between them. For planting in May, compared to March, the difference reached 4.766 and 1.558 t ha^{-1}, respectively. This is due to the fact that at the beginning of its development the days are short and cold, and the plants received little luminous intensity, reducing its development.

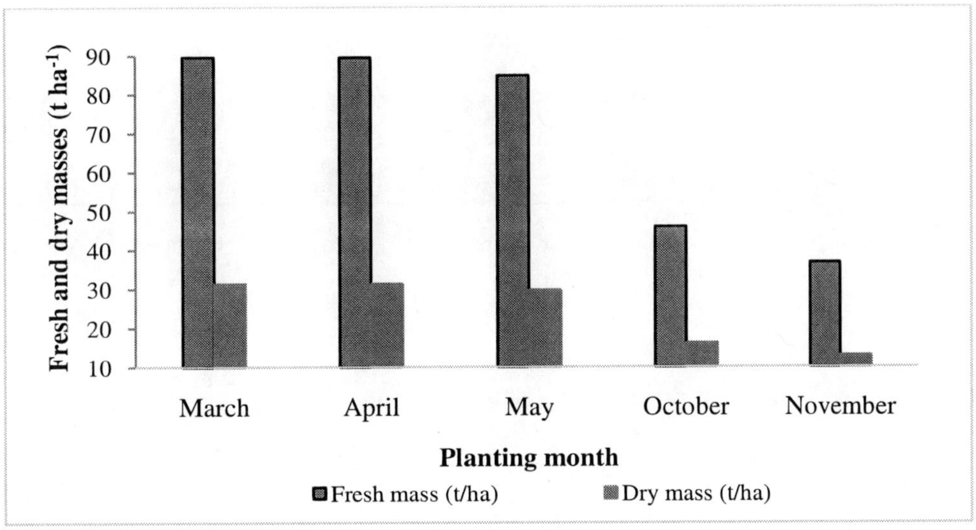

Figure 4 – Production of fresh and dry mass per cycle, according to the planting month.

The sugarcane presents expressive photosynthetic efficiency, mainly in the use and the rescue of CO_2, it is adapted to the high luminous intensities and high temperatures, even with relative water shortage. However, the crop needs large amounts of water to meet its water needs, since only 30% of its total biomass is represented by dry matter, reaching up to 70% water content, depending on the phenological stage (Segato et al., 2006).

Pelo modelo também foi estimada a produtividade/água evapotranspirada (Figura 5), onde o plantio no mês de abril foi o de melhor resultado para esta relação, com 2,6 Kg para cada m³ de água evapotranspirada. O mês de maio, mesmo com valores do ciclo incompleto, obteve resultado semelhante ao mês de março.

The model also estimated the water use efficiency (Figure 5). Planting in April showed the best result for this relation, with 2.6 kg for each m³ of evapotranspirated water. The month of May, was similar to March, even with incomplete cycle values.

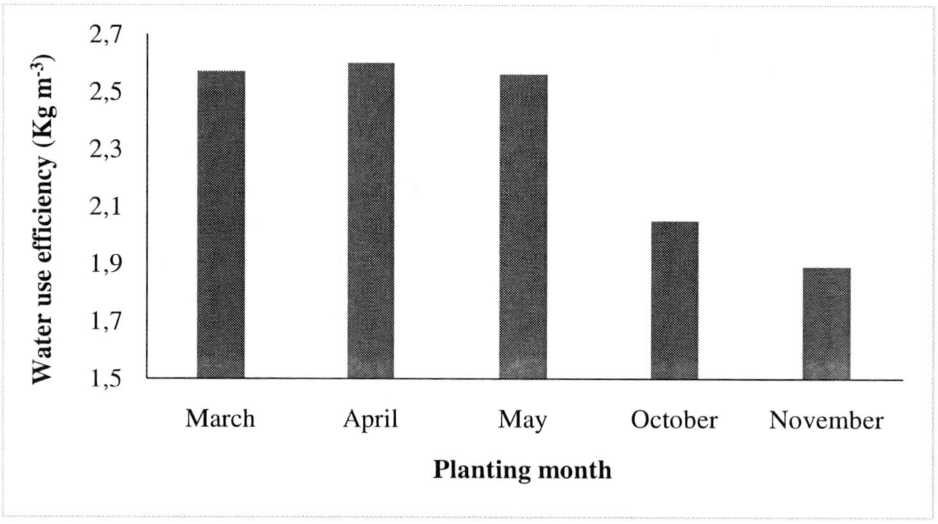

Figure 5 – Water use efficiency for the cycle of each month of planting.

CONCLUSIONS

Planting in April showed the smaller depth of irrigation water need, given by the greater rainfall in the period evaluated. In addition, the simulation of planting in April presented the best water use efficiency. Planting in May can be risky, because the environmental conditions at the beginning of the development of the crop are not favorable.

ACKNOWLEDGMENT

To Ifes *campus* Santa Teresa for the financial, operational and infrastructure available for the experiment and for providing scholarship to the students. To Prodif-Ifes for the financial support in the trip to present this paper.

REFERENCES

Hsiao, T.C., L.K. Heng, P. Steduto, B. Rojas-Lara, D. Raes, E. Fereres. 2009. AquaCrop - The FAO crop model to simulate yield response to water: III. Parameterization and testing for maize. Agronomy Journal. 101:448-459.

Lima, T.A.D., E.C. oliveira, R.P. Posse, A.A. Pires. 2013. Desempenho do modelo AquaCrop na estimativa da produtividade do feijoeiro (*Phaseolus vulgaris* L.) sob diferentes sistemas de cultivo em Colatina – ES. In: VII Jornada de Iniciação Científica, Desenvolvimento Tecnológico e Inovação do IFES. Serra, ES, Brazil.

Marin, F.R. 2005. Árvore do conhecimento – Cana-de-açúcar. AGITEC – Agência Embrapa de Informação Tecnológica. Also at http://www.agencia.cnptia.embrapa.br/gestor/cana-de-acucar/arvore/CONTAG01_10_711200516716.html

Maule, R.F., J.A. Mazza, G.B. Martha JR. 2001. Produtividade agrícola de cultivares de cana-de-açúcar em diferentes solos e épocas de colheita. Scientia Agricola. 58(2): 295-301.

Marafon, A. C. Análise Quantitativa de Crescimento em Cana-de-açúcar: uma Introdução ao Procedimento Prático. 1 ed. 2012. Embrapa Tabuleiros Costeiros, Aracaju, SE, Brazil.

Pimentel, C. 2004. A relação da planta com a água. Seropédica: Edur, pp. 191.

Proater – Programa de Assistência Técnica e Extensão Rural 2011 – 2013, São Roque do Canaã. INCAPER, 2011.

Raes, D., Steduto, P., Hsiao, T.C., Fereres, E. 2009. AquaCrop - The FAO crop model to simulate yield response to water: II. Main algorithms and software description. Agronomy Journal, 101(3):438-447.

Ramesh, P., M. Mahadevaswamy. 2000. Effect of formative phase drought on diferente classes of shoot, mortality, cane atributes, yield and quality of four sugarcane cultivars. Journal Agronomy & Crop Science. 185:249-258.

Segato, S.V., C.F.M. Mattiuz, A.E. Mozambani. Aspectos fenológicos da cana-de-açúcar. In: Atualização em produção de cana-de-açúcar. Piracicaba, SP, Brazil. 2006. pp. 19-36.

Vieira, G.H.S., E.C. Mantovani, G.C.Sediyama, F.T.Delazari. 2014. Productivity and industrial yield of sugarcane as a function of irrigation depths. 30(1):55-64.

A New Approach to VFD System Specifications and Rebates

Kyle Feist, MS, PE, CID, CAIS

Irrigation Training & Research Center (ITRC), California Polytechnic State University (Cal Poly), San Luis Obispo, CA 93407-0730. kfeist@calpoly.edu

Randall Cole, CEM

Pacific Gas and Electric Company, 245 Market Street, Rm 692D, San Francisco, CA 94105, randall.cole@pge.com

Charles Burt, Ph.D., P.E., CID

Cal Poly ITRC, San Luis Obispo, CA 93407-0730. cburt@calpoly.edu

Abstract. *Variable frequency drives (VFDs) are used to provide flexibility and efficiency enhancements for agricultural pressurized irrigation systems. Standard 6-pulse VFDs commonly used for agricultural pumps can also create real problems with the pump motor and nearby electrical systems, unless mitigated by special equipment.*

With support by the California Polytechnic State University (Cal Poly) Irrigation Training and Research Center (ITRC), Pacific Gas and Electric Company (PG&E) recently began a new rebate program for installing VFD systems for pressurized irrigation. A major component of the new rebate program is the requirement to comply with a detailed set of minimum VFD system design and installation specifications.

It was recognized that specifications were needed for farmers to receive comparable quotes for a good-quality VFD system, without needing to know all the technical details. Instead of a more typical qualified product list, which requires continuous time and effort to maintain, the specifications focus on minimum performance-based criteria and therefore provide flexibility for the VFD system designer. This paper provides an overview of the specifications' content and rebate program implementation.

Keywords. Agriculture, pumps, pump drives

Introduction

Variable frequency drives (VFDs) provide speed adjustment for three-phase alternating current (AC) irrigation pump induction motors. VFDs are typically sold as part of a package, or VFD system.

A VFD system is defined as the combination of a VFD module and other supplemental components. Common component examples include:

- An enclosure housing some or all of the components (including the VFD itself). Good enclosures are rated to provide a specific level of protection from physical damage or dust and moisture ingress.
- Cooling devices or sub-systems. Operating a VFD creates heat that must be dispersed, so that component temperature ratings are not exceeded.
- Devices designed to mitigate a variety of potential problems, as detailed in the next section.
- Convenience-related items that enable advanced operational features or improved user interfaces.

A conceptual diagram of a VFD system is shown in Figure 1.

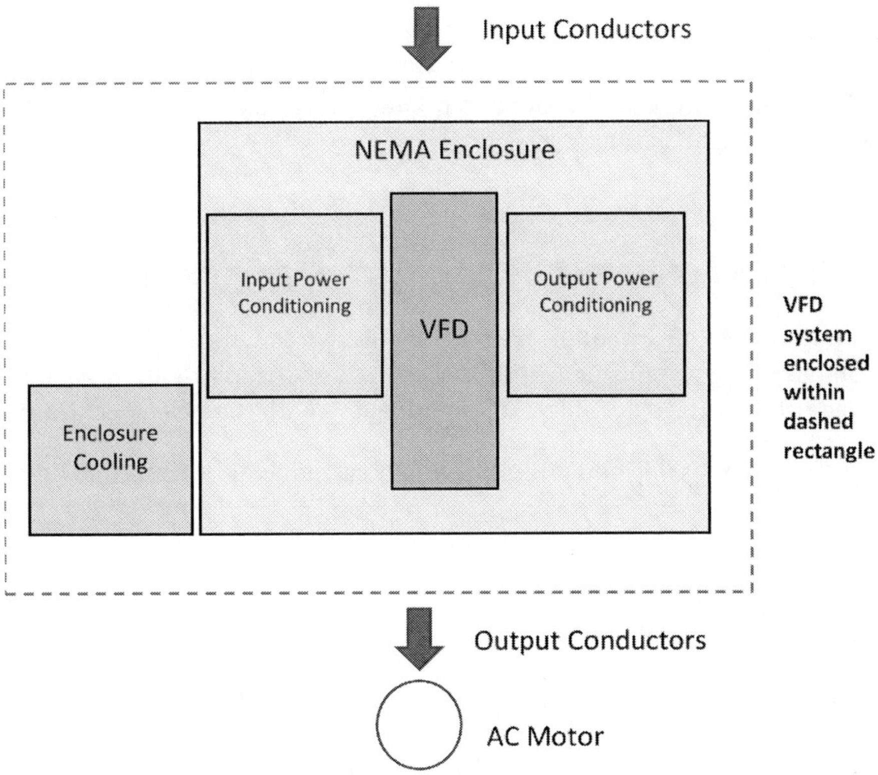

Figure 1. Major potential components of VFD systems

Other key points include:

- VFD systems are typically purchased in one of two ways:
    - Through a distributor of a major VFD manufacturer as part of a pre-packaged system with a variety of ordering options
    - Through an Underwriters Laboratory (UL) panel shop that constructs custom VFD systems
- The VFD module, typically provided by major manufacturers, is only one part of a VFD system.
- There are usually a number of options or add-ons that can be provided in the VFD system. Some of the add-ons can be considered essential for the customer's application, while others may be simply for convenience.
- VFD systems are typically designed by technical VFD experts and licensed electrical engineers.

Incentivizing VFDs: A First Attempt

Pacific Gas and Electric Company (PG&E) took the initial approach of providing a general rebate to those installing a *VFD* on pressurized irrigation systems. The initial rebate was $40 per VFD horsepower, without many qualifications. The rebate value was computed as the historical average value of the first year energy efficiency gains (energy savings in kilowatt hours (kWh) and demand reduction savings in kilowatt (kW)) from over 500 paid custom VFD energy efficiency applications, multiplied by $0.09 per kWh and $150 per kW. The rebate value was subsequently reduced to $30 per VFD horsepower due to incentive budget concerns.

PG&E experienced extensive participation in the initial rebate program through the efforts of PG&E customer representatives, irrigation system owners, and with the support of enterprising irrigation dealers that helped customers fill out applications. In terms of participation level, the initial rebate program was a success. However, it became clear that the lack of specifications or qualifications for rebated VFD installations presents risks for the two primary stakeholders: PG&E, and the farmer or VFD owner. These risks are discussed in the next section.

Rethinking the Rebate Program

PG&E contracted with the Cal Poly Irrigation Training and Research Center (ITRC) to develop minimum specifications for on-farm VFD systems delivering water to pressurized irrigation systems. Preliminary discussions identified risks, as listed in Table 1, to be mitigated with the new specifications.

Table 1. Potential risks of purchasing and operating a VFD system without specifications

Risk Category	Affecting Farmer	Affecting Others Nearby	Affecting PG&E
Uncertainty in what is included with the VFD system, or the expected performance	X		X
Paying more for less energy efficiency or other performance characteristics	X		
Power quality degradation (including harmonic distortion) emanating on-farm and going back into the grid	X	X	X
Decreased pump motor life due to voltage spikes (additional heat) and bearing damage	X		
Electromagnetic and radio frequency noise that can interfere with the performance of nearby sensors and radio systems[1]	X	X	

[1] While this may sound exotic, real problems can happen when operating any of the following close to the VFD: drones, remote monitoring or control systems, and local automation based on analog sensors. The magnitude of the risk depends on installation-specific details.

While some of these risks are well-documented in publicly available technical articles, these topics are rarely discussed during the bidding or purchasing process of VFD systems within the agricultural market. Additional objectives of the VFD systems specifications included:

- Providing minimum performance-based requirements, that are intended to be:
    - Flexible, such that future advancements in VFD technology and components will not require rewriting to the specifications. However, that does not mean the specifications are unchangeable.
    - Understandable and achievable by trained technical sales staff, without necessarily requiring a licensed electrical engineer or extensive details of the infrastructure situated on the primary transformer winding (the utility-owned equipment).
    - All-inclusive for the design, installation, field configuration and documentation of a basic VFD system installation.
- Eliminating the need for the farmer and vendor to know all the technical VFD details.
- Establishing a relatively simple means for the utility to administer a complex incentive program.

New Rebate Program

A new VFD rebate program based on compliance with the ITRC specifications was rolled out on Jan 1, 2018. The program provides the following rebates:

- For a 75 horsepower (HP) motor or smaller: $60 per motor HP plus $2,000
- For a greater than 75HP motor: $60 per motor HP (booster pumps are limited to a maximum of 150 HP and well pumps are limited to a maximum of 600 HP)
- For VFD incentive applications that do not meet the ITRC specifications, customers can still receive $20 per motor HP (booster pumps are limited to a maximum of 150 HP and well pumps are limited to a maximum of 300 HP)

The specifications and PG&E rebate program details are hosted on the ITRC website at www.itrc.org/vfd. Participation in the rebate program also includes key procedural requirements and record documents as follows:

- The certification number or Underwriters Laboratory (UL) File Number, proving that the VFD system designer is listed by a certifying agency for the design and fabrication of VFD system equipment.
- Two documents (or invoices) with a written statement proclaiming that the VFD system meets the specification requirements regarding both design and installation.

Major Initial Specifications

A non-exhaustive list of unique and important specifications includes:

- There is a minimum efficiency (the ratio of power output to power input) for both the VFD and the complete VFD system.
- The VFD system must be designed and fabricated by a UL-listed panel shop for UL508C VFD systems.
- New VFD-driven motors must be designed for VFD applications and be equipped with shaft grounding rings and insulated bearing carriers, depending on the application and horsepower.
- A minimum level of VFD input surge protection is required.
- Some harmonic mitigation is required, depending on the size of the motor. Additionally, there exist multiple options for the level of compliance, or the specific equipment used.
- Application-specific RF-shielded VFD cable must be used.
- Specific installation and configuration tasks are listed in the specifications, including a minimum amount of documentation and training that must be supplied to the owner.

Initial Reactions and Lessons Learned

Reactions and feedback from VFD vendors and farmers were received in two ways. Multiple workshops were held with farmers and VFD system vendors, and an online question-and-answer platform was provided from the specifications website.

In addition to the expected level of general hesitation to accept change, some important insights were gained based on valuable feedback from farmers and vendors:

- Some vendors are unaware of the complexity and technical details of the products they sell.

- Nuisance VFD system tripping is occurring at unexpectedly high levels, either from poor VFD input power quality (voltage spikes, sags, harmonics or phase imbalances) or insufficient cooling and maintenance.
- There is a pattern of premature motor failures in new VFD system installation and retrofits. While many VFD vendors are uncertain as to the exact cause of the problem, some examples of "current industry standards" can be part of the problem. Major examples include:
    - Selecting new pump motors, or keeping existing motors that are not designed for VFD applications and have insufficient insulation ratings.
    - Under-sizing the pump motor so that normal operations require that it operate overloaded (at a service factor of 1.1 or greater), in relatively hot environments, without shading.
    - Neglecting to follow VFD manufacturer recommendations in the selection and installation of VFD to motor cabling.

It appears that the new specifications will help with many of shortcomings of "current industry practices".

Future Work

Based on a balance of continued feedback and new information, it is anticipated that the specifications will be modified over time. It is also anticipated that the specifications may motivate VFD vendors to consider alternative design options so that the performance and reliability of on-farm VFD systems improve over time and keep pace with market demands and technological advances.

Conclusion

Prior to the project, publicly available performance-based specifications for agricultural VFDs did not exist. As such, the specifications provide a number of direct and indirect benefits to the pressurized irrigation industry:

- A list of topics that can be discussed and considered with VFD vendors, and ultimately modified if needed – even for VFD system purchases outside of PG&E's service area.
- Simplification of the process of ordering a VFD, both for farmers and irrigation/pump dealers, because the key requirements are already taken care of. All that is left is identifying add-on, convenience items to meet customer-specific needs and preferences.

Furthermore, the new rebate program, combined with the new specifications, is expected to:

- Enhance energy efficiency and flexibility for on-farm pressurized irrigation systems by promoting VFD use. Moreover, the additional flexibility provided by the VFD enables farmers to more easily experiment with advanced irrigation and chemical/fertilizer injection strategies.
- Improve the reliability of pump operations in the future by:
    - Reducing future power quality degradation that would likely to have occurred without the specifications.
    - Providing incentives for good design and installation practices, such as proper motor selection, grounding, etc.
- Improve the conditions for the adoption of other advanced on-farm technologies such as drones, instrumentation, remote monitoring and automation by reducing future electromagnetic noise and radio frequency interference.

Presentation to Irrigation Association Conference, 2018, Long Beach, California.

Robert L Welke - Australia

Qualifications:

- Diploma, Mechanical Engineering
- Irrigation Agronomist (Irrigation Australia)
- Irrigation Auditor, Landscape (Irrigation Australia)
- Indentured Fitter and Turner

Affiliations:

- Member: Irrigation Australia Ltd (IAL) since 2006
- Member: Pumping Industry Australia (PIA)

Address:

- Windsor Gardens (Adelaide), South Australia, AUSTRALIA

Email/Web:

- rob@talle.biz
- www.talle.biz

Abstract:

Testing pumps to determine pumping energy efficiency results in only part of the pumping energy efficiency answer.

This paper explores the "unseen and often unrecognised" areas of pumping energy efficiency, namely those associated with the hydraulic efficiency of the pipeline delivery system.

Traditional energy efficiency audits have concentrated on testing the pump's absolute efficiency as well as identifying any OFF-BEP (Best Efficiency Point) operation.

This approach ignores the "elephant in the room", the hydraulic efficiency, which in turn is solely a function of pumped head.

Examining hydraulic efficiency includes any items which will result in a reduction of friction head, such as pipeline optimisation, predictable pipeline performance deterioration, filter configuration, excess residual head and others.

Pumping system energy efficiency incorporating hydraulic efficiency can be optimised either in the design stage, which results in a "fit for purpose" pumping system or resulting from an audit of an aged pumping system, in which case some capital expenditure may be required to optimise pump performance, paid for with pumping electricity savings.

The author has extensive experience in this field, having identified over $600,000 worth of achievable annual pumping electricity costs over the last 6 years on 13 irrigation systems in Australia, with an average return on investment (ROI) of 2.0.

The presentation will show case some of the findings of these audits.

Introduction

Testing irrigation pumps is the time-honoured method of establishing the pumping energy efficiency.

This involves measuring the flow rate, pump head and input energy, usually electrical kW, to establish a pump hydraulic efficiency.

There are two parameters that are sought in a pump test:

1) The absolute pump efficiency, a measure of the mechanical condition of the pump. With the efflux of time, pumps wear and become less efficient, therefore require periodic testing.
2) The point at which the pump is working relative to its Best Efficiency Point (BEP). Pumps operating at a duty too far left or right of BEP will be less efficient, even though the pump is in top condition.

There is a third crucial parameter which is not regarded as part of pumping energy efficiency, and yet it has a profound influence on pumping costs.

That is pipeline hydraulic efficiency (and associated equipment). Items such as pipeline friction, pipeline ageing, filters, hydraulic valves, emitters and residual head, in fact any component which results in a friction head component.

And it doesn't stop there. If an irrigation system is deficient in residual head, low emitter CU's and DU's will prevail, meaning that the irrigation scheduling co-efficient will be raised to compensate, resulting in up to 20% more water pumped. That's 20% more pumping energy used!

It is the authors experience, after 50 years in pumping and hydraulics, having conducted over 1300 pump tests and dozens of pipeline friction tests, that there is as much pipeline hydraulics energy efficiency losses existing in irrigation systems as there is pumping hydraulics losses.

This presentation highlights some of the authors findings.

Theory

The formula[1,2] for calculating the annual cost of electrical energy for irrigation pumping is:

$$\$/yr = \frac{2.73 \times \textbf{(pumped head in metres)} \times (\$/kWh) \times (ML/yr)}{\textbf{(Pump } \eta\textbf{)} \times \textbf{(motor } \eta\textbf{)}}$$

This can also be expressed as:

$$\$/yr = \frac{3.15 \times \textbf{(pumped head in feet)} \times (\$/kWh) \times (MgUS/yr)}{\textbf{(Pump } \eta\textbf{)} \times \textbf{(motor } \eta\textbf{)}}$$

Accordingly, the principle variable parameters for pumping energy efficiency are:

- Pumped head
- Volume pumped
- Pump efficiency
- Motor efficiency

(Pump efficiency and Motor efficiency: covered in testing pumps)

Pumped head is one of the principle parameters for measuring pumping energy efficiency

However, pumped head is **the** principle parameter for measuring pipeline hydraulic efficiency.

Pump hydraulic efficiency

Pump efficiency is defined by three essential components[1]:

- Pumped head (metres/feet)
- Pumped flow rate (litres per second/gpm)
- Electrical energy input (Electrical kilo Watts [EkW])

Since the concept and practice of pump testing irrigation pumps is universally accepted and well understood, and because of time limitations of this presentation, only assessing pipeline hydraulic efficiency will be discussed.

Pipeline hydraulic efficiency

Since pipeline hydraulic efficiency is essentially the friction of the pipeline delivery system, hydraulic efficiency includes, but not limited to:

1) Pipe friction optimisation
2) Pipe performance deterioration
3) Filter losses
4) Excess Residual head

We are going to examine these applications of hydraulic efficiency and quantify the contribution that they make to pumping energy efficiency.

1) Pipeline friction optimisation

All pipes have friction losses, which equate to energy costs in pumping. Keeping these costs to a minimum but at the same time minimising pipe costs is a process called optimisation.

Therefore, assessing a pipe's friction losses is critical to determining if the correct size pipe has been used, and assessing if it is economical to duplicate or replace a pipe to reduce pumping costs.

This process is ideally made in the initial design process, where the higher capital costs of a larger pipe selection will be more than be offset with lower pumping energy costs.

The process can also be made as a result of a field audit, however, the economics are less attractive since the cost of the optimised pipe solution is actually added to the original pipe cost.

Both processes involve using software which can calculate Nett Present Value (or amortised costs) for the electricity costs of a pipeline given a range of operating parameters.

Example

Compare the following two pumping systems: (NOTE: numbers rounded/estimated)

Both systems: 1 km (0.6 mile) pipe, pumped flow rate 50 l/s (792 USgpm) and 200ML (53 MgUS)/yr

	System 1	System 2
Pipe size	150NB (6")	200NB (8")
Pipe cost (est)	$42,000	$55,000
Friction loss	45m (148 feet)	11m (36 feet)
Pump cost (est)	$25,000	$15,000

Electricity cost/yr	$8,800	$2,200
Electricity cost/15 yrs	$163,000	$40,000
Total LLC pipe+pump+15yr/elect	$230,000	$110,000

The larger pipe/smaller pump combination saves $120,000 over 15 years.

This is an example of pipeline optimisation. Ideally, more calculations should be carried out with other pipe diameters, to achieve the lowest Life Cycle Costing (LCC) for the pipe+pump+15yr/electricity cost.

2) Pipe performance deterioration

The day after an irrigation pipe is placed into service, its performance has already commenced to deteriorate. Pipes which pump raw river, dam water or recycled water are particularly prone to a steady drop in performance due to biofilm build-up, especially in warm weather.

New PVC and Poly pipes above 6" (150NB) have a Hazen & Williams (H&W) C value (friction factor) of around 155. It's a reasonable chance that within a year, the pipes' C value may be down to 140 (10% drop) and continue downwards thereafter. The hotter the climate, the worse the effect, as biofilm thrives with warmth.

Glacial melts are not immune to this. In New Zealand South Island (lat 45 deg), glacial flour is a common component of glacial river water and is known to significantly deteriorate pipe performance over time[(3)].

What that means is that over time, there's higher friction than originally designed for these pipes.

So, either there will be a deficit in operating pressure (lower irrigation efficiency) or if the pumping system has a VFD and surplus HQ capacity (unlikely), higher operating costs as the pump is ramped up to overcome additional pipe friction.

Water Authorities in Australia, such as Goulbourn-Murray Water recommends derating PVC and Poly pipe performance 10% at design stage to allow for pipe performance deterioration.

South Australia's SA Water has traditionally designed pipeline systems off the River Murray with low range C Values as low as 110[(4)].

The author's experience would indicate that a 20% derating after 10 years is reasonable when using unchlorinated river or dam water, or recycled water.

A 20% reduction in H&W C Value means that for the same head loss, 20% less flow can be pumped, or for the same flow to be maintained, approx. 40% additional friction losses will occur.

Cases of up to 60% pipeline performance reduction over 10 years have been found due to iron hydroxide or contamination from river borne Bryozoa, aquatic invertebrate animals, typically about 0.5 millimetres (0.020 in) long [(5)], both of which attach to the inside pipe wall.

The alternative to allowing for pipeline performance degradation is to design pigging entry and exits infrastructure into pipeline systems, predicted to suffer from performance degradation. This is now becoming a more common practice in Australia with those designers "in the know".

Case Study 1

"There's a wheel barrow in my pipeline"[6]. This is a colloquialism meaning there's an unexplained head loss in my pipeline. An associate approached the author in 2000 saying there was a wheel barrow in his pipeline and asked the author to find it!

His pumping system delivering recycled water through a 450NB PVC (18") pipe 10km (6 mile) long to a storage dam had sustained a 10% increase in pumped head and 15% increased energy cost over 18 months.

A detailed pipeline friction audit conducted by the author determined that the loss in pumping energy efficiency was only attributed to the increase in friction in the pipeline, caused by a build-up of biofilm.

The pipeline was pigged, which then restored the pumping energy efficiency to like new.

This experience led to a permanent pigging entry point being installed at the pumping station.

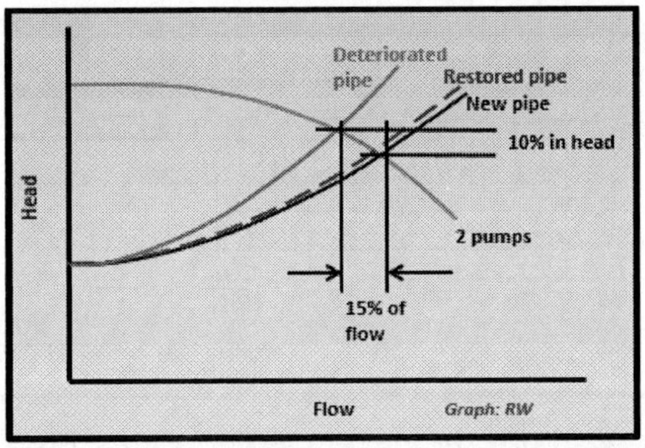

Figure 1: A new recycled water irrigation pumping system at Christies Beach, SA, in 1999 suffered a 15% electricity cost increase ($2,500) over 18 months. The cause was found to be biofilm build-up in the pipeline. Pigging the pipe restored its performance. This chart shows the performance reduction in terms of the increase in system curve steepness, resulting in a reduction of pump output.

Case Study 2

The author conducted a detailed pressure and flow audit on a turf farm[7] pumping and pipeline delivery system, feeding river water at 28 l/s to a 200m (660 feet) long lateral move.

The pump was costing about $12,500 /yr in electricity.

The pumping system had been twice audited over the last 5 years by a "qualified" irrigation auditor and given a clean bill of health.

The results from the authors audit was that the pump, despite being 30 years old, was well maintained and was 65% efficient, down from 75% but due entirely from operating too far left of BEP.

The pipeline returned a friction value of H&W C Value = 80, down from new pipe approx 155. Pigging the pipe revealed that it was severely affected with iron hydroxide.

Despite iron hydroxide water usually having a tell-tale brown water stain, the water was absolutely clear with no staining apparent.

Over its 15 years of service, it appeared that, as the pipeline deteriorated in performance, the pump was progressively speeded up by changing the belt drive ratio, there being plenty of power available and the flows were reduced by variously configuring the type of irrigation with reducing flow rates. However, no one identified the deteriorating pipe performance as the problem.

The additional pumping electricity costs due to the increased pipe friction loss was $5,200/yr, or 43%. Amortised at 3.5% with an electricity price index of 7%pa, that's equal to $58,000 over 15 yrs. That was the same as the cost of replacing the pipeline and fitting a VFD to the pump!

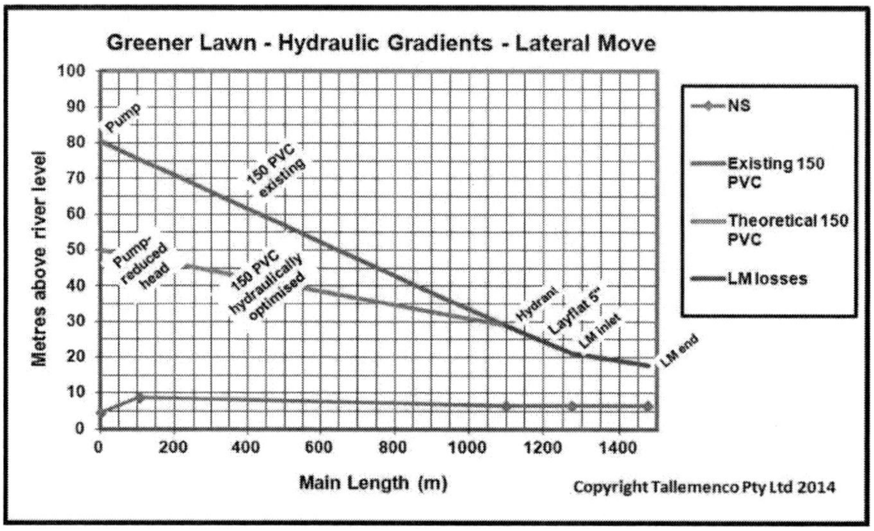

Figure 2: Hydraulic grade lines shows both the new pipe performance (in green) and existing performance measured on site (in red) of the 1km (0.6mile) of 150NB (6") PVC pipeline on a turf farm the author audited in NSW in 2014. The additional electrical energy due to the clogged pipe, amortised over 15 years, was equivalent to replacing the pipe and fitting a VFD.

3) Filter losses

The selection of a filter that's too small for the job (lower capital cost) will result in a higher head loss and therefore a higher energy cost. In addition, the filter will backflush more frequently, resulting in a shorter life. The ultimate cost will likely be many times the initial cost saving.

A filter that is too large or unsuitable for the job will result in significantly higher backflush flow rates, which can damage pumping equipment and result in low irrigation efficiency, hence higher pumping costs.

<u>Cast Study 3:</u>

A recycled water irrigation pumping system designed and commissioned by the author (as an employee of Hydrotech Aust) was installed in an irrigation system at Sanctuary Cove, Gold Coast, QLD, in 1997.

It was designed to operate ultimately with 2 pumps and two filters but was installed in Stage 1 with only one pump and one filter. The client later added a second pump, but did not add the second filter, to save costs.

The author, in the capacity of an independent consultant, was called to site about 10 years later, to discover that the single filter had a history of wear and malfunction. The addition of the second filter would have averted this situation.

Figure 3: A recycled water irrigation pump station was installed initially with one filter, one pump. Later, the second pump was added, but not the second filter. This resulted in significant maintenance and early failure of the single filter. In addition, when the second pump was installed, the clean water head loss across the filter increased from 1m (3') to 4m (13'), a 3 metre (10 feet) head loss increase, costing approx $5,000 over 15 yrs in pumping electricity costs.

Case Study 4

A three pump recycled water pumping station was designed and assembled by the author in 1999 (whilst an employee of Hydrotech Aust Pty Ltd), to specifications for a contractor acting on behalf of a golf course in central NSW. The pumping system was installed by the contractor, but no commissioning was requested.

The author, out of personal interest and passing by the golf course 11 years later, called in to see how the unit was performing.

The Superintendent, eager to vent his frustration, explained that the pumps and VFD's had suffered significant failure over the 11 years.

A brief visual audit (free of charge) established that (unbeknown to the author at the time the pump system was supplied) four x 48" media filters were deployed after the pumps. The filter backflush flow rate added 50% addition flow rate over and above the pumps' maximum output. As no pressure sustaining valve was ever installed, the pumps spent much of their life running off the end of their curve and cavitating.

This is a case of simply the wrong filter selection, and no local knowledge to recognise the problem that it produced.

The energy efficiency cost of this disaster can only be imagined. It could be measured in terms of pump down time labour, pump and VFD repairs, plus a significant amount of additional pumping to overcome the extremely poor CU's and DU's which would have resulted from irrigating with virtually no field pressure for the duration of 4 x 48" media filters backflush, probably 40 minutes each backflush cycle.

The water source was a recycled pond adjacent to the pumping station and was overloaded with algae, meaning that the filter likely backflushed frequently.

The author's offer to the Local Council (who administered the golf course) to rectify the design was rejected.

Figure 4: This collage of photos shows a three pump system teamed up with a massive 4 x 48" media filtration system, requiring a flow output 50% in additional to the pump system capacity during backflush, but which was not allowed for. With no pressure sustaining valve, the result was that the pumps and VFD's suffered periodic failure, not to mention the significant field pressure loss during backflush and hence low CU's and DU's to the golf course it served.

4) Excess residual head

All emitters require a nominal head pressure to correctly operate. This value is always defined by the emitter manufacturer. To ensure this head is always available, a margin is usually built in to the pumped head in the design stage, usually up to 5 metres head (16 feet), to ensure that there is always sufficient operating head at the emitter.

However, some irrigation systems operate with excessive residual heads, resulting in excess energy consumption.

(Probably many more operate with less head than required, but that's a story for another presentation!)

Case Study 5

The author was commissioned by a Victorian Govt Dept to undertake audits[8] on a sample of vegetable farms in the Gippsland, VIC.

A pump supplying water from a river system delivered water to a 200m (660 feet) long Lateral Move (LM). The audit revealed that the surplus head at the end of the lateral move was 11m (36 feet). This is 6m (20 feet) more than is required.

Considering the total pumped head required for this LM was only 38m (125 feet), the excess represents a 16% energy efficiency impost. This amounted to $680/yr electricity costs, or $12,900 over 15 yrs.

A simple impeller trim would have optimised this LM's pumping energy efficiency, costing about $1,000, with a ROI of 1.5.

A VFD fitted to this pump would have optimised its energy efficiency, with a cost of $5,000, or a ROI of 7.3 years.

Additional excess head losses (7m or 23 feet) were also identified in the layflat hose supplying the LM, contributing even more energy efficiency losses, at 18%.

The pump itself was measured at 68% efficient, deemed not worth any attention.

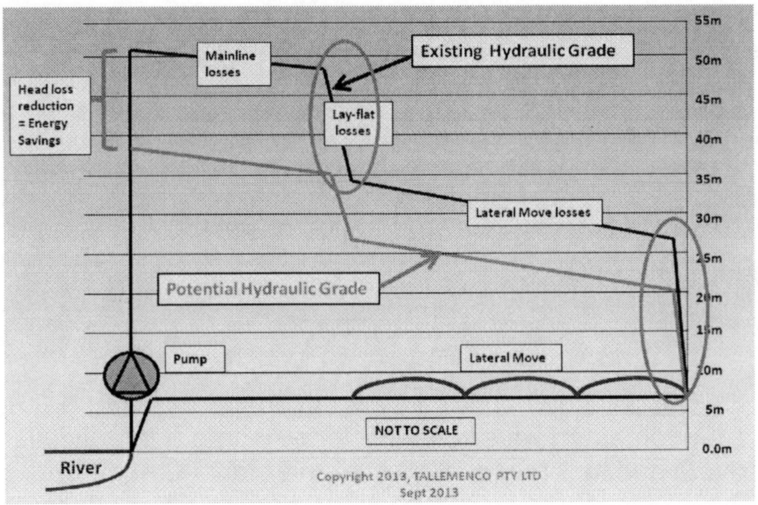

Figure 5: Hydraulic Gradients (HG) above indicates just where the head losses were occurring on this lateral move irrigation system. The black HG line indicates the existing system, the red HG line indicates an energy optimised system. Energy savings of $680/yr (or 16%) were achievable by reducing the LM's excess residual head simply by trimming the pump impeller.

Conclusion

Testing pumps to determine pumping energy efficiency is important, but only represents part of the pumping energy efficiency story in an irrigation system.

Pipeline energy efficiency is the "elephant in the room" representing a major component of pumping energy efficiency.

Components such as pipes being too small, pipe performance being degraded over time with silt and biofilm, and poorly sized irrigation system components such as filters, hydraulic valves and emitters all contribute to significant pumping energy in-efficiencies.

References

(1) http://www.talle.biz/data.html
(2) http://www.talle.biz/units04.pdf
(3) Anecdotal, feedback from Tallemenco Pumping training course participants in South Is., NZ.
(4) Reference drawing EWS 3691/74: Berri DWS, Low Level System 'F' Units
(5) https://en.wikipedia.org/wiki/Bryozoa
(6) http://www.talle.biz/wheelbarrow.pdf
(7) http://www.talle.biz/turfenergy.pdf
(8) http://www.talle.biz/gippslandenergy.pdf

Manage Irrigation for Energy Conservation & Load Shifting

Charles Hillyer

Irrigation for the Future, 424 W 16TH ST, CHICO, CA 95828,
charles.c.hillyer@gmail.com

Marshall English

Irrigation for the Future, 424 W 16TH ST, CHICO, CA 95828,
Marshall.English@oregonstate.edu

Abstract. This paper deals with development of a decision support system (DSS) for conjunctive management of energy and water use efficiency in irrigated agriculture. The primary goal of this work is to motivate and facilitate energy demand management (energy load shifting) to improve grid stability, reduce costs and facilitate use of more sustainable sources of energy. Energy and water conservation *per se* are secondary goals. The project is funded by the California Energy Commission under the EPIC program.

Keywords. Irrigation Management, Load Shifting, Deficit Irrigation, Proportional Deficit, Demand Response Programs

Introduction

Irrigated agriculture in parts of California is an energy intensive industry. One farm we are working with provides a rather startling example. Irrigating 360 acres of cotton with ground water from deep wells, the farm must lift approximately 500,000 tons of water more than 1,000 feet annually. That energy intensity attracts the attention of utilities for potential energy load shifting to mitigate problems associated with the highly variable demands on the electrical grid.

Utilities have devised incentive programs for energy load shifting in irrigated agriculture. The project reported here, funded by the California Energy Commission, is designed to motivate and facilitate farms' participation in those programs. Initially this project focused on two general strategies, *Time of Use* (TOU), for which utilities charge higher rates for electrical energy used at times when energy demands on the grid are high, and *Demand Response* (DR), which offers incentive payments for temporarily curtailing pumping if and when requested by the providing utility at times of peak demand. As this project has evolved we have found increasing interest in a third strategy, participation in a *Transactive Energy* market through which the farm decides when to buy pumping energy based on a rate schedule that varies with grid demand.

To take advantage of these incentive programs it is necessary to align irrigation schedules as closely as possible with anticipated intervals of lesser energy demands on the grid. That makes irrigation scheduling a more challenging, two dimensional problem, managing both water and energy use conjunctively. Rather than simply tracking soil moisture and irrigating at management allowed depletion, the irrigator must schedule irrigations ahead of time, sometimes well into the future, in anticipation of times when irrigation is likely to be curtailed. A more comprehensive understanding of crop water availability in spatially variable fields will be needed to ascertain whether to accept or decline a curtailment request. And the irrigation schedule will need to be updated quickly when curtailments occur.

The project is built upon an existing Decision Support System (DSS) developed originally with USDA, BPA and Oregon State funds that is used to design optimal, farm-specific water use strategies, translate those general strategies into detailed seasonal irrigation schedules and track and update those schedules to adhere to the chosen strategy as circumstances change during the season. That DSS is being augmented with new software elements to more easily align irrigation schedules with grid energy availability and cost.

Eighteen months into the project the DSS is near completion. Preliminary use of the software has revealed useful insights into the merits of the different energy load shifting strategies mentioned above.

The original decision support system (IMO)

A panel of nationally recognized experts in agricultural irrigation, commissioned by a consortium of utilities to prepare a 'Road Map' to advance efficiency in irrigated agriculture, stated that "Irrigation efficiency is ultimately determined by management, and good management depends on soft technologies (software, modeling)" (English and Evans, 2013). This project, promises to be a major contribution to development of such soft technologies.

IMO was specifically designed to provide the following capabilities:

- *Crafting optimal irrigation strategies in advance of the season*: IMO is used to first guide advance development of research-based water use strategies to target optimal soil moisture levels for the different stages of crop development, specifically including regulated deficit irrigation (RDI) strategies. It then generates detailed irrigation schedules to implement those strategies based on historic local weather and subject to the specific circumstances and operational constraints of individual farms. For purposes of the present project the schedule to be developed can be adjusted to minimize energy costs and/or capitalize on incentive payments for energy load shifting.
- *Tracking and projecting crop water availability as a season progresses.* IMO tracks variable soil moisture patterns in heterogeneous fields and projects field-wide crop water availability weeks into the future enabling farm managers to determine whether and when crop water availability is sufficient to tolerate pumping curtailment while still ensuring that the irrigation schedule will achieve targeted soil moisture levels.
- *Automatically generating revised irrigation schedules.* IMO enables rapid updating of the seasonal irrigation strategy to accommodate changing circumstances; in this application that capability facilitates revising the scheduling strategy to accommodate DR curtailments that have occurred or in anticipation of events that may occur in the near future while still maintaining the intended water use strategy.

Proportional Deficit

Implementing deficit irrigation in the context of energy management programs requires an irrigation strategy that applies a minimum yet still economically viable amount of water. Additionally, the strategy must be flexible enough to accommodate the uncertainty inherent in load shifting programs. The IMO system implements a deficit irrigation strategy proposed by University of California Cooperative Extension researchers. The strategy is based on the concept of Proportional Deficit wherein the amount of water applied is a fraction (or percentage) of the crop evapotranspiration (ETc) that occurred since the last irrigation. Typical irrigation practice in micro-irrigated almond orchards in California is to irrigate every day (or ever few days). The Proportional Deficit strategy is easy to implement because the grower only needs to track ETc for a day or two and application depths are simply a percentage of the ETc.

Use of the IMO system provides two default strategies for Almonds. The first strategy, proposed by David Doll of UCCE, has five distinct stages. This strategy uses a gradual drawdown to 50% of plant available water so that mild stress begins at onset of hull split (approximately the first week of July). This moisture level is maintained with 100% ETc replacement until 75% of hull split, followed by greater than 100% ETc replacement in anticipation of harvest. A brief dry down period occurs prior to harvest and is followed by 100% ETc replacement for the remainder of the season. The advantage of this strategy is that it provides precise control of stress during hull split however, the manager must have some means to measure stress (e.g. stem water potential) during the critical periods. The target level of stress might be based on stem water potential or some other measure of crop water status or crop stress. Because of the uncertainties involved in carefully managing mild stress there would be some advantage to applying lighter, more frequent irrigations. That would make it easier for the farm to respond quickly if/when stress reaches some allowable limit. It could also increase flexibility to respond to DR/ADR energy load shifting incentives.

The second strategy is more prescriptive and is derived from FAO66 and David Goldhammer's recommendations for RDI of almonds (Goldhammer, 2012). The strategy is expressed as a series of dates and percentages. Two sets of values, targeting either 30 or 35 inches of application, is shown in the following table.

Table 1 Second Proportional Deficit strategy

30 inches			**35 inches**	
70%	until May 1		**75%**	until July 1
50%	until July 1		"	"
25%	until July 15		**50%**	until July 15
100%	until Aug 15		**100%**	until Aug 15
60%	until October		**75%**	until October

The figure below illustrates a target pattern of optimal soil moisture prescribed by the UCANR advisor, the first of the above general strategies. The general pattern can be adjusted to higher or lower total water use, maintaining the same seasonal pattern but with different levels of total seasonal water use.

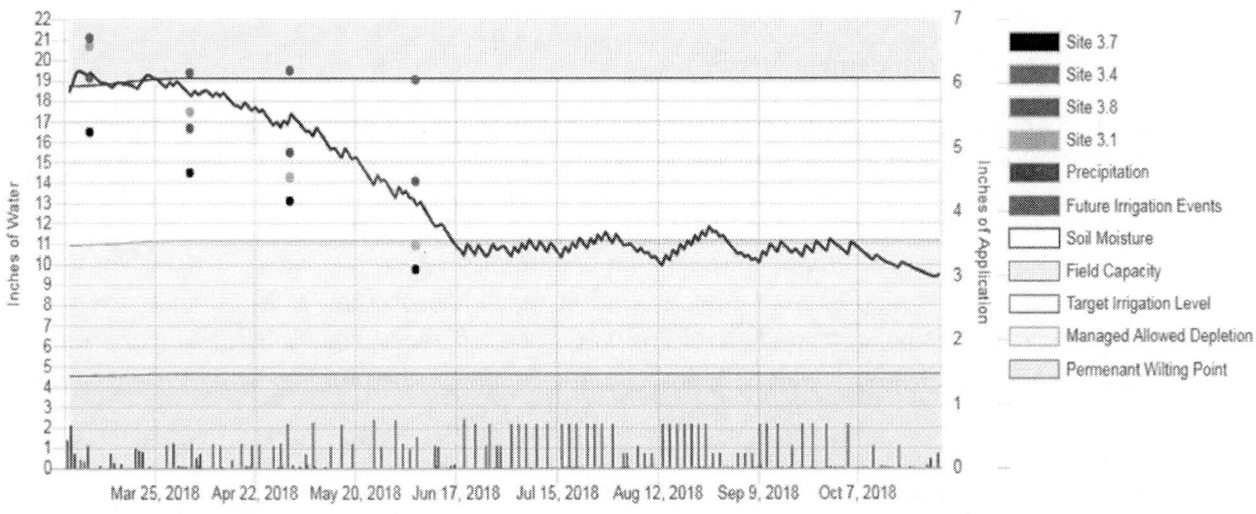

Figure 1 Proportional Deficit strategy

Planning for Proportional Deficit irrigation is more complicated than implementing it. IMO supports the strategy by providing a special interface that allows the user to interactively change the percentage of ETc replacement at various stages during the season. Figure 2 shows a screen shot of the prototype interface. The upper chart is a graph of soil moisture produced by the IMO simulation system. The lower chart is the sequence of % ETc replacement values. The user can click & drag on the point to adjust the values. After each adjustment, the simulation results are updated. The planning process involves interactively adjusting the ETc values until reaching an acceptable soil moisture pattern. During the season (e.g. after a DR event), the user can repeat the planning process using actual weather and irrigation data collected during the season.

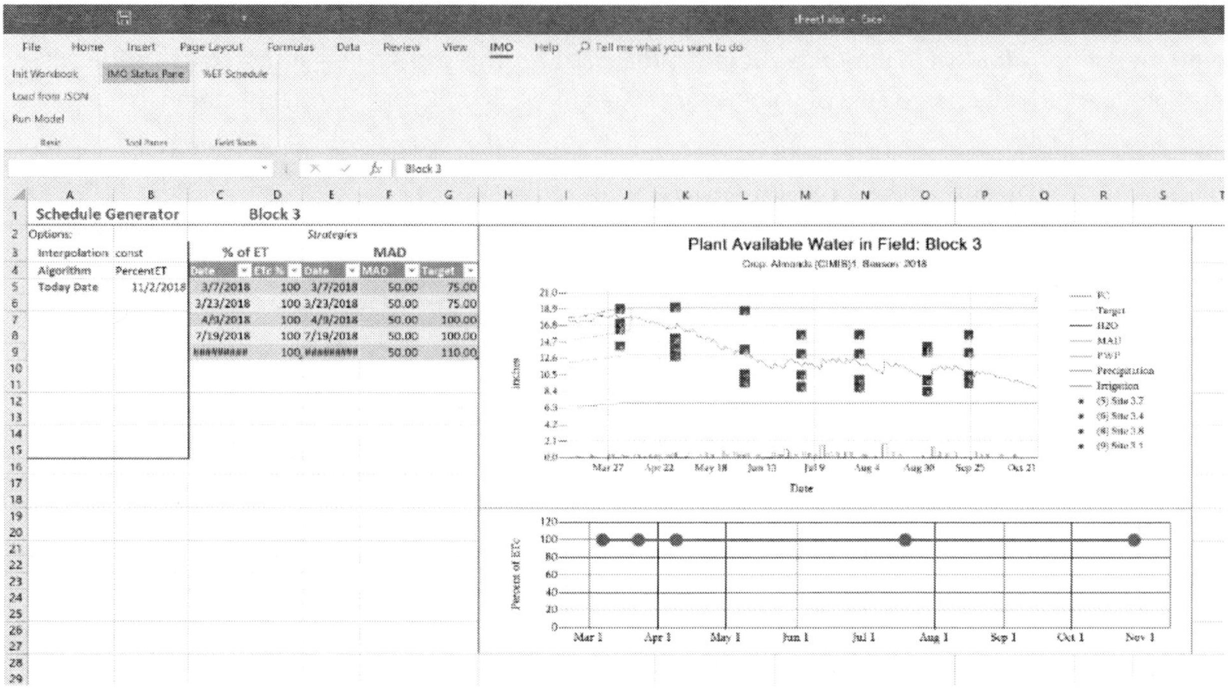

Figure 2 Screenshot of the prototype planning tool

Demand schedules: the water demand pattern

The following sections outline characteristic patterns of irrigation water use and energy costs which the decision support system is designed to accommodate.

Seasonal water use patterns are illustrated below for almonds in the southern San Joaquin Valley of California. A generic seasonal pattern for full irrigation of almonds is shown in Figure 3. However this pattern is not immutable. Some degree of reduced water use, an RDI strategy, may also increase net income by reducing operational costs and increasing water use efficiency. Figure 4shows an alternative pattern recommended when a farm water supply is limited to 60% of the nominal water requirement (from Goldhammer, 2012).

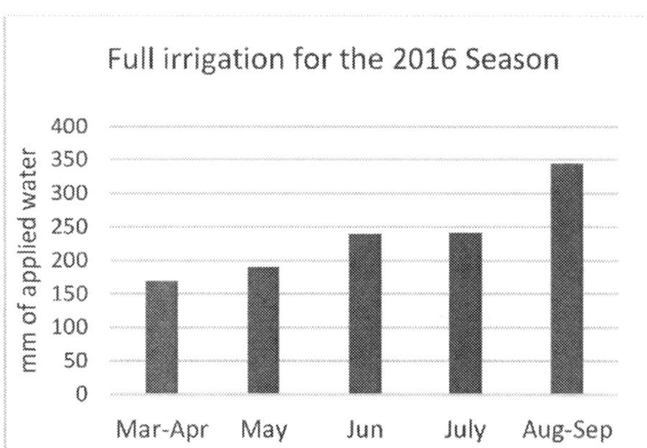

Figure 3 pattern for full irrigation of almonds

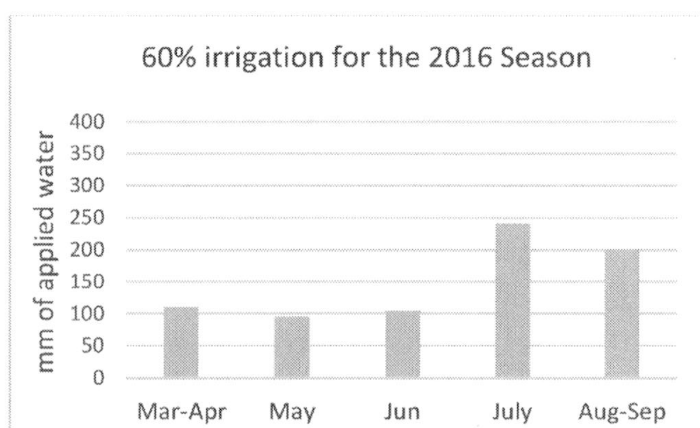

Figure 4 pattern for 60% irrigation of almonds

The energy demand pattern

Three temporal patterns of grid energy demand are of interest here; the seasonal pattern of energy demand, the diurnal pattern of a typical day during the irrigation season, and the emerging pattern of renewable energy generation.

The complex dynamics of grid energy demand and supply cannot be adequately presented within this paper. We will use daily and seasonal patterns of the dollar value assigned to energy conservation to illustrate the patterns of grid energy demand that are relevant to agricultural users.

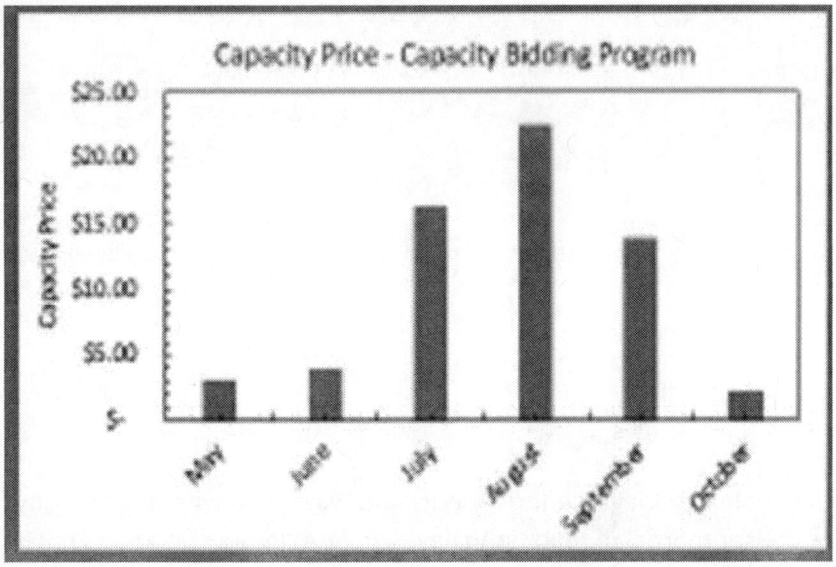

Figure 5 Monthly capacity price incentive factor

Seasonal patterns of grid energy costs:

These can be illustrated by a proxy variable, the monthly capacity payments offered for curtailments of pumping energy use by PG&E. The DR incentive program offered by PG&E computes a 'capacity' payment based on the product of the kilowatts of power to be curtailed (the capacity) multiplied by a 'capacity factor' that varies monthly depending on the relative demands on the grid for each month. The capacity factors range from $3.00 per kWh in May to $22.00 in August. As indicated, decisions to irrigate in July through September may need to account for these prices.

Diurnal patterns of grid energy costs:

One type of diurnal rate schedule is a two-tiered schedule. PG&E offers $0.19 /kWh for off peak hours and $0.45/kWh for peak periods for peak demand times (noon to 6:00 PM). Another typre of rate schedule proffered by Southern California Edison (SCE) varies with time of day and, equally significantly, with daily high temperatures (Figure 5). The various lines indicate rates for day when temperatures range between 81 and 84 degrees, 85 to 90 degrees and so on. The TOU rate offered by PG&E is shown as well for comparison. So, for example, at 3:00 PM (15:00) on a day when temperatures exceed 95 degrees the energy use rate will be $2.50 /kWh under the real time pricing schedule, whereas with the PG&E summer rates the price would be $0.45 /kWh.

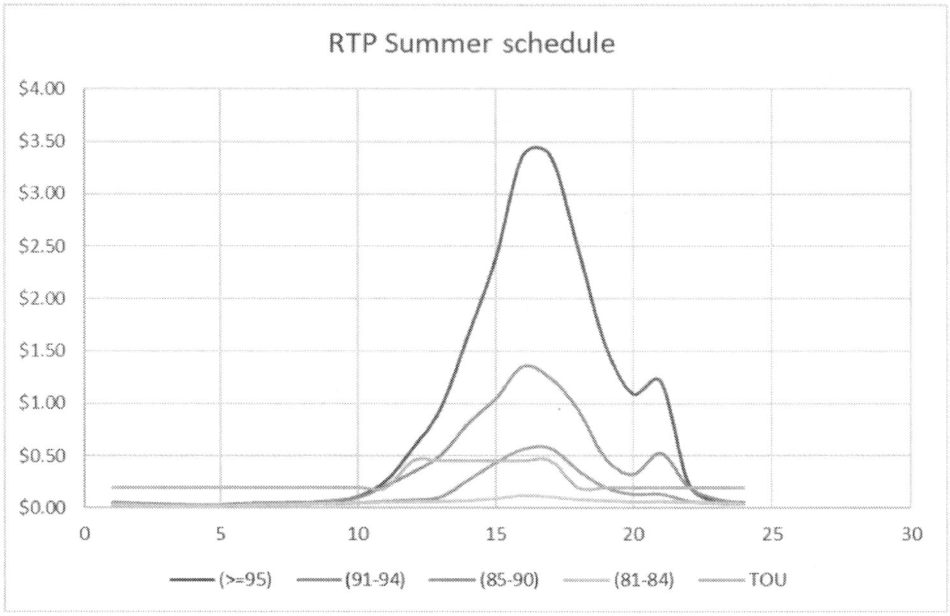

Figure 6 hourly rates for Real Time Pricing for Southern California Edison in 2018

Illustrations

We will illustrate conjunctive management of energy and water use with a case study from work with a cooperating farm in the southern San Joaquin Valley. The farm in question had limited access to water, so the seasonal water use could only provide 60% of full irrigation for almonds. The IMO system was therefore used to develop the deficit irrigation strategy proposed by University of California Cooperative Extension which was shown earlier. The farm stipulated that they prefer to irrigate weekly and prefer 24 hours set times in a three-set rotation.

IMO was used with to derive a detailed irrigation schedule to produce the target soil moisture pattern following the farm operational preferences as closely as possible. That schedule can then serve as a baseline for comparing alternative irrigation strategies, particularly including DR/ADR based strategies for a pumping head (TDH) of 200 feet and a field size of 160 acres. The bar chart below illustrates the

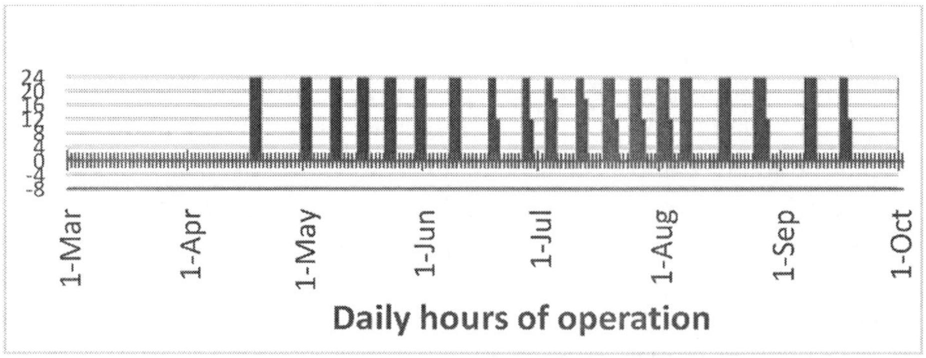

Figure 6 Baseline irrigation schedule for sample analysis

schedule as a bar chart showing the prescribed dates and number of hours for each irrigation during the season. As indicated, the pumping energy cost in this baseline case would be $36,910.

The system was then run again but with the irrigation schedule modified to follow a TOU schedule based on the two tier PG&E rate schedule ($ 0.45 per kWh for peak rate pumping and $ 0.19 /kWh for off-peak pumping). The pumping costs would be reduced by $9,900, a 27% saving.

The resulting irrigation schedule, illustrated in Figure 7, would involve more prolonged irrigation events, but the applied water would still be sufficient to maintain the intended pattern of crop water availability. However, a fundamental issue with TOU scheduling is that the delivery system capacity is generally reduced by the number of hours curtailed each day. In this case

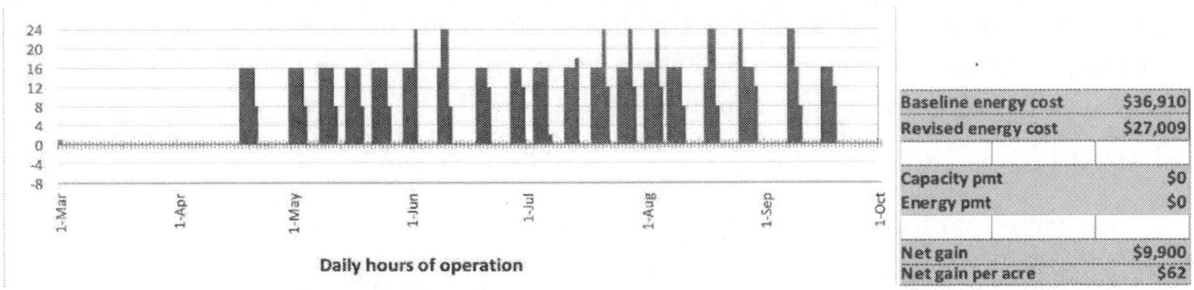

Figure 7 TOU irrigation schedule for sample analysis

The analysis was then repeated a third time for a DR strategy, superimposing the DR schedule actually called in that region in 2017 as a test case, assuming the farm accommodated every DR request and using the capacity incentive payments illustrated in Figure 4. An additional component of the incentive payment called the *Energy payment* added to the total incentive. The DR events are shown as negative hours (inverted red bars) in Figure 8. The effect on pumping costs would be unchanged since the pumping would be delayed but not eliminated. However the incentive payments would provide an offsetting payment of $2817.

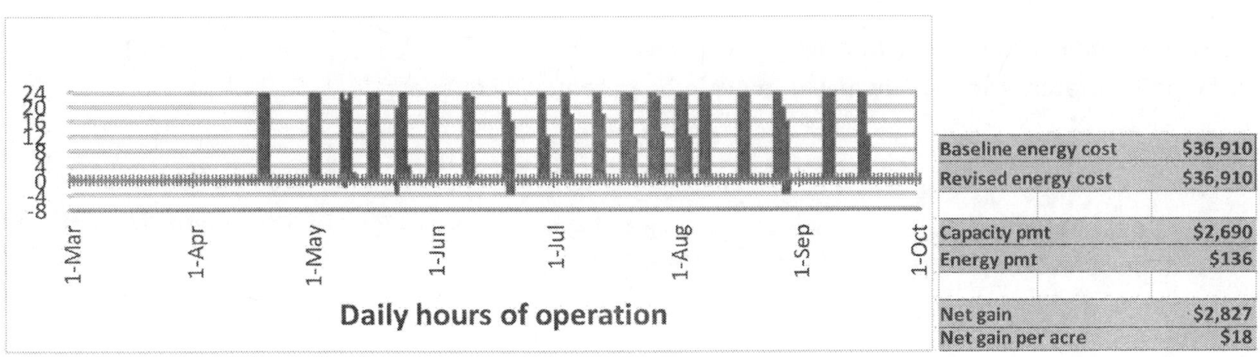

Figure 8 DR irrigation schedule for sample analysis

Transactive energy

Finally, we will compare TOU and DR incentives with a transactive energy strategy involving a different farm, another almond orchard in Merced County. The comparison is for a single irrigation event, a three day rotation of 28 hour sets. The TOU and DR strategies were based on the hours of curtailment and incentive payment rates outlined in the previous example. However, in this case, pumping costs were based on 360 kW pumping capacity for a 312 acre orchard. The cost of a single irrigation cycle using those rates and hours were:

Baseline, no load shifting	\$6420
TOU based on PG&E rates	\$4790

The analysis was then repeated using the RTP price schedule from SCE shown earlier (Figure 5). The farm was assumed to curtail pumping during the schedule to shut down during the six or eight highest cost hours from the RTP schedule. The cost of a single irrigation cycle for three different temperature days using SCE summer RTP rates (optimized for time of day) were:

Moderate temperature day (<80)	\$1189
High temperature day (85 90)	\$1847
Very high temp day (>95)	\$3063

Though the transactive rate schedule offers the lowest cost strategy in all three cases, it should be noted that if, for any reason, the farm should elect to irrigate during peak price periods that advantage would be lost, and in fact the cost for a very high temperature day would be:

\$9,948

Conclusion

This project described in this paper is still ongoing. Eighteen months into the project the DSS is near completion. Preliminary use of the software has revealed useful insights into the merits of the different energy load shifting strategies mentioned above. Complete results from the 2018 growing season will be presented at the 2018 Irrigation Association Education Conference.

Transactive energy strategies offer the greatest savings, but if for any reason a farm is forced to depart from the TE strategy the penalty may outweigh the potential incentives.

REFERENCES

- Christian-Smith, J.; H Cooley; Peter Gleick; 2012; "Potential water savings associated with agricultural water efficiency improvements: a case study of California", Water Policy 14 (2012) 194-213
- BPA, 2012; Web-based Advisory Program for Energy Conservation in Irrigation; Bonneville Power Administration; Final Report to: Bonneville Power Administration; From: Oregon State University
- English, Marshall and Charles Hillyer, 2016, *MANAGING FOR DEFICIT IRRIGATION*; Proceedings, USCID Conference, ***Water Shortages and Drought: From Challenges to Solutions,*** San Diego, May 17-20, 2016
- English, Marshall and Robert Evans, 2013; *A Road Map for Energy and Water Efficiency in Irrigated Agriculture; Guiding principles for transforming the market.* February, 2013. Prepared for the Northwest Energy Efficiency Alliance by Irrigation for the Future
- English, M.J., Solomon, K. H., and Hoffman, G.J. 2002. "*A paradigm shift in irrigation management.*" Invited paper, Jubilee Year Edition, J. Irrigation and Drainage Engineering, ASCE. Vol. 28, No 5.
- English, M.J. and S.N. Raja, 1996; "*Perspectives on Deficit Irrigation.*" Agricultural Water Management, Vol. 32, (1996) 1-14.
- English, M.J. *Deficit Irrigation; An Analytical Framework.* (1 of 2) ASCE, Journal of the Irrigation and Drainage Engineering, Vol. 116, No. 3, pp 399-412, 1990.
- English, M.J., J.T. Musick and V.V.N. Murty. 1990, *Deficit Irrigation;* In*: Management of Farm Irrigation Systems*; ASAE, St. Joseph, MI; 1990.
- Olsen, D, Arian Aghajanzade and Aimee McKane, 2015; "Opportunities for Automated Demand Response in California Agriculture", Lawrence Berkeley National Laboratory LBNL-175486
- English, M.J., Solomon, K. H., and Hoffman, G.J. 2002. "*A paradigm shift in irrigation management.*" Invited paper, Jubilee Year Edition, J. Irrigation and Drainage Engineering, ASCE. Vol. 28, No 5.
- English, M.J. and S.N. Raja, 1996; "*Perspectives on Deficit Irrigation.*" Agricultural Water Management, Vol. 32, (1996) 1-14.
- Geerts, s> and D. Raes; " Deficit irrigation as an on-farm strategy to maximize crop water productivity in dry areas; Agricultural Water Management, V 96 (9), pp 1275-1284; Sept, 2009
- Goldhammer, D., 2012; *Almonds* (pp 358-373); IN; *Crop yield response to applied water*, by: Steduto et als, FAO Irrigation and Drainage Paper 66, Food and Agriculture Organization of the United Nations, Rome
- Hillyer, C., Sayde C. (2010) A Web Based Advisory Service For Optimum Irrigation Management. In: 5th National Decennial Irrigation Conference, Phoenix, AZ: ASABE
- Keller, Jack, and Ron D. Bliesner; 1990; *Sprinkle and Trickle Irrigation*; Blackburn Press

Engineering Adaptations Optimizing IOT in Agriculture

Rob Hibbs, Agricultural Engineer, MeasureTek, Albany, OR; rhibbs@measuretek.com

Abstract

The Internet of Things (IOT) is changing agriculture with a higher quantity of sensor inputs and control outputs. This is due to lower per-unit costs and greater reliability of data transmission. This comes at a critical time, when agricultural automation can assist in lowering labor demand/costs, as well as improve resource utilization efficiencies, especially water.

IOT has the unique capability to utilize and combine data from non-proprietary and different resources, such as point-specific data, public or government data, farm and crop data. This data can combine to not only display data, but to display and report on actionable task items to provide real-time control or make real-time decisions and alarming.

Irrigation control is now more than a time clock, smart control enables modifications to schedules based on any monitored variables, ETc, soil moisture, temperature, etc. Many forms of control can now be accomplished remotely, including irrigation scheduling and VFD automation. The capability of remote control is successfully enabled, because the monitoring allows reliable feedback to actions in the field.

Distributed (wireless) sensors are available at an attractive price range, that allows for system design utilizing an effective number of representative sensors over a larger acreage.

Keywords: Control, Internet of Things, IOT, Monitoring, wireless, measure, agriculture, water, efficiency, automation, remote control

Introduction

There is a transformation of monitoring and control happening in Agriculture. The cost and complexity of sensors and control systems are coming down, while the quantity of available hardware options are increasing. This is happening at the same time the application needs are increasing with the high cost and availability of labor is high. And the cost effectiveness, and/or flat-out requirement, for smart utilization of resources such as water and fertilizer is ever increasing.

IOT, 'Internet Of Things', simply refers to a collection of physical devices and systems that are internet enabled, and have the ability to operate and interact with other devices, computers and/or people. IOT devices and systems can range from very simple to complex, but the underlying premise is that IOT systems inherently provide more ability (sensors and control points) at a much smaller cost. IOT in Agriculture, being a conservative industry, lags behind the industrial sector in widespread adoption of the technology; but is rapidly growing into the space.

Wireless, distributed, sensors are changing the scope and footprint of what can realistically be installed in the field due to lower costs and increased data transmission distances.

Utilized appropriately, the agriculture industry and growers are increasingly realizing the benefits of the monitoring and control systems in increasing productivity and efficiencies of operations.

Target Goes From Management to Operations

Historically, monitoring and control systems were installed on a limited basis to provide feedback to management on operation effectiveness.

For example, a soil moisture sensor or two would be put in the field and results would be stored and analyzed by a manager who knows the specifics about the resident crop, soil type and irrigation. This analysis by the manager, due to time constraints, would happen once a week, month, at the end of the season, or even not at all.

An IOT System has the inherent ability to pull data from several different resources, such as VFDs, water level, soil data, crop data, weather data, irrigation data, all coming from different sources, allowing the pertinent and actionable items to be pinpointed and displayed real-time, provide control feedback, alarms and reporting.

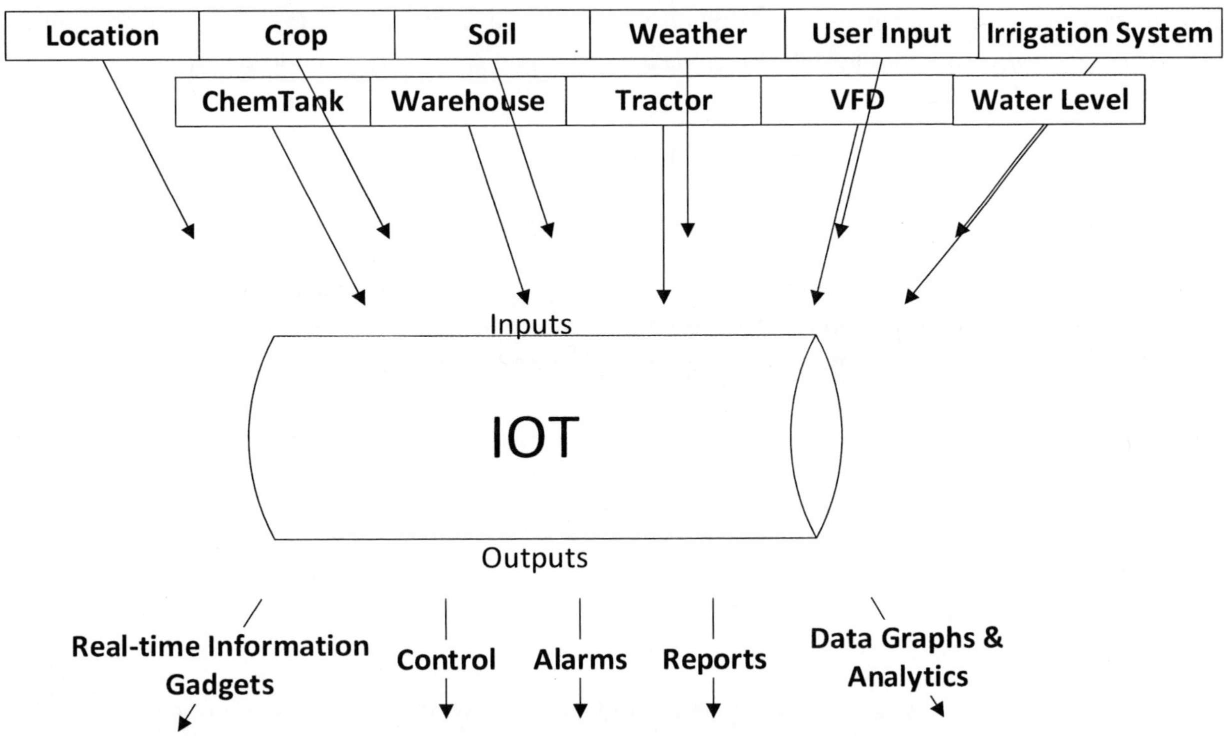

Figure 1. IOT System Overview

Unlike an all-encompassing, massive, proprietary or government computer program, the context, data and interpretation can come from disparate sources. This is an advantage to the industry and to the growers. Weather data can come from public networks if available, or not; irrigation data can come from the irrigation designers, sensor data can come from many different manufacturers' products, forecast data can come from private or public networks.

The outputs can be simple data gadgets, complex control, alarms based on data or pest reports; the organization is able to decide the priorities, information and outputs that provide the most cost-benefit for the organization.

A temperature alarm has been around for decades, but now that alarm can come in the form of a text message, email and voice phone call in English and/or Spanish, and have a history associated with who responded and when. How about a similar alarm that calls you to say "your irrigation is scheduled, but is not happening"? This is taking alarms to the next level, not based on data only but on information regarding your irrigation system.

The most beneficial use of the IOT system is therefore the irrigator, or the applicator, or the agronomist, as the data and/or control is to help and enable the operator do their job more efficiently and effectively. The system also provides an invaluable teaching tool for the manager in training, improving operations and catching problems before they become large.

Irrigation Control Has Changed

Labor costs, or lack of available labor, combined with limited or restricted water and pesticide use has driven the need for automated monitoring and control. The days of setting an irrigation schedule for a month are over. Irrigation zones have become smaller for better, more efficient, irrigation.

Wireless Control

Wire and trenching are expensive and a continual problem with wire theft, breakage, and remapping. Wireless valves are now robust, and make the 16-port controller obsolete, now the industry has a 100+ port controller. To effectively manage this new irrigation system, some supporting information for each valve such as location, crop, and soil is necessary. An irrigator may be able to program each valve from a keypad at the pump panel, or from a screen by valve, but this is not the best utilization of the irrigator's time and talent. An irrigator gains knowledge and capability when he can focus on zones, crops and soils; rather than Valve 67, 72 & 84.

Extend High-End Labor Resources

Farms are diverse and often geographically separated, combine that with the requirement of conserving and limiting resources; this means the irrigator and manager's time is a precious resource and will continue to be. A simple display of current status, highlighted maps with irrigation zones, allows the irrigator to view status of multiple irrigation systems in minutes. The irrigator can locate issues or problems in a daily or twice daily system review from his phone and go directly to the problem or dispatch staff with direction.

Figure 2. Map View of Flow Data

Growers have utilized the savings in time to extend the irrigated acres that an irrigator can cover, as well as leverage their time to concentrate on irrigation efficiencies and soil moisture.

The irrigator has the ability to view and download schedules from anywhere in the world with today's technology, no need to travel to the site; saving time and resources.

VFD's Are Here to Stay

VFD's offer sustained performance with continual changes in irrigation control adjustments. Valves can be added, turned off, or sustained based on maintenance, soil moisture, temperature or climate control; the VFD allows this to function without interruption.

However, the VFD has operating parameters, settings and faults that are normally reviewed, adjusted or reset by an experienced source such as a distributor, supplier or manager. IOT allows for the display and adjustment of these parameters remotely, saving grower and distributor's costs in deploying service technicians, and more quickly solving grower's issues.

Smart Irrigation Schedules

A time-based control system is the baseline but is not adequate for the long term. It can be made adequate currently by having someone continually looking at the data and changing the operating schedule. Or, the system can have the built in capability to adjust schedules via data and information.

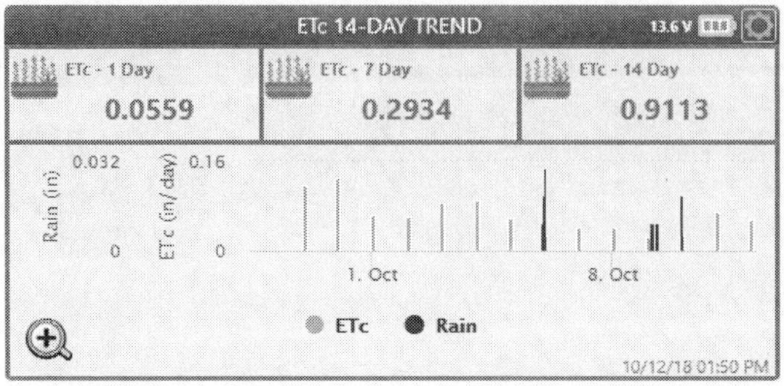

Figure 3. Daily ETc for the Specific Crop

An IOT based controller has many inputs from which control can be based. The same controller that operates irrigation, can turn on overhead for frost control, and in the heat of the summer turn on irrigation for climate control to reduce pest incidence or sunburn.

Of the early adopters, there is not one grower who has not changed their irrigation practices based on utilization of the new control capabilities. They learn of the cause-effect relationships and adjust accordingly.

Distributed (Wireless) Sensors

Sensors are now available that will communicate directly with the cloud, and not tied to a traditional power-hungry radio network utilizing dedicated, non-licensed frequency bands.

This brings the sensor cost down significantly, as the infrastructure of battery, modems and A/D conversions is minimized.

The system integrator now has more ability to effectively install a representative number of sensors in the field to get desired outcomes and analysis, rather than monitoring a few select locations.

Information Not Just Data

The tabular report, or graph of data, may be OK for regulators or managers hind-sight view; but this is not enough for operational personnel to make decisions. The data needs to be current and combined with supporting information and data such that operational personnel can perform actions based on the displayed information.

A 'Report' is largely replaced by a 'Gadget' in IOT terms. A gadget has the following attributes:

- Information displayed and updated at a frequent interval
- Utilizes supporting information and data
- Can be moved around, or grouped, with similar or related gadgets

As an example, a soil moisture gadget shown in Figure 4 can show much more than soil moisture, since the gadget knows the irrigation system, crop and soil type. An operator may not understand the meaning of 25% volumetric soil moisture in a particular field and crop, but the gadget clearly shows the location specific 'upper' and 'lower' bound of irrigation, as well as the cause and effect of his applied irrigation.

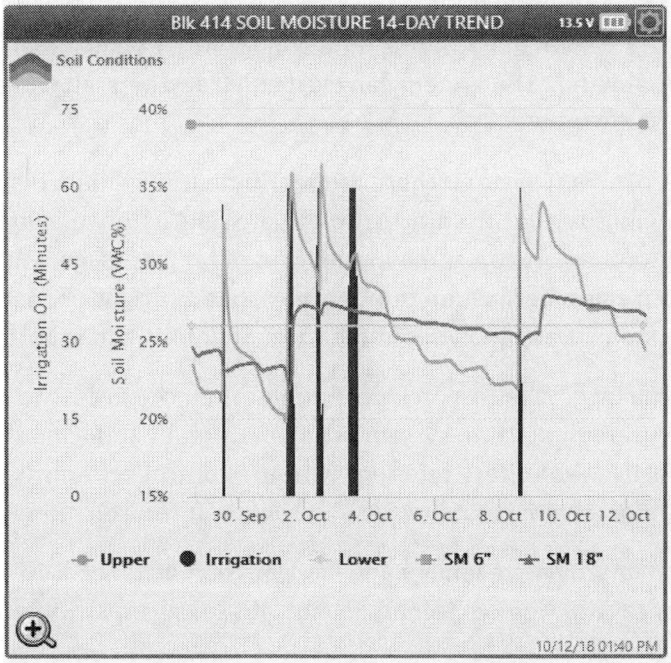

Figure 4. Soil Moisture Gadget showing volumetric soil moisture, irrigation on time, field capacity and Maximum Allowable Depletion(MAD)

Another corresponding gadget, shown in Figure 5, utilizes weather, crop and irrigation data to calculate suggested irrigation on-times to replenish 1-day or 7-day ETc. This operator knows from this gadget that he needs to run his irrigation system for 3 hours to replenish the last 7 days water use via ETc.

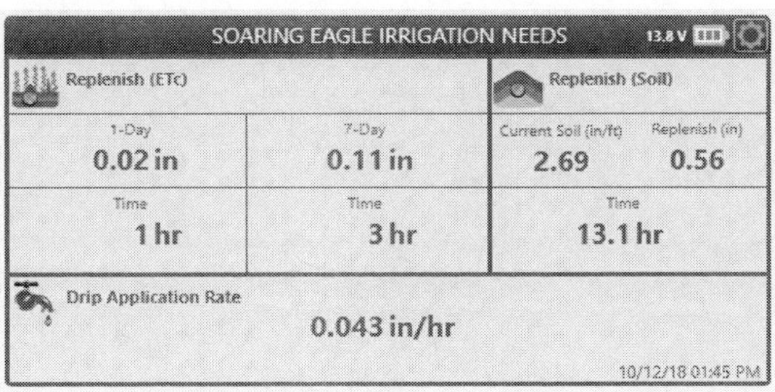

Figure 5. Irrigation Needs Gadget

Another real-time calculation shown in Figure 5 is the length of irrigation time needed to replenish the soil profile to field capacity. These are real-time calculations and data available for the operator and manager that is specific to their operation and management.

Conclusion

IOT in agriculture is here to stay, it is needed by the industry to increase effectiveness and efficiencies. Expanding existing labor's production and increasing water/nutrient efficiencies are the driving forces.

Enabled is a non-proprietary system of gathering data and information in a manner that can be utilized specifically for the grower at hand. The system can start small and be scaled in accordance with the needs and requirements of the grower.

There are several outputs of an IOT system: there are real-time 'smart' data displays called Gadgets, which combine raw data with site specific information to provide actionable information. Alarms, not just of raw data, but of system errors can come in the form of text message, email, and voice in English or Spanish. The system can alarm not just on temperature, but of irrigation system variables, pest alerts, and more in one system. Data analytics in the form of graphs and reports are standard.

Control systems provide the biggest return on investment, as remote access to monitored variables allow personnel to concentrate directly on system problems, greatly reducing the time allocated to troubleshooting and travel between sites. Inherent with the control systems is Smart Control, where control is effected by monitored variables, such as ETc, temperature, soil moisture, etc

Due to decreased cost and improving reliability and transmission distance. Distributed (wireless) sensors are enabling a new view into what can be accomplished with sensors distributed throughout the farm.

It is important to list the priorities and desired outcomes of any IOT system at the beginning, largely concentrating on daily, routine action items that aid in production, efficiencies or solving problems.

Incentives and Challenges for Agricultural Energy Use and Grid Services

Collin English
Irrigation for the Future
collin.english@gmail.com

Arian Agahjanzadeh
Lawrence Berkeley National Labs
aaghajanzadeh@lbl.gov

ABSTRACT

In this paper we analyze the 2018 irrigation season for three different farms using three different demand management options (Baseline, Time of Use, and Demand Response) and compare the profit potential for each management approach. We also identify critical challenges to making irrigated agriculture a resource for demand management and electricity grid and next steps and possible technological developments to improve the economics of agricultural load management.

KEYWORDS: Agricultural energy use; Pumping costs; Energy use efficiency; Solar; Grid services; Demand management; Demand response; Time of use; Transactive energy; Real-time pricing; deficit irrigation; Evapotranspiration

INTRODUCTION

On-farm energy use is an increasingly important cost and variable irrigators must balance against their water availability, yield targets and production schedules. Declining surface water and lower water tables in many western states are increasing the energy intensity and therefore the cost of irrigation. At the same time, due in part to the low cost of land, agricultural areas are emerging as major producers of clean energy. In California, for example, over the last five years solar production has jumped from 5.5 million net MWh to 28.5 million MWh, with the largest volume of solar generated in counties with some of the largest volume of agricultural production.

As the electricity grid evolves to include more distributed sources of energy generation, it needs flexible loads to maintain its stability. Twenty-nine states, Washington, D.C., and three territories have adopted a Renewable Portfolio Standard (RPS). Top irrigated agricultural states are among those with ambitious RPS targets such as California (50% by 2030), Texas (10 GW by 2025), Kansas (20% by 2020), and Montana (15% by 2015). In order to meet those RPS goals, large quantities of solar and wind electricity need to be integrated into the electricity grid.

Unlike conventional thermal power plants (e.g. natural gas and coal), these new generation sources are inherently intermittent and not always available. Until large scale battery storage becomes widely available at a low enough cost, many renewable energy sources are only available during when their energy source is available -- solar generation occurs during daylight hours, wind generation is only available when winds are blowing. This intermittency of renewables, their lack of dispatchability, and currently high cost of energy storage makes planning and managing the electricity grid more difficult.

With intermittent generation sources, managing energy demand is a critical part of maintaining grid stability. Historically, generation would be modulated to match demand, but with intermittent supply from renewables now it's the other way around. Demand is altered to match generation. Agricultural loads, with their large magnitude and uniformity, have the potential to satisfy a significant portion of that critical resource.

DEMAND MANAGEMENT

The term demand management encapsulates any mechanism through which and end users' demand is modified in order to benefit the grid. In return, end users are compensated either through incentives or lower rates. The most common demand management strategies are Time of Use (TOU) pricing and various forms of Demand Response (DR) or Automated Demand Response (ADR). An emerging approach which may be better suited to address the future grid needs is Real-Time Pricing (RTP) through a Transactive Energy (TE) market mechanism.

In this section we discuss each of these mechanisms and identify some of the benefits and drawbacks of each. This is followed by cost benefit analysis of each demand management approach at a farm setting – and how much things will vary from one farm to another. Three farms, each with different pumping demands and irrigation management constraints, are analyzed for baseline energy use and three demand management scenarios: TOU, DR, and RTP. All data is from the 2018 growing season in California's San Joaquin Valley.

Time of Use (TOU)

Time of Use pricing aims to re-shape the underlying load profile through long duration price responses that are announced months ahead of time and change seasonally. TOU rates are typically higher during peak hours (hours with highest demand for electricity) and lower during off peak hours. Figure 1 shows an example of TOU rates for agricultural accounts in the Pacific Gas and Electric (PG&E) territory in California.

Figure 1: Time of Use Rates for PG&E Agricultural Customers (Based on AG4A rates per kWh 9/1/18)

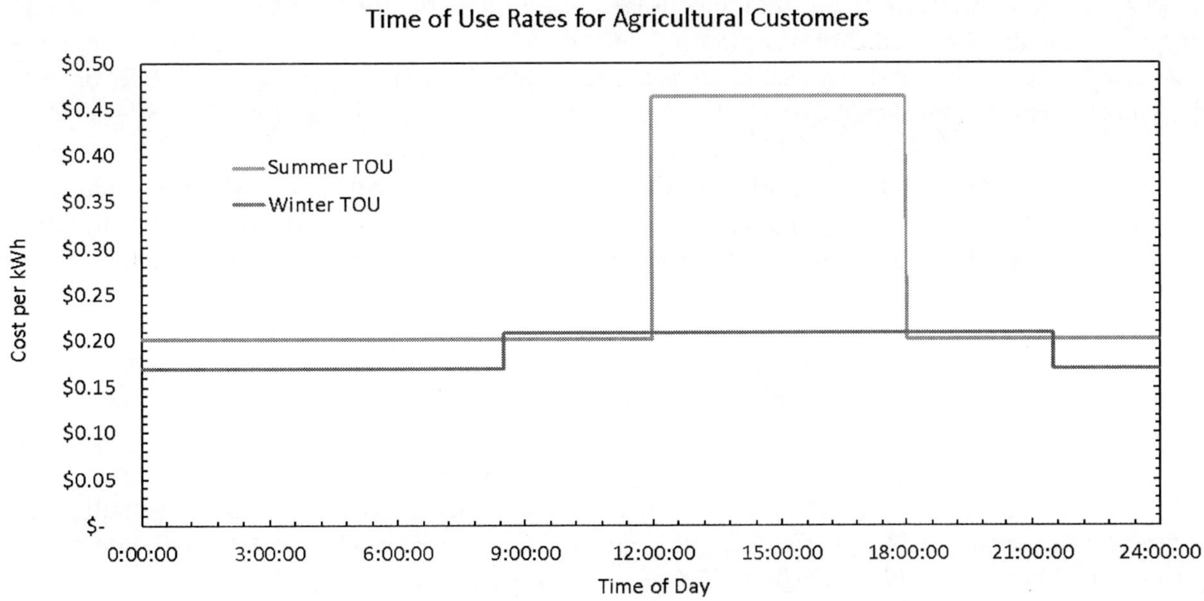

TOU has historically been the most cost-effective option for modifying load shapes. A recent study conducted by LBNL for California Public Utilities Commissions (CPUC, 2016) estimates that time of use pricing is particularly cost effective because few site-level technologies are required for enablement. While the load reduction at any given site is typically small, the breadth of participation – possibly for a variety of rates schedules that currently exist – we can see substantial statewide effect/benefit. With 60% of all growers in California taking part in some form of TOU scheduling, this is by far the most widely used demand management program in agriculture. With the goal being to avoid peak energy rates, there are many different approaches.

Pros of TOU:

- Low cost of implementation
- No need for program management, it's completely up to the customer to modify load based on the rate

Cons of TOU:

- Rates need to be constantly updated to reflect the latest grid status
- Requires largest numbers of customers to respond to the rates in order to achieve its goal

- Predictable outcome. In aggregate, result from a TOU rate is predictable

- Irrigation systems during peak season may require schedules shift to irrigate during peak hours.

Demand Response (DR/ADR)

DR and ADR is referred to strategies that encourage end-users of electricity to shed or shift their load for economic or reliability reasons. DR is commonly referred to strategies that are implemented manually, and ADR refers to automatically adjusting loads based on the signals received from the utility or the Independent System Operator (ISO).

DR and ADR programs are often supported by third party aggregators who act as an intermediary between the farm and the utility, vetting the farm's energy use history and setting up the contracts and hardware that might be required to monitor and control pumps and power supplies, enabling the farm to profit from the curtailment calls. Curtailments are typically announced in advance by as much as 24 hours or as little as 30 minutes. With DR, the curtailment call comes to the grower asking if they can shut down their pumps for a set amount of time lasting anywhere between 30 minutes and 4 hours. With and ADR program, unless the grower opts out ahead of time, the pump is controlled automatically.

With most DR and ADR programs, growers sign agreements to take part in a minimum number of curtailments during the season. Failure to fulfill these agreements results in penalties -- penalties vary and are determined in a case by case basis but in general failure to curtail up to 60% of the nominated capacity, will result in penalties as high as 60% of the capacity payment (PG&E, 2013).

Pros of DR/ADR:

- Ensuring grid reliability when demand exceeds generation
- Cost and environmental benefits by avoiding expensive and polluting peaker plants form going online
- If done correctly can result in cost saving for the customer

Cons of DR/ADR:

- Complex utility programs
- DR/ADR and TOU are in conflict (doing TOU lowers DR potential)
- Baselining issues with DR programs. 10-day baseline causes farm to not be compensated based on their actual load shed/shift
- Running DR programs are costly for the utilities
- Grid needs have evolved but the framework of utility DR programs have stayed the same

- Without automation, responding to DR is a huge undertaking, especially in a farm setting

- Short notification periods with penalties to follow if the end user misses the notification

- Long gaps in irrigation for cotton, of ten days or more, create a negative incentive for DR (see graphic)

Real-Time Pricing and Transactive Energy (RTP/TE)

Real-Time Pricing and Transactive Energy are demand management for the "smart grid." Both provide the customer with greater transparency and control over energy pricing opportunities, allowing the user to find energy pricing that more effectively meets their needs. Both rely on "smart metering" and interactive smart grid infrastructure. Both provide the customer with more control. And both avoid the baseline problem of DR. Transactive Energy also provides pricing and control systems for distributed energy resources (DERs), micro-grids, and peer-to-peer exchange. TE is designed for the current trend of increasingly distributed, renewable energy generation, and to provide greater economic and management stability in a future of increasingly dynamic energy markets. RTP tariffs are offered by some utilities. TE provides a platform through which loads and electricity generators can directly communicate. As a market mechanism, TE is still in the demonstration and testing phase.

Pros of RTP/TE:

- No need for program management; similar to TOU everything is left up to the customer

- The rates are much more dynamic and allows the utility to adjust rates in real-time to reflect the status of the grid (allows for pricing of congestion or greater availability, for example)

- TE enables decentralized, local, peer-to-peer, and micro-grid energy exchange

Cons of RTP/TE:

- Requires high level of automation and communication (e.g. Smart grid infrastructure; OpenADR communication protocols)

- Almost all RTP rates offered through utilities are experimental

- TE as a control and pricing mechanism is still in an experimental stage

- Unlike DR, end users can plan ahead of time; or adjust operations in real-time
- Unlike DR, No penalties for participation or opting out
- Unlike DR, there is no program management required
- Avoids the baseline problem of DR
- Low cost (or no cost) to implement (once Smart grid or OpenADR are established)
- If proven successful, TE rates need to be approved by the Public Utilities Commission, which could be a significant hurdle

ANALYSES OF THREE FARMS

We tracked the 2018 irrigation season for three cooperating farms in California's San Joaquin valley, including their energy use requirements, irrigation strategy (irrigation dates, amounts and set times) and the resulting soil moisture profiles, each farm's unique irrigation system, pumping demands (measured as total dynamic head), soil moisture holding capacity, climate and weather, and seasonal crop water demands. Using this information as a baseline, we modeled each site and analyzed three different demand management scenarios (TOU, DR, and RTP), and calculated the irrigation timing strategies necessary to meet each demand management approach, and the cost of energy for each strategy based on the unique constraints of each location.

Gustine, CA

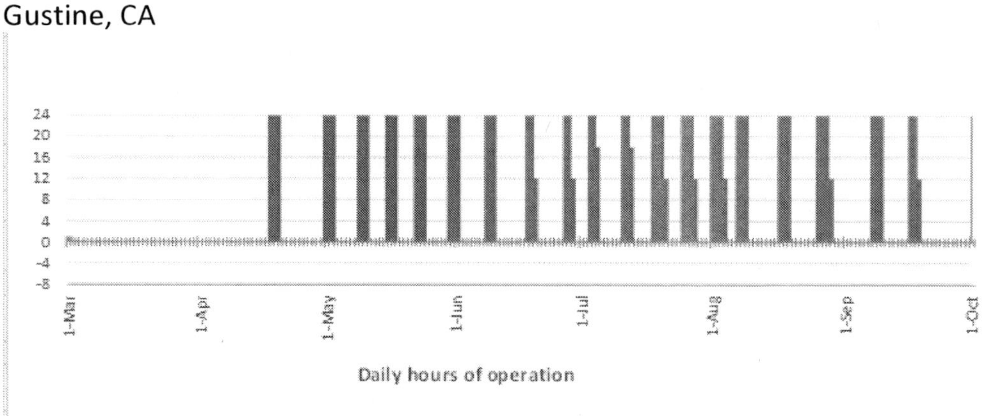

TDH	85
Application rate (in/hr)	0.064
Acres	92
Power (kW)	21
Sets	3
Baseline energy cost	$9,020
Revised energy cost	$9,020
Capacity pmt	$0
Energy pmt	$0
Net gain	$0
Net gain per acre	$0

The Gustine site is a small family almond orchard (92 acres) on soils that are two types of sandy loam and a significant amount of river rock. Their pumping lift is minimal (TDH = 85) as their water comes directly from a canal beside the orchard. The blue bars in the graph, above, show the irrigation times and amounts of their 2018 season. Using this as a baseline, the following two graphs show alternative irrigation timing strategies for TOU and DR. Their baseline energy costs is: $9,020, or $98 acre.

The TOU strategy (below) generated for the Gustine site, shifts the start and stop times to avoid peak hours, producing a savings of $2,419, or $26 acre.

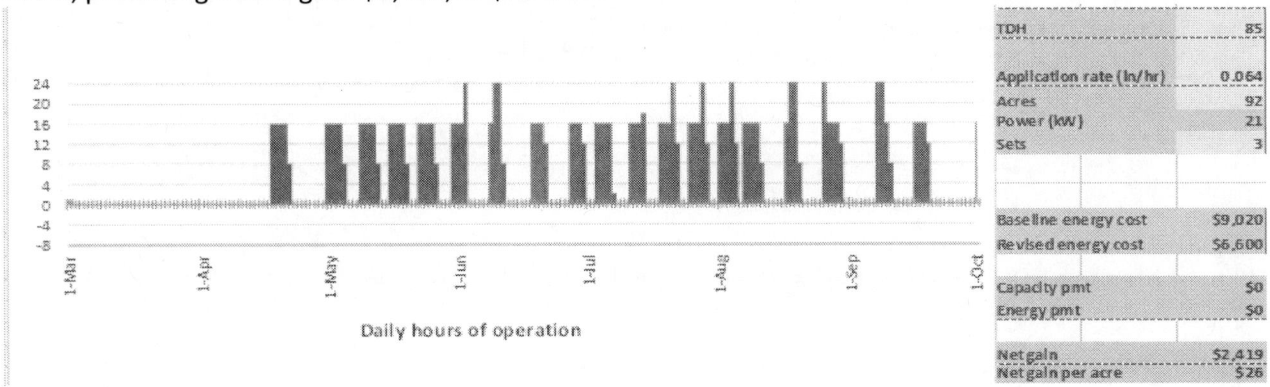

The next graph, below, looks at the economic potential of following their standard schedule while using a DR program at the same site. The red bars below the irrigation times indicate actual DR calls during the 2018 season. Had the irrigator shut down with each DR call, they would have realized a savings of $691, or $8 an acre.

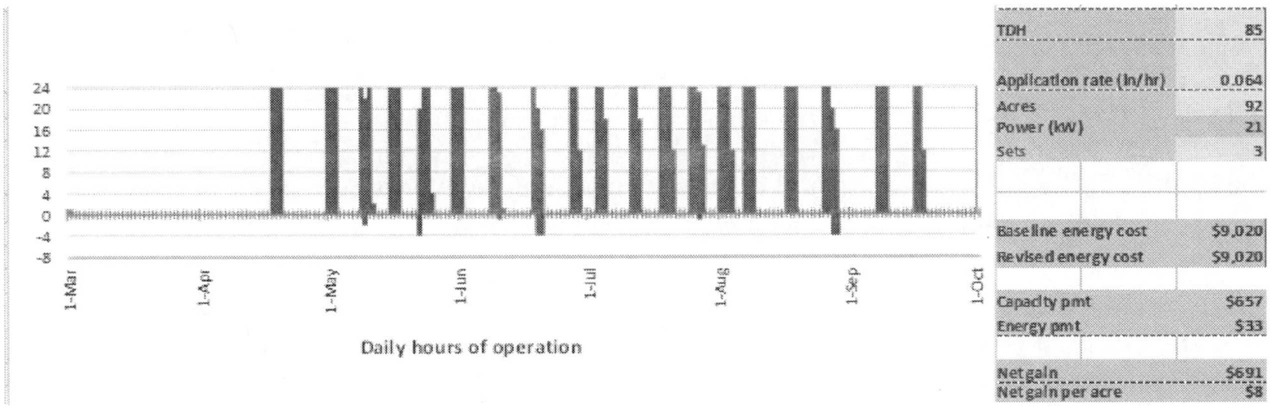

Chowchilla, CA

The Chowchilla site is one field (almonds) of a much larger ranch, using deeper wells at a much higher energy cost. There are a range of 10-15 different soils across the ranch. The irrigation system uses fan jets. The graph, below, shows their 2018 season, with a baseline energy cost of $113,127, or $362 per acre.

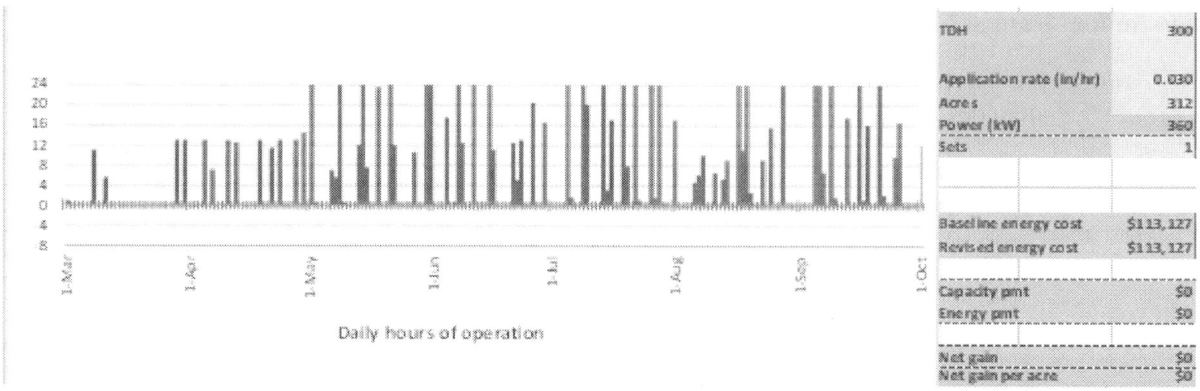

When shifted to a TOU schedule, below, the Chowchilla site could see a savings of $28,390, or $91 acre.

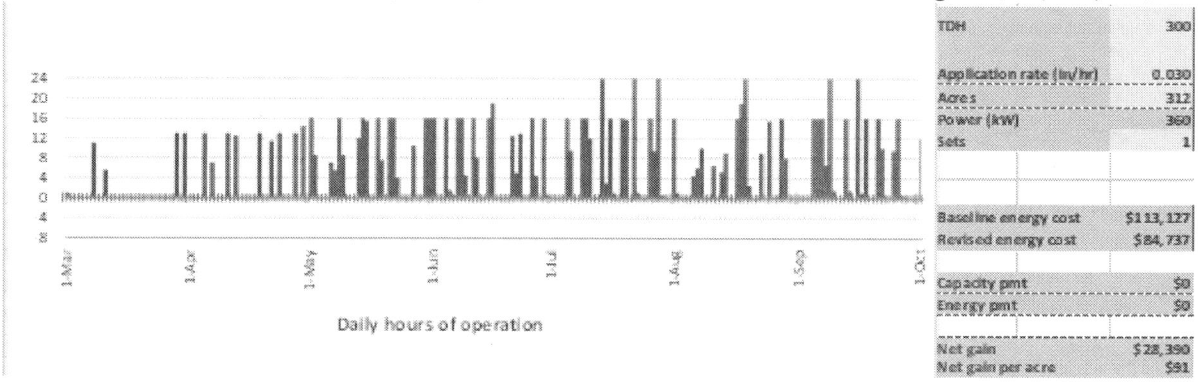

If irrigators at the Chowchilla site had taken advantage of DR events during the 2018 season, below, they would have realized a savings of $5,663, or $18 acre.

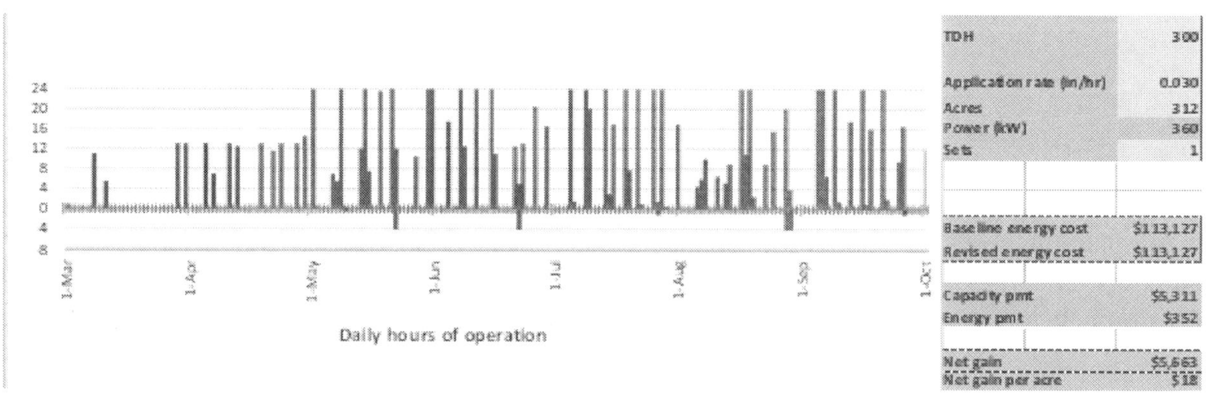

Five Points, CA

This site, below, differs from the other two in a couple significant ways. The soils are heavy loams with much higher holding capacity. The irrigation system is subsurface drip. And the crop is Pima cotton. Their baseline energy cost for 2018 was: $118,743, or $742 acre.

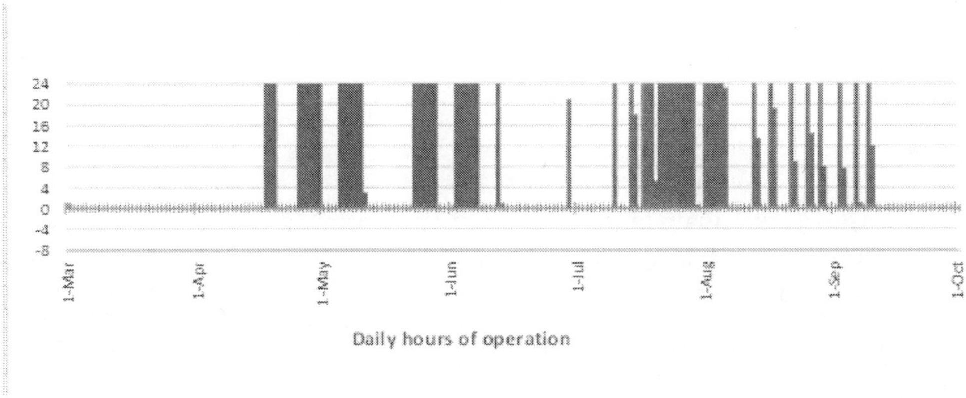

TDH	1000
Application rate (in/hr)	0.042
Acres	160
Power (kW)	287
Sets	3
Baseline energy cost	$118,743
Revised energy cost	$118,743
Capacity pmt	$0
Energy pmt	$0
Net gain	$0
Net gain per acre	$0

Using a TOU schedule, below, they could have saved $28,061, or $175 acre.

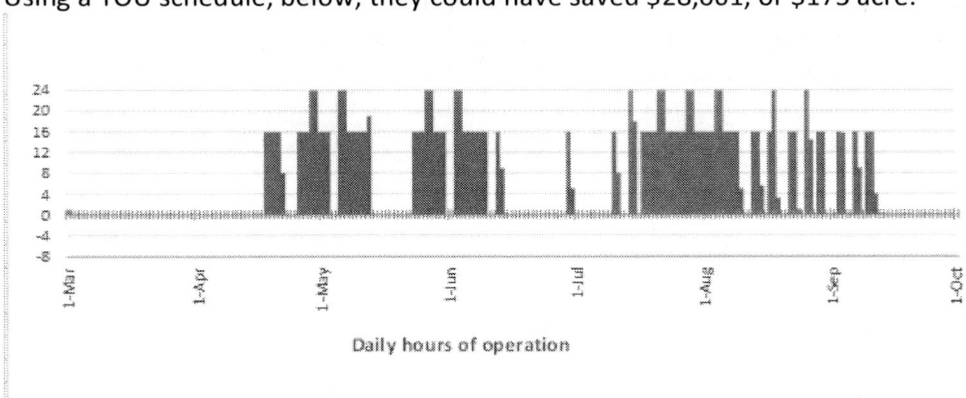

TDH	1000
Application rate (in/hr)	0.042
Acres	160
Power (kW)	287
Sets	3
Baseline energy cost	$118,743
Revised energy cost	$90,682
Capacity pmt	$0
Energy pmt	$0
Net gain	$28,061
Net gain per acre	$175

The analysis for a DR strategy for this site produced an interesting finding. Guided by the soil moisture conditions and crop stress, the irrigator was able to pause between irrigations for more than ten days. As a result, when the second DR event was called, under the 10-day baseline rules of the DR program, they would incur a financial penalty had they chose to respond to the DR curtailment request. As a result, their savings using DR would have been, $10,484, or $66 acre.

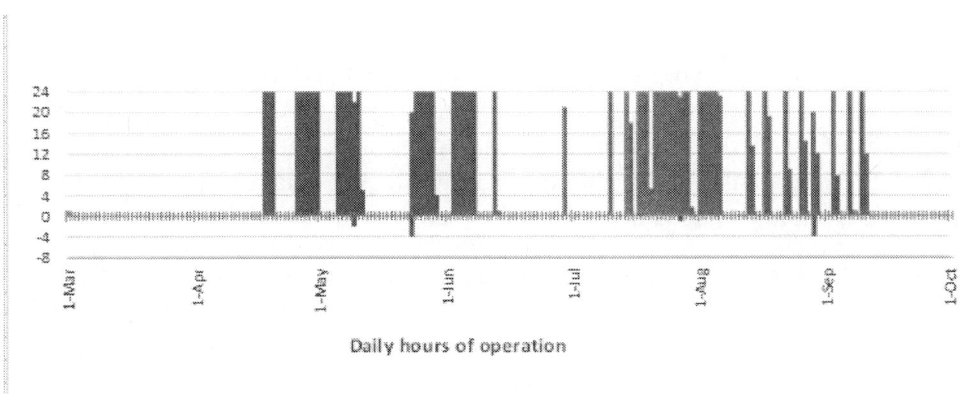

TDH	1000
Application rate (in/hr)	0.042
Acres	160
Power (kW)	287
Sets	3
Baseline energy cost	$118,743
Revised energy cost	$118,743
Capacity pmt	$10,272
Energy pmt	$213
Net gain	$10,484
Net gain per acre	$66

SUMMARY OF THREE FARMS

This analysis suggests that TOU strategies for demand management are more profitable for irrigated agriculture than DR programs. This also supports the larger trend in irrigated agriculture towards adoption of TOU demand management programs and suggests why DR programs have seen less adoption. This analysis is based on the soils, seasonal weather, age of the orchard and irrigation system constraints at this site. While each farm represents a different irrigation and energy use management challenge, there are some clear benefits for irrigators using TOU strategies. DR is more complicated, less profitable, and costlier to implement and manage than TOU.

CHALLENGES

For irrigated agriculture to successfully participate in these programs and provide value both to the grid (as change in demand) and the farm (as financial incentives), several barriers need to be addressed: (i) Automation; (ii) Irrigation timing and system constraints; (iii) DR programs are complex and poorly aligned for agriculture.

Automation

If irrigated agriculture is to work seamlessly with energy markets, agriculture needs to be integrated with the smart grid in ways and with technologies unique to agriculture. Many irrigation systems still require someone to visit the pump and manually turn it on or off. In many cases data about the field -- flow meter and pressure data -- are also collected manually. To make irrigated agriculture a more flexible, adaptable and profitable participant in energy markets, irrigated fields need to communicate directly with the grid.

From the utility side, connecting agriculture to the smart grid means pumps equipped with smart meters and automated start and stop controls to enable the pump to shut down quickly. But for an irrigator, a market signal to the smart meter on the pump is only a small piece in a chain of evaluations, decisions and actions that must be made about the irrigation system (valves, gates, filters) and crop health (soil moisture profile, stem water potential, crop stress, weather forecasts). For irrigated agriculture to optimize the economics of their energy use and work profitably with the grid, automation solutions should go beyond the smart meter to include remote control of valves, gates, and filter stations, as well as the automated transmission of field data (flow meter and pressure, soil moisture, climate, weather, crop stress), and software to track, interpret and generate decisions based on this data. The smart meter for an irrigation pump should connect to a smart meter for the crop and irrigation system.

Irrigation Timing and Irrigation System Constraints

Crop yields are directly proportional to the amount of water a crop receives and uses over the season. The standard measurement of this relationship between water and yield is evapotranspiration (ET). How easily an irrigator can participate in a demand management program, such as TOU or DR, depends on the limits of the irrigation system and the irrigators ability to keep up with ET. Because many irrigation systems are designed to meet ET during peak season, the opportunity to stop irrigations during the hottest times of the year is limited, but not impossible.

For irrigators to work with demand management with irrigation systems at full capacity during peak season requires the ability to predict how each irrigation will shape the of soil moisture profile going forward to meet crop water demand and crop stress. This level of precision makes it possible to do two things: (i) identify irrigation strategies below full ET --deficit irrigation; and (ii) estimate the impact, positive or negative, of different irrigation timing strategies -- including deficit irrigation -- on crop stress and yield outcomes. This will enable the irrigator to more easily see the potential to get behind in ET and the risks to the crop of reducing an irrigation, running different combinations of irrigation intervals (for TOU), or moving an irrigation to a different time to profit from demand peaks (for DR and TE).

Risk to Pumps and Demand Charges

This analysis did not look at the risk to pumps or the potential for demand charges that might be encountered by following DR or TOU programs.

Deep groundwater wells in particular can be very expensive, and thought to have a limited number of starts/stops before damage or failure. If the expected value of the well's damage (failure cost * failure probability) is higher than the incentives to participate, then participation is financially unwise. Even moderate or shallower wells can incur some damage from increased start/stops, and so can the pumps, unless they have soft start/stop capabilities (Olsen, et. al., 2015).

DR Programs are Complex and Poorly Aligned with Agriculture

Currently a suite of DR incentive programs are offered through the utilities and third-party aggregators. However, they are not well suited for adoption by irrigated agriculture for two reasons. First, they are complex to set up, manage and verify. And second, they are poorly aligned with actual farm practices.

Mechanisms through which a farm can be approved for DR, participate in DR events, and receive compensation are complex. Most farms lack the in-house expertise for going through the entire process without the help of external consultants or DR aggregators.

In addition to DR program complexities, DR programs are poorly aligned with actual farm practices. Stringent baseline energy use requirements put in place by the utilities deter, and even disincentivize, agricultural customers from participating. Specifically, the use of a 10 day baseline in order to determine incentive payments, based on the past 10 days of the the participant's energy use. If there is no energy used for the 10 days prior to the DR event, the DR incentives will be significantly reduced, and the participant will not be appropriately compensated for the load shed provided.

CONCLUSION

- TOU demand management programs offer the most profit to the irrigator, with the least amount of complexity; but calculating, planning and implementing full season TOU strategies comes with technical challenges, including irrigation system limitations and the ability to identify timing strategies that meet crop water requirements mid-season.
- DR program complexity and incentives are poorly aligned with the needs of irrigated agriculture, in some cases incentivizing growers to avoid providing loads even when they can, due to the 10 day baseline requirement.
- Demand management programs, in general, present technical challenges for the irrigator because many irrigation systems are designed to meet crop water demands at peak ET without regard for shutdowns.
- To engage with the dynamic, variable market signals RTP and TE make possible, irrigators participating in demand management need ways to precisely calculate how each irrigation will shape their soil moisture status so that they can (i) plan a season using the most cost effective energy timing schedule; and (ii) quickly evaluate how a change in their irrigation schedule presents a risk to the crop and the best way to rebalance their irrigation times to profit from market signals. This sort of crop water status monitoring and forecasting tool would benefit TOU and DR programs as well.
- Real time Pricing and Transactive Energy markets suggest even greater profit potential for irrigators, but demonstrations of TE markets working with irrigated agriculture are needed.

- TOU offers a no cost approach in reshaping load shapes. Historically, TOU has observed the highest level of participation by end users. However, TOU rates are not dynamic and they only change seasonally. Existing TOU rates, put in place by California utilities over a decade ago, define the hours of noon to 6pm, as peak periods. Those hours are now the same hours with highest level of solar generation, which is why utilities have proposed adjusting TOU periods to later in the day (4pm-9pm) to mitigate the steep ramps in demand during the evening hours.
- Dedicated tools, such as a decision support system (DSS), for monitoring the status of crop could enable irrigated agriculture to more profitably and effectively support demand management and
- ADR/DR may not be as profitable as TE/TOU
- The infrastructure needed for a full TE implementation will directly benefit DR.

NEXT STEPS

- Developing DSS to coordinate management calculations
- Evaluate the benefits of automated data / better field monitoring
- Evaluating incentives of TE for farm energy management

ACKNOWLEDGEMENTS

We would like to thank our cooperators who have helped us better understand the challenges faced by irrigated agriculture; the California Energy Commission for supporting our research and analyses; and the University of California Cooperative Extension for advice and assistance developing our model of crop growth cycles.

REFERENCES

Alstone; et al., 2017. Charting California's Demand Response Future. Final Report on Phase 2 Results 2025 California Demand Response Potential Study. Lawrence Berkeley National Laboratory. Berkeley, CA.

Olsen, et. al., 2015. Opportunities for Automated Demand Response in California Agricultural Irrigation. Lawrence Berkeley National Laboratory. Berkeley, CA. Also available at: http://eta-publications.lbl.gov/sites/default/files/opportunities_for_automated_demand_response_.pdf

Evaluating the Operation of Residential WIFI Based Irrigation Controllers

Charles Swanson, Extension Program Specialist II -Landscape Irrigation
Texas A&M Agrilife Extension Service
2117 TAMU, College Station, Texas 77843-2117

Dr. Guy Fipps, Professor & Extension Agricultural Engineer
Texas A&M Agrilife Extension Service
2117 TAMU, College Station, Texas 77843-2117

Abstract

Smart controllers for residential end users have been available for over 20 years. Originally these controllers either used on site climate sensors or communicated via signal service providers (such as pager technology) to determine local evapotranspiration or plant water requirements. As technology in general has evolved over the last 20 years, so have residential smart irrigation controllers. Many manufacturers now market WiFi based irrigation controllers than can incorporate data from a growing number of weather networks to manage local irrigation needs. In 2017 a study was begun to evaluate the effectiveness of six commercially available wifi based irrigation controllers. Controllers were programmed to irrigate the same type of landscapes and bench tested simultaneously from the summer through the fall and winter of 2017. This paper compares controller performance and hopes to guide end users to selecting a wifi based controller.

Background

A testing facility was developed in 2008 to evaluate the performance of smart irrigation controllers. Most of the controllers evaluated between 2008 and 2014 used an onsite weather sensor to calculate irrigation schedules or adjust a pre-programmed irrigation runtime. Since then, manufacturers have advanced their smart controller technologies to incorporate WiFi technologies that allow residential irrigation controllers to access an array of internet based weather networks. As many new residential home constructions now include wireless networking, more homeowners and contractors are adopting wireless internet based controllers because of their ability to access local weather data as well as enable remote access and management of the irrigation controller through smart devices such as cell phone, tablets and other Internet of Things (IoT) Devices.

As the number of residential WiFi based irrigation controller's increase in the marketplace, homeowners and contractors have a difficult decision of choosing a controller. To help address these concerns, in the summer of 2017, six commercially available WifI controllers were selected to be evaluated that were being marketed and installed in Texas. These controllers are identified in Table 1.

Table 1. Residential Wifi Controllers

Controller	Manufacturer	Model
A	Hunter	Hydrawise-HC
B	Rainbird	ESP-TM2 LNK Wifi
C	Rachio	Generation 2
D	Skydrop	Halo Controller
E	RainMachine	Touch HD
F	Orbit	B-HYVE

Evaluation Methodology

The six controllers were installed in a laboratory and connected datalogger (See Figure 1). Each controller was programmed for six different virtual landscape zones that varied in plant water requirements, soil type, root zone depth and sprinkler type (precipitation rate). The virtual landscape is defined in detail in Table 2. However all controllers varied in how they were programmed. Some controllers asked specific questions about the landscape during the setup process while other controllers required the user to program in a "peak" irrigation schedule. For this purpose, a peak irrigation schedule was calculated using historic ETo data for College Station, Texas. The details of the calculated peak irrigation schedule are shown in Table 3. As part of this evaluation, no rain shut off device was connected to the controllers to evaluate their ability to manage rainfall.

The datalogger recorded when each station turns on and off. Post testing analysis calculated irrigation runtimes and irrigation amounts were determined using the precipitation rate defined in the virtual landscape. Irrigation amounts over various time periods were compared to a daily calculated soil moisture balance using ETo data from the TexasET Network (http://TexasET.tamu.edu) for College Station, Texas. A comparison was also made to weekly irrigation watering requirements (7 day water balance).

Figure 1. Controller installation and Setup in the Lab

Table 2. The virtual landscapes as defined by Representative Texas Landscapes

	Station 1	Station 2	Station 3	Station 4	Station 5	Station 6
Plant Type	Flowers	Turf	Turf	Groundcover	Small Shrubs	Large Shrubs
Plant Coefficient (Kc)	.8	.6	.6	.5	.5	.3
Root Zone Depth (in)	3	4	4	6	12	20
Soil Type	Sand	Loam	Clay	Sand	Loam	Clay
MAD (%)	50	50	50	50	50	50
Adjustment Factor (Af)	1	.8	.6	.5	.7	.5
Precipitation Rate (in/hr)	.2	.85	1.4	.5	.35	1.25
Slope (%)	0-1	0-1	0-1	0-1	0-1	0-1

Table 3. Texas Landscapes, Peak Watering Schedule for College Station

	Station 1	Station 2	Station 3	Station 4	Station 5	Station 6
Plant Type	Flowers	Turf	Turf	Groundcover	Small Shrubs	Large Shrubs
Peak Month (July) ETo, in	7.10	7.10	7.10	7.10	7.10	7.10
Weekly ETc, in	1.28	.77	.58	.40	.56	.25
Plant Available Water, in	.23	.57	.70	.45	1.70	3.50
# Of Irrigation Days Per Week	7	2	2	2	1	1
Days	Every Day	Mon/Thurs	Mon/Thurs	Mon/Thurs	Mon	Mon
Total Runtime Per Day	54	28	12	24	96	12
Cycles Per Day	2	2	2	2	2	2
Runtime Per Cycle	27	24	6	12	48	6

Results

Figures 2 through Figure 5 show the amount of irrigation applied per controller for each station during the evaluation period. To simplify analysis, data was graphed seasonally for typical Summer, Fall and Spring periods common in College Station, Texas.

Additionally in Figures 6-10, the irrigation pattern of three zones was graphed to evaluate the seasonal performance of the controllers. The three zones selected represent significant differences in plant water requirements and required irrigation frequency. Controller D is not listed for this analysis as the datalogger recorded no operation of the controller between 6/12/18 and 8/24/18. It is uncertain at this time whether this was datalogger hardware related or a problem with the controller operation.

Figure 2. August –September 2017 Analysis

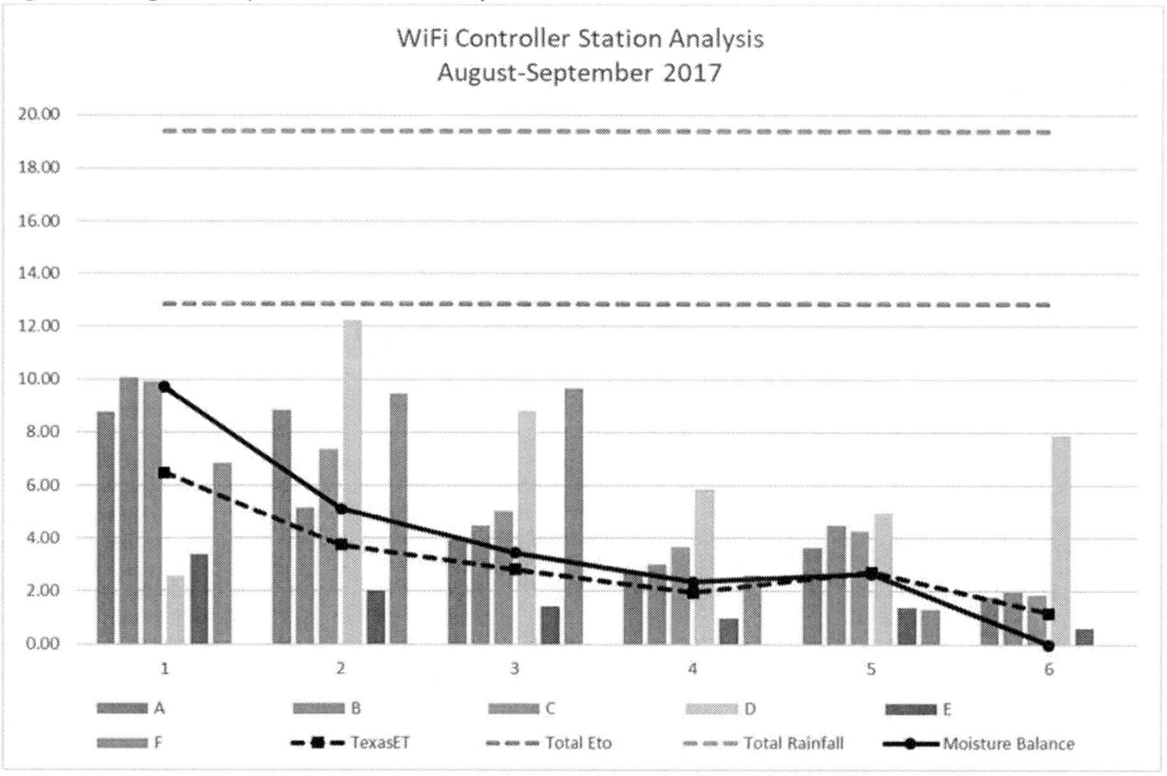

Figure 3. October – December 2017 Analysis

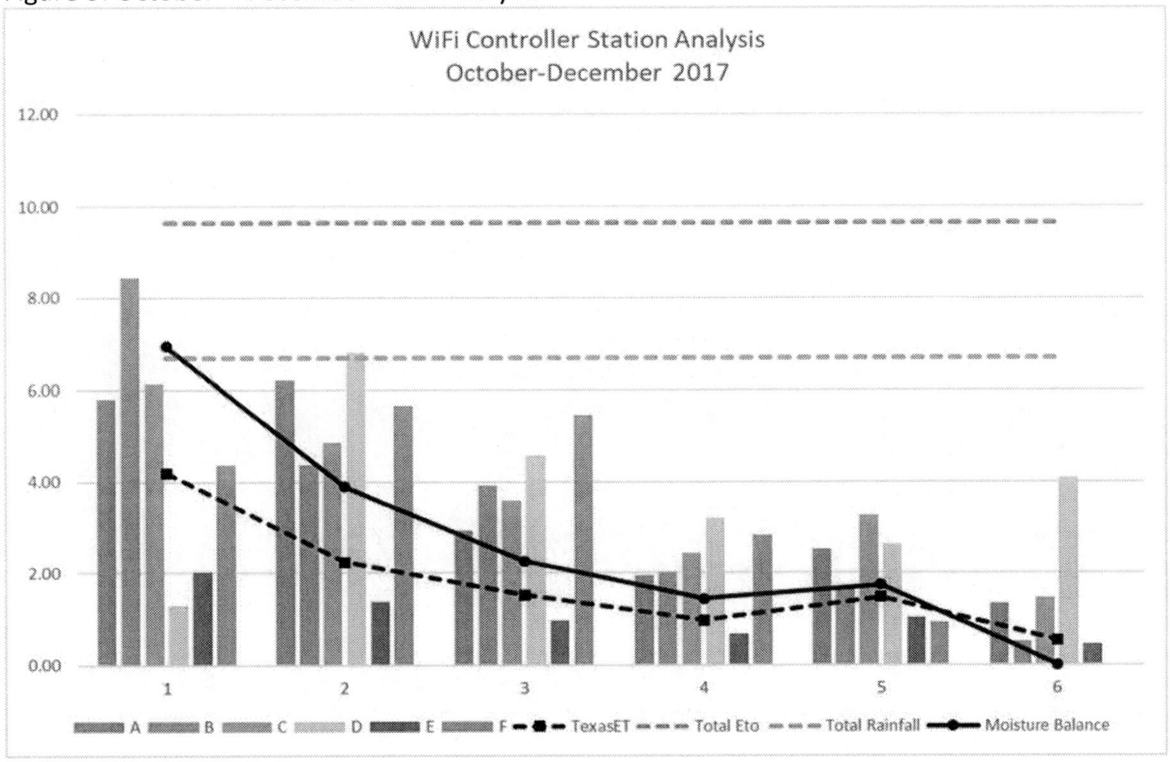

Figure 4. March- April 2018 Analysis

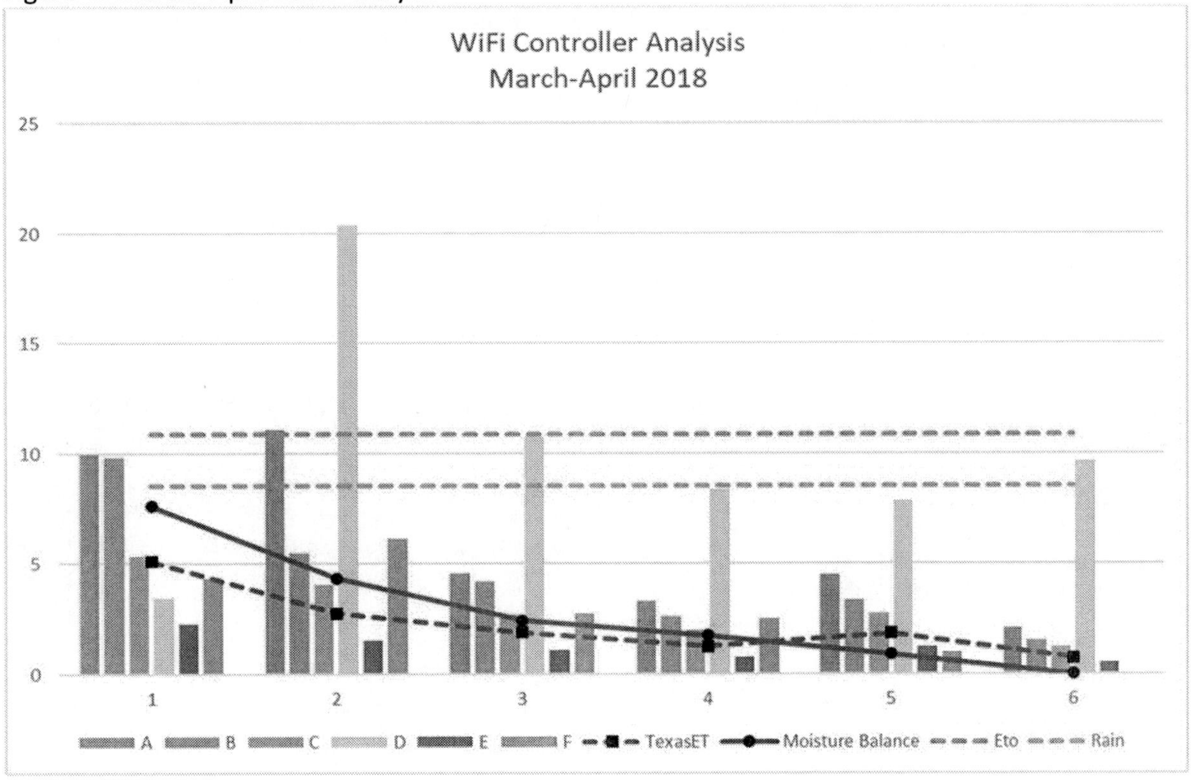

Figure 5. May- August 2018 Analysis

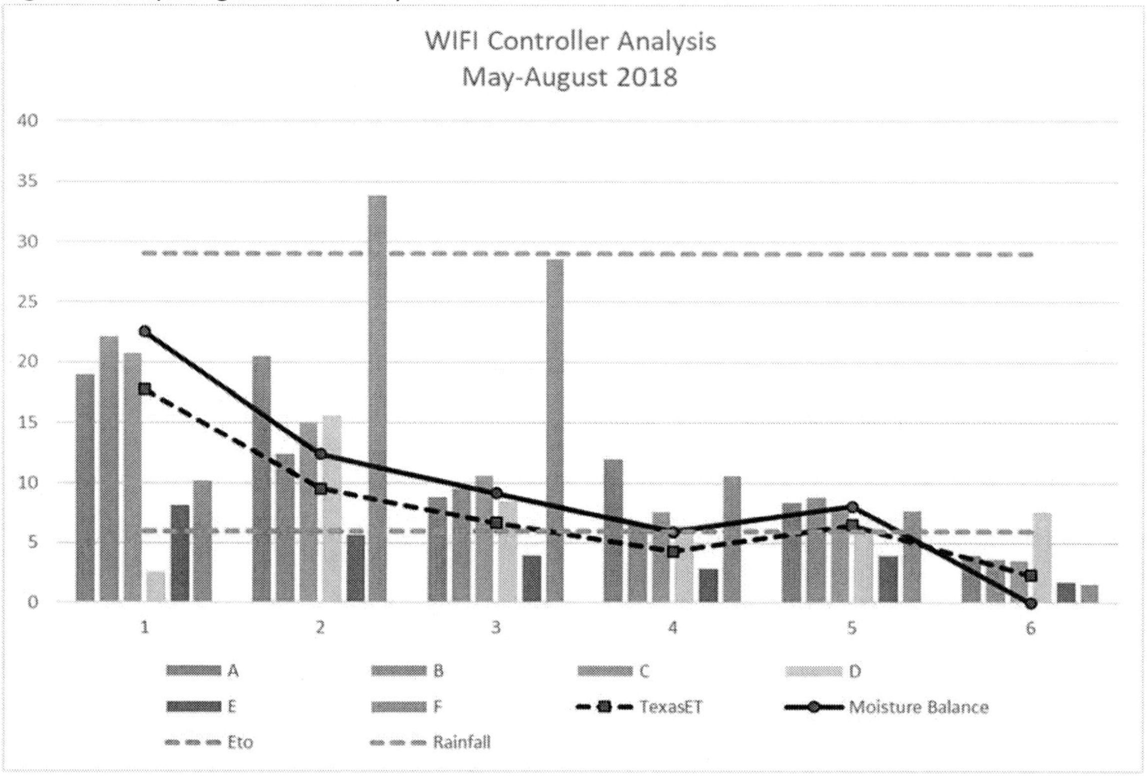

Figure 6. Controller A Irrigation Runtime and Frequency

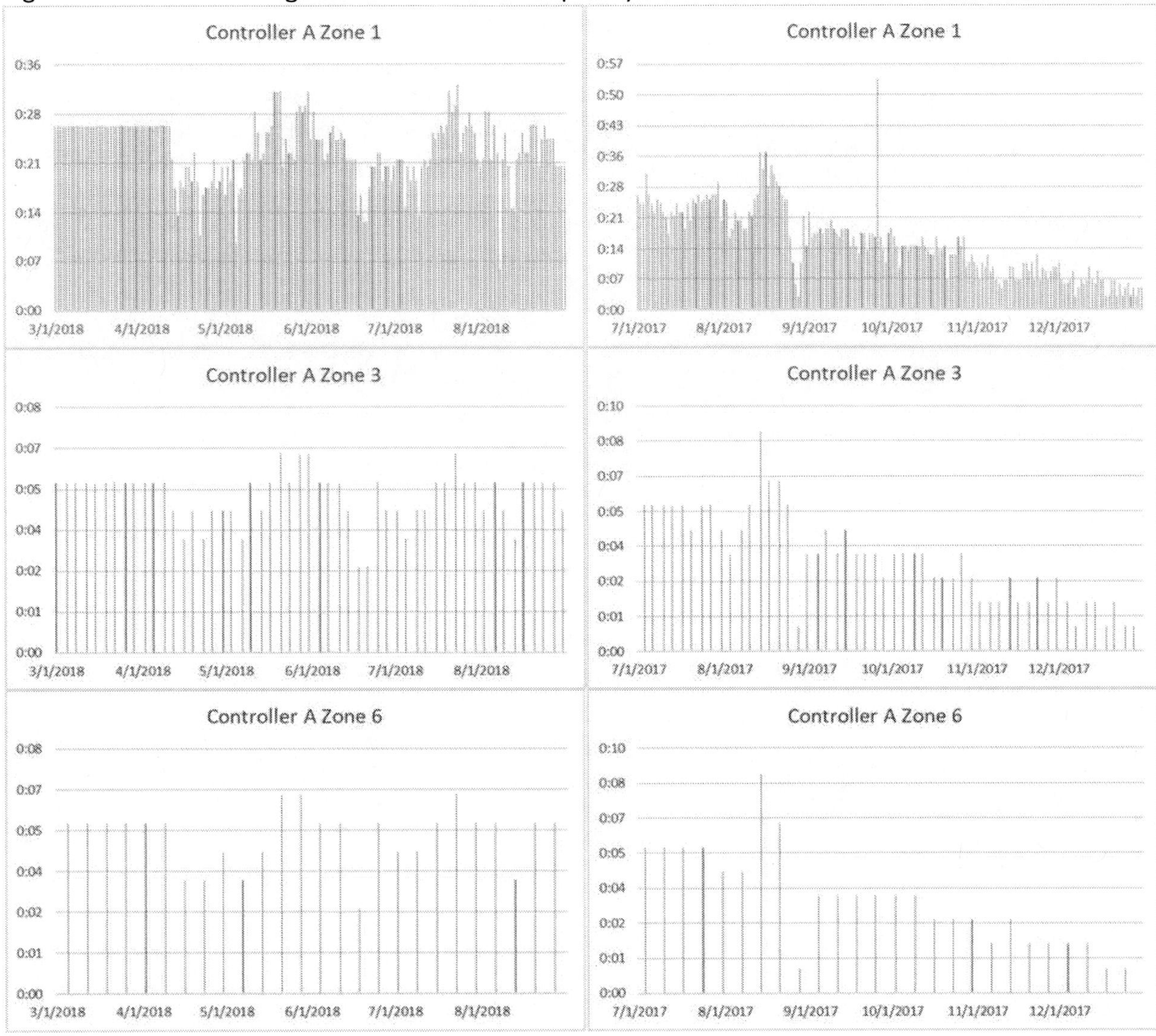

Figure 7. Controller B Irrigation Runtime and Frequency

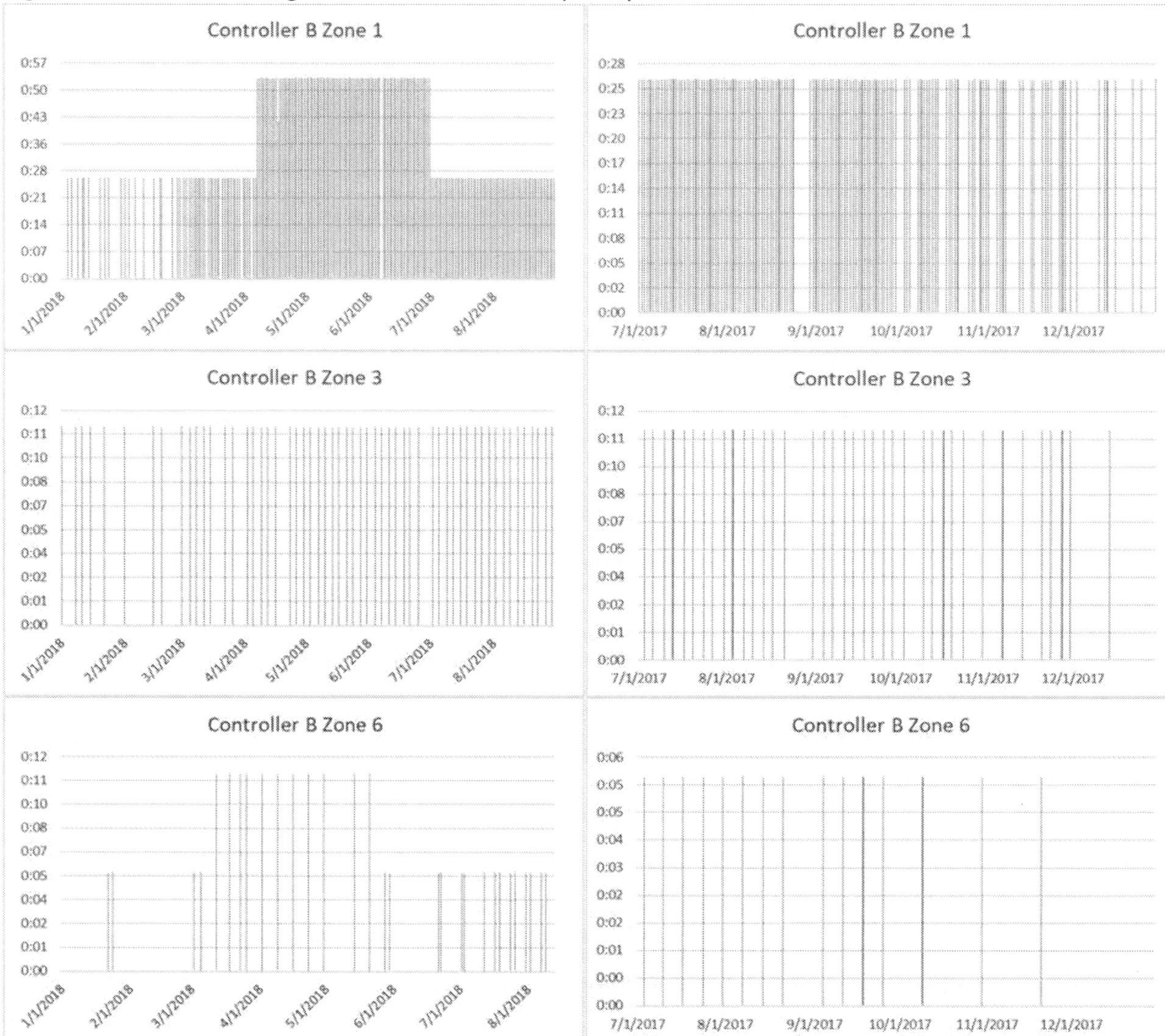

Figure 8. Controller C Irrigation Runtime and Frequency

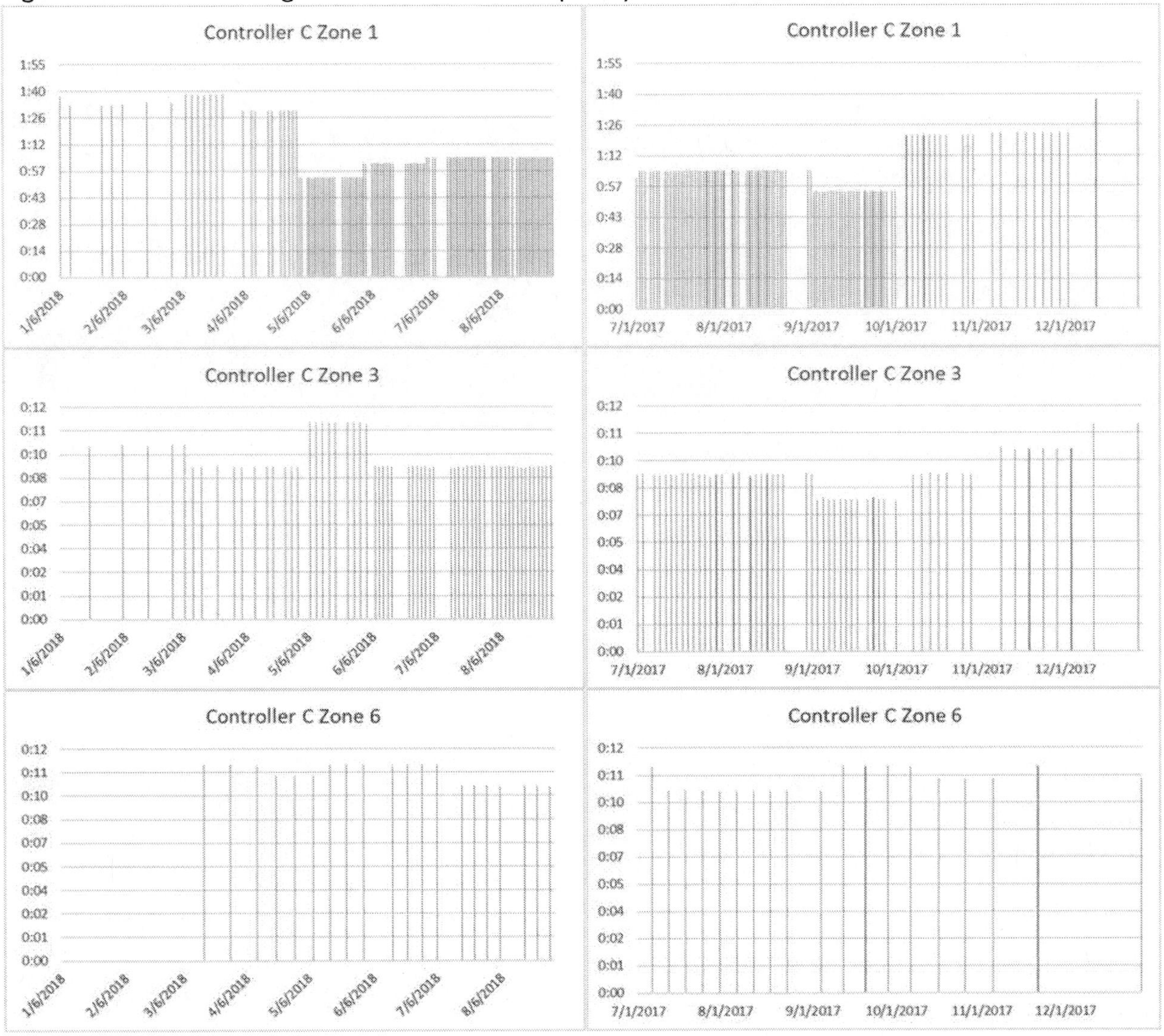

Figure 9. Controller E Irrigation Runtime and Frequency

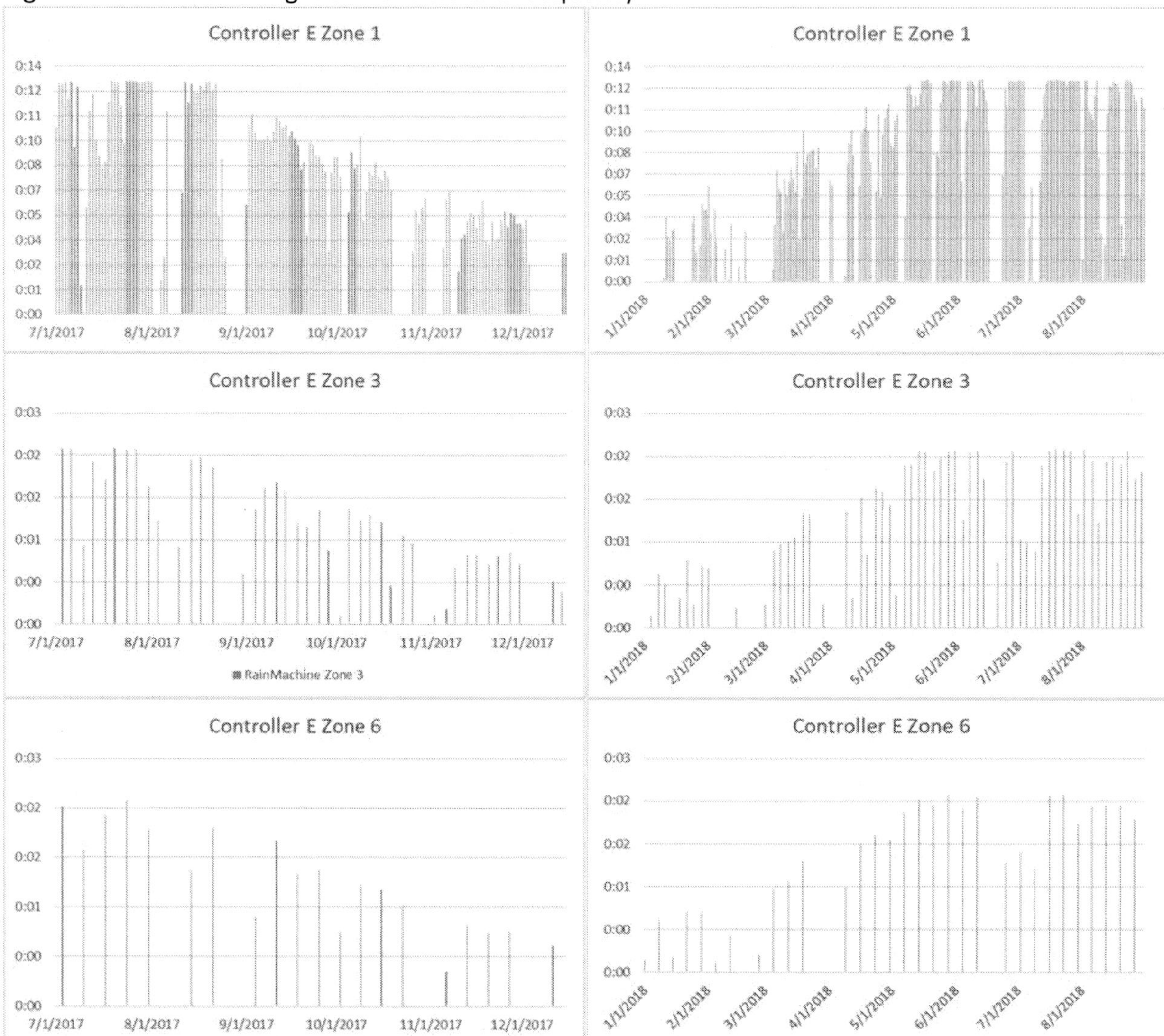

Figure 10. Controller F Irrigation Runtime and Frequency

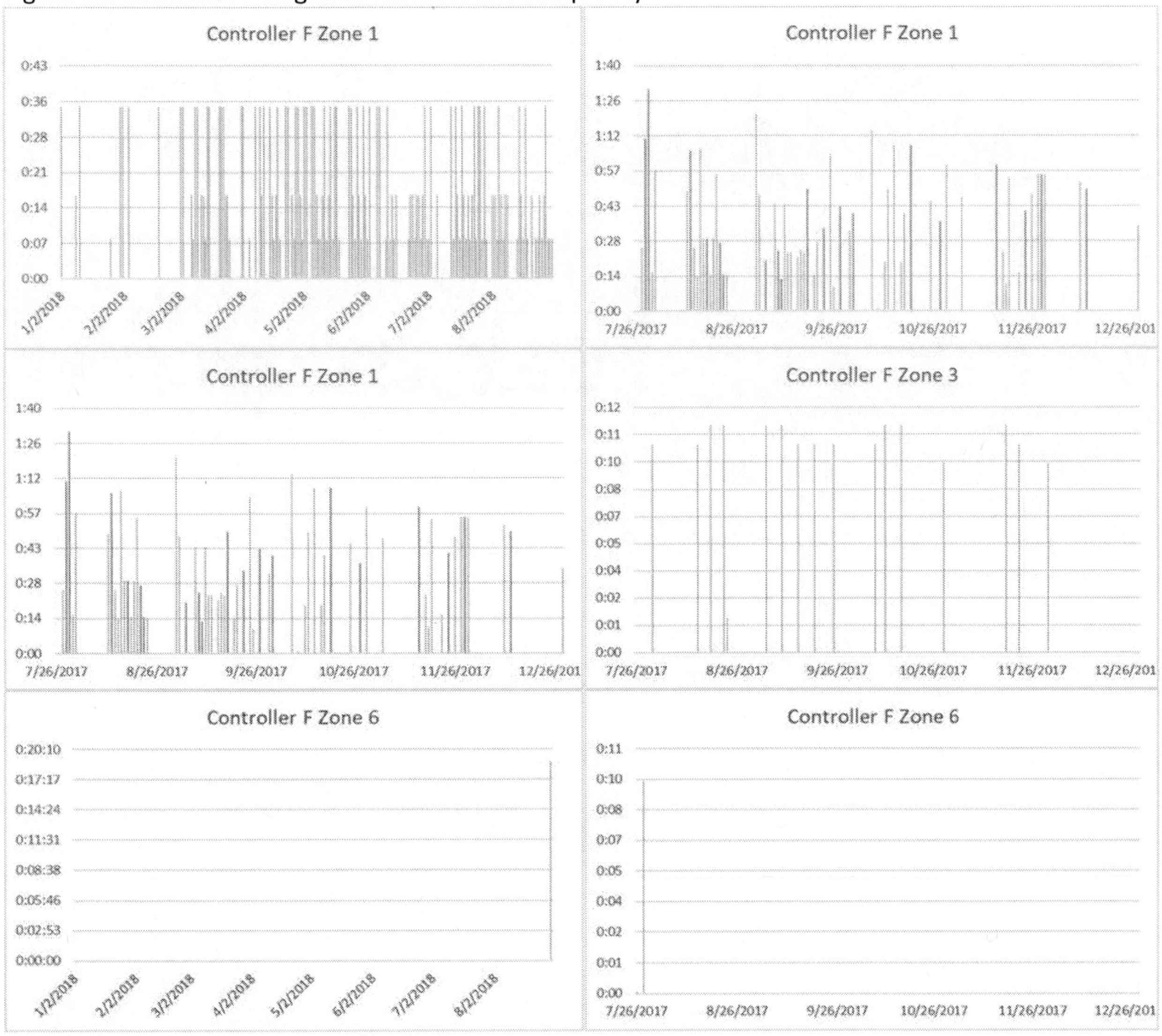

Summary

Performance of all controllers varied throughout the evaluation period. Based on the first year of performance, most WiFi based irrigation controllers showed that they can do an effective job of managing the seasonal water requirements of landscapes compared to a fixed conventional time based irrigation controllers. However, some controllers did irrigated much more or less than was determined based on the daily soil moisture balance. Analysis shows some controllers change the irrigation runtime as water requirements vary but that some controllers only adjust the frequency of irrigation as seasonal water requirements change. Controllers also showed a positive response to rainfall events even though no rain shut off device was connected to the controllers. Future continued evaluation will help determine the water conservation potential of WiFi based residential irrigation controllers. Additionally, future evaluation with the inclusion of a rain sensor could also help prevent controllers from irrigating any excessive amounts.

Median Retrofit Projects

Malarie Gotcher
Water Conservation Manager
City of Oklahoma City

Squeeze Every Drop.com

Why Medians?

- Highly visible across the city and large amounts of runoff and overspray onto the street.
- Typical for residents to blame the city for water waste
- Great opportunity to demonstrate water savings to contractors and for future median irrigation installations

2

Partners

OKC Beautiful

Local Keep America Beautiful Affiliate non-profit with a mission to enhance the appearance of Oklahoma City through education, programs, and community engagement.

Urban Lawn and Landscape

Oklahoma City irrigation contractor who maintains many of the medians. Provided labor to install.

Pressure-regulated spray head retrofit projects

- Three medians in Oklahoma City
  - Cooper Median (Median 1)
  - Gaylord Median (Median 2)
  - Newmark Median (Median 3)
- Sprays on each median were retrofitted with pressure-regulated versions
- "Real World" scenario

4

Median 1 - Pressure

Completed in January 2016

- Tall fescue, trees and shrubs
- Six zones
  - 169 heads
  - ~7,400 ft^2

Pre-install zone pressures

Zone	Pressure (PSI)
1	60
2	58
3	65
4	60
5	60
6	60
Average	60.5

Before installation - Median 1

Median 1 – Water Consumption

Completed in January 2016

Measured pressure in February 2016, all zones were operating at 30 psi.

Year	Total Water Consumption (thousand gallons)	Rainfall Total (inches)
2013	286	50.5
2014	536	31.9
2015	255	51.8
2016	187	29.7
2017	193	51.1
Average	291	43

Median 1 – Water Consumption

Completed in January 2016

Year	Total Water Consumption (thousand gallons)	Rainfall Total (inches)
2013	286	50.5
2014	536	31.9
2015	255	51.8
2016	187	29.7
2017	193	51.1
Average	291	43

Water Use
Completed in January 2016

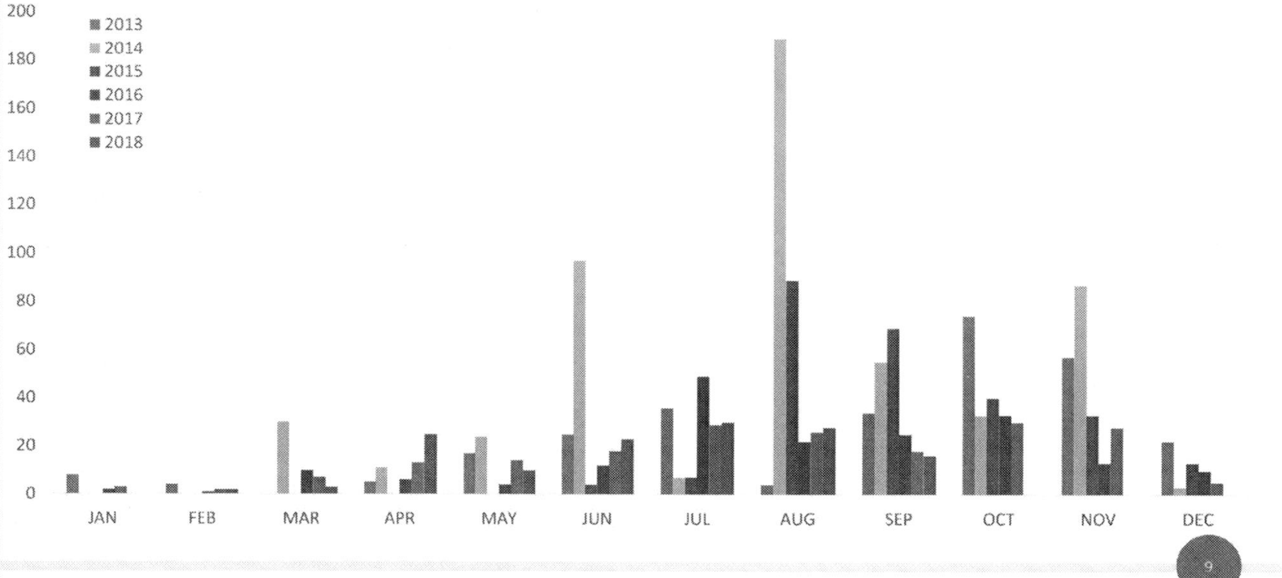

October 17, 2018 – Follow-up

Median 2 - Pressure

Completed in August 2016

- Nine zones (171 heads, ~11,107ft^2)
- Tall fescue, shrubs, and trees

Zone	Pressure (PSI)
1	60
2	60
3	60
4	60
5	60
6	60
7	51
8	60
9	62
Average	59

Median 2 – Water Consumption

Completed in August 2016

Year	Total Water Consumption (thousand gallons)	Rainfall Total (inches)
2013	167	50.5
2014	547	31.9
2015	390	51.8
2016	321	29.7
2017	257	51.1
Average	336	43

Median 2

October 17, 2018, Follow-up

Water Use Median 2
Completed in August 2016

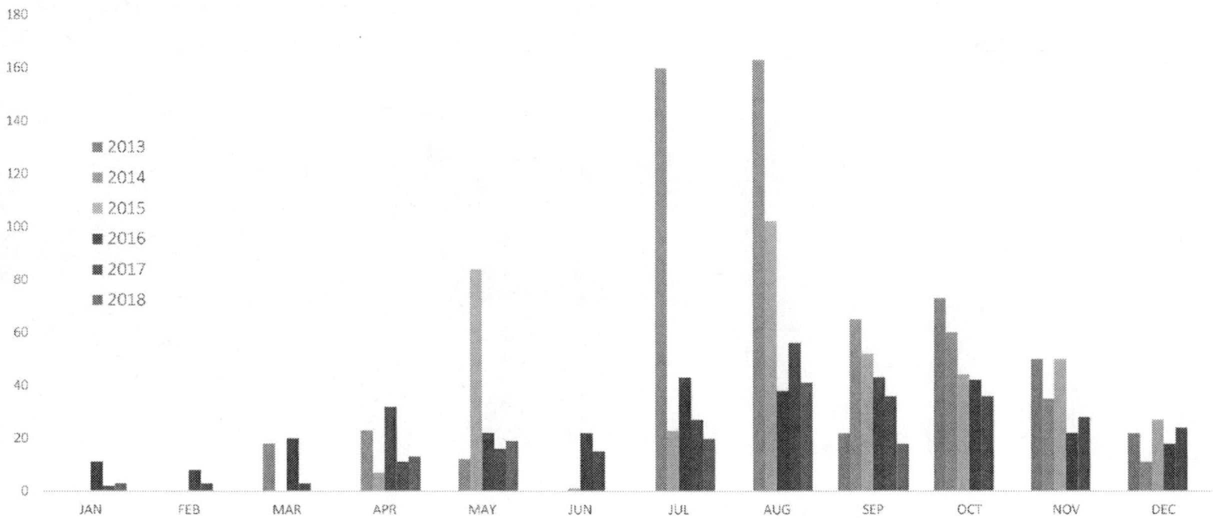

15

Median 3 - Pressure

Completed in July 2017

Pre- and Post Audit Completed

- Average DU improved from 0.45 to 0.69

Seven zones (180 heads, ~32,264ft^2)

Zone	Pressure (PSI)
1	55
2	55
3	70
4	70
5	33
6	55
7	63
Average	57

Median 3 – Water Consumption

Completed in July 2017

Year	Total Water Consumption (thousand gallons)	Rainfall Total (inches)
2013	-	50.5
2014	547	31.9
2015	390	51.8
2016	321	29.7
2017	257	51.1
Average		43

Water Use – Median 3
Completed in July 2017

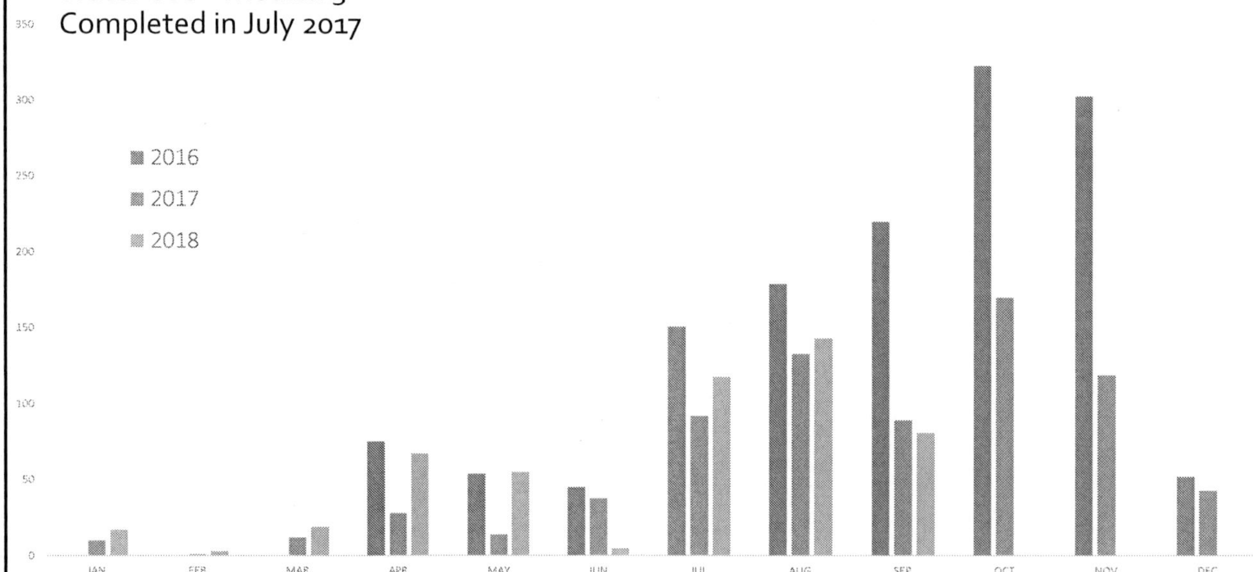

18

Median 3

Sprinkler Zone*		Rotary DU_{lq}	Spray DU_{lq}	Precipitation Rate (in/hr)	Pressure (psi)
1&2	Pre-Install	0.44	-	0.59	55
3&4		0.61	-	0.67	70
5		-	0.26	2.20	33
6		-	0.37	0.68	55
7		-	0.59	2.00	63
1&2	Post-Install	0.70	-	0.59	65
3&4		0.67	-	0.48	65
5-7		0.71	-	0.44	40

*Zones 1 and 2, 3 and 4 were overlapping zones that supply to one area

Zones 5 through 7 were combined as zone 5 because of the lower-flow

What did we learn?

- Need more information
  - Pre-Audit
  - Post-Audit
  - Flow test at the meter
- Regular maintenance is key to ensure water savings past the installation

Thank you
Malarie Gotcher
Water Conservation Manager
City of Oklahoma City
Utilities Department
Malarie.Gotcher@okc.gov
405-297-3380
21

Actual rain sensor dry out times
compared to
estimated soil dry out times
Bernardo Cárdenas and Michael D. Dukes
Agricultural and Biological Engineering Dept.,
University of Florida
Irrigation Show and Education
Conference
Long Beach, CA
December 4, 2018
UF UNIVERSITY of FLORIDA

UF UNIVERSITY of FLORIDA

Timer

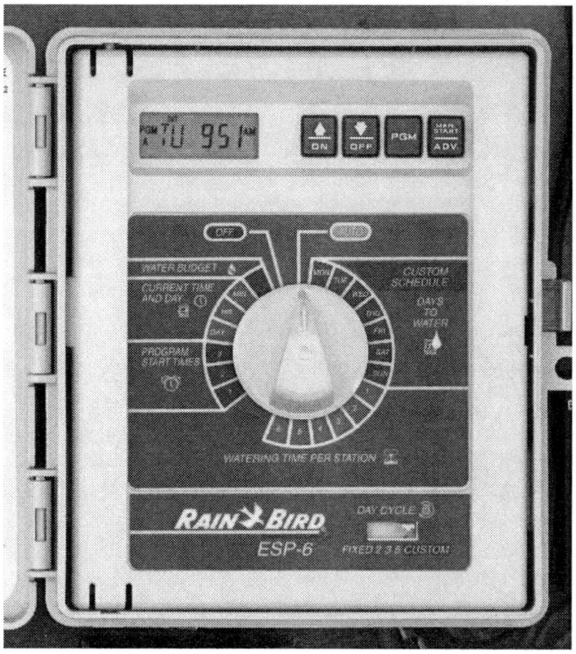

Rain Sensor (RS)

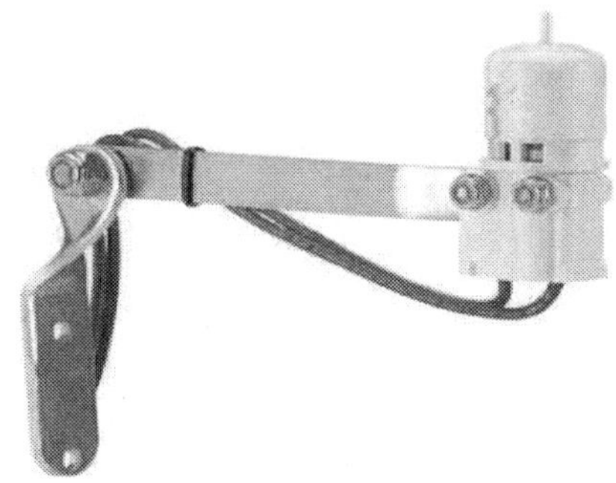

- **RS can bypass the timer settings, when there has been sufficient rain.**

UF | UNIVERSITY of FLORIDA

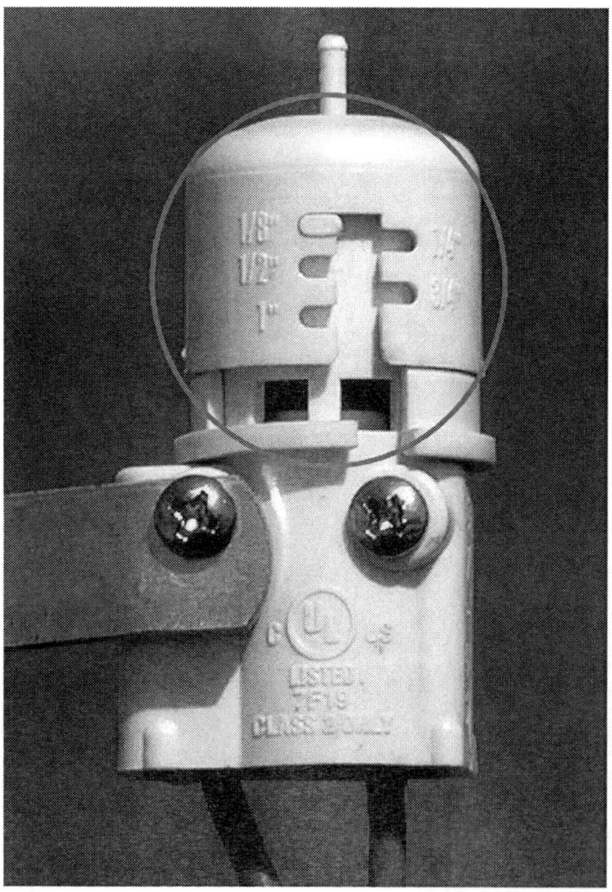

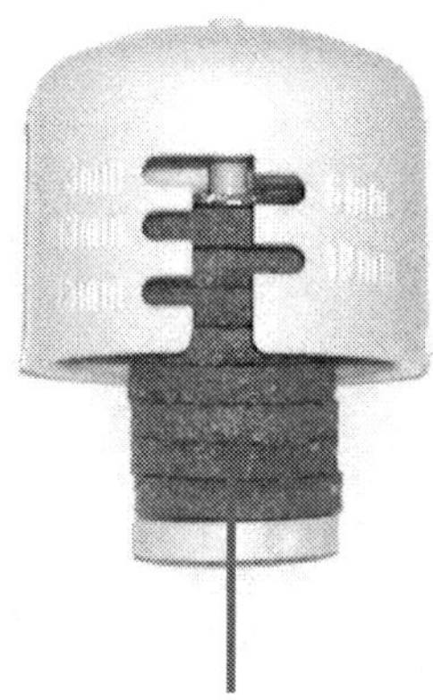

Expanding Disks

UF | UNIVERSITY of FLORIDA

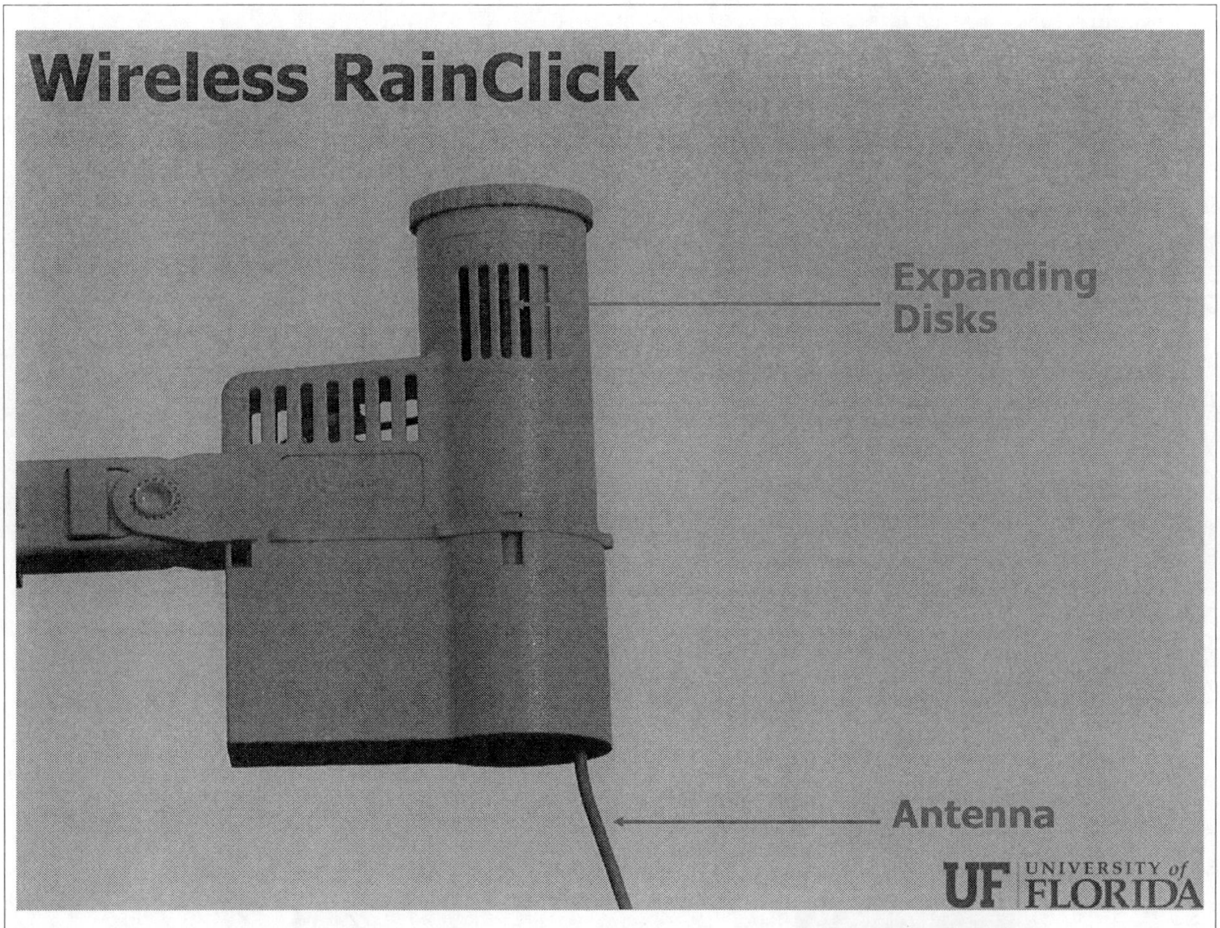
Wireless RainClick
Expanding Disks
Antenna
UF UNIVERSITY of FLORIDA

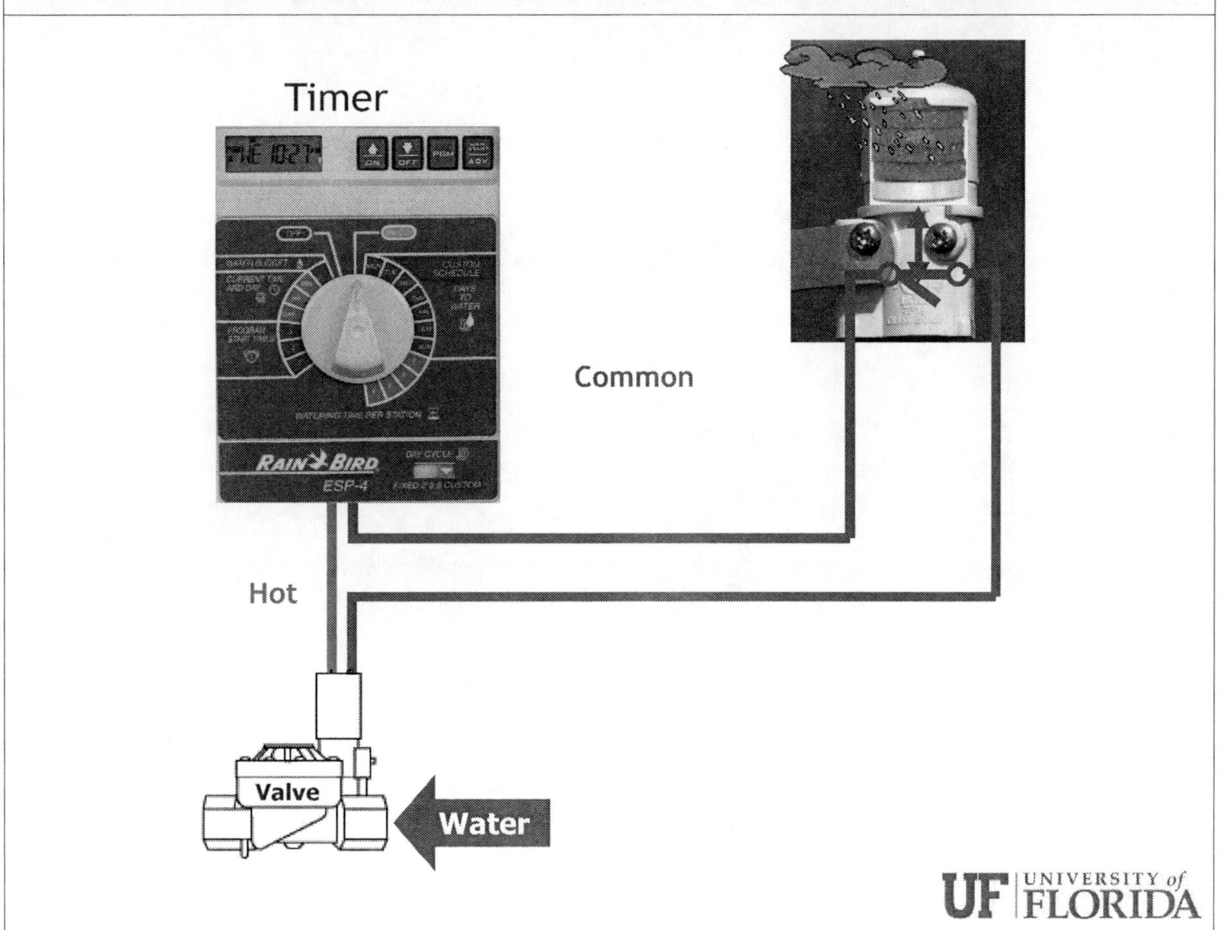
Timer
Common
Hot
Valve
Water
RAIN BIRD
ESP-4
UF UNIVERSITY of FLORIDA

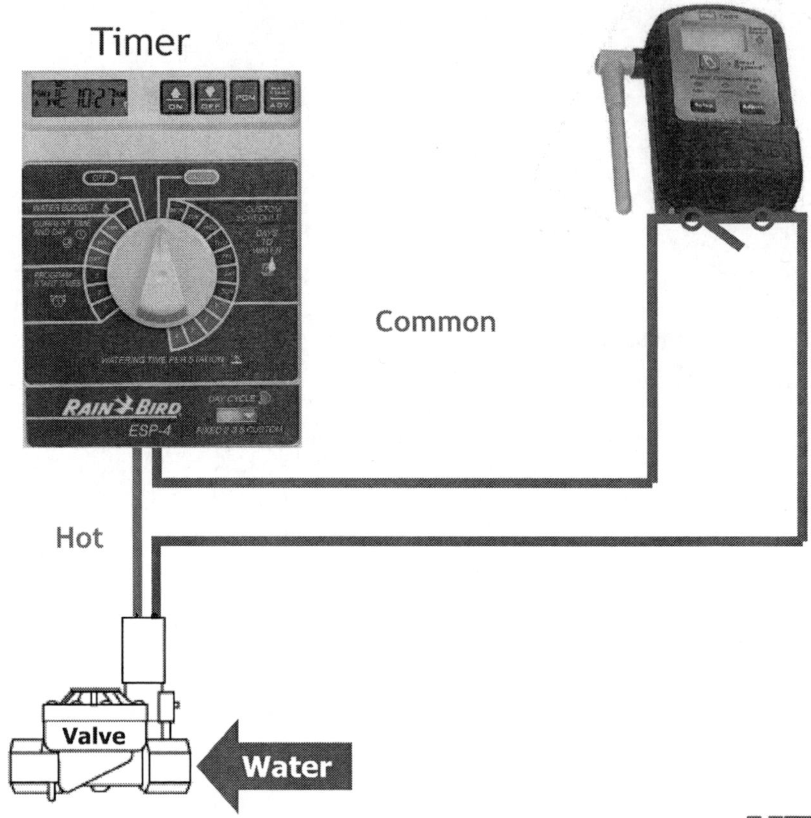

UF | UNIVERSITY of FLORIDA

Question

Do the RS dry out periods match the soil dry out periods?

UF | UNIVERSITY of FLORIDA

OBJECTIVES

a) Determine the dry out periods of 2 RSs

b) Estimate the dry out periods of 3 soil textures through a soil water balance model

c) Compare a) vs b)

UF | UNIVERSITY of FLORIDA

UF | UNIVERSITY of FLORIDA

MATERIALS AND METHODS

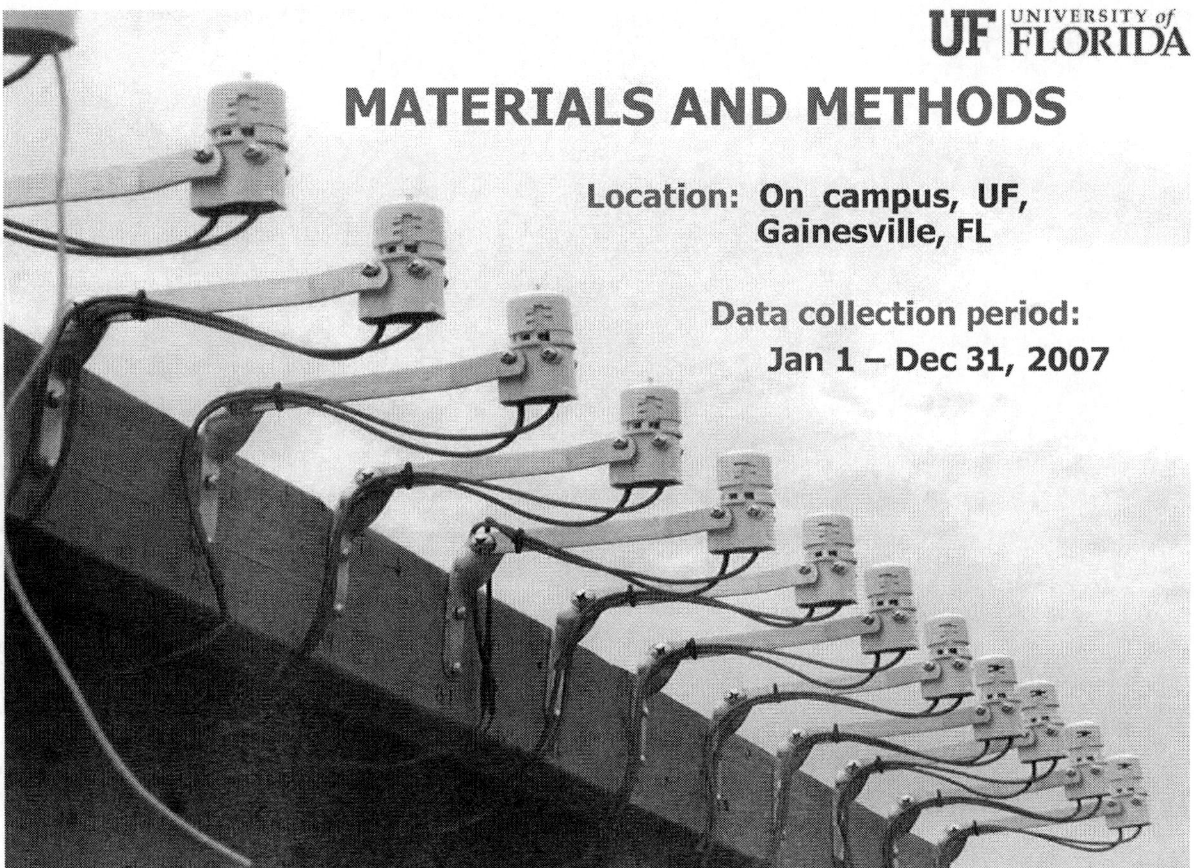

Location: On campus, UF, Gainesville, FL

Data collection period: Jan 1 – Dec 31, 2007

UF UNIVERSITY of FLORIDA
Hunter RainClik

UF UNIVERSITY of FLORIDA
Toro TWRS
MUX

Weather Station

Tipping bucket (0.25 mm)

Data logger

(m/d h:m:s)

UF UNIVERSITY of FLORIDA

Weather Station

Weather data

↓

Software Ref ET

UF UNIVERSITY of FLORIDA

WAVE to simulate hourly soil water balance

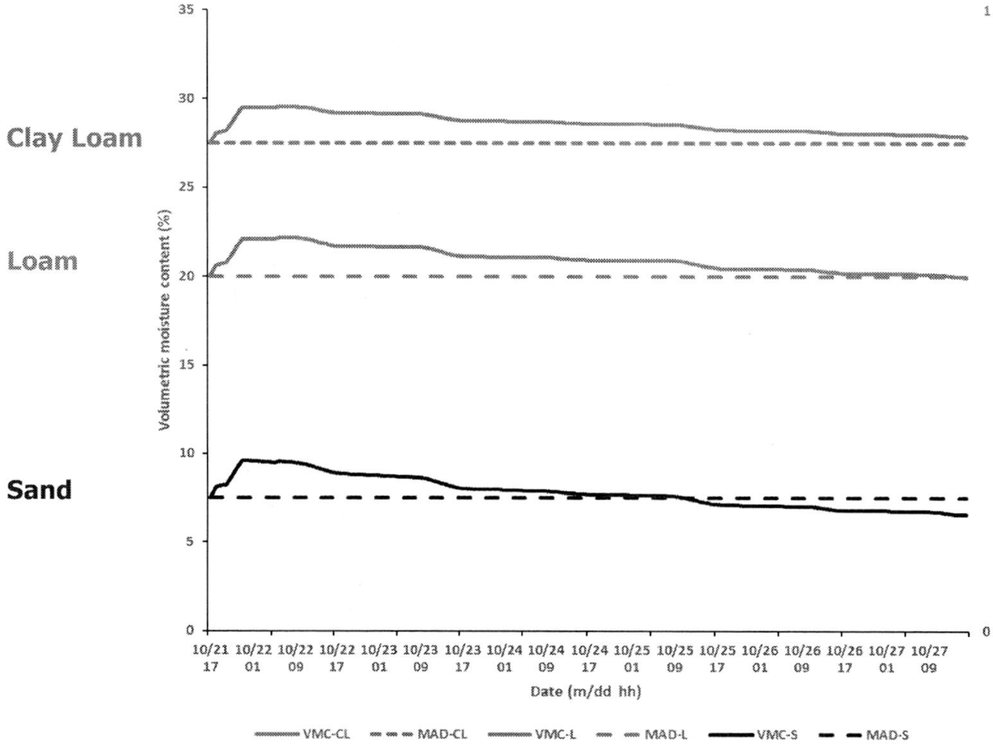

UF UNIVERSITY of FLORIDA

WAVE to simulate hourly soil water balance

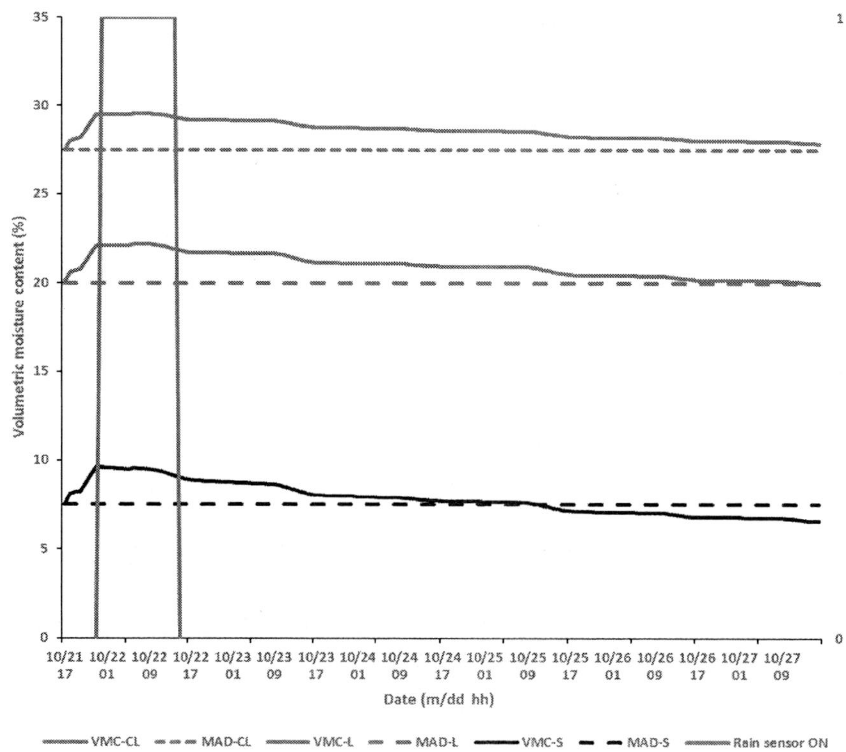

UF UNIVERSITY of FLORIDA

RESULTS AND DISCUSSION

UF UNIVERSITY of FLORIDA

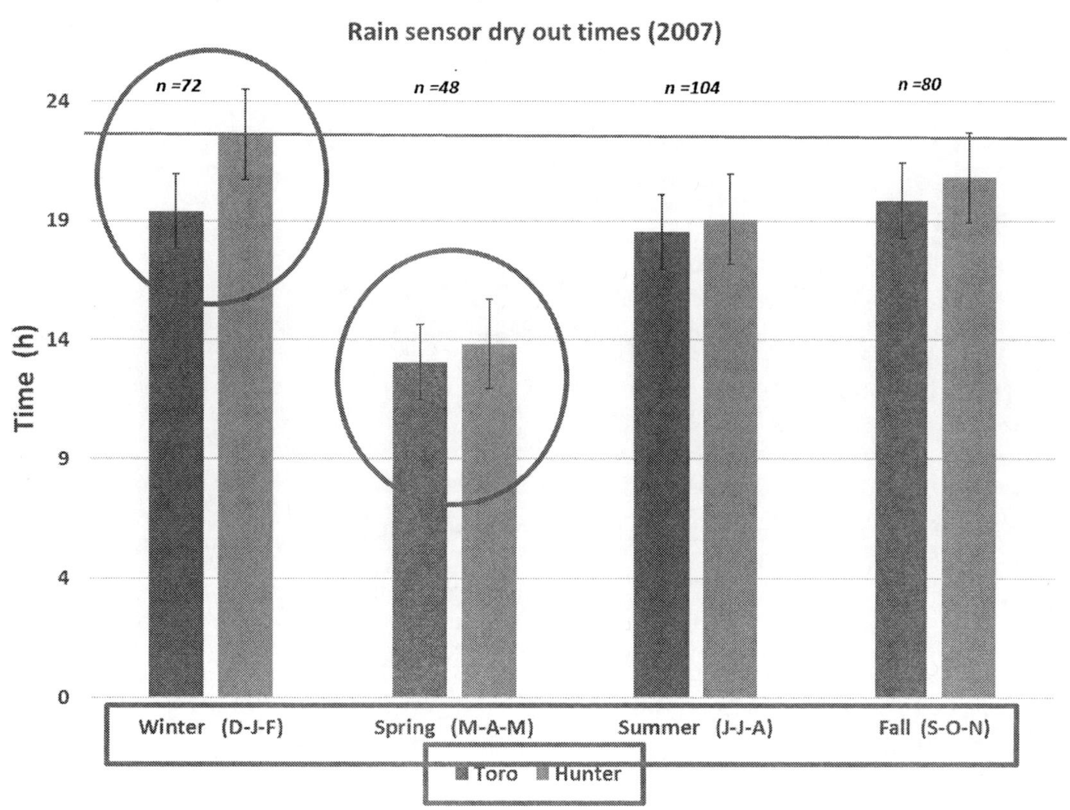

Standard error bars

UF UNIVERSITY of FLORIDA

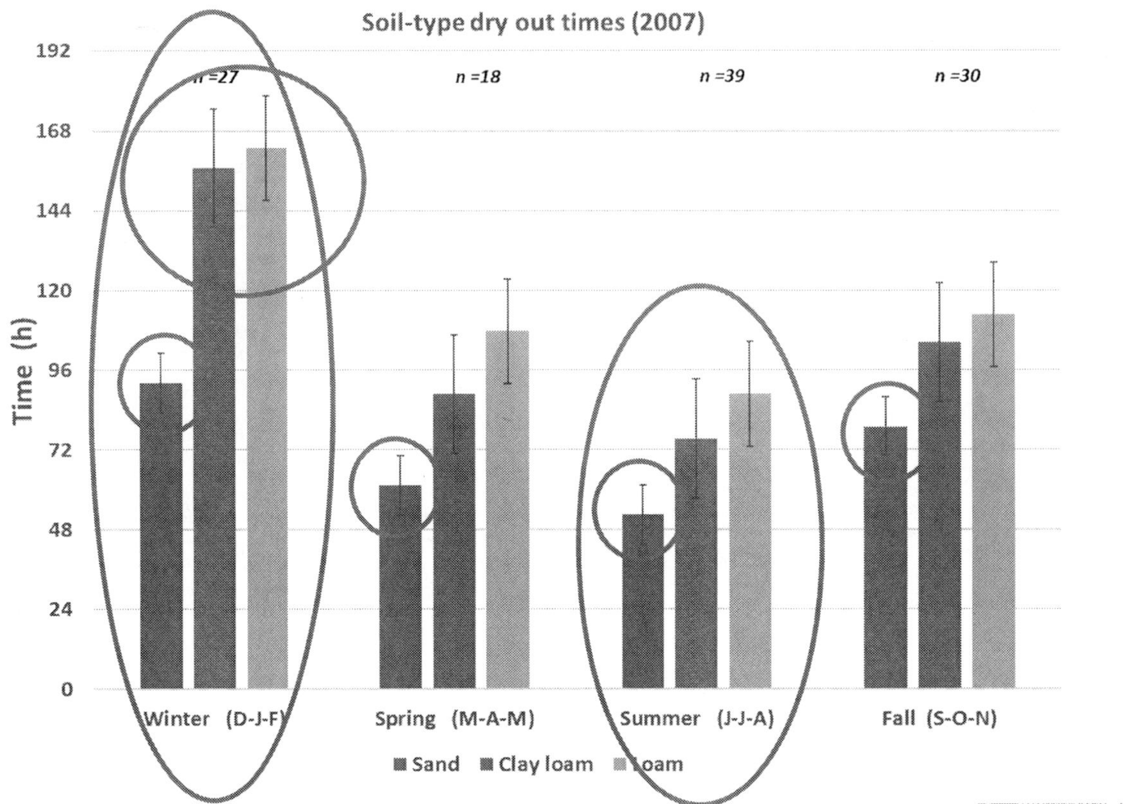

Standard error bars

UF UNIVERSITY of FLORIDA

Time (hh:mm)	Rain (mm)
18:00	1.5
19:00	0.3
20:00	0.3
21:00	1.3
22:00	1.3
23:00	1.3
Total	**5.8**

No runoff

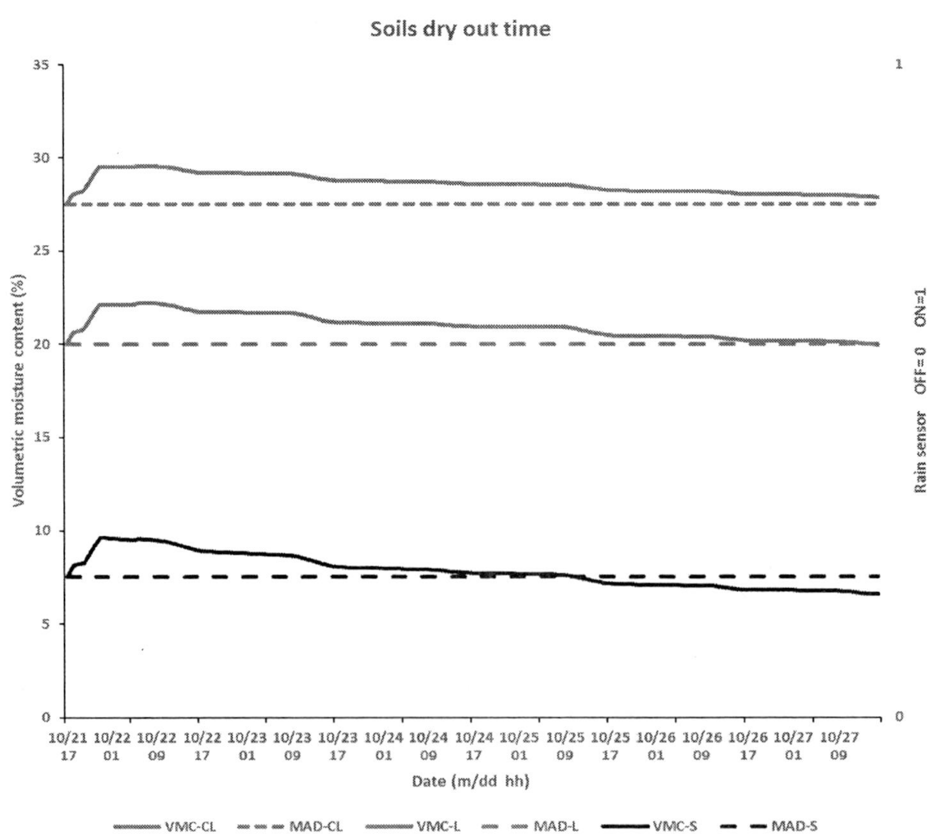

UF UNIVERSITY of FLORIDA

Time (hh:mm)	Rain (mm)
11:00	14.2
12:00	0.3
13:00	
14:00	7.9
15:00	
16:00	
17:00	14.7
Total	**37.1**

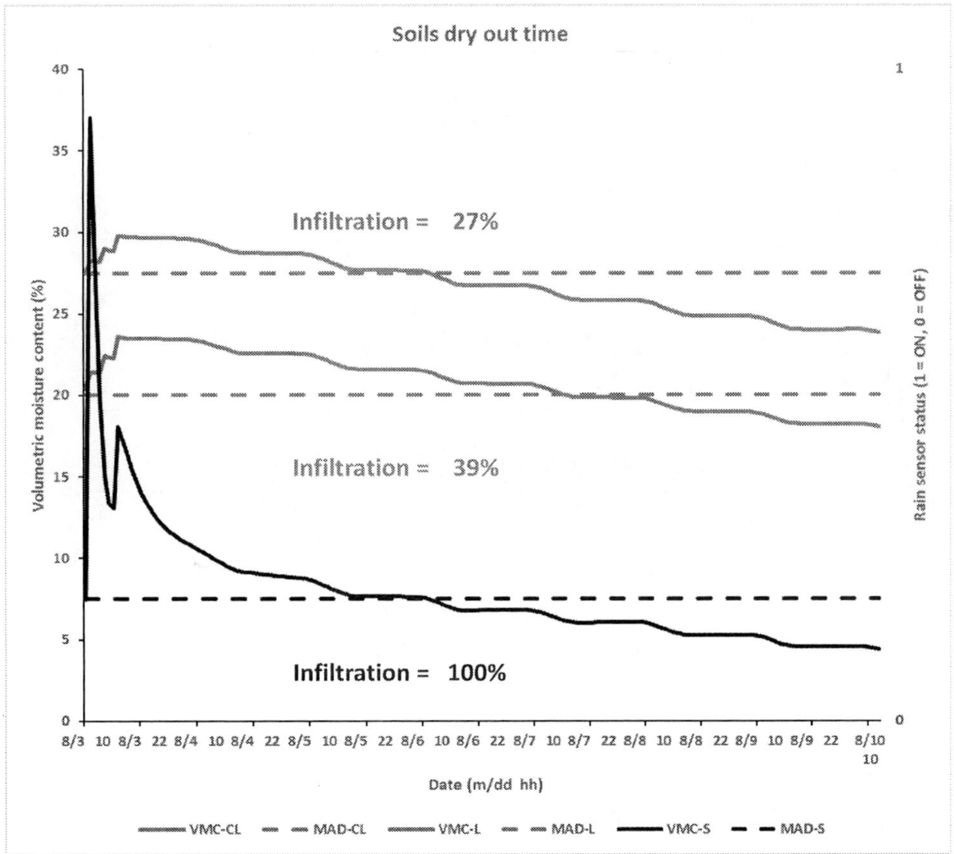

UF UNIVERSITY of FLORIDA

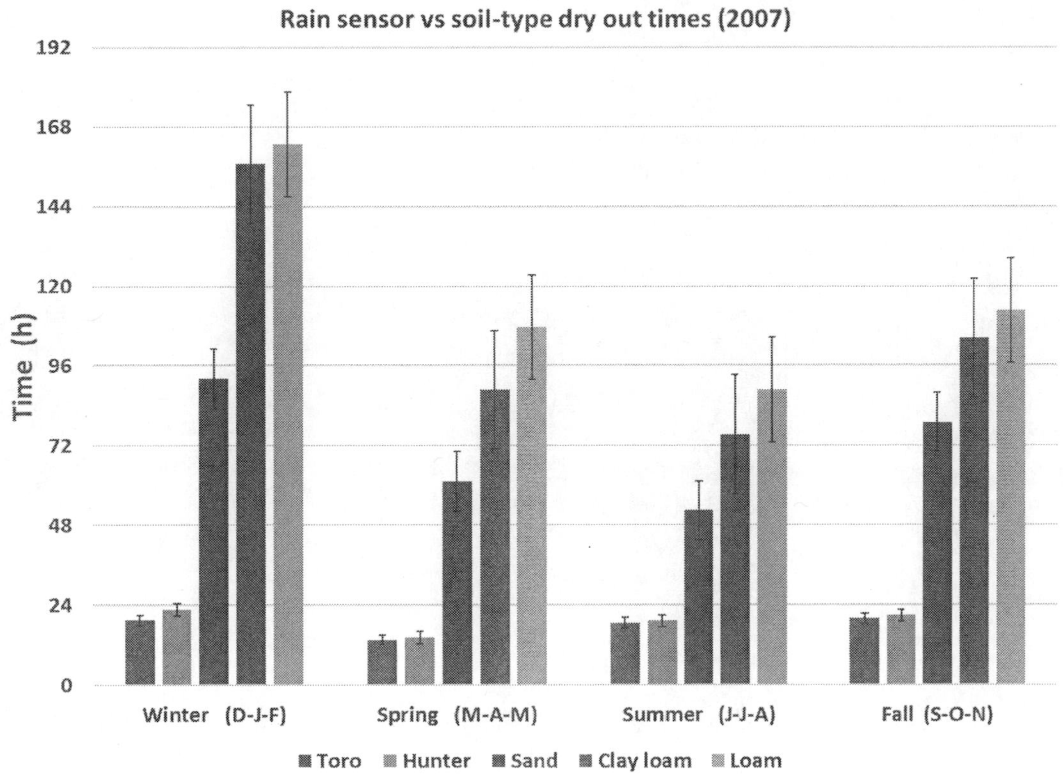

Standard error bars

UF UNIVERSITY of FLORIDA

CONCLUSIONS

- **The dry out periods of the tested RSs were shorter than those of the different soil types modeled**

- **RSs: limited usefulness under Florida conditions**

UF | UNIVERSITY of FLORIDA

Estimating Annual Water Demands from Irrigation Flow Rates

Kent Sovocool, Senior Conservation Research Analyst, MAS, CLIA

Southern Nevada Water Authority, P.O. Box 99956, Las Vegas, NV 89193-9956

kent.sovocool@snwa.com

Abstract.

Landscape design is widely recognized as influencing sites' water demands. While continually developing, "green" efficiency programs, standards and codes now often include various design criteria for new landscapes with a goal of promoting water efficiency. These are typically one of two types, water budgeting-based or founded on restrictions on plantings. Both of these have pros, but also significant cons, and are not easily worked into a framework that relies on auditors with limited site visit time and lack of expertise in evaluating plantings. Ideally a metric that integrates aspects of landscape design in an easy, robust, repeatable manner that is linked to water use is needed.

In some startling findings, the Southern Nevada Water Authority (SNWA) demonstrates that for single-family home sites, all of these considerations can be considered summarized within a single metric that expresses and, even predicts within a range, likely annual irrigation use. Totalizing flow rates of all stations and normalizing this and comparing this to submeter usage reveals this relationship in the Southwest.

Even more surprising, the relationship holds nationally when data from the Residential End Uses of Water Version 2 is examined. For a recent specification, RESNET's HERS_{H2O} water efficiency rating, *this has been developed and introduced as an optional method for demonstrating outdoor efficient design and is a points opportunity in that standard. For this purpose, the measure is area normalized and defined as the Residential Irrigation Capacity Index (RICI) where each change in RICI equates to a 10% change in outdoor use relative to a baseline condition.*

In this presentation, the author explains the development of this new metric and its advantages in rating the water efficiency of new homes' landscape designs and irrigation systemsfrom a water conservation perspective.

Keywords.

Turf/Landscape (Residential), Water Manager, Water Provider, Water Conservation, Design, Auditing, Standards, Water Budget, RICI

Introduction.

There are, at present, several organizations working on "green" (high-efficiency / better for the environment / healthier) standards, specifications, and code language for new buildings and development. One of these is the Residential Energy Services Network (RESNET), a 501 (C)(3) membership corporation that makes standards relating to building efficiency and associated rating standards. It is probably best known for its core Home Energy Rating Standards (HERS). RESNET is in the process of developing a water component to its ratings in partnership with the International Code Council (ICC). This was known as the Water Efficiency Rating Index (WERI) though recently RESNET has named it $HERS_{H2O}$ which serves to emphasize that the index is designed to build on and work in concert with the HERS index system (RESNET News, 2018). It also helped to practically differentiate it from another similarly named standard that has been developed. In addition to RESNET's partner, the ICC, major contributors to the effort include the Natural Resources Defense Council (NRDC), EPA's WaterSense Program, and KB Homes. While the focus of this manuscript is on the Residential Irrigation Capacity Index (RICI), and RICI can standalone as a distinct topic from $HERS_{H2O}$, there is no denying that the index nature of the rating provided the constraints and the necessity for RICI's invention which was also reliant on availability and analyses of large sets of field data by others, including the author.

The need for a new way of estimating future irrigation demand when auditing a new home.

First, an understanding of part of $HERS_{H2O}$ is required. $HERS_{H2O}$ is a true mathematical index with a "starting point" of 100, which is representative of a reference home's 2006 water usage, where the reference home is a theoretical similar construct. Decreases in water use are directly represented by the score. For example, a home with a $HERS_{H2O}$ score of 60 uses 40% less water that the 2006 reference home. It should be noted that the reference home and the rated home have similar assumptions (i.e. for example, a 2500 square foot home on a 4200 square foot results in the reference home having the same considerations). The $HERS_{H2O}$ model in entirety is sophisticated relative to the more normative manner typically used in efficiency standards and specifications that are intended to improve water efficiency in new buildings. While an in-depth examination of the functions of the model is beyond the scope of this manuscript, looking at some excerpts is necessary to understand how and why the Residential Irrigation Capacity Index (RICI) was developed.

Here are excerpts of Sections 3.0 and 5.0 of the RESNET $HERS_{H2O}$ Technical Guidelines (RESNET, 2018) pertaining to calculations of outdoor use:

3.0 Determining the Outdoor Reference Home Annual Water Use (in thousands of gallons per year). The reference home outdoor annual water use shall be calculated using the following two equations

If the rated home has a netET of less than 12 inches/year ***OR*** *the rated home has an automatic irrigation system, use Equation 1.*

Equation 1: $\left[\frac{exp(A)}{1+exp(A)}\right] * 1.18086 * [2.0341 * netET^{0.7154} * Ref\_Irr\_Area^{0.6227} + 0.5756 * ind\_Pool * netET]$

If the rated home has a netET of greater than 12 inches/year ***AND*** *the rated home does NOT have an automatic irrigation system, use Equation 2.*

Equation 2: $\left[\frac{\exp(B)}{1+\exp(B)}\right] * 1.22257 * [1.4233 + 0.6311 * \mathrm{netET} + 0.9376 * \mathrm{Ref\_Irr\_Area}]$

Either equation shall be constrained as follows:

IF

Rate_Irr_Area < Ref_Irr_Area

THEN

Ref_Out= equation 1 or 2 (as identified above) equation 1 (Using Rate_Irr_Area and pool indicator=0) equation 1 (with Ref_Irr_Area and pool indicator=0)

AND

Outdoor Reference Home GPD shall never be lower than equation 2

Where:

3.1 Reference Irrigated Area. Reference irrigated area shall be calculated as:

IF

the lot size of the rated home is < 7,000 ft²

THEN

$$Ref\_Irr\_Area = \text{Lot\_Area} * (0.002479 * \text{Lot\_Area}^\wedge 0.6157)$$

IF

The lot size of the rated home is ≥7,000 ft² Then

*Ref_Irr_Area = lot_area * 0.577*

Where:

Ref_Irr_Area= The size of the landscape that receives supplemental water in the reference home

Lot_Area= The size of the lot on which the rated home is being constructed

And,

5.0 Determining Outdoor Rated Home GPD. The rated home outdoor GPD shall be calculated as:

If the rated home has an automatic irrigation system

$$\left[\frac{exp(A)}{1 + exp(A)}\right] * 1.18086 * \left[2.0341 * netET^{0.7154} * Rate_{Irr_{Area}}{}^{0.6227} + 0.5756 * ind_{Pool} * netET\right]$$

Constrained such that:

Outdoor Rated Home GPD shall never be lower than

$$\left[\frac{\exp(B)}{1 + \exp(B)}\right] * 1.22257 * [1.4233 + 0.6311 * \text{netET} + 0.9376 * \text{Rate\_Irr\_Area}]$$

Where:
Exp(A)= exponent of [1.4416 + 0.5069 * (Rate_Irr_Area/1,000)]
Exp(B)= exponent of [0.6911 + 0.00301 * netET * (Rate_Irr_Area/1,000)]
Irr_Area= The size of the landscape that might receive supplemental water in the rated home
netET= The annual historic sum of mean reference evapotranspiration minus the mean precipitation for all months that evapotranspiration exceeds precipitation
ind_Pool= Indicator representing the presence or absence of a swimming pool

Ref_Irr_Area= The size of the irrigated area in the reference home, calculated in accordance with section 3.1
Rate_Irr_Area= The size of the irrigated area in the rated home

(Though somewhat different from the expressions under consideration at the time the early draft standard was first presented to the RESNET Outdoor Subcommittee, the currently available version on RESNET's web are restated here to avoid causing confusion.)

While there are several assumptions that could be discussed about the calculations, the RESNET Outdoor Subcommittee which was made up of several stakeholder parties, including it should be noted, the IA, came to focus on what variables were important in the equations predicting outdoor use. In brief, the model treats outdoor use as a function of whether or not an automated system is installed, Net ET, irrigated area, and pool area (if a pool is installed at the time of construction, a rare occurrence).

No one on the Outdoor Subcommittee felt such a narrow focus was sufficient to estimate water use in irrigated landscapes. Anyone that has taken the IA's Certified Landscape Auditor course or even just informally compared use with plant type immediately recognizes that the plantings and irrigation system absolutely impact a site's water use. Landscape design, including irrigation was missing from the considerations.

While a number of creative ways to address irrigation and landscape design have been developed for various specifications and, in some places, these have even been codified, the approaches tend to fall into one of two major categories:

- The site is subjected to some kind of landscape water budget which, at least in theory, helps drive design decisions about features, plantings, and irrigation towards reducing water use relative to what would have been normally done.

OR

- The site is subjected to limitations on certain types of high water use plantings and/or other high-volume irrigation or features. In the case of plantings, ornamental turfgrass is typically at least one of the plantings that is restricted in use. Savings result because past development norms emphasized these higher use areas of plantings, features, etc.

In an ideal world, these alternatives should result in similar savings if proper care is taken in developing them. In practice, most green industry and product trade associations tend to support the budget-based approach while most governmental agencies (that are trying to conserve water) tend to favor the planting restrictions or budget-based with some modifications that act as a backstop to maintain integrity of the statute (i.e. in reality, a less explicit restriction).

The Outdoor Subcommittee began to consider landscape design but was confronted with some difficult challenges:

- The HERS auditors are not landscape specialists and would not have the necessary knowledge of plant water use and site characteristics to do a budget.
- Even if they were trained in this, there really isn't time to do a budget in the time allotted for audits.
- Turf area provisions work but they are unpopular and putting them into the standard would draw significant opposition from the landscape industry and trade associations.
- While available residual area to landscape around the house can be typically broken into rectangular shapes for area size estimation, the landscape plantings themselves are often non-standard in shape. While this non-orderly arrangement is no doubt more natural appearing than totally uniformly-spaced plantings, it would translate into lots of area estimation work.
- Commonly the plantings are one of the last things installed. This presents a significant practical barrier to implementing either of the archetypical paths.

The Subcommittee recognized that builders understandably want relatively new, attractive, fresh plantings to accompany their product from a customer preference and relatively little leaf litter. Where builder provided irrigation systems typically are provided, these are often installed in advance of the plantings. The group considered how this status and these challenges could be overcome and landscape and irrigation design could be incorporated into $HERS_{H2O}$.

Research to the rescue.

From the mid-1990s to the early 2000s, SNWA ran a research study (Sovocool and Morgan, 2005) that involved submetering discrete areas of landscaping for purposes of quantifying per unit area application of water volumes, among other things. While the focus of that research was confirming and quantifying savings of water efficient landscaping, the author noted, first informally, that annual per unit area use through the submeter could often within a range, be approximated if one simply looked at the average irrigation station flow rates. An example from one of the manuscripts associated with that work (Sovocool and Rosales, 2001), in this case for xeric areas only, appears below:

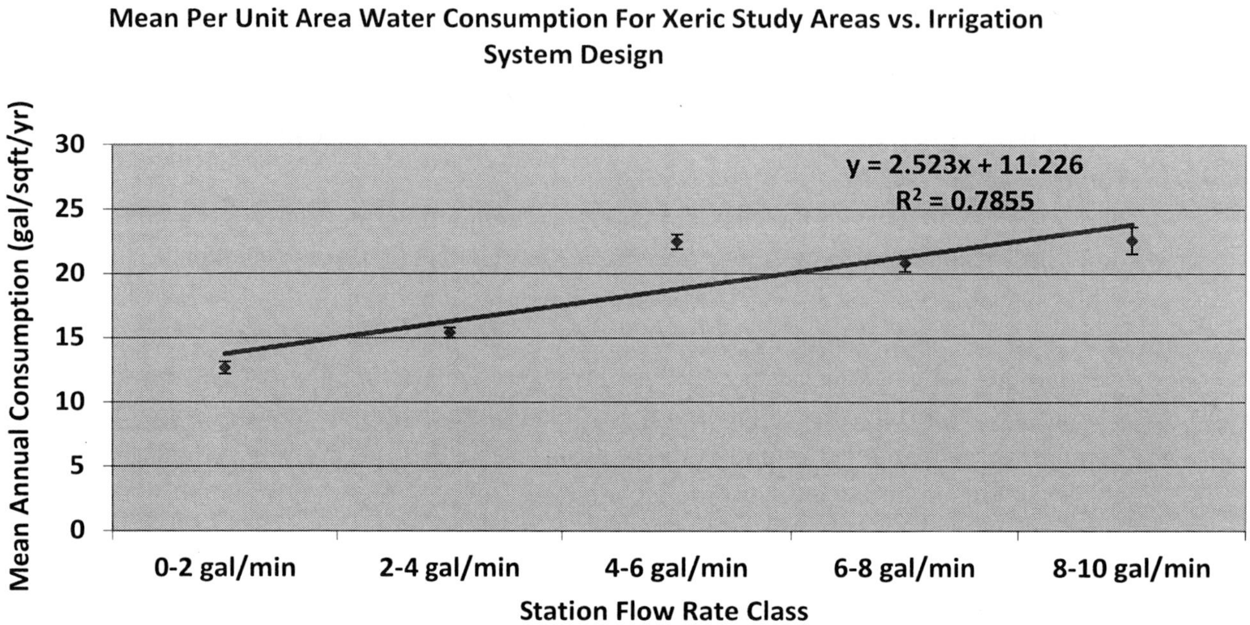

While xeric application volume are is best explained with multiple regression, the graphic demonstrates that annual usage on a per square foot basis tended to increase and was correlated, with increasing station flow rates in the monitored areas.

The finding was significant enough that it was incorporated into the final report linear regression model that quantified known sources of variability in xeric areas' use (Sovocool and Morgan, 2005). The correlation was scientifically quantifiable, interesting, and helped to inform SNWA about how savings would vary for more lush landscaping with lots of emitters. Beyond this though, nothing further was done with the finding for many years with the author assuming it was interesting nuance probably confined to the Southwest.

During the discussions of incorporating design into $HERS_{H2O}$ by the Outdoor Subcommittee, the research SNWA had done, including the observation that use could be estimated by station flow rates came to light. The author and the chair of the Subcommittee, Doug Bennett, decided to further explore the phenomenon as an avenue for overcoming the identified challenges.

The next logical analyses were to examine larger datasets for all types of landscaping to see if this trend endured or was only landscape-type specific or a local phenomenon. It was realized that some kind of index needed to be created to accomplish this. A scatterplot exploring correlation between area normalized summed flow rates (i.e. the Index) and annual use for 1211 data points across all local landscape types was developed and is shown below.

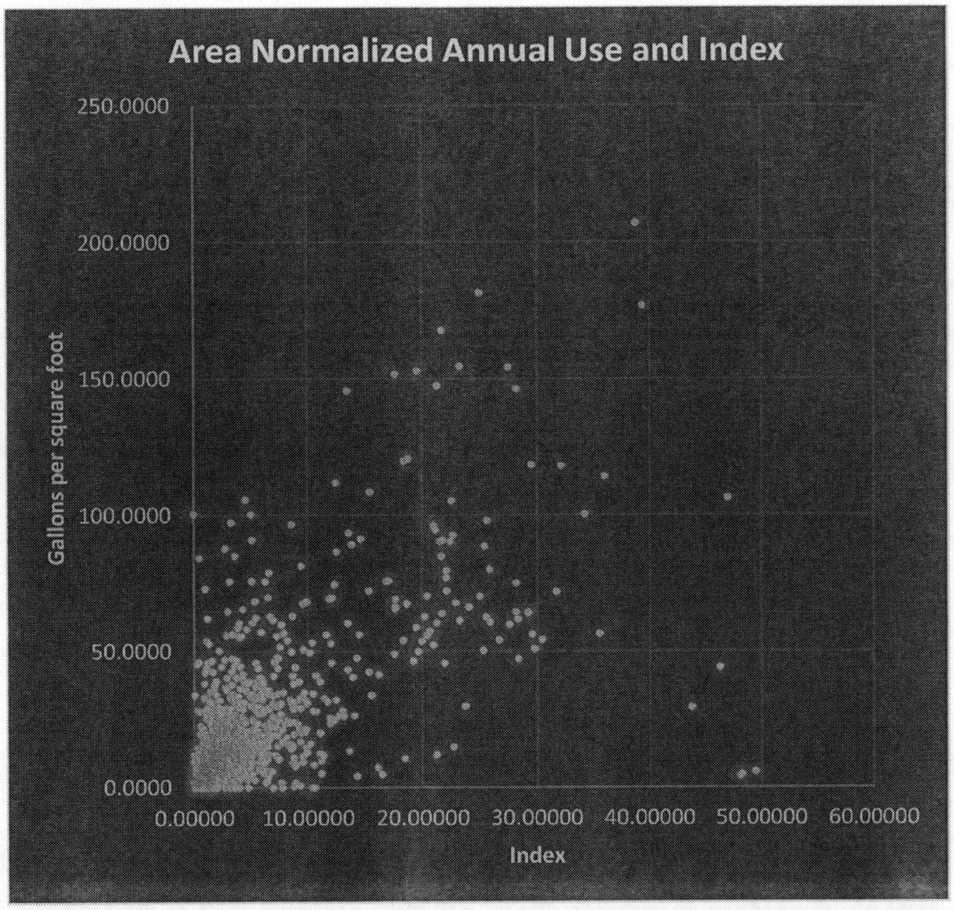

A relationship appeared to exist, though above an index value of 20 the pattern looked more random. A regression summary shows that a fitted line would have a correlation coefficient of .410 for this data. That it wasn't a near fit was not surprising as there are numerous variables that impact outdoor irrigation. What was surprising was how strong the relationship is to this one variable (t=29.00; p<000001). Within a range, the concept of using an irrigation capacity based-index to approximate use appeared plausible.

Invention of the Residential Irrigation Capacity Index (RICI).

This finding was reasonable enough that the author formulated what he called a Residential Irrigation Capacity Index or, "RICI". He defined this, as the totalized flow rates in gpm, from discrete flow rate field measures of each irrigation station installed, divided by the area served by these stations (i.e. irrigated area) times 1000. While defining RICI as an independent variable is a simplification of reality, for purposes of understanding its use, this is a helpful designation. Annual irrigation, that is water application, to that irrigated area in gallons per year is then the dependent variable. For the discretely metered landscapes in southern Nevada, this graphic demonstrates the model relationship between RICI and submetered use for RICI values of less than 20, of which the great preponderance of landscapes fell and over which the relationship is linear:

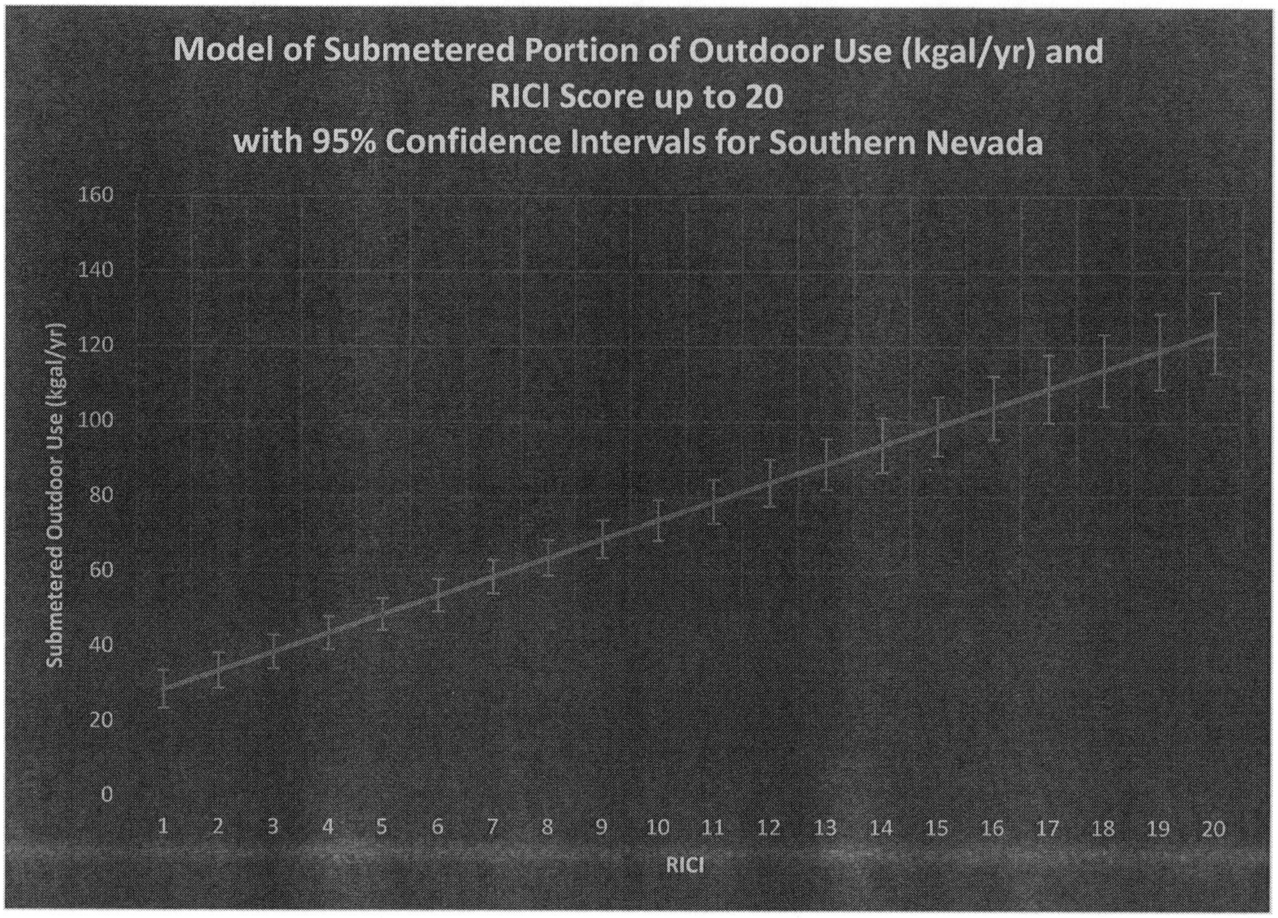

In southern Nevada a one unit change in RICI, above, is equivalent to 10.6% change in outdoor irrigation. Still, these were just local findings for southern Nevada.

Simultaneously though, the EPA under an effort being led by Jonah Schein, was reviewing outdoor irrigation use as they looked at the same challenges. EPA was on the Outdoor Subcommittee and a major partner in developing $HERS_{H2O}$ and the supporting equations and calculation tools. Specifically, they were utilizing trace flow records and other data that were collected and used in the Residential End Uses of Water Study, Version 2 (DeOreo et. al., 2016), abbreviated as REUWS II hereafter. Trace flow records are characterized as having very high-resolution use data, here having and interval of 10s between reads. In REUWS II the trace flow data was used to characterize and quantify how water was used at homes, both indoors and outdoors. EPA queried all homes with automated irrigation and recorded irrigation events and then proceeded to assemble the flow rates data by assuming that flow values within 15% of one another were from the same irrigation zone. Informed of SNWA's analyses, they compiled the flow rates data per property and used the irrigated area values as determined in the REUWS II research to calculate RICI in an analogous manner to SNWA.

The similarities in the outcomes were impressive. Not only did the distributions appear highly similar, the statistics appeared to align as well as is briefly summarized below:

Analytical Similarities between Local and Nationwide Findings Relating to RICI

SNWA service area (local to southern Nevada using submeter data)	EPA (using nationwide REUWS II data and trace flow data)
Distribution clumped towards low values end.	Distribution clumped towards low values end.
Average RICI = 4.97	Average RICI = 4.90
If average assumed as baseline threshold, 76% of homes would make threshold (i.e. They have RICI scores lower than this.).	If average assumed as baseline threshold, 73% of homes would make threshold (i.e. They have RICI scores lower than this.).

Based on the similarity in findings, the EPA provided their RICI values from the national REUWS II data set to the author and they had a dialog about whether RICI could successfully meet the need and at what level a threshold RICI should be set. An analysis of a histogram of the RICI values and the average RICI score above led them to suggest a baseline RICI score of 5:

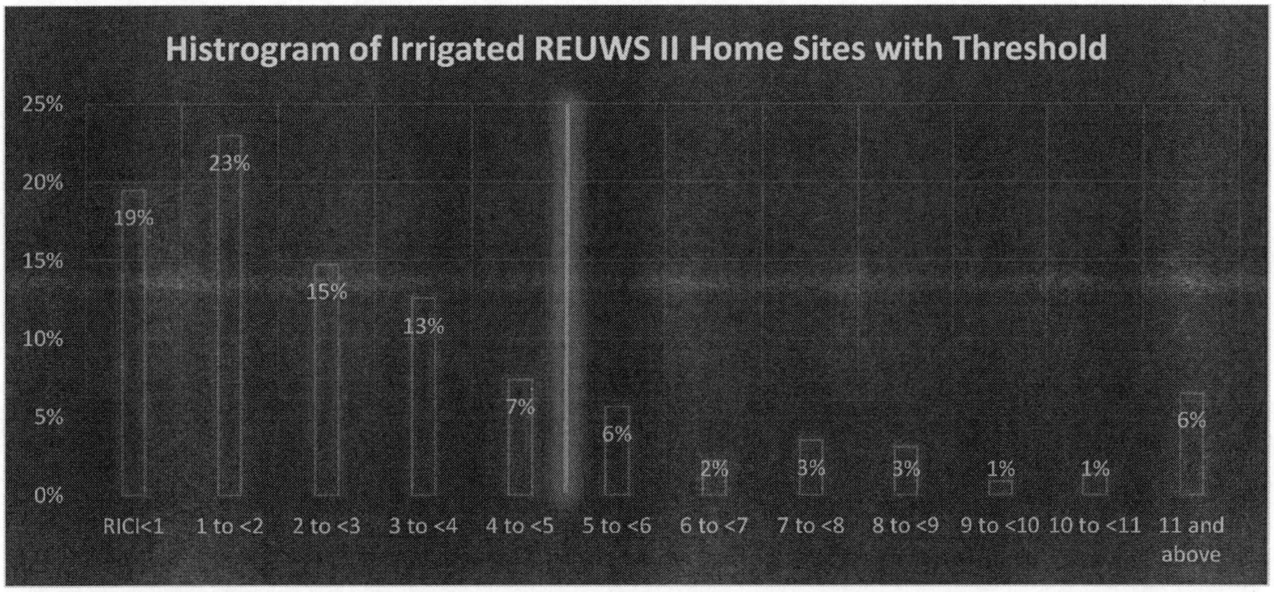

EPA felt a slight conservative rounding down of the expected yield from what SNWA observed was in order so a 10% reduction in outdoor use was to be assumed for each one-unit reduction in RICI score that a builder achieved. EPA also felt most comfortable with treating RICI as an option that a builder could turn on or off given it is a truly new approach to estimating future outdoor irrigation volumes. The author concurred as he agreed it's important to evaluate a new metric such as RICI carefully. The only practical way to see how it works is to have builders try it out and generate sufficient sample for analyses. Starting with making it optional essentially eliminates the risk to builders of RICI causing harm and thus makes it an enticing option to consider.

SNWA and EPA brought the concept to the Outdoor subcommittee and it was unanimously endorsed as the ideal solution to the challenges identified. It ultimately went into, and has remained contained within, the version of $HERS_{H2O}$ current as of this writing (RESNET, 2018). The metric has been included as an option in the standard and is mathematically incorporated into associated calculation sheets. It appears formalized in $HERS_{H2O}$ as:

5.2 Residential Irrigation Capacity Index (RICI). In rated homes where documentation is provided, a RICI may be calculated as

$$RICI\_rat = \frac{sum\ of\ flow\ (gpm)\ of\ all\ irrigation\ valves}{ft^2\ irrigated\ area}\ X\ 1{,}000$$

5.2.1 Applying RICI. A rated home where documentation for a RICI is provided may adjust the portion of water use associated with irrigation (less the water use associated with pools) in the rated home's outdoor gpd by 10% for every point from a baseline RICI (RICI_ref) of 5.

The selected RICI threshold of 5 means that nationally the majority of homes (77%) would be expected to be advantaged by RICI and that is only if they installed landscaping and irrigation systems as they have in the past. It is to be expected that a higher percentage would be below 5 given several parts of the country have in place some kind of provisions intended to promote more sustainable landscape installations by builders. Finally it should be again noted that if a builder was not going to be advantaged by using RICI, then they would simply forgo the option to apply it.

Measuring the Residential Irrigation Capacity Index (RICI).

Determining a RICI score is easy and requires minimal training and field supplies.

1. **Determine the installed irrigated area**. This can be done in the field by breaking up the installed landscape into rectangles and figuring out areas of the rectangles. Because plantings individually are not being counted rather the entirety of the landscape being irrigated is, typically all angles are right angles. An alternative to field measurement for larger lots would be to examine plans or aerial images showing the areas served by irrigation. In most new homes this is simply the area of the front-yard.
2. **Turn off all water sources and taps**. The test requires running stations so don't turn off water at the meter, rather assure all taps and appliances that could use water are off and that the irrigation controller will not run any scheduled programs that may interfere with the test. After everything is off, open the meter box lid and check that there is no movement in the meter. Be careful not to inadvertently disconnect or damage any of the utility's infrastructure including remote reading devices. This is also a de facto test for leaks. Leave the meter box lid open for step 3.
3. **Set the controller to provide sufficient pre-sampling equilibrium time for each station.** Each station needs to be run separately with the other ones off. Set the irrigation controller to manually run the station to test for a long enough period for the auditor to get from the controller to the meter, for the valve to fully open, and for air in the system to be purged. There are distinct audio cues for both the valve opening and the air spitting out of the system. The

amount of time to wait varies for a number of reasons including the components and pressures and is typically less than a minute for a spray system to sometimes several minutes for a drip station. Practice will better allow the auditor to estimate how long to wait before taking readings in a given area and development.

4. **Measure the flow rate.** This is done by recording the volume of water that passes through the meter in a defined amount of time. With a stopwatch or countdown timer, observe how much water passes through the meter in 30 seconds and multiply this by 2 to get gallons per minute (gpm). For most home meters gallons are marked on the meter face with each rotation equaling 10 gallons for meters displaying in gallons. If a meter reads in cubic feet, a conversion must be done by multiplying the cubic feet by 7.48 to change the value into gallons terms. Turn off the station after measurements are completed.
5. **Repeat steps 3 and 4 to record gpm for each station.** Remember to do each station individually and check for no movement between runs that would indicate use during the test. It is best to cross correlate with the controller which station is which where many stations are present.
6. **Get the RICI score.** Sum together all the flow rates and divide this number by the landscape area in square feet. Finally multiply this number by 1000 to get the RICI value.

What Residential Irrigation Capacity Index (RICI) Means.

What does RICI reveal about irrigation system and landscape design? At its most basic level it demonstrates that higher use is associated with higher capacity to use a resource. This is why lighter sports cars still use more gas than subcompacts with less powerful engines. It is why usually a high butane gas grill burns through a gas cylinder more quickly than a low output grill. And, more to the point, it's why homes that have a 3/4" service use more water on average than those with a 5/8"meter. In each of these cases more use on average is to be expected while recognizing for any given case it is not a guarantee.

Part of this is that, where a human is involved, there is a behavioral tendency to use capacity to whatever extent it exists. To a degree, for irrigation, taking some control away from the person could save water and the $HERS_{H2O}$ rating to its credit has added to the outdoor model a 15% decrease if builders provide a smart controller. This links also to education and time when considering irrigation. For example, a homeowner may set a station of microsprays for an hour because they treat it identically to a point drip system they have because hey lack education about it. Thus the higher output system results in higher use than it should. But clearly more than just human behavior is involved.

RICI integrates the major inputs that would otherwise go to water budgeting schemes in a way that is landscape size neutral and, uniquely, that is relatively unbiased with respect to where the site is located. Higher demands whether from plant type, planting density, or other factors it seems, based on the analyses by SNWA and EPA, to be reflected in the capacity of the irrigation stations installed. While this seems entirely logical given these are specified systems, somehow measuring the capacity of a system as a means to estimate future demands has not to date been considered to the author's knowledge. It may be that the tendency to focus on the controller and the user has caused a bit of to neglect of an obvious predictor of use. And the other paths worked have worked until the aforementioned challenges related to practical and timely auditing of new irrigated landscapes by non-horticultural experts was encountered. As with other advancements, it was necessity and, here, large samples of powerful datasets, that revealed the correlation and powered the development of RICI.

Is RICI really a long-sought third way to promote water conservation in landscape design that doesn't involve prescriptive turf limitations or complicated water budgets? The analyses alluded to herein suggest the answer could well be "yes". Only time and further use of RICI within the $HERS_{H2O}$ rating system, and possibly in other efficiency initiatives and studies, will tell us if this path works as well as it appears to and is embraceable by those that currently are limited to the two standing approaches for promoting efficient design.

Conclusions.

The desire to consider landscape design and challenges encountered during development of that preference for RESNET's $HERS_{H2O}$ standard has resulted in innovation of a remarkable and simple metric for estimating future demands in builder installed irrigated landscapes at new homes. The Residential Irrigation Capacity Index (RICI) is based on summing together flow rates for irrigation stations installed at the time of construction and normalizing this for area served. This new RICI score is:

- Based on state-of-the-art research examining real-world correlation between the capacity of residential irrigation systems to deliver water and water consumption.
- Developed based on a national data set in addition to development local to southern Nevada. It is anticipated to work nationwide.
- Currently in $HERS_{H2O}$ as an option. Each one unit decrease in RICI from a baseline value of 5 is associated with a 10% decrease in outdoor use.
- Normalized on a per unit area basis, meaning homes with a diversity of landscape sizes and configurations are treated in an apples-to-apples manner.
- Not a metric that requires specialized knowledge of plant water use or taxonomy and does not rely on variable, mutable and unadjudicated assumptions about plant water use in a given area.
- Is not dependent on plantings being installed in time for the audit.
- Not a trade industry disliked turf limitation.
- Easy, straightforward and can be done by anyone. The sampling method is rapidly learned and RICI audits are quickly completed in the field.
- A metric that integrates not just irrigation capacity, but the contributions to water use resulting from species composition and density as well.

RICI may represent an innovative approach to evaluating new landscapes for future water use levels and is just starting to be tried out. Future findings will further qualify the interest and success of this novel approach for considering landscape design for water conservation.

References.

DeOreo, W., P. Mayer, B. Dziegielewski, J. Kiefer. 2016. Residential End Uses of Water, Version 2, pp. 363. Water Research Association. Denver, CO.

RESENT HERSH20 Technical Guidelines. February 19, 2018. Residential Energy Services Network (RESNET). Accessed October 9th 2018. http://www.resnet.us/blog/wp-content/uploads/2018/06/Draft-HERS_H2O_Guidelines_FINAL-2018.02.19.pdf.

RESNET News. 2018. RESNET Announces Kick-off of HERSH20 pilot phase. June 20th 2018. Residential Energy Services Network (RESNET). Accessed October 10th 2018. http://www.resnet.us/blog/resnet-announces-kick-off-of-hersh2o-pilot-phase-2/.

Sovocool, K. and M. Morgan. 2005. Xeriscape Conversion Study Final Report. Southern Nevada Water Authority. Las Vegas, NV.

Sovocool, K. and J. Rosales. 2001. A five-year Investigation into the potential water and monetary savings of residential xeriscape in the Mojave Desert. Presented at the American Water Works Association Annual Conference and Exposition, Washington D.C. American Water Works Association, Denver, CO.

5+ Years of Testing Smart Irrigation Controllers in Single Family Homes

Irrigation Show and Education Conference
December 4, 2018

Bernardo Cárdenas & Michael Dukes

Agricultural & Biological Engineering Dept.
University of Florida/IFAS

UF UNIVERSITY of FLORIDA

Central Florida Water Initiative

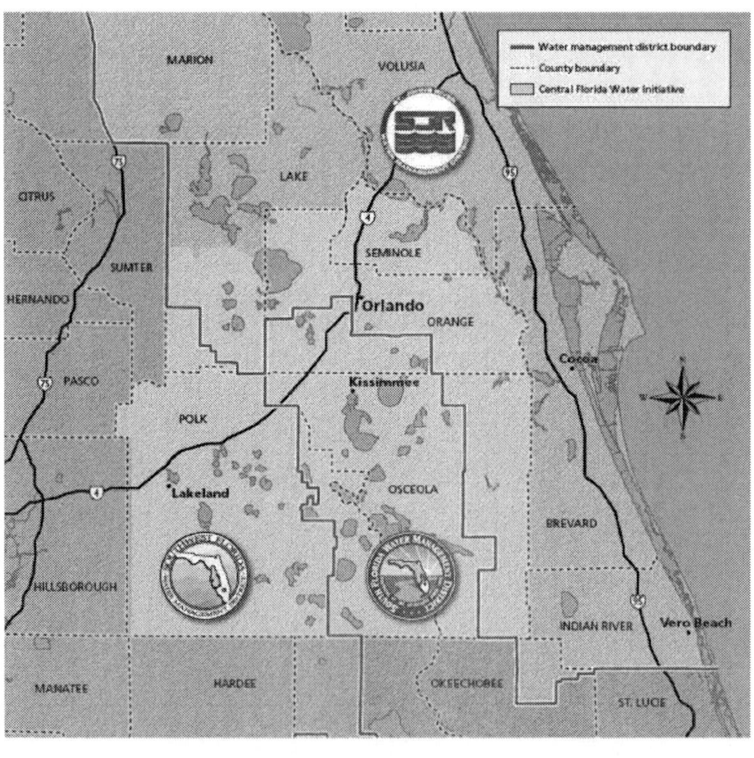

UF UNIVERSITY of FLORIDA

Testing Smart Irrigation Controllers (SICs)

Questions

- Can ***SICs*** help conserve irrigation water in homes?
- How much water can they save?
- Would those savings have a negative impact on the turf grass quality?
- Are SICs reliable for a mid/long term period?

UF | UNIVERSITY of FLORIDA

Objectives

- Evaluate if two types of smart controllers could reduce irrigation application of “excessive” irrigators
- Compare the water applied to a theoretical irrigation requirement
- Determine the significance of water savings

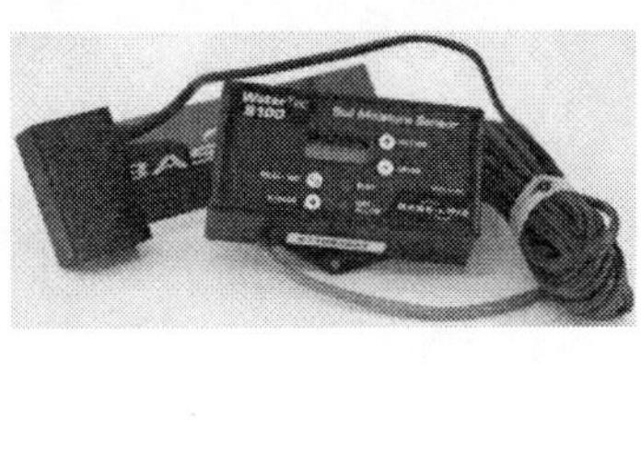

UF | UNIVERSITY of FLORIDA

Materials and Methods

Selection of Cooperators (excessive irrigators)

OCU sent to UF historical billing info

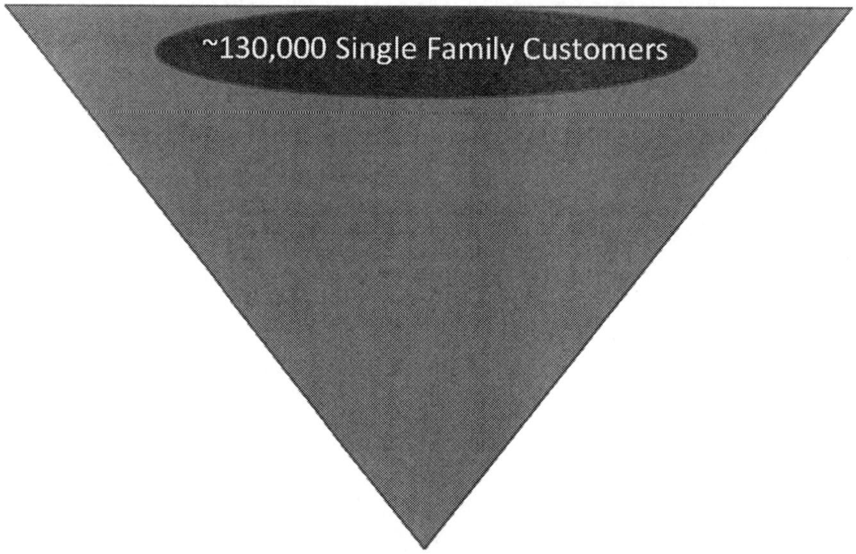

UF | UNIVERSITY of FLORIDA

Materials and Methods

Selection of Cooperators (excessive irrigators)

- Estimated Irrigation

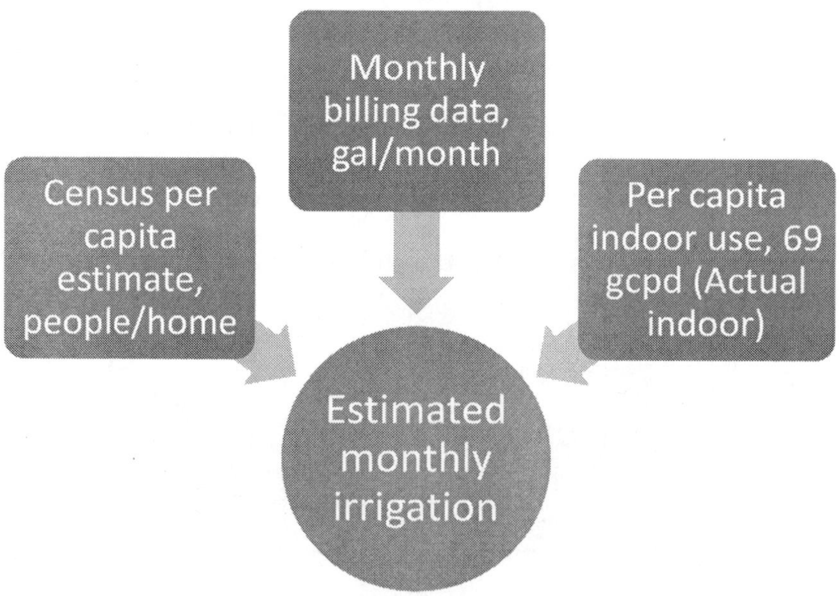

UF | UNIVERSITY of FLORIDA

Materials and Methods

Selection of Cooperators (excessive irrigators)

- Estimated daily ***Gross Irrigation Requirement*** (GIR)

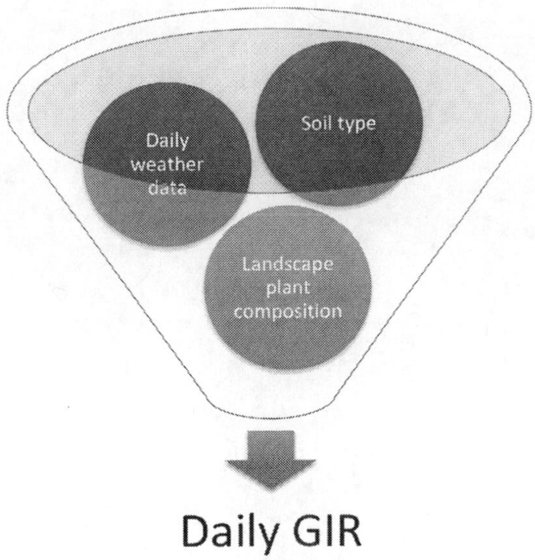

UF | UNIVERSITY of FLORIDA

Materials and Methods

Selection of Cooperators (excessive irrigators)

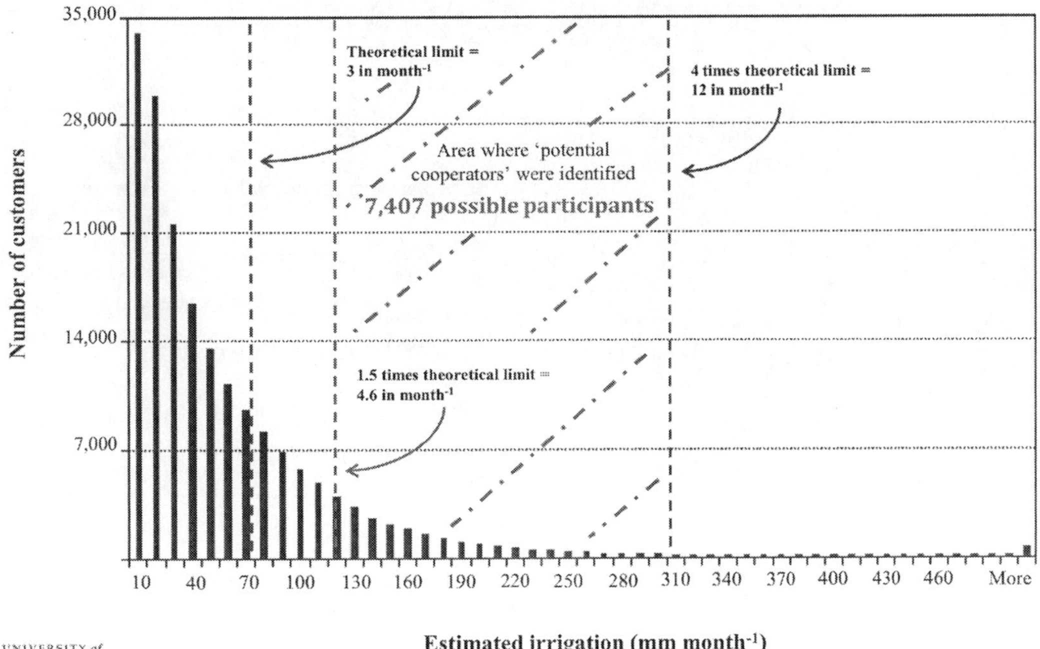

UF | UNIVERSITY of FLORIDA

Materials and Methods

Selection of Cooperators (excessive irrigators)

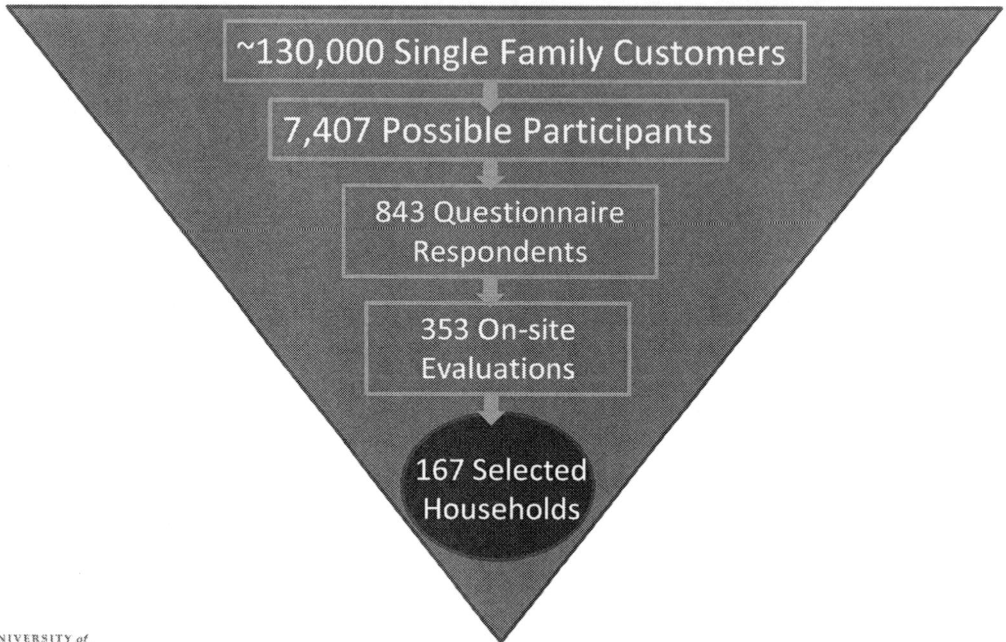

UF | UNIVERSITY of FLORIDA

Materials and Methods

Treatments and Installation

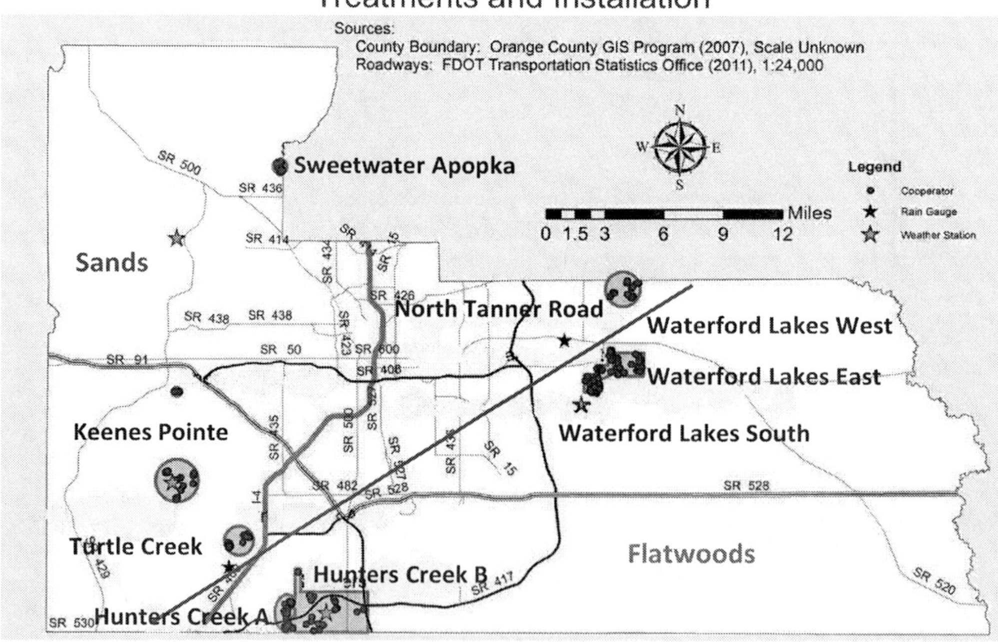

UF | UNIVERSITY of FLORIDA

Materials and Methods

Treatments and Installation

Treatment	ET	ET+OPT	SMS	SMS+OPT	MO
Smart Irrigation Controller	Rain Bird ESP-SMT 		Baseline WaterTec S100 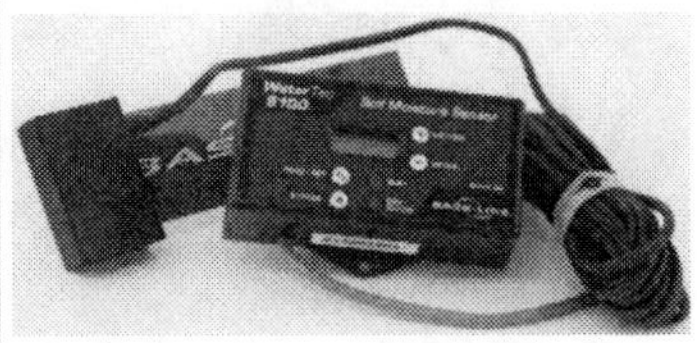		--
Schedule	7 d/wk	3 d/wk	7 d/wk	3 d/wk (2/d)	2 d/wk
Programmed	Contractor	UF Site-specific settings	Contractor	UF (0.25"/event)	N/A

UF UNIVERSITY of FLORIDA

Materials and Methods

Treatments and Installation

Treatment	ET	ET+OPT	SMS	SMS+OPT	MO
Smart Irrigation Controller	Rain Bird ESP-SMT 		Baseline WaterTec S100 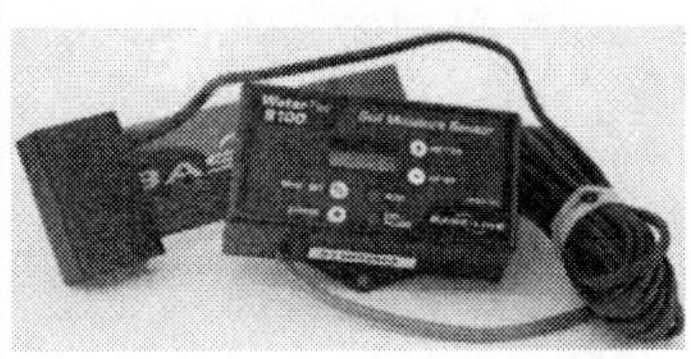		--
Locations Installed	7	9	7	9	9
Number Installed	28	38	28	38	35

UF UNIVERSITY of FLORIDA

Materials and Methods

Treatments and Installation

- OPT Treatments:
  - Five minute Tutorial
  - Educational Brochure on controller features

UF | UNIVERSITY of FLORIDA

Materials and Methods

Treatments and Installation

- All homes got:
  - Dedicated irrigation meters
  - Backflow devices
  - Minor repairs by contractor
  - Automatic Meter Recording devices (AMRs)
    - Records hourly irrigation volumes
    - Bi-monthly downloads

UF | UNIVERSITY of FLORIDA

Materials and Methods

Turf quality

- Measured seasonally
- Scale: 1 - 9

2

5

9

UF | UNIVERSITY of FLORIDA

Materials and Methods

Data collection & Statistical analysis

- Data Collection Period:

  Nov 2011 – Feb 2017 (62 months)

- Weekly irrigation application
  - Fixed effects of treatment, soil type, and rainfall
  - Random effects of location and week
- Tests treatment differences
- Tests significance of soil type
- Means procedure

UF | UNIVERSITY of FLORIDA

Results

UF | UNIVERSITY of FLORIDA

Weekly irrigation application

Sand locations

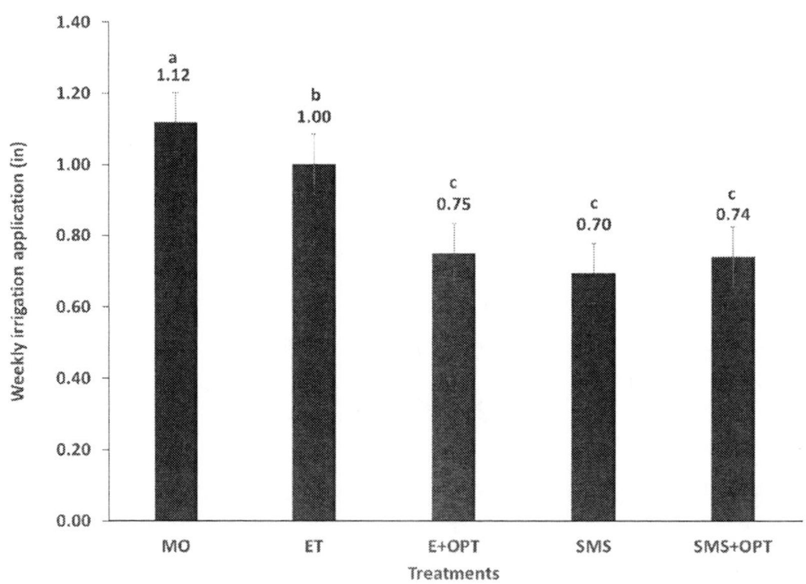

UF | UNIVERSITY of FLORIDA

Weekly irrigation application
Flatwoods locations

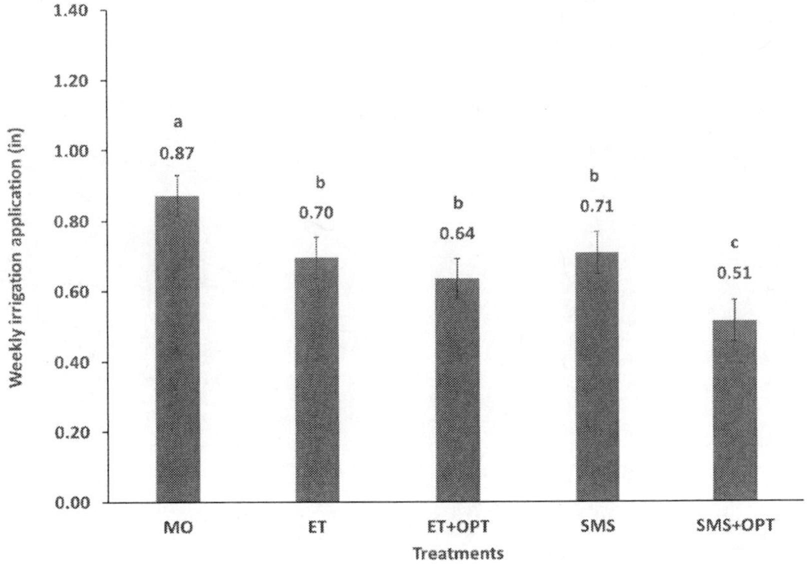

UF | UNIVERSITY of FLORIDA

Weekly irrigation application
Both soil types

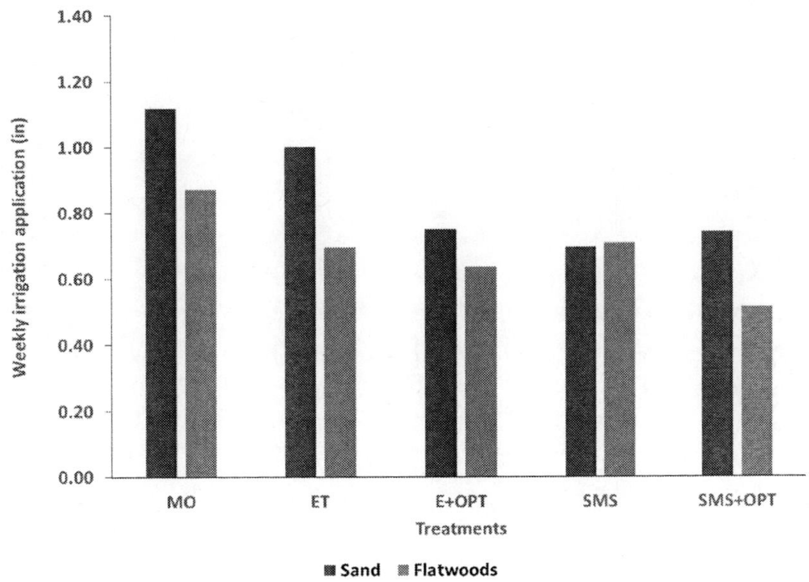

UF | UNIVERSITY of FLORIDA

Cumulative irrigation vs irrigation requirement
Sand locations

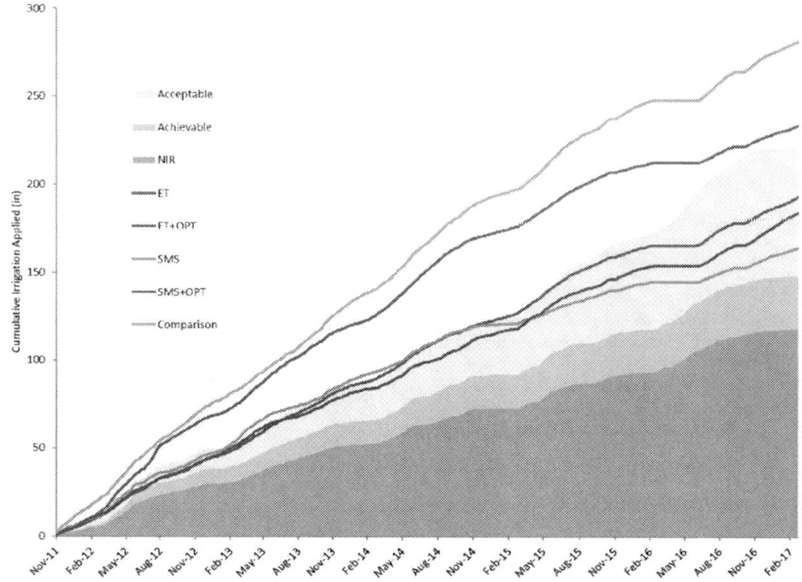

UF UNIVERSITY of FLORIDA

Cumulative irrigation vs irrigation requirement
Flatwoods locations

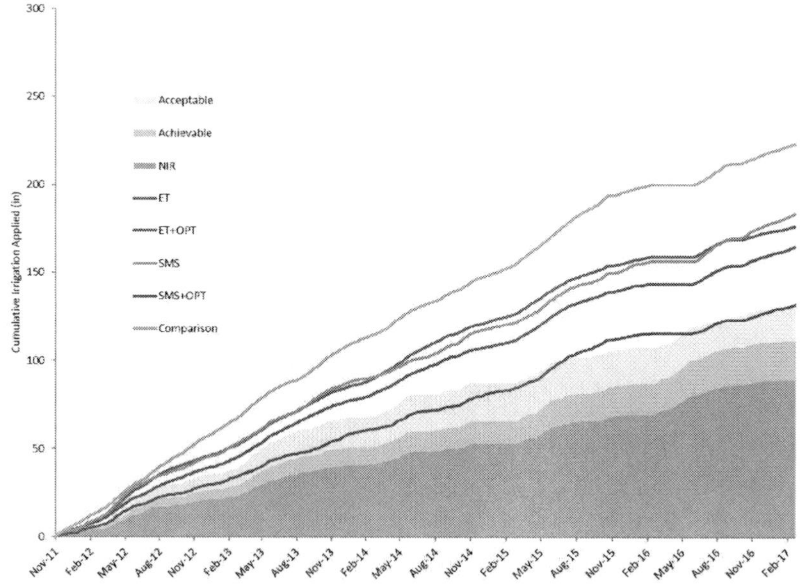

UF UNIVERSITY of FLORIDA

Results
Turf quality

- Almost every home averaged above a 6.2 rating
- During the whole study time frame
- No TQ differences between treatments

UF UNIVERSITY of FLORIDA

Conclusions

- After 62 months: all treatments with SIC significantly decreased irrigation compared to MO

ET	: 19%	ET+OPT	: 32%
SMS	: 30%	SMS+OPT	: 43%

- No difference on turf quality between treatments
- Water savings achieved did not result in a negative turf quality impact.

UF UNIVERSITY of FLORIDA

Conclusions

- These results demonstrate the ability of SMSs and ET-controllers to regulate irrigation based on real-time soil moisture/weather conditions, on the tested soils.

UF UNIVERSITY of FLORIDA

Acknowledgements

- Orange County Board of County Commissioners, Water Research Foundation, St. Johns River Water Management Districts, South Florida Water Management District
- Stacia Davis, Michael Gutierrez
- Terri Thill, James Williams

UF UNIVERSITY of FLORIDA

VOCs and Solvent Cements

The role of VOCs in solvent welding thermoplastic pipes & fittings

By Matthew Chang

June 8, 2018

455 W. Victoria Street, Compton, CA 90220 U.S.A. • www.weldon.com

Introduction

Volatile Organic Compounds ("VOCs") are found in a wide variety of everyday products including solvent cements, which are used to weld thermoplastic pipes and fittings. The use of solvent cements will inadvertently release VOCs, which cause smog and indoor air pollution. This has led agencies such as southern California's South Coast Air Quality Management District ("SCAQMD") and the California Department of Public Health ("CDPH") to develop and establish VOC guidelines to improve air quality for the environment. This white paper will define what VOCs are and explain the different VOC guidelines used to reduce their impact on the environment. Finally, the white paper will discuss the future of VOC regulations and solvent cements.

What are VOCs?

VOCs are carbon-containing compounds that easily evaporate at ambient temperature and pressure from solids or liquids. In indoor environments, excessive amounts of these compounds can cause adverse health effects. VOCs released outdoors can cause hazardous air pollutants such as airborne particles and ground-level ozone.

Solvent Welding and VOCs

The source of VOCs in solvent cement are the different type of solvents used to help soften the surface of thermoplastic pipes and fittings in preparation for the welding process. During storage, the VOCs in the solvents are safely contained inside the solvent cement cans. VOC evaporation mainly occurs during the solvent welding process when the applied layer of solvent cement is exposed to air. However, little to no evaporation occurs once a pipe joint is welded as the remaining VOCs are trapped inside the welded joints.

VOCs and Outdoor Air Quality

The South Coast Air Management District (SCAQMD) Rule 1168 was adopted in 1989 to protect outdoor air quality against smog by setting limits for VOC content in solvent cements.[1] This is one of the most stringent air quality regulations in the U.S., setting a standard adopted by regulatory agencies in many other states across the country.

In response, the solvent cement industry has been active in providing products that minimize VOCs. For instance in 1991, Weld-On introduced the first low VOC solvent cement in the market in response to growing air quality concerns. In 2009, Weld-On became the first company in the industry to offer a complete Low VOC product line of solvent cements and primers that meets the VOC content limits established by SCAQMD Rule 1168 and phased out products that have VOCs above these limits.

All solvent cements currently in the market must meet SCAQMD Rule 1168 to be considered "Low VOC". The current VOC content limits set by SCAQMD Rule 1168 are:

Products	Maximum VOC content (gram/liter)
PVC Solvent Cement	510
CPVC Solvent Cement	490
ABS Solvent Cement	325
ABS to PVC Transition Cement	510
Primer	550

VOCs and Indoor Air Quality

To address concerns regarding indoor air quality, the California Department of Public Health (CDPH) created the CDPH Standard Method for the Testing and Evaluation of Volatile Organic Chemical Emissions from Indoor Sources Using Environmental Chambers.[2] This is one of the most widely used standards to evaluate products for low VOC emissions in a controlled environment.

In the CDPH test, VOC emissions are measured in a controlled testing chamber after conditioning of the test specimen for 10 days. The gathered emissions results are then compared against a list of allowable concentration limits of different VOCs. LEED v4 currently sets concentration limits for 33 individual VOCs and for total VOCs (TVOCs)[3]. TVOCs is the cumulative concentration of measured VOCs in a sample.

The following table clarifies the differences in VOC guidelines between SCAQMD Rule 1168 for VOC content and CPDH Standard Method for VOC emissions:

	Outdoor Air Quality SCAQMD Rule 1168	Indoor Air Quality CDPH Standard Method
Purpose	To reduce **outside air pollution** by limiting the amount of VOC content in a product	To evaluate VOC emissions from low-emitting VOC products in an **indoor environment**
What is measured?	VOC content of a product in grams per liter	VOC emissions in an indoor chamber test under defined temperature, relative humidity, and ventilation
Who performs the test?	Evaluation is done by the solvent cement manufacturer	Evaluation is done by a 3rd party laboratory.

Measuring only VOC content is not the most accurate way to evaluate indoor air quality since there is no direct correlation between VOC content in a product and VOC emissions indoor over time. VOC emissions are more difficult to measure as a number of factors can affect the accuracy. The CDPH Standard Method minimizes these variabilities by measuring emission in a controlled environment. For these reasons, LEED v4 of the US Green Building Initiative specifies limit values for both VOC content and VOC emissions in order to obtain LEED credit for Indoor Environmental Quality.[3]

Future of VOC regulations

New regulations by SCAQMD will further reduce VOC content limits for solvent cement by 2023. Solvent cement manufacturers are already working on new products that can meet the new regulations while ensuring product stability and performance compliance. The following are the new proposed VOC content limits:

	Maximum VOC Contents (gram/liter)	
	Current	2023
PVC Solvent Cement	510	**425**
CPVC Solvent Cement	490	**400**
ABS Solvent Cement	325	325
ABS to PVC Transition Cement	510	**425**
Primer	550	550

Future VOC rules and regulations might phase out specific VOCs that have a much higher toxicity or other potential adverse environmental impact. Not all VOCs are created equal and their impact on the environment depends on several variables such as unique chemical reactivity and volatility. For instance, some VOCs form ozone at a much slower rate and are considered minimally reactive compared to VOCs that react at a much faster rate. This could lead to the development of new products that have reduced emissions of the most reactive VOCs while meeting the performance standards required by the market.

Conclusion

The presence of VOC containing materials in solvent cement is essential to perform its primary function of fusing thermoplastic pipes and fittings. However, a small fraction of VOC evaporation occurs during the application process whenever the solvent cement is exposed to air. Over time, the industry has made considerable effort in minimizing the impact of solvent cement VOCs in compliance with rules and regulations such as SCAQMD Rule 1168 for VOC content and CDPH Standard Method for VOC emissions. In addition, more stringent VOC rules and regulations coming in the near future could lead to the development of a new generation of low-emitting VOC products that can perform as well as traditional products on the market today while prioritizing the reduction of the most harmful VOCs.

References

1. South Coast Air Quality Management District. 2005. Rule 1168 – Adhesive and Sealant Applications.
2. California Department of Public Health. 2010. Standard Method for the Testing and Evaluation of Volatile Organic Chemical Emissions from Indoor Sources Using Environmental Chambers, v 1.2 Sacramento, CA: CDPH, Division of Environmental and Occupational Disease Control,
3. U.S. Green Building Council. 2016. LEED v4. Rating System Selection Guide. Washington, DC: USGBC, http://www.usgbc.org/guide/bdc

IN-STEM FLOW REGULATION

Abstract

The author is an experienced landscape and irrigation licensed contractor with almost 60 years of active participation in the industry. In 2002, the author invented and patented the "Little Valve" technology for pop up and above-grade spray heads in order to make the maintenance of spray nozzles and subsequently, rotating, multi-stream nozzles faster, easier and much more efficient. Up to this present time, this was and is the first significant change to sprinklers in almost 60 years. In follow-up testing beginning at the Center for Irrigation Technology, it was discovered that the "Little Valve" technology offered unique water and labor saving features, notably, during installation and maintenance. This was followed by testing showing increased distribution uniformity of conventional 15" spray nozzles adjusted to 15', 12', 10' or 8' when compared to MPR nozzles designed to perform at these same distances at 30 psi and tested also at 55 p.s.i. In addition, the technology made it possible to eliminate over-spray and high pressure misting at pressures up to 165 psi, giving the installing or maintenance contractor the ability to precisely control the distance of throw of any individual head.

In the 2010 – 2011 period, the water-saving characteristics of the technology were further confirmed under the auspices of the Innovative Conservation Program (ICP) of the Metropolitan Water District of Southern California (MWD). The audited trials showed an average water savings of 30.2% with the "Little Valve" technology. This, in turn, led MWD to add the "Little Valve" components to its commercial rebate program. When the Water Management Usage Division of the Municipal Water District of Orange County published its "Cost Effective Analysis – Existing Programs" report, the "Little Valve" components were identified as the most cost-effective water-saving technology available in the marketplace.

A Lecture on the Many Benefits of In-stem Flow Regulation (IFR)

Author: Ted Sirkin, Irrigation Designer & Consultant
California Landscape Contractor's License No. 217789 (1963)
Owner: Valvette Systems, Woodland Hills, CA

Prior to the In-stem Flow Regulator's introduction, it was virtually impossible to remove and replace nozzles without one person at the controller or the RC valve and another at the sprinkler to remove and re-install the nozzle OR for one service person to go back and forth to the valve or controller two or three times, shutting off the system, then going back to the problem sprinkler, all of which was very time-consuming. The only other method for one person was and is using remote-control units, which were then just gaining popularity but were often unreliable and in other cases were in the shop for repair or had just simply been forgotten when the crews left in the morning. Even today, remote control units can be time-consuming in their operations meaning slow to turn on or shut off a valve, and when connecting up to a controller. Range and interfering structures or topography is another inhibition facing remote control units. For the long-range units, even today cost can be a deterring factor, as well as the need to connect wires to each and every station on a controller. Yet, back at that time, there simply was no other way for one person working alone to be able to shut off a sprinkler in the middle of its operation for servicing clogged nozzles, flushing out dirty lines and nozzle change-outs.

Hence, In-stem Flow Regulation (IFR) was initially created to make sprinkler head maintenance and installation easier by providing a means for one person to approach a pop-up or above-grade shrub adapter sprinkler with a clogged nozzle or desiring to change out or replace an existing nozzle, to be able shut off the water to that one particular sprinkler all without shutting off the other sprinklers within the same valve system and upon completing the task, be able to turn the one sprinkler back on and move onto the next problem. Another major problem that always occurs when one is servicing pop-up sprinklers is when the nozzle is removed, the pop-up riser stem will slide down back into the pop-up body with or without the water being on. In the case of the sole existing IFR presently in the marketplace, the adjustment screw and its surrounding material provide a surface for the incoming water to keep the pop-up elevated while the nozzle is being worked on, hence rarely, if ever, does the pop-up stem retract until the RC Valve has been shut down.

The first testing on In-stem Flow Regulators was performed at the Center for Irrigation Technology and was conducted by Ed Norum, a former director and long time consultant with the C.I.T. The finished results were titled "Detailed Analysis of the Testing Results for LittleValve Sprinkler Parts. [1] The testing centered on the water and labor savings IFRs provided under a variety of normal field circumstances with simulations for both new installations and post-installation maintenance work. In the majority of trials, significant water and labor savings were achieved versus non-LittleValve sprinklers and parts. In the 'Maintenance' trials, savings of 66% less labor and 36% less water were reported. See the full Report at ValvetteSystems.com in the sub-section "Test Data & Water Savings Reports" under the 'Think Water Conservation' tab.

There is an interesting aside to this discussion of the C.I.T. testing: It was not until 6 or 7 years later that this author started touting the 'pressure regulating' benefits of these products as he did not yet have a good grasp of that benefit the products offered. It was a work in progress over the first 5 years wherein the water saving benefits started taking form. The only awareness that the author had about pressure regulation during that time period was that even if the pressure was 165 psi, the LittleValve, consisting of a 3/8" diameter stainless steel set screw introduced into a reduced-diameter water passageway of a sprinkler part, would, in fact regulate water flow allowing a 15-ft nozzle to be reduced down to about 6 feet, allow for 98 -99% shut down of a sprinkler for maintenance purposes while others on the same line were still operating, save water by eliminating overspray onto streets, sidewalks and buildings and when using a 15-ft nozzle at distances of 14 feet or less all misting/fogging would go away and cease to exist. In fact, as opposed to 5-ft and 8-ft nozzles, which cause misting and fogging sometimes at even low pressures, the more the radius distance is lowered with 15' nozzles connected to IFRs, the less possibility of misting/fogging.

However, Mr. Norum was aware of it and made this notation in Page 4 near the end of the final analysis: "If pressure regulators are not used, then the headloss introduced by the Little Valve is beneficial in reducing the operating pressure to values closer to the range of operating pressures under which spray heads were designed to operate. Further, the adjustable feature of the LittleValve allows it to introduce the additional headloss required to match the supply pressure to the design pressure."

The benefits of using primarily 15-foot nozzles and avoid using 5', 8', 10' and 12' nozzles became evident with the 2004 testing conducted by Joe Kissinger, a certified irrigation auditor and a part time irrigation teacher At Cal-Poly, Pomona, working under Dr. Joseph Hung, Professor Emeritus, as the Technical Adviser. Kissinger brought Dr. Hung onto the team to help determine why he was saving water during the testing. At that time, Dr. Hung, the no. 1 irrigation hydraulics expert in the western USA, was

[1] LittleValve is the registered name for this technology manufactured and distributed by the inventor. However, in 2011, the Metropolitan Water District of Southern California (MWD) gave it the generic name of In-Stem Flow Regulator when it became the first and only sprinkler part on the MWD's rebate list

able to figure out that the reason LittleValve parts saved water is because a chamber is created between the IFR plug area and the nozzle, hence the water swirling around the inside of the chamber eventually goes up into the nozzle but with considerable less turbulence than being shot straight into the nozzle through a pop-up stem or sprinkler riser. The water entering into the nozzle with less turbulence allows the water to exit the nozzle more efficiently. Through their testing, Dr. Hung and Joe Kissinger were able to determine that the chamber caused the 15-foot nozzle to emit water with higher uniformity than the same nozzle, at the same pressure as on a regular sprinkler.

Then in 2010 -2011, those benefits became further memorialized in the trials conducted under the auspices of the Innovative Conservation Program (ICP) of the Metropolitan Water District of Southern California (MWD) in cooperation with the U.S. Department of Reclamation.[2] Four of the five one-year long trials were devoted to recording monthly water usage in turf and shrub areas with 8 separately-metered RC valves using Neptune meters from 1" to 1 ½" to record flow separately for each valve. One valve in each trial measured the water usage of X number of IFR sprinklers, mostly pop-ups whereas the other valve in that same trial had an equal number of non-IFR sprinklers. All the sprinklers and all the nozzles were brand new. Two of the trials were in turf locations and the other two were in shrub areas. One turf trial and one shrub trial was in the Los Angeles Dept. of Water & Power service area (LADWP) and one turf trial and one shrub trial was in the Las Virgenes Municipal Water District (LVMWD). Each month the author, accompanied by the conservation guru from each water district, would go to the trial sites in their respective districts and read the meters together. This was done to assure the veracity of the results through 3rd-party verification of all meter readings.

After the meter readings at each test site, a very important process would take place. The water agency guru and the author would take a substantial number of soil probes at each test area and closely examine the samples in order to jointly agree on the water times and frequency of each station for the ensuing one-month period. The watering times and frequencies of each station were then adjusted accordingly. In the opinion of this author, nothing can beat a 3-dimensional examination of the soil moisture.

Both turf trials and one of the two shrub trials took place in 12-ft wide median strips. The sprinklers were properly spaced on 12-foot centers and alternate-spaced on both sides of each median. In these 3 IFR areas, new 15-ft Rain Bird nozzles were used and in the 3 non-IFR zones, new 12-ft Rain Bird nozzles were used. That means the nozzles in the IFR areas did NOT provide matched precipitation. Yet, an examination of pages 9, 10 and 12 of the MWD testing, the 3 pages devoted to the results of these 3 trial areas, shows that the water savings of non-MPR nozzling connected to IFR sprinklers exceeds the water savings expected from the use of MPR nozzles. The percentage of water savings in the 3 trials range from 22.7% up to 33.5%. When the results of the non-median shrub area trial are included, the average water savings of the four IFR zones compared to the non-IFR, MPR zones reduced water use by 30.23%. (See page 13 of MWD report.)

One of the most notable findings that came from monthly evaluation of soil probe samples at the test sites is that the droplets from 15-ft nozzles are larger in diameter and heavier than the droplets from the 12-ft nozzles; and even more so when compared to droplets from 10-ft and 8-ft nozzles. The author and the water conservation gurus deduced that the larger droplets penetrate the ground deeper, diffuse throughout the soil more widely and certainly resist wind drift much better. This is a good place to note that due to the water savings results from the IFRs participation in the Innovative Conservation Program, the IFR sprinkler parts are the ONLY actual sprinkler parts on the MWD's commercial rebate list. The other major items associated with sprinklers on that list are high-efficiency nozzles and high-tech 'Smart' controllers.

---

[2] The MWD is the nation's largest water purveyor of treated water serving over 19 million customers over 5 counties via 26 member agencies..

There are, of course, some exceptions to using just 15-foot nozzles. Rain Bird and Toro both make flat spray nozzles. The writer is not aware of a designated flat spray nozzle being made by Hunter, K-Rain, Weathermatic or Hit. As mentioned above, 15-foot nozzles can be brought down to about a 6-foot radius and do a good job of watering when properly spaced. However, the question is, "What nozzle do I use if I want to water less than 6 feet or even lower." That's where the flat spray nozzles come into the picture. Connected to IFRs, Rain Bird's 8-foot Flat, one of Rain Birds grey-colored specialty nozzles, and Toro's Black Flat – that has no radius designation – are simply the world's BEST micro-spray applications anywhere and with the best looking patterns. If you seek a low-volume micro-spray sprinkler with a distance range of 6 feet down to 2 – 2 ½ feet, you will find that those two nozzles will do the watering job you want with little or no maintenance problems normally encountered with all other micro-spray application products. And to top it off, all pressure-regulation concerns normally associated with micro-spray sprinklers are non-existent.

The next group of nozzles benefiting from IFRs is a boon to most irrigators in that it solves a problem heretofore not really solvable. We're talking side strip nozzles. In most cases, side strip distances are 4-5 feet in front of the nozzle and 15 feet to the left and 15 feet to the right. The single biggest problem with side strips is when you adjust them down to lower than about 9 feet left-right; the nozzle ceases to emit water out in front of itself. It basically becomes a center strip nozzle at that point, which doesn't respond well to being adjusted via the little screw on top.

Hunter and Rain Bird's rotating side strip offerings provide a much different watering pattern than the spray-type side strips do. Hence, excluding those rotating models, there is one spray-type side strip that is different from all the others and that is Hunter's SS530. Its difference is that it is a "5-by" side strip rather than a "4-by". Hence, when connected to an IFR, the SS530 can be taken down to as low as 2 – 3 feet left/right and there still is water coming out in front of it. Because of that feature, the SS530 can actually be turned into another micro-spray nozzle similar to the Toro and Rain Bird Flat sprays only with a more elongated spray pattern.

To help get past that problem of a "4-by" becoming a center strip nozzle, Toro's newer line of Precision Spray Nozzles (PSN) has two spray-type side strip offerings: 4 x 30 and 4 x 18. The 4 x 18 can be used for tighter areas than can be served by a typical 4 x 30.

But now with In-stem Flow Regulators, all side strips can be controlled so that the installer no longer has to live with a 30-foot wide pattern. Side strips can now easily be taken down to 10 feet left/right before water stops coming out in front and oftentimes eliminate water being thrown onto the front entryway of a home, apartment or over-spraying onto a parking lot. It should be noted, however, that with IFRs, one gains a lot more precision placement of the water with spray-type side strips than with the rotating models.

As a side note, be aware that when encountering a planter area that is less than two feet wide, this IFR irrigator has learned that he can only use Rain Bird's 15SST for those areas. When turning the IFR down to confine the pattern within the two feet, the nozzle will act as a center strip but the side throws can be better contained within the 2-foot width than other side strips.

A final note about side strips deals with those occasions when, for various reasons, side strips need to be mixed in with rotating nozzles on the same valve system. Excluding the two 5-foot rotating side strip nozzles on the market, under normal circumstances we have learned that only the PSN 4 x 30 side strip nozzle can be mixed in with other rotating nozzles on the same system and be brought down to about 8 – 10 feet left/right or lower but at the sacrifice of no water out in front. The reason is simple: The PSN side strip is the only spray-type that has flow rates similar to those of all the rotating nozzles. If spraying the side strip less than 5 feet left/right, then Hunter's SS530 can generally be used with rotating nozzles.

But, now those rotating nozzles will be discussed in more detail because In-stem Flow Regulators are making a very large impression in conjunction with rotary or rotating nozzles. Of the four major rotary nozzles on the market with which this author is familiar, the IFR works "magic" on two of them but still provides benefits to all four.

Just for clarity, even though many of you here already know this, the stated low-end radius distance for MPs is 6' on the new 800 model and 8 feet on the 1000 model, K-Rain's minimum distance is 13 feet, Rain Bird's new RVAN14 minimum radius distance is 8 feet and that is for adjustable models only as the full circle minimum is 17 feet. And Toro's PRN has a minimum distance of about 14 feet if the pressure is at 30 psi. If the inlet pressure is higher, the minimum distance increases.

There are two major differences between the Toro and Rain Bird offerings versus the Hunter and K-Rain models. The first difference is the Toro rotating nozzle and Rain Bird's adjustable rotary nozzle have an arc range of 45° to 270°. With one minor exception, the other two have a minimum arc of 90°. The big difference insofar as IFRs are concerned is in the inner construction of the nozzle. When one turns down the radius of a Hunter or K-Rain rotary below its stated lower distance, the mechanism stops turning as it basically needs a certain amount of water flow to keep it rotating.

That is NOT the case with Toro or Rain Bird rotating nozzles when paired with In-stem Flow Regulators because the operating mechanism of those latter two function more like a gear-driven sprinkler. Hence, by having an IFR under a Toro or Rain Bird rotating nozzle, you can reduce the radius way, way beyond its stated minimum radii and that's where the magic comes into the picture.

Recently, Toro has been experimenting with IFRs and after examining the results of pairing their Precision Rotating Nozzle (PRN) with the IFR, Toro has found that it has the largest distance range of any rotary nozzle on the market. It has also learned that it only needs one size of PRN. Generally speaking, the maximum distance of a PRN is around 20 feet up to about 25 feet depending on pressure. However, when looking at minimum distances, no other rotary nozzle can compare with the PRN when it's paired with an IFR. When connected to an In-stem Flow Regulating sprinkler, the PRN distance can easily be brought down to 18 inches or even lower, regardless of pressure. That means the IFR gives the PRN a range of 18 inches all the way up to 25 feet. That's a huge radius-distance range.

Following close by is Rain Bird's 17-24 foot model. The 17-24's ability to reach 20 – 25 is also affected by pressure. However, regardless of pressure, when paired with an In-stem Flow Regulating sprinkler, the 17-24 can easily be brought right down to 4 – 5 feet. Hence, an IFR gives the 17-24 a range of 4 feet up to 25 feet, also a big radius-distance range of coverage. Under 4 feet, it pretty much stops rotating, but effectively, IFRs make Rain Bird's other two sizes obsolete.

Radius reduction of a rotary nozzle without an IFR in the picture can be tricky. Hunter's 1000 model and its newer 800 model each have a shorter minimum radius of 8 feet and 6 feet respectively. However, where water pressures exceed 55 – 60 psi, it is not likely one will easily or always be able to get down to those two radius distances. Hence, Hunter promotes pressure-regulating stems partly as way to accomplish the minimum distances; but even then, it's still not always easy to do especially if you have a system designed with high volume in mind. Here in Southern California where water pressures tend to be on the high side, this author, who is also a licensed landscape and irrigation contractor, has oftentimes struggled to bring down that 1000 model to below 9 feet - even with an IFR, although other contractors have told him that with IFRs, they are able to do so easily. Consequently, with Hunter or K-Rain rotating nozzles, when you turn down the adjustment screw - or with an IFR - to that particular point, generally in line with the manufacturer's minimum distance calculation, there simply is not enough water flow to keep it turning, hence the nozzle stops rotating and just gurgles at you.

There is an exception to the previous point, though, and that is the MP Rotator 3000 model. The minimum stated distance for the MP3000 is 22 feet. The In-stem Flow Regulator available today is able to bring the 3000 lower distance range down to 17 feet, adding another 5 feet of range to that nozzle.

Another big benefit for rotating nozzles is that some of them have tiny little screws that one must use to control the distance when not connected to an IFR part. At this moment, most of us would be thinking of Hunter's MP Rotator line. The tiny, little screw atop the rotating nozzle is recognized by most contractors as being on the troubling side. Regardless of what nozzle you have, rarely do you ever touch the screw atop the nozzle again when you have IFRs underneath. In almost all cases, all distance adjustments and most on/off control are done via the IFR even though the IFR only shuts down the sprinkler 97%-99% but not 100%. It doesn't need to shut off the sprinkler 100%; the RC or Remote Control valve needs to do that.

Additionally, IFRs oftentimes are the only method that allows for sufficient shutdown of a sprinkler with rotating nozzles AND for any and all above-grade shrub rotors. Basically, that is something you cannot do to any of them without shutting down the RC valve. This is not referring to just being able to turn off a rotating nozzle or Hunter's I-20s, Rain Bird's 5000 series or Toro's T5s or T7s for nozzle change-outs, rather it is a reference to being able to shut off the water to the rotor itself for replacement of the entire rotor or rotor nozzle for line flushing, if necessary.

However, when it comes to above-grade rotors, there is one huge benefit of IFRs that just can't be beat. The only way to reduce the spray distance of a rotor is to use a smaller nozzle or adjust the diffusion screw. When you engage the diffusion screw into the rotor stream, the distribution uniformity of the stream is affected; distribution uniformity decreases while the precipitation rate increases dramatically because the same flow rate is now distributed over a smaller area. What makes IFRs and above-grade rotors so much more efficient than just a standard shrub rotor is that you maintain distribution uniformity and reduce the flow rate at the same time. You insert the desired nozzle into the rotor, set your arc but then you set your distance with the IFR screw. You now have the freedom to adjust that distance to its maximum that the pressure and the nozzle will allow all the way down to the OFF position. Your distance can be dialed in like a guided missile. They're also very handy when servicing and/or replacing defective rotors by eliminating countless trips up and down a hillside slope. It should also be noted that ¾" IFR parts under rotors substantially reduce misting/fogging commonly seen on many hillsides in Southern California. They also prevent water hammer from damaging shrub rotor cases or the internal gear drives on slopes because each IFR helps protect the shrub rotor from incoming, high-pressure surges.

That takes us to other styles of In-stem Flow Regulators. The author assumes that future In-stem Flow Regulator products will likely have the on/off control located on the sides of other sprinkler parts in a fashion more or less similar to the existing IFRs in the marketplace. Because of that, IFRs are not confined to just being appurtenances to pop-up stems. In addition to being available presently in both male and female shrub adapters, IFRs are also available in certain plumbing fittings commonly used in irrigation. In the future, the author can imagine them in other fittings or simply in sections of pipe. The fittings in which they are presently available now are ½" and ¾" couplings – female thread at both ends, and ½" and ¾" riser extenders – female thread at one end and male thread at the other.

To be clear, the ½" couplings are necessary for use on above-grade shrub adapters where multi-stream, rotating nozzles are the nozzle of choice. The filter screens of these nozzles are a requisite to be installed with the nozzles. Until there is a different design for IFR shrub adapters, just use an IFR coupling assembled with a minimum 1" or 2" long ½" diameter nipple topped with an ordinary shrub adapter in order to accommodate the rotating nozzle and its filter screen.

Above-grade shrub adapters constitute somewhere between 10% -15% of all sprinklers in the USA. Most, of course, are found in the Sunbelt sections of the country. Pressure regulation in shrub adapters seems to be more difficult to achieve than in pop-up sprinklers where it is now becoming very common. In shrub adapter systems it is a rarity, partly because that apparatus is expensive and somewhat bulky, hence obtrusive.

But the problem with pressure regulation in shrub adapter types is completely eliminated with In-stem Flow Regulating sprinkler parts because every IFR part on the market today is pressure regulating and is simply unaffected no matter how high the pressure.

You may have noted that it has been mentioned several times that the minimum distances on certain rotating nozzles can be achieved regardless of pressure. Not knowing how future IFRs will be designed, we can only relate to those out in the marketplace now. Today's In-stem Flow Regulators simply do not care how high the pressure is. The sprinkler body may care, the RC valve may care, the sprinklers lines themselves may care. But the IFR stems do not care. It's not being said that there have not been cases where the IFR stems came apart due to high pressure, but that it has never been brought to this writer's attention. IFRs have been installed on jobs with pressures as high as 175 p.s.i. with no ill effects on the pop-up stems having been seen.

What is amazing to see is to have 100 – 110 psi or more on a given site, put a 15-foot nozzle onto an IFR pop-up, an IFR shrub adapter or an IFR coupling/shrub adapter assembly, turn on the sprinklers and then with the IFR in place see a substantial reduction in misting/fogging, then take the distance down to under 14 feet and watch as ALL of the misting and fogging totally disappears, and stays that way even as you turn the IFR screw slowly down to the off position.

An excellent visual example of this is on the author's website: watersavingsprinklers.com. Under the tab titled "Gallery & Videos", there are 11 pages of videos and pictures of the In-stem Flow Regulators in conjunction with Toro's Precision Rotating Nozzles. To get a clear picture of the ability of IFRs to override all high pressure, we suggest a visit to Page 6 and see the 'before' and 'after' videos and pictures. Page 6 features the sprinklers at the San Diego Maintenance yard of the California Transportation Dept. (Cal-Trans) where the pressure is 135 p.s.i. It is impressive to see nothing but fogging before the IFR change-out and then see it all gone after the change-out.

Because of the IFR's ability to override all high pressure and the control IFRs have over the water flow, IFRs are now being adapted to replicating many of the advantages of 'Drip' systems. It just takes the right nozzle or something that can replace the nozzle, i.e, a threaded cap with holes drilled in it, to act as an above-grade, point source watering application. Several cities and Cal-Trans, itself, are getting away from 'Drip' tubing and its accompanying maintenance and going to pin point watering devices using ½" pvc piping to bring water directly to the plant. Although more expensive, these systems are quite trouble-free, are much sturdier, and easier to service. One new model using In-stem Flow Regulation determines water flow based on the distance from the sprinkler to the plant and can be brought down to .05 gpm with the IFR. Present-day stream bubblers coupled with IFRs can accomplish many of the same benefits.

Within this lecture, 4 different testing events are noted. In truth, 18 total testing and trial events have taken place validating the water saving benefits of In-Stem Flow Regulators. Even back in 2008 before Rotating nozzles had established the foothold they have now, IFRs were proving that they were going to be a vital cog in the water conservation arena. The fourth testing event is an interesting trial called "Agoura Hills Trial No. 1". It started out on March 10, 2008 and has updates up to December 30th of that year. The trial is full of data that was confirmed by the Parks Manager for the city of Agoura Hills and led to the eventual changing out of almost every sprinkler in the city to IFRs. It is worthwhile to note that almost all the nozzles used in this trial and the subsequent city-wide change-out were 15-foot nozzles even though most of the city median strips are only 10-feet wide.

On the websites noted within, all 18 tests and trials can be seen under the sub-section called "Test Data & Water Savings Reports", which is a sub-section within the 'Think Water Conservation tab'. Directly under the Water Savings Reports, the graph referred to in the next paragraph below can be seen. Again the websites are 'watersavingsprinklers.com' or 'valvettesystems.com'.

In closing, another document is being presented which it is believed this audience will find interesting. It is called "Cost Effective Analysis – Existing Programs". It was prepared sometime in the last few years by the director of the Water Management Usage division of the Orange County (Calif) Municipal Water District, a member agency of the MWD and itself a re-seller of water to smaller water companies within Orange County, a very populous county in Southern California.

The Analysis features 30 water conservation methods/technologies/products. They are arranged in the order from least-to-most of how many dollars it takes to save one acre-foot of water – as you know, about 325,000 gallons. As it turns out, the In-stem Flow Regulator is at the top of the list in water-conservation affordability. Using IFRs costs $91.00 to save that acre-foot. No. 4 out of 30 on that list is almost double the cost of IFRs at $180 to save an acre-foot. And so we irrigators can gloat, in last place: No. 30, the least cost-effective methodology, the author's personal least favorite - the MWD's Turf Removal Program, coming in at $1,679 to save that one acre-foot of water.

Conclusion

Hence, our first conclusion is that In-stem Flow Regulating sprinklers are the proven, most cost-effective water-conservation products on the marketplace. We follow that by stating unequivocally that the key to IFRs is knowing what nozzle to use and that list is short. We can also conclude that all of the existing IFR models are pressure regulating and wasteful misting/fogging is brought totally under control. Further, because of the individual on/off control at each sprinkler, the labor and water savings to install and maintain new or existing sprinkler systems is incomparable when In-stem Flow Regulators are on the scene. Lastly, no sprinkler product on the market allows for the dialing in of the radius distance more accurately and effectively as an IFR part, which not only saves water on the merits but also reduces or eliminates run-off, which is such a wasteful practice, it is now illegal in California.

References

Ed Norum. Center for Irrigation Technology. 2002. Detailed Analysis of the Testing Results for LittleValve Sprinkler Parts. Pg 4, Sect. C.

Joe Kissinger and Dr. Joesph Hung. 2004. Comparing the Water Application Uniformity of 15' Rain Bird Nozzles Adjusted-down by LittleValve Sprinkler Parts & Fittings Versus Standard Rain Bird Nozzles: 5' - 8' - 10' - 12' – 15' (under the title "uniformity Charts". Pp 1 -10.

Metropolitan Water District. May 15, 2010 - May 15, 2011. EXECUTIVE SUMMARY OF YEAR-END FINAL RESULTS - ICP Water Conservation Program. Title: Water Savings Study With LittleValve Sprinkler Parts. 3rd party water district personnel: Robert Estrada - L.A. Department of Water & Power; Scott Harris – Las Virgenes Municipal Water District. PP 9, 10, 12 and 13.

In coop w/ Frank Gonzalez, Parks Manager, City of Agoura Hills, CA. Mar – December, 2008. WATER SAVINGS PROGRESS REPORT DUE TO MODIFICATION OF SPRINKLER HEADS WITH "LITTLE VALVES" IN MEDIAN STRIP ON KANAN RD, AGOURA HILLS, CALIFORNIA. Title: **AGOURA HILLS TRIAL NO. 1** With Updates to December 30, 2008. PP 1 – 6

Joe Berg, Director: Water Management Usage division of the Orange County (Calif) Municipal Water District. "Cost-Effectiveness Analysis – Existing Programs", a graph from power point presentation to various water managers in Orange County. 2014

Acknowledgements

Michael G. Baron, National Specifications Manager for Water Management Products for The Toro Company, with over 38 years of experience in the landscape and turf irrigation industry.

Managing Irrigation to Water Budgets in San Diego County

Carlos Michelon*, BSCE
San Diego County Water Authority, 4677 Overland Ave., San Diego, CA 92123, cmichelon@sdcwa.org
(*WSLEP Program Co-Manager)

JoEllen Jacoby, RLA, CLIA, QWEL, WWLP
Registered Landscape Architect, JoEllen.jacoby.baugh@outlook.com

Kyrsten Burr-Rosenthal, BA, MPA
San Diego County Water Authority, 4677 Overland Ave., San Diego, CA 92123, kburr@sdcwa.org

Alyssa Barman, Student Intern
San Diego County Water Authority, 4677 Overland Ave., San Diego, CA 92123, abarman@sdcwa.org

Abstract. *From 2014 to 2016, the San Diego County Water Authority conducted a grant-funded program aimed at improving water use efficiency in large urban landscapes by means of hardware upgrades, water budgeting, water use monitoring, and workforce training. The results were impressive, with an average reduction in water use across 20 participating sites that exceeded 30%. This technical paper and presentation will discuss program protocols; descriptive statistics of hardware enhancements deployed; workforce training; and highlights of program results, including potential program enhancements.*

Program implementation was a collaborative effort with the California Landscape Contractors Association (CLCA), leveraging the expertise and resources of CLCA's Water Management Certification program to empower participating contractors to successfully manage mature landscapes to a water budget.

Keywords. Contractor, Water Manager, Water Provider, Turf/Landscape (Commercial), Audit, Conservation, Distribution Uniformity, Scheduling, Pressure Regulation, Training, Water Budget.

Executive Summary

With a Proposition 50 Water Use Efficiency grant from the California Department of Water Resources, the San Diego County Water Authority (Water Authority) set out to implement a small-scale demonstration program to showcase the effectiveness of irrigation equipment upgrades when coupled with irrigation management training. Participating landscape maintenance contractors registered in the program with the goal of achieving a twenty percent (20%) reduction in water use relative to a three-year water consumption baseline at twenty commercial landscape sites. The program was a Water-Energy Nexus (WEN) partnership between Water Authority and San Diego Gas and Electric, leading to water and embedded energy savings through effective irrigation management and high-efficiency hardware upgrades. The program was designed to be consistent with CLCA's Water Management principles. Participating sites received financial incentives to evaluate irrigation systems, upgrade equipment and manage water consumption within a water budget. The results demonstrated the water conservation effectiveness of a comprehensive approach that integrated site management, equipment upgrades, and contractor training. The average reduction in water use across twenty participating sites exceeded thirty percent (30%).

Background

In 2008-2009, the Water Authority and San Diego Gas & Electric (SDG&E) successfully carried the first iteration of this type of pilot program. This a performance-based pilot program, entitled the Managed Landscapes Program (MLP), was designed to improve the outdoor water-use efficiency of a variety of large landscapes, which included homeowner associations, apartments, and commercial properties. Through that approach, a single vendor deployed proprietary irrigation control technology and water management services, achieving significant water savings. At the conclusion of the successful MLP, there was interest in implementing a follow-up pilot program to test the viability of a de-centralized approach (involving multiple contractors) and relying on non-proprietary industry best practices. The Water Authority and SDG&E, as part of their ongoing WEN partnership, collaborated from 2013-2015 to implement the Water Savings Landscape Efficiency Program (WSLEP), which included program modifications based on lessons learned from the evaluation of the MLP.

The global objective for WSLEP was to achieve a twenty percent (20%) reduction in water use relative to a three-year water consumption baseline at twenty commercial landscape sites. WSLEP goals included engaging landscape contractors and their existing customer base (pool of candidate sites) to identify high-water use sites, deploying a limited range of proven irrigation technology retrofits, and following specific irrigation management protocols.

Financial incentives were provided to participating landscape contractors for fulfilling specific program requirements and meeting pre-determined performance objectives (See Table 1. Schedule of Incentives). WSLEP was jointly administered by the Water Authority and SDG&E, in collaboration with CLCA. The Water Authority was the program lead for program design, technical oversight, and hardware financial incentives. SDG&E's was the program lead for contracting and administration. Administration of program implementation was outsourced to CLCA. Implementation activities included participant enrollment, workforce development, site assessments, installation verification, incentive disbursement, reporting, and a final determination of each site's irrigation management performance.

Table 1. Schedule of Incentives

Activity	Incentive Amount
Site Participation Enrollment	$100 per site
Landscape Audit, Water Budget Calculation and Water Savings Estimate	$500 per site
Executed Site Agreement	$100 per site
Hardware Upgrade Allowance	$5,000 per site maximum (supplier paid directly)
Installation Labor Allowance	$5,000 per site maximum
Water Management Services	$1,200 ($100 per month per site)
Customer Satisfaction Survey	$100 per site
Performance Fee (for 20% savings or better)	$1,700 per site
Total Fees	**$13,600**

Program Protocols

Selection of sites. The program hinged on the careful selection of sites. The landscape contractor and the site owner were required to jointly enroll to participate. Landscape contractors agreed to follow the principles of the CLCA water management program. The typical site was serviced by dedicated landscape meters with an approximate irrigation area of about four acres (1.62 ha). Some larger sites were treated as separate sites for administrative purposes. For eligibility, prospective sites needed to demonstrate the potential for water savings of at least twenty percent (20%). A site's plant coverage could be a mix of turf and shrubs. However, this program did not allow plant replacement (e.g. turf conversion) to occur during the participation period. Twenty sites managed by nine contractors were ultimately enrolled, totaling 100.6 acres (40.73 ha). See Table 2. Breakdown of Project Types with Total Acreage, for a breakdown of project types and associated acreage.

Table 2. Breakdown of Project Types with Total Acreage

Site Type	Qty.	Total Acres (ha)
Multi-family (SFR HOA and apt/condo)	14	63 (25.50)
City Park	5	32 (12.95)
Commercial Complex	1	6 (2.43)
Average Site Acreage		**5 (2.02)**

Landscape Irrigation Audit: The potential for water savings was determined through an irrigation audit, comparing historical water use to projected water use as calculated by the Maximum Applied Water Allowance (MAWA), defined by California's Model Water Efficient Landscape Ordinance (2010 version). An audit was performed on each property by a Certified Landscape Irrigation Auditor (CLIA). This free service was provided as part of a separate program entitled WaterSmart Checkup, which is jointly funded by the Water Authority and its member agencies (outside the scope of the WSLEP grant).

The MAWA calculation is based on the factors defined in the Model Water Efficient Landscape Ordinance of 2010 (State of California, Department of Water Resources). MAWA relies on evapotranspiration (ET_o) rates to determine the water needs of various plant types (plant factor) across California's diverse climate zones. It also takes into consideration the state's required level of irrigation efficiency. The ratio of plant factor to irrigation efficiency is known as the ET Adjustment Factor (ETAF). In accordance with the 2010 MWELO, the ETAF factor for existing landscapes was 0.8 (80% of ET_o) and 0.7 for new landscapes. WSLEP targeted participation by established landscapes. WSLEP's requirements were designed to help these sites achieve the same performance levels required of newly designed landscapes. For this reason, a .7 plant factor was used in WSLEP's MAWA calculation to determine the water budget. See Figure 1. Typical Audit Map and MAWA Calculation.

The audit identified the location and serial number of water meters, a measured service area for each meter, plant material and irrigation types, existing irrigation problems and recommendations for improved efficiency. Historical water-use was provided for each meter by the water agency serving each site. Historical water use was compared to each site's calculated MAWA. The irrigation audit helped to identify potential irrigation equipment upgrades. The historical water use baseline and the MAWA budget were used as a screening tools to validate the water savings potential for each site. While the original expectation was to enroll only sites that demonstrated at least a twenty percent (20%) potential savings based on MAWA, actual values for enrolled sites ranged between two percent (2%) and thirty seven percent (37%) with an average theoretical water savings projection of fourteen percent (14%). See Table 3. Water Savings Potential.

Figure 1. Typical Audit Map and MAWA Calculation

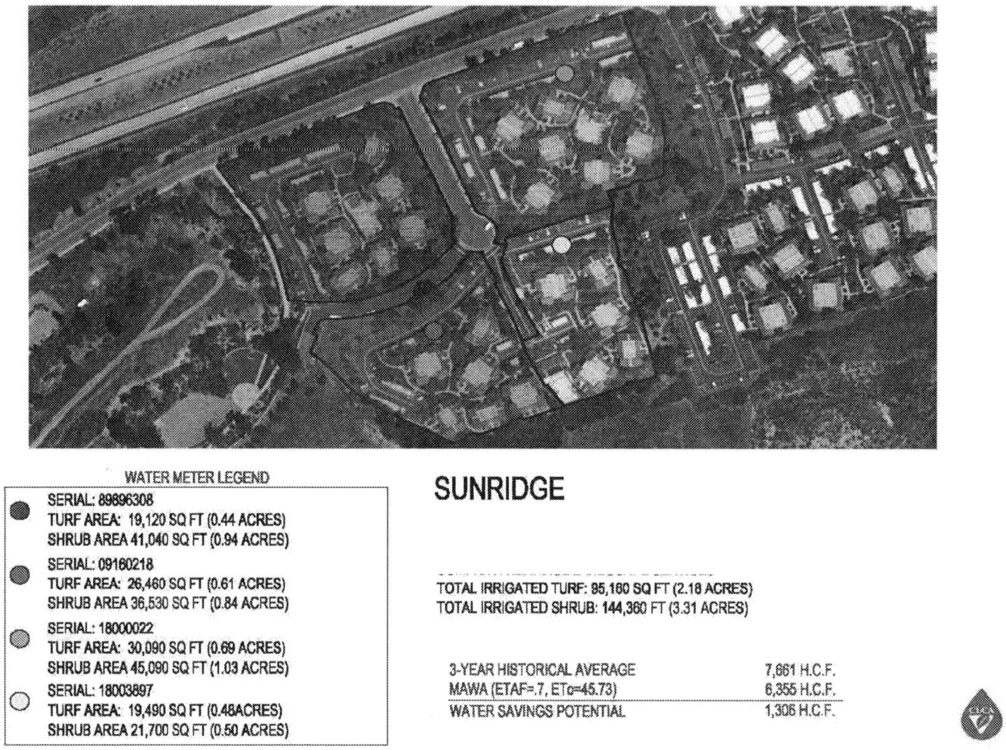

$$MAWA = (ET_o - Eppt)(0.62)[(0.7)(LA)+(0.3)(SLA)\}$$

Table 3. Water Savings Potential

	3-Year Historical Avg. HCF (m3)	MAWA Budget HCF (m3)	Water Savings Potential (%)
Average of 20 sites	5,156 (14,600)	4,331 (12,264)	14%

Equipment Upgrades: Each contractor was required to recommend irrigation equipment upgrades from a prioritized list that included pressure regulation, distribution uniformity, water loss (leak repair) and scheduling components. With a maximum budget of $5,000 per site for the equipment upgrades, the materials list recommended by contractors sought to maximize water savings. A designated irrigation supplier provided quotes for the recommended equipment. These were reviewed and approved by the Water Authority and SDG&E prior to purchase. All equipment was sourced from a single supplier. This allowed the supplier to directly invoice SDG&E and contractors did not incur upfront expenses. To further support the contractor, a $5,000 incentive was provided toward labor for the installation of the equipment. Hardware upgrades averaged $4,665 per site or $927 per acre (.4 ha) with tax. (Figure 2).

Figure 2. Average Equipment Expenditure per Site and per Acre

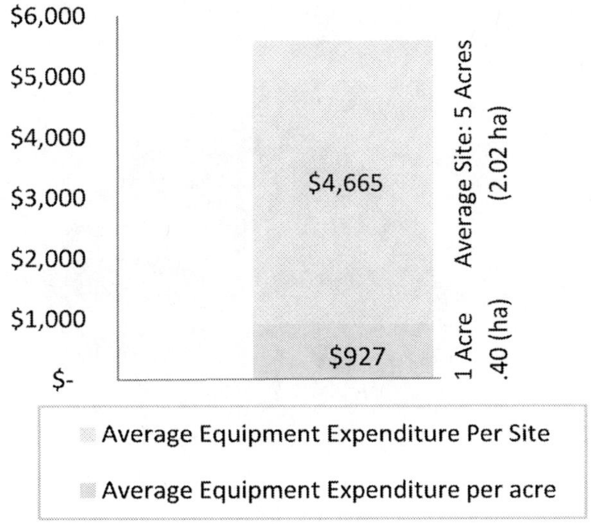

Each site invested in pressure regulation at one of three entry points: the backflow, the valve or the head, as appropriate. The equipment purchases broke down into four broad categories: Distribution Uniformity (DU)/Irrigation Efficiency (IE); Pressure Regulation; Leak Repair/Prevention; and Scheduling. With the first three categories representing ninety-seven percent (97%) of all purchases. The categories and general item descriptions are presented in Table 4. Equipment Categorization by Dollar Amount.

Table 4. Equipment Categorization by Dollar Amount

General Item type	Category	Pre-tax Category Total
Drip Rotator Nozzle Rotor Sprinkler	DU and/or IE	$ 29,145.01
Check Valve Flow Sensor Misc. pipe, fittings, wire, glue, boxes Remote Control Valve Sprinkler w/ check valve Swing Unit Valve	Leak Repair/Prevention	$ 27,215.45
Press Reg Rotor w/ Press Reg Sprinkler w/ Press Reg	Pressure Regulation	$ 27,352.12
Controller Components Rain Sensor	Scheduling	$ 2,219.40
Pre-tax TOTAL		**$ 85,931.98**

Figure 3. Equipment Category by Percent, illustrates the distribution of the broad categories and shows the focus of the WSLEP pilot program on pressure regulation at thirty-two percent (32%), distribution uniformity and/or Irrigation Efficiency at thirty-four percent (34%) and leak prevention/repair at thirty-two percent (32%).

Figure 3. Equipment Category by Percent

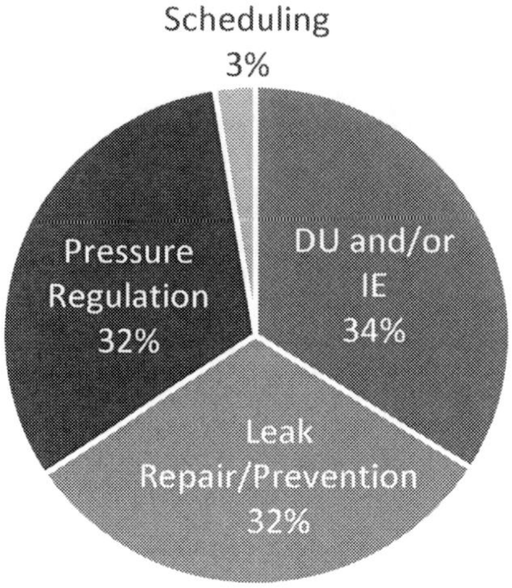

Workforce Training: A significant component of the WSLEP was the promotion and expansion of water management skills among landscape contractors. CLCA was tasked with training of the landscape contractors according to the Water Management Certification Program principles. The certification requires managing a landscape site at or below 80% of MAWA for one year. CLCA used its own database and reporting procedures to collect periodic water use reports that would enable them to track water use performance at each site.

Some of the participating contractors already possessed a Water Management certificate, while others took advantage of the program's training and technical support to further develop their irrigation management expertise and credentials. Each contractor had the necessary experience to identify needed irrigation upgrades to improve their site's irrigation performance and the potential for savings through careful water budgeting. All participants not already certified as a Water Manager received the standard one-day certification course. The need for additional support and education became apparent as the program progressed. A supplemental eight-hour specialized training was provided for all participants by Blue Watchdog Conservation, Inc. covering in depth concepts like water budget calculations, water scheduling, soil infiltration rates, runoff, controller programming and pressure issues. In addition, one on one coaching was provided, as needed, in the field and via telephone.

At the start of the project three contactors had their CLCA Water Manager Certification, including two that had attained 'Expert' level. By program end, the ranks of Certified Water Managers grew by an additional four managers, making a total of seven out of nine participants that held a certification. One of the new members has even gone on to attain Expert status.

Training in Action:

> Contractor, James Cothrine of Earthwise Industries shared his experience participating in WSLEP. Cothrine had two sites that suffered from significant distribution uniformity and pressure regulation issues. He started WSLEP without a certification in irrigation management. He was aware that to perform effectively, his sites would need to improve distribution uniformity and pressure regulation. WSLEP provided his sites the necessary financial assistance to purchase and install equipment deemed necessary to upgrade the system. He also benefitted from WSLEP's training and guidance to complete the CLCA Water Management Certificate, using the participating sites as his examples of water budgeting. Prior to participating, Cothrine had heard of MAWA and water budgeting, but had never applied it to any of his existing landscapes. He felt that MAWA only worked in a perfect world, not the variable and inconsistent reality of an older irrigation system. Using MAWA as a baseline, the sites started with a 3% (average) potential for water savings. At the end of the program a 40% (average) savings was achieved. Cothrine is now a Water Manager and applies his WSLEP training and experience to other sites under his management.

Maintenance Period: After completion of all irrigation repairs, participating sites entered the maintenance period. Water use performance was recorded for one year. Contractors were required to take monthly meter readings (bi-weekly recommended), maintain the irrigation systems and adjust schedules to maximize performance. No plant material changes (e.g., turf conversion) was allowed during the WSLEP program. To ensure high quality maintenance and a positive outcome for participating sites, a customer satisfaction survey of property owners was conducted at the end of the maintenance period.

A pre-existing CLCA Water Manager dashboard was adapted for the specific needs of this pilot program, including the ability to record baseline water use information and to compare it to on-going water use. Contractors were responsible for uploading monthly meter readings to the CLCA Water Manager dashboard. CLCA provided oversight and management of the site data. Contractors had the responsibility to program irrigation controllers appropriately for plant type and weather conditions with the aim of achieving water savings while ensuring positive landscape performance.

Program Results: The results of the WSLEP can be measured in terms of customer satisfaction and water savings. The definition of program success – and the long-term viability of the WSLEP approach – required both elements to be achieved concurrently. An anticipated program risk was the possibility of a statistical reduction in measured water use at the expense of plant health and aesthetic appearance. The desired target was to realize a significant reduction in water use while ensuring a high level of customer satisfaction.

Each customer (property owner or representative) completed a satisfaction survey at the end of the program. Program satisfaction was measured across fifteen criteria. Program-wide, survey responses were highly favorable. Customers were pleased with the level of funding support provided, which allowed them the opportunity to upgrade irrigation systems. Respondents generally understood the goals of the program and the potential implications for their sites' actual performance. Customers were satisfied with their contractors' implementation of WSLEP's irrigation management protocols. WSLEP water saving goals were achieved overall and participants found their landscape quality to be the same or better than prior to program participation. Seventy-nine percent (79%) of customers would participate again, if given the opportunity. On a scale of 1-10 (10 being Extremely Satisfied) the average response was nine (9).

Table 5. Customer Satisfaction Survey Results

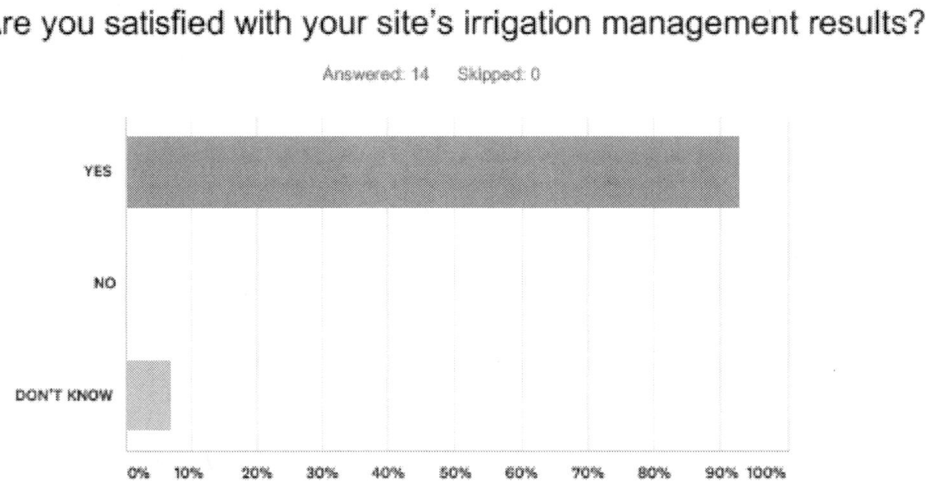

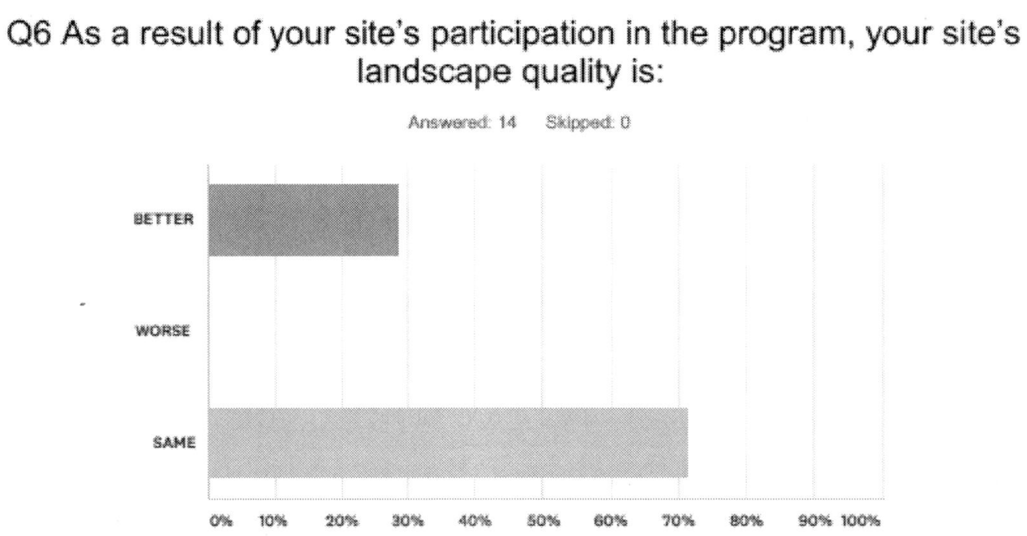

The results of actual water saved are presented below in Table 6. The raw consumption data was adjusted for weather. For the Pre-Use (baseline consumption), the quantities were weather normalized relative to a three-year average of the variance above and/or below the average ET_0. Similarly, the Post-Use consumption was weather normalized to reflect the actual ET_0 during the twelve-month period of the WSLEP. Weather-normalized differences ranged from 2.1 percent to 65 percent savings with an average savings of 30.7 percent and a median value of 35.75%. Fourteen of the twenty participants achieved the goal (minimum 20% savings) and received the final incentive of $1,700. Overall weather-normalized water savings totaled 30,939 Hundred Cubic Feet (HCF) (87,608.5 m3).

Table 6. Results of Water Use Before and During the Water Based Budget Period

WSLEP (Industry-Based Water Budget Program)										
	Raw Consumption Data				Weather Normalized Data					
Site #	Pre-Use 2011-2013 Average HCF (M3)	Post-Use Year 2014/2015 Average HCF (M3)	Raw Difference HCF (M3)	Raw Difference Relative (%)	Pre-Use Weather Effect (%)	Post-Use Weather Effect (%)	Pre-Use Weather Normalized HCF (M3)	Post-Use Weather Normalized HCF (M3)	Weather Normalized Difference in Use HCF (M3)	Weather Normalized Difference in Use, Relative (%)
1	2,710.0 (7,673.9)	2,588.0 (7,328.4)	-122.0 (-345.5)	-4.5%	0.90%	-0.10%	2,686.1 (7,606.2)	2,589.7 (7,333.2)	-96.43 (-273.1)	-3.60%
2	3,370.3 (9,543.6)	2,379.0 (6,736.6)	-991.3 (-2,807.1)	-29.40%	1.30%	6.90%	3,326.2 (9,418.7)	2,215.4 (6,273.3)	-1,110.83 (-3,145.5)	-33.40%
3	3,172.3 (8,983.0)	2,080.0 (5,889.9)	-1092.3 (-3,093.1)	-34.40%	1.30%	6.90%	3,130.8 (8,865.4)	1,937.0 (5,485.0)	-1,193.86 (-3,380.6)	-38.10%
4	5,435.7 (15,392.2)	4,719.0 (13,362.7)	-716.7 (-2,029.4)	-13.20%	0.90%	-0.10%	5,387.7 (15,256.3)	4,722.1 (13,371.5)	-665.68 (-1,885.0)	-12.40%
5	9,661.0 (27,356.9)	5,828.8 (16,505.3)	-3832.2 (-10,851.5)	-39.70%	0.50%	9.00%	9,613.9 (27,223.5)	5,306.3 (15,025.8)	-4,307.63 (-12,197.8)	-44.80%
6	4,093.0 (11,590.1)	2,662.0 (7,537.9)	-1431.0 (-4,052.1)	-35.00%	0.50%	9.00%	4,073.1 (11,533.7)	2,423.4 (6,862.3)	-1,649.69 (-4,671.4)	-40.50%
7	6,849.0 (19,394.2)	4,311.0 (12,207.4)	-2538.0 (-7,186.8)	-37.10%	0.50%	9.00%	6,815.6 (19,299.6)	3,924.5 (11,112.9)	-2,891.09 (-8,186.7)	-42.40%
8	6,018.0 (17,041.1)	4,052.0 (11,474.0)	-1966.0 (-5,567.1)	-32.70%	0.50%	9.00%	5,988.7 (16,958.1)	3,688.8 (10,445.5)	-2,299.92 (-6,512.6)	-38.40%
9*	4,017.3 (11,375.7)	3,647.0 (10,327.2)	-370.3 (-1,048.7)	-9.20%	0.50%	5.10%	3,997.8 (11,320.5)	3,461.2 (9,801.0)	-536.54 (-1,519.3)	-13.40%
10*	3,218.3 (9,113.2)	2,895.0 (8,197.7)	-323.3 (-915.6)	-10.00%	0.50%	5.10%	3,202.7 (9,069.0)	2,747.5 (7,780.1)	-455.12 (-1,288.8)	-14.20%
11	5,687.3 (16,104.6)	4,483.0 (12,694.4)	-1204.3 (-3,410.3)	-21.20%	1.30%	6.90%	5,612.9 (15,894.0)	4,174.7 (11,821.4)	-1,438.20 (-4,072.5)	-25.60%
12	4,656.3 (13,185.2)	3,578.0 (10,131.8)	-1078.3 (-3,053.5)	-23.20%	1.30%	6.90%	4,595.4 (13,012.7)	3,331.9 (9,434.9)	-1,263.46 (-3,577.7)	-27.50%
13	4,359.0 (12,343.3)	2,757.8 (7,809.2)	-1601.2 (-4,534.1)	-36.70%	0.50%	9.00%	4,337.8 (12,283.3)	2,510.6 (7,109.2)	-1,827.19 (-5,174.0)	-42.10%
14	3,845.0 (10,887.8)	2,456.2 (6,955.2)	-1388.8 (-3,932.6)	-36.10%	0.50%	9.00%	3,826.3 (10,834.9)	2,236.0 (6,331.6)	-1,590.25 (-4,503.1)	-41.60%
15	3,192.0 (9,038.7)	1,930.0 (5,465.2)	-1262.0 (-3,573.6)	-39.50%	0.50%	9.00%	3,176.5 (8,994.8)	1,757.0 (4,975.3)	-1,419.46 (-4,019.5)	-44.70%
16	2,514.0 (7,118.9)	2,433.0 (6,889.5)	-81.0 (-229.4)	-3.20%	1.30%	6.90%	2,481.1 (7,025.7)	2,265.7 (6,415.7)	-215.42 (-610.0)	-8.70%
17	3,712.7 (10,513.2)	1,353.0 (3,831.3)	-2359.7 (-6,681.8)	-63.60%	1.30%	6.90%	3,664.1 (10,375.6)	1,260.0 (3,567.9)	-2,404.13 (-6,807.7)	-65.60%
18	8,298.7 (23,499.3)	5,504.0 (15,585.6)	-2794.7 (-7,913.6)	-33.70%	0.50%	9.00%	8,258.2 (23,384.6)	5,010.6 (14,188.4)	-3,247.64 (-9,196.3)	-39.30%
19	7,661.0 (21,693.5)	6,026.0 (17,063.7)	-1635.0 (-4,629.8)	-21.30%	0.50%	9.00%	7,623.7 (21,587.9)	5,485.8 (15,534.1)	-2,137.87 (-6,053.8)	-28.00%
20	9,021.7 (25,546.6)	9,359.0 (26,501.7)	337.3 (-955.2)	3.70%	1.30%	6.90%	8,903.6 (25,212.2)	8,715.4 (24,679.3)	-188.23 (-533.0)	-2.10%
Total	**101,492.7** (287,395.3)	**75,041.8** (212,494.7)	**-26450.8** (-74,900.4)	**-26.10%**			**100,702.2** (285,156.9)	**69,763.5** (197,548.2)	**-30,938.66** (-87,608.5)	**-30.70%**

Note: Sites 9 & 10 had meter replaced. Old meters tend to under-register consumption. Data from 2015 only

Note: First order weather correction (ETo-0.25*Precipitation)

(M3) = Cubic Meter (2.83168466 M3=1 HCF)

Program Challenges and Suggested Enhancements for Future Rounds

Challenges:

- Identifying sites: The ability to find and vet appropriate sites was an initial challenge. In some cases, contractors may hesitate to propose sites that are currently under their management despite having a water savings potential. The general concern is that a significant water savings potential might be perceived by the owner as a sign of current under-performance by the contractor.
- Water management skills and establishing baseline data: Participating contractors had a very diverse range of skills. Determining prior water use baselines, accurately measuring irrigation areas, and calculating accurate water budgets were challenging tasks for most participants. In a few cases, meter issues and confusing water utility consumption records added to these challenges, causing pre-use consumption baselines to be inaccurate. These were later identified and adjusted. In some cases, area measurements were not associated with the correct meter which also impacted performance against the reported MAWA budgets. CLCA retained an independent service provider, Blue Watchdog Conservation, to provide technical assistance and quality control on WSLEP records. This added resource ensured consistent and accurate records across the program.
- Although directed to carefully identify needed equipment in advance to minimize the number of transactions necessary with the irrigation supplier, many contractors had returns, exchanges and multiple purchases as part of their projects. Reconciling these transactions became a heavy burden on the sponsoring agencies, ultimately requiring third party assistance.
- Payment schedules for incentives were impacted by the review process and contractors desired a quicker payout for installation expenses incurred.
- MAWA water budgets did not seem to fit real world conditions in some situations.

Future Program Enhancements:

- Plan timeline to allow for careful screening and selection of sites and completion of audits.
- Increase upfront training to ensure the contractors have the knowledge and tools to effectively participate in the program.
- Prompt contractors to thoroughly plan for equipment requirements and to avoid exchanges. This will avoid delays in the approval process and minimize processing time.
- Modify the incentive schedule to ensure payouts are issued more expediently.
- Modify the CLCA water budget tool to incorporate actual ET_o .
- Improve water use data by installing flow sensors at each meter, or, use automatic meter readings where installed by participating utilities.

Conclusions

The WSLEP pilot program has successfully demonstrated the use of water management protocols based on California's MWELO for managing the water use at large, established landscapes. Thus, MWELO's methodology is as relevant to the design of new landscapes as it is to the ongoing management of existing landscapes. WSLEP areas of emphasis included pressure regulation; distribution uniformity; leak repair; and smart controllers along with workforce training and water management. This comprehensive and integrated approach to irrigation management of mature landscapes has proven to be an effective and scalable methodology that can be replicated as needed to help meet local and regional water conservation goals.

WSLEP implementation coincided in part with periods of state-mandated water use restrictions in California. Such restrictions can be very challenging to the green industry. WSLEP has demonstrated a business-friendly model that has the potential to help sites comply with such requirements while also generating additional voluntary water savings. WSLEP's training activities demonstrated the importance and effectiveness of workforce development initiatives on topics such as water budgeting, area measurement techniques, reading and recording water use via the water meter, scheduling controllers and regular system adjustments. Such training empowers contractors to achieve water-saving targets while preserving, and even enhancing, landscape quality. Water budgets are an essential tool for effective landscape irrigation management.

On average the program exceeded its stated goal of a 20% reduction in water use. WSLEP validated the premise that a well-trained workforce, coupled with pressure regulation and distribution uniformity improvements can deliver significant water savings. Should the program be implemented again in the future, the use of flow sensing equipment and/or submeters is highly recommended to help automate the WSLEP's extensive data collection needs.

Acknowledgements

The authors gratefully acknowledge the following individuals and organizations for their contributions to this program:

The California Public Utilities Commission; the California Department of Water Resources (funding made possible in part by a Proposition 50 grant administered by DWR); Jeff Alexander (Program Co-Manager) of San Diego Gas & Electric; David Silva (Implementation Administrator) of the California Landscape Contractors Association (CLCA); the San Diego Chapter of CLCA; and Patrick Crais of Blue Watchdog Conservation, Inc. (training, technical assistance and quality control); all participating property owners and landscape contractors, especially James Cothrine (WSLEP participant) of Earthwise Industries for sharing his WSLEP insights and experience.

The WSLEP program was implemented in close collaboration with the following member agencies of the Water Authority: City of Carlsbad, City of San Diego, Helix Water District; and Olivenhain MWD.

References:

Chesnutt, Tom, June 2017, A & N Technical Services, Inc., DROP Grant Program Process Evaluations, San Diego County Water Authority.
San Diego County Water Authority, June, 2017, Proposition 50 Urban Drought Assistance Grant 2008 Drought Response and Outreach Program (DROP), Final Report.
Model Water Efficient Landscape Ordinance, 2009, State of California, Department of Water Resources, https://water.ca.gov/LegacyFiles/wateruseefficiency/docs/MWELO09-10-09.pdf

Appendix A Results of Water Use Before and During the Water Based Budget

WSLEP (Industry-Based Water Budget Program)

Raw Consumption Data					Weather Normalized Data**					
Site #	Pre-Use 2011-2013 Average (HCF)	Post-Use Year 2014/2015 Average (HCF)	Raw Difference (HCF)	Raw Difference Relative (%)	Pre-Use Weather Effect (%)	Post-Use Weather Effect (%)	Pre-Use Weather Normalized (HCF)	Post-Use Weather Normalized (HCF)	Weather Normalized Difference in Use (HCF)	Weather Normalized Difference in Use, Relative (%)
1	2,710.0	2,588.0	-122.00	-4.5%	0.9%	-0.1%	2,686.1	2,589.7	-96.43	-3.6%
2	3,370.3	2,379.0	-991.33	-29.4%	1.3%	6.9%	3,326.2	2,215.4	-1110.83	-33.4%
3	3,172.3	2,080.0	-1092.33	-34.4%	1.3%	6.9%	3,130.8	1,937.0	-1193.86	-38.1%
4	5,435.7	4,719.0	-716.67	-13.2%	0.9%	-0.1%	5,387.7	4,722.1	-665.68	-12.4%
5	9,661.0	5,828.8	-3832.17	-39.7%	0.5%	9.0%	9,613.9	5,306.3	-4307.63	-44.8%
6	4,093.0	2,662.0	-1431.00	-35.0%	0.5%	9.0%	4,073.1	2,423.4	-1649.69	-40.5%
7	6,849.0	4,311.0	-2538.00	-37.1%	0.5%	9.0%	6,815.6	3,924.5	-2891.09	-42.4%
8	6,018.0	4,052.0	-1966.00	-32.7%	0.5%	9.0%	5,988.7	3,688.8	-2299.92	-38.4%
9*	4,017.3	3,647.0	-370.33	-9.2%	0.5%	5.1%	3,997.8	3,461.2	-536.54	-13.4%
10*	3,218.3	2,895.0	-323.33	-10.0%	0.5%	5.1%	3,202.7	2,747.5	-455.12	-14.2%
11	5,687.3	4,483.0	-1204.33	-21.2%	1.3%	6.9%	5,612.9	4,174.7	-1438.20	-25.6%
12	4,656.3	3,578.0	-1078.33	-23.2%	1.3%	6.9%	4,595.4	3,331.9	-1263.46	-27.5%
13	4,359.0	2,757.8	-1601.20	-36.7%	0.5%	9.0%	4,337.8	2,510.6	-1827.19	-42.1%
14	3,845.0	2,456.2	-1388.80	-36.1%	0.5%	9.0%	3,826.3	2,236.0	-1590.25	-41.6%
15	3,192.0	1,930.0	-1262.00	-39.5%	0.5%	9.0%	3,176.5	1,757.0	-1419.46	-44.7%
16	2,514.0	2,433.0	-81.00	-3.2%	1.3%	6.9%	2,481.1	2,265.7	-215.42	-8.7%
17	3,712.7	1,353.0	-2359.67	-63.6%	1.3%	6.9%	3,664.1	1,260.0	-2404.13	-65.6%
18	8,298.7	5,504.0	-2794.67	-33.7%	0.5%	9.0%	8,258.2	5,010.6	-3247.64	-39.3%
19	7,661.0	6,026.0	-1635.00	-21.3%	0.5%	9.0%	7,623.7	5,485.8	-2137.87	-28.0%
20	9,021.7	9,359.0	337.33	3.7%	1.3%	6.9%	8,903.6	8,715.4	-188.23	-2.1%
Total	**101,492.7**	**75,041.8**	**-26450.83**	**-26.1%**			**100,702.2**	**69,763.5**	**-30938.66**	**-30.7%**

** Sites 9 & 10 had meter replaced. Old meters tend to under-register consumption. Data from 2015 only*

*** First order weather correction ((ET$_o$ -0.25*Precipitation)*

Student Water Managers

Christopher J Eppler, Energy Engineer, BS Geology and Masters in Energy Engineering, IA Certified Irrigation Designer, IA Certified Irrigation Contractor, IA Certified Irrigation Auditor
Canyons School District, 9150 South 500 West, Sandy, Utah 84070-
Christopher.eppler@canyonsdistrict.org

Abstract

Properly trained high school students can become water managers extending a program's ability to gather data to more optimally program irrigation controllers. This saves limited water, critical school district funds, and provides students competitive summer employment. The process of gathering required data to create irrigation schedules can be time-consuming or low-priority for smaller staffs at school districts, universities, cities, and municipal parks. Properly educated and supervised students can identify and document needed information to create accurate program schedules. Students are taught using The Irrigation Training & Research Center's online classes. Students find real-world application for their math, computer and physical science learned in school and develop skills with a potential crossover in future jobs. It deepens their understanding of large landscape actual water needs. We partnered with the Utah Water Conservancy District who provided an annual $10,000 grant to hire students the first two summers. Based on the 2015 field season, students projected a savings of 24,740,616 gallons using their data, combined with historical evapotranspiration rates. After the 2016 irrigation season, 19,677,361 gallons were actually saved. Not reflected in this is a defective meter that could potentially have increased the saving of another eight million gallons.

Canyons School District
Students Projected 2016 Savings 24,740,616 Gallons per Year for 13 Elementary Schools

	ET (Gallons)	Current (Gallons) 2015	Projected (Gallons) 2016	Savings (Gallons)
Copperview Elem	7,667,264	18,678,355	10,106,160	8,572,195
Altara Elem	6,100,975	11,300,499	8,013,288	3,287,211
Park Lane	4,393,367	7,593,547	5,213,760	2,379,787
East Midvale Elem	4,385,881	6,865,000	5,274,000	1,591,000
Sunrise	6,210,451	11,054,957	9,408,941	1,646,016
Bella Vista Elem	7,115,450	9,807,321	8,278,597	1,528,724
Edgemont Elem	6,608,033	9,636,464	8,148,988	1,487,476
Willow Canyon	6,234,342	8,886,565	7,584,081	1,302,484
Crescent Elem	4,759,622	6,647,620	5,735,902	911,718
Granite	4,580,366	5,487,379	4,623,329	864,050
Silver Mesa	5,302,668	6,468,917	5,777,058	691,859
Brookwood Elem	4,350,255	5,762,636	5,497,130	265,506
Quail Hollow	4,783,421	6,791,727	6,901,798	-110,071
Total Savings	61,724,449	90,276,715	74,800,153	**24,417,955**

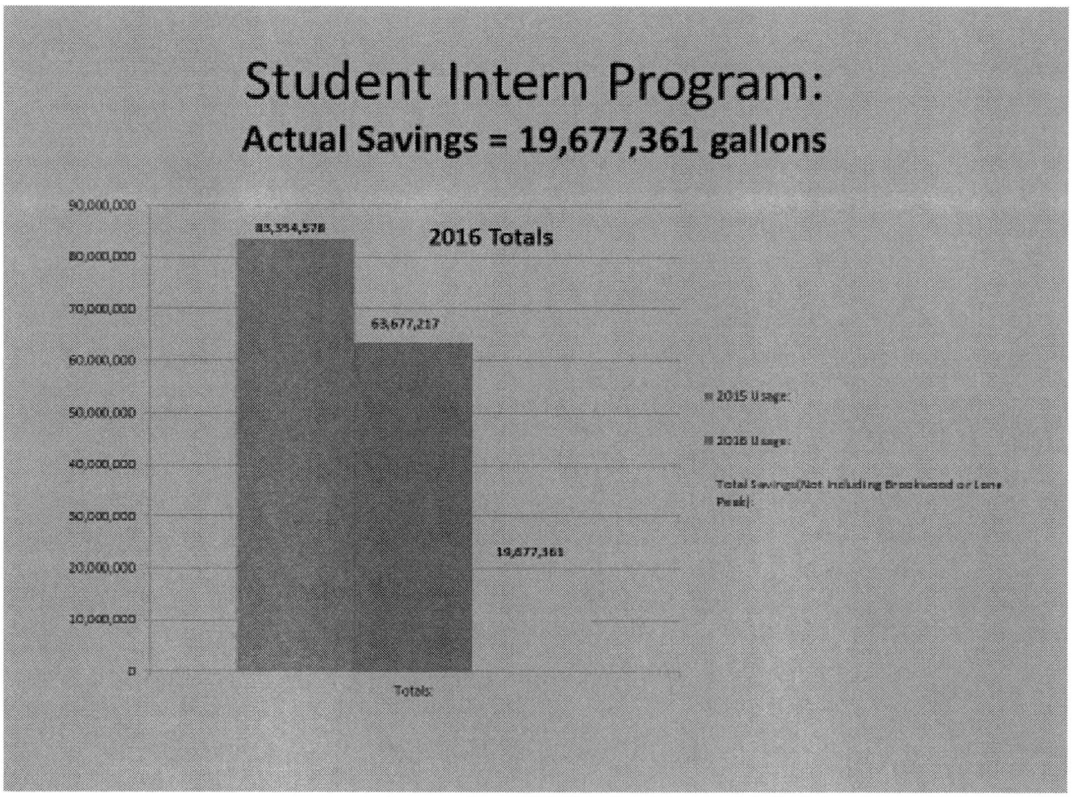

Students

Like hiring any employee, selecting the right students can determine the success of the program. After four years implementing the program, my greatest success has been hiring high school juniors or seniors who come recommended by the Advanced Placement Math department. During a student's interview, a fuller explanation of the responsibilities of the position are discussed. For our program, 6–8 qualified students are optimal balancing my time to instruct and supervise them. I work to create an enjoyable work experience and am flexible with anticipated summer vacation schedules, typically about two weeks off. I have found that structuring the work to be Monday through Thursday, 9 to 2:00 p.m. with a half hour lunch*, allows a balance that keeps youth engaged and focused. Teams work in pairs to collect field data and input into spreadsheets. Once a school is completely evaluated, teams are reassigned new partners. I find this keeps productivity high, and allows different opportunities for students to learn to work together. Field work is done in the morning before the heat of the day and classroom work is done during the afternoon. Students provide their own transportation, logging mileage for reimbursement. Cell phones are allowed for communication and safety. I have found students have been respectful with personal cell phone use while working. Rarely do I need to ask for phones to be put away. (*check with your HR on local employment regulations)

Curriculum

The Irrigation Training & Research Center (ITRC) online classes comprise the curriculum for the program including Basic Hydraulics; Basic Soil-Plant Water Relationships; Distribution Uniformity and Precipitation Rate; Evapotranspiration; and Irrigation System Components.

In the Basic Hydraulics class, the concepts of flow, velocity, and pressure are explained so that students develop an understanding of how water moves in an irrigation system. Students learn about characteristics of pipe size and about how the pipe's physical properties impact water flow.

The Basic Soil-Plant Water Relationships class lays the foundation for principles related to soil moisture content using a sponge to represent soil. Topics covered are soil saturation, field capacity, permanent wilting point, available water holding capacity, management allowable depletion, and soil moisture depletion. It explores the different soil types such as clay, silt, sand, loam and the management allowable depletion (MAD) of each.

Distribution Uniformity Lower Quarter (DUlq) enables students to understand system performance, and how evenly plants throughout a zone are watered. DUlq is used for scheduling irrigation and determining the amount of time the system should run. The students take the average of the lower quarter water applied and divide it by the average of the whole area applied. A catch canister test is performed to determine the DUlq.

The students learn to calculate the precipitation rate (PR) or identify the PR using a catalog. PR is the rate at which irrigation water reaches the ground surface, usually measured in inches per hour. They learn the difference between flow and PR. Additionally, they come to understand the relationship between soil, slope and the precipitation rate, along with the concept of cycle and soak to overcome potential over watering. In order to maintain a healthy turf, students discover that sometimes the zone's lowest precipitation rate must be used to irrigate. Matched precipitation is taught to be the objective of each zone.

Evapotranspiration (ETo) allows the students to understand the combined effect of plant transpiration and evaporation from the soil and plant leaves. Students learn that a landscape coefficient needs to be multiplied by the ETo in order to determine the plant water requirement ETl. The landscape coefficient is comprised of three things: plant type (tree, shrub, turf, flower), plant density (how closely plants are spaced together in an area), and microclimate. Finally, students ascertain daily, weekly and monthly ETo rates collected from the controller and from the Utah Department of Natural Resources, along with how to use ETo in the calculation of their run times.

The effective rainfall is the fraction of total rainfall that is actually stored in the root zone. Rainfall can be sometimes ignored, especially when there is a minimal recorded amount.

Prior to the class, few if any of the students have had any experience installing or working on sprinkler systems, so I include a class on system components. This allows them to learn the various irrigation hardware and to begin visualizing the layout of a designed system. Students are not asked to repair or install any irrigation components.

Field and Classroom Procedure

Step 1: Prior to sending students on location it is helpful to notify the custodian that students will be working on the property. Working in pairs, students arrive at a school and introduce themselves to the custodian. They politely explain what they will be doing and when both parties are comfortable, they begin evaluating the sprinkler system. Students first, determine sprinkler type, nozzle size, arc (full, half, etc.), and establish how many heads are in each zone. Any broken, missing, bent, or poorly adjusted sprinkler heads are reported so that a work order can be submitted to the maintenance department. The students responsibility is to collect data not to repair. I ask the students to note landscape material: turf, trees, shrubs or flowers. After an initial observation of the process, students measure the root zone, using either a soil probe or vertical slice of the soil. Students then categorize the area's exposure; full sun, partial sun or shade. Student take steps to identify the soil type by estimating whether it is clay, silt, sand or loam. Proper soil identification includes taking a soil sample to the local horticulture extension office or university for lab analysis. The square footage of the landscape area is determined from Google Earth.

Step 2: Utilizing field data, students enter this into an excel spreadsheet with columns for zone, zone location, number of heads per zone, head type, arc, nozzle, head gpm, zone gpm, psi, catalog or calculated head and zone precipitation rate. The precipitation rate, as well as the gallons per minute, are determined by nozzle color or orifice size. When a zone seems poorly designed, a catch canister test provides a more accurate precipitation rate and DUlq. When pressure looks abnormally low or high, a pitot tube is occasionally used. The students historically have struggled to get accurate psi readings with the pitot tube. Until a zone has a matched precipitation rate, I recommend to the students to input the lowest PR. By doing this we prevent under watering and damage to the turf. I know that this is not ideal water conservation, but landscape costs can be exorbitant and children can get hurt if the play areas become too hard and turf bare. Fortunately, I have been able to get the sport field heads matched within a year or two.

Student Intern Program

Step 2: Populate spread sheets with Zone Data

Site:	Albion			Date:	7/11/2018						
Small Rotors	Use 45 psi										
Rotors	Use 60 psi						Total				
Fixed	Use 30 psi					Head	Head	Zone		Head	Zone
Zone	Location	head #	Head type	Arc.	Noz.	GPM	GPM	GPM	PSI	PR	PR
1	E. Parking Edge	1	Hunter I-40	Q	Grey	15.7	15.7	187.7		2	0.66
		1	RB 8005	Q	Brown	15.9	15.9			1.48	
		3	RB 8005	Q	Teal	14.3	42.9			1.48	
		1	Hunter I-40	H	Grey	15.7	15.7			1	
		1	RB 8005	H	Blue	17.8	17.8			0.81	
		2	Hunter I-40	Q	No noz	23.9	47.8				
		1	RB 8005	H	Brown	15.9	15.9			0.72	
		1	RB 8005	H	Tan	12	12			0.66	
2	E. Field W.	4	RB 8005	F	Teal	14.3	57.2	89		0.37	0.36
		2	RB 8005	F	Brown	15.9	31.8			0.36	
3	Next to 2	3	RB 8005	F	Teal	14.3	42.9	86.7		0.37	0.33
		2	RB 8005	F	Brown	15.9	31.8			0.36	
		1	RB 8005	F	Tan	12	12			0.33	
4	Next to 3	3	RB 8005	F	Teal	14.3	42.9	86.7		0.37	0.33
		1	RB 8005	F	Tan	12	12			0.33	
		2	RB 8005	F	Brown	15.9	31.8			0.36	
5	Next to 4	4	RB 8005	F	Teal	14.3	57.2	89		0.37	0.36
		2	RB 8005	F	Brown	15.9	31.8			0.36	
6	Next to 5	3	RB 8005	F	Teal	14.3	42.9	90.6		0.37	0.36
		3	RB 8005	F	Brown	15.9	47.7			0.36	
7	Next to 6	1	RB 8005	F	Brown	15.9	15.9	87.4		0.36	0.36
		5	RB 8005	F	Teal	14.3	71.5			0.37	
8	Next to 7	5	RB 8005	F	Teal	14.3	71.5	87.4		0.37	0.36
		1	RB 8005	F	Brown	15.9	15.9			0.36	
9	Next to 8	5	RB 8005	F	Teal	14.3	71.5	87.4		0.37	0.36
		1	RB 8005	F	Brown	15.9	15.9			0.36	

Step 3: The students calculate the ideal zone run time, run times per cycle, and irrigation days per month. I have taken the Irrigation Association Audit worksheet and made it into an excel sheet for each zone. I have made the sheet interactive so as I change values, the sheet updates. The exercise allows each student to have an understanding of how DU, PR, root zone, and soil type affect the irrigation schedule. As values change for each of the zone characteristics, so does the run time.

Student Intern Program

Step 3: Calculate Ideal Zone Run Time

Project Name: Altara Elementary
Location:
Date:
Soil Type: Sandy Loam

Plant Water Requirement		April	May	June	July	August	September	October
		16.07	4.12	3.25	3.05	5.32	8.07	35.47
Plant Material								
Reference Period								
Reference ET	in. per mo.	1.28	4.99	6.32	6.75	6.2	2.55	0.58
Landscape Coefficient		0.7	0.7	0.7	0.7	0.7	0.7	0.7
Allowable Stress								
Plant Water Requirement	inches	0.896	3.493	4.424	4.725	4.34	1.785	0.406
Irrigation Water Requirement								
Precipitation Rate	in. per hr.	0.58	0.58	0.58	0.58	0.58	0.58	0.58
Distribution Uniformity		0.6	0.6	0.6	0.6	0.6	0.6	0.6
Irrigation Water Requirement	inches	1.49	5.82	7.37	7.88	7.23	2.98	0.68
Total Run Time per Period	minutes	91.43	356.43	451.43	482.14	442.86	182.14	41.43
Scheduling Requirements								
Root Zone Soil Type								
Available Water	in. per hr.	0.12	0.12	0.12	0.12	0.12	0.12	0.12
Active Root Zone Depth	inches	8	8	8	8	8	8	8
Plant Available Water	inches	0.96	0.96	0.96	0.96	0.96	0.96	0.96
Allowable Depletion		0.48	0.48	0.48	0.48	0.48	0.48	0.48
Irrigation Days per Period	days	1.87	7.28	9.22	9.84	9.04	3.72	0.85
Total Run Time per Day	minutes	48.98	48.98	48.98	48.98	48.98	48.98	48.98
Run Time per Cycle	minutes	19.63	19.63	19.63	19.63	19.63	19.63	19.63
Cycles per day								
Corrected Run Times	minutes							

Step 4: The students create an irrigation schedule. Included in this schedule are the zone run times, the days between irrigation cycles, and the gallons used for each cycle. The students can then sum the gallons for a complete cycle and estimate annual irrigation. These calculations are based on historical ET.

Since 2017, the controllers are operating using daily ET along with the information gathered in the field and input into the step 2 spreadsheet. I continue to have the students calculate a run time schedule as a learning experience.

Student Intern Program:

Step 4: Implement a Calculated Schedule or Water to Evapotranspiration Rate using Calculated Precipitation Rates

Evapotranspiration is the sum of evaporation and plant transpiration from the Earth's land and ocean surface to the atmosphere.

Altara											
		GPM	Gallons per cycle	Time	April	May	June	July	August	September	Ocotober
zone:	description										
a-2	East field center	75.8	8,461	111.63	16.07	4.12	3.25	3.05	3.32	8.07	35.47
a-3	West curb east field	139	9,016	64.86	16.07	4.12	3.25	3.05	3.32	8.07	35.47
a-4	adjacent to a-2	154	9,357	60.76	16.07	4.12	3.25	3.05	3.32	8.07	35.47
a-5	adjacent to a-4	93.6	7,131	76.19	16.07	4.12	3.25	3.05	3.32	8.07	35.47
a-6	adjacent to a-5	52.5	4,941	94.12	16.07	4.12	3.25	3.05	3.32	8.07	35.47
a-7	void										
a-8	Along 110 south east side	86.5	2,386	27.59	16.07	4.12	3.25	3.05	3.32	8.07	35.47
a-9	Along 110 south west side	75.1	2,072	27.59	16.07	4.12	3.25	3.05	3.32	8.07	35.47
a-10	Front west corner	32.85	906	27.59	16.07	4.12	3.25	3.05	3.32	8.07	35.47
a-11	West front perimeter fence	66.3	1,829	27.59	16.07	4.12	3.25	3.05	3.32	8.07	35.47
a-12	West side building	39.8	1,098	27.59	16.07	4.12	3.25	3.05	3.32	8.07	35.47

Step 5: Students calculate an estimated monthly water use amount from historical ETo, DU, and landscape acreage and compare it to the billed water use for the site. Inside water use is taken into consideration. First, the landscape acreage is multiplied by 27,154 gallons to determine the gallons of water for 1 inch of water over the landscape site. Second, the monthly ETo is divided by a determined distribution uniformity percentage, to get a value for the required irrigation in inches for that month. Third, the required irrigation for each month is multiplied by the gallons for the landscape area. The calculated value is the total gallons estimated for that month. This value is compared to the actual usage. The difference between the two represents an estimation of how close the program irrigated to ETo.

Student Intern Program

Step 5: Compare Estimated Turf Water Requirement to Actual Water Use

Altana Elementary			
Custodian		Elisa Rodriguez	
Sq. Footage		270,508	
Acres		6.21	
1 Acre Inch		27,154	gallons
Site Acre In.		168,626	gallons
Estimated Inside Usage		40,000	gallons
	2015	2015	2016
	ET	Actual Usage	New Usage
January	40,908	40,908	40,908
February	40,000	40,000	40,000
March	47,631	47,631	47,631
April	518,683	502,645	376,742
May	268,554	504,831	1,344,874
June	1,468,611	2,007,911	1,723,708
July	1,106,626	2,873,401	1,779,832
August	1,165,084	2,337,438	1,779,832
September	996,457	2,082,594	671,891
October	448,422	863,140	206,371
November			
December			
Total Gallons	6,100,375	11,300,409	8,015,288

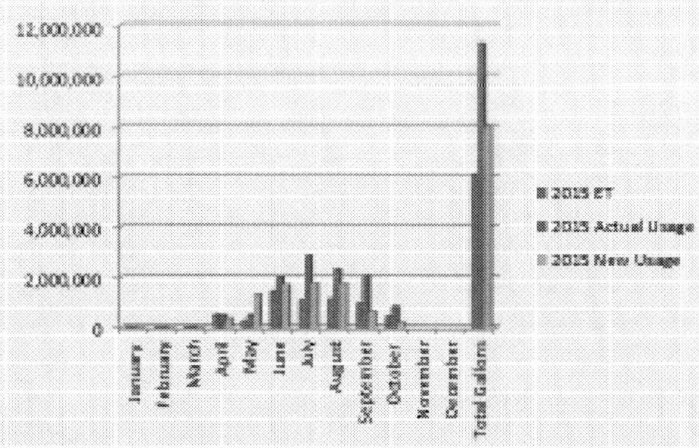

Conclusion:

Canyons School District has found sufficient value in the student water management program to employ students as summer employees. Students fill the void of needed staff to oversee the landscape irrigation. I realize that the program has to make assumptions in order to oversee the many large sites, but the students are all instructed to follow proper water management protocol learned through the Irrigation Training & Research Center. The program no longer applies for, or is dependent on, grants from the The Central Utah Water Conservancy District. Canyons School District now employs the students. By identifying interested, committed students with a willingness to learn, a productive and effective program can be built.

References

Irrigation Training & Research Center, 2018. Basic Hydraulics, Basic Soil-Plant Relationships, Distribution Uniformity & Precipitation Rate, Evapotranspiration, Irrigation System Components at https://www.irrigation.org/IA/Education/Online-Learning/ITRC-Classes/IA/Education/ITRC_Classes.aspx?hkey=de197fa6-967d-47c8-810d-5176ee292c54

The Irrigation Association, 2000. Certified Irrigation Designer Reference Manual

The Irrigation Association, 2002. Certified Irrigation Contractor Workbook

Irrigation Training and Research Center, 2002. Landscape Water Management Principles

Rain Bird, 2018. Landscape Irrigation Products Catalog

Hunter, 2018. Product Catalog

Allan Hancock College Overcomes Drought Challenges

Erin Hagen, Product Specialist
Calsense, 2075 Corte Del Nogal, Suite P, Carlsbad, CA 92011, erinh@calsense.com

A.J. van de Ven, President
Calsense, 2075 Corte Del Nogal, Suite P, Carlsbad, CA 92011

Abstract. The Governor of California declared a State of Emergency in January 2014 in response to the prolonged drought conditions plaguing the State. All Californians were asked to reduce water use by 20%. Allan Hancock College began evaluating ways to conserve water to comply with the emergency and help manage both their labor and water costs.

One of the top priorities during the evaluation was to maintain the playability of the sports fields which was challenging due to their constant use, resulting in a limited water window.

To overcome their challenges, the staff installed the latest in cloud-based irrigation technology including smart irrigation controllers, on-site weather devices, and master valves and flow sensors to detect and shut down leaks and breaks, manage flow, and record water usage. The controllers automatically operated to the full capacity of the system by dynamically determining how many valves could be energized at a time based on actual measured flow rates.

The result was a savings of over 40% of their historical water use while maintaining the sports field turf quality. Furthermore, the labor savings was calculated at 3-hours per staff member, per week, time previously spent on manual watering and programming their previous controllers. The water and labor savings paid back the cost of the control system in less than one year.

Keywords. Manufacturer, Turf/Landscape (Commercial), Conservation, Evapotranspiration, Soil Moisture, Sustainability, Water Budget, Groundwater, Controllers, Turf/Landscape Central Control, Turf/Landscape Smart Controller, Flow & Water Meter, Water Resources Management

Background

In response to the prolonged drought affecting the State of California, Governor Jerry Brown declared a State of Emergency in 2014. His proclamation "that conditions of extreme peril to the safety of persons and property exist in California due to water shortage and drought conditions with which local authority is unable to cope" was primarily driven by 2014 being the driest year on record, diminishing water supplies. The resulting regulation mandated that Californians reduce water usage by 20 percent.

Allan Hancock College is in northern Santa Barbara County and is part of the California Public Community College system. The college serves the community well with over 98% of its students coming from the local area. It is one of the five best community colleges in the state and has four locations in Santa Maria, Lompoc, Solvang and at Vandenberg Air Force Base. The college is well known for its accomplished athletic programs including intercollegiate football, basketball, soccer, tennis, baseball, softball, golf, volleyball, cross country, and track and field. The college offers opportunities for those who want to begin a bachelor's degree (university degree), earn an associate degree (two-year degree), prepare for a career, or upgrade skills. The college offers degrees and certificates in more than 100 areas of study. Allan Hancock College is well-known for its English as a second language program, its professional theatre program, and for providing superior support services for its students, including counseling and tutoring.

Although Allan Hancock College was not mandated to achieve the State-wide conservation goal of 20% because the Santa Maria Valley Groundwater Basin's storage was plentiful, the college felt the need to reduce water usage to comply with the local and state restrictions and protect their water supplies for future generations. The College also saw this as an opportunity to reduce their consumption in the face of continued increases to their water costs, thereby saving money.

Challenges

The primary challenge for Allan Hancock College was the maintenance and care of its sports fields. The quality and playability of the turf needed to be maintained, despite the ongoing drought. The need for regular, consistent watering was further hampered by the fields' constant use, both day and night. This prevented the maintenance staff from being able to effectively manually water the fields and their existing irrigation system was unable to apply the necessary water during the allotted time at night.

Although the local water utility had groundwater water available without restrictions, the supply was still limited, and the availability window was short. Furthermore, the City of Santa Maria, in which Allan Hancock College is located, has an average annual rainfall of only 9-inches, further requiring management of the groundwater in the event the basin is not recharged fully and the aquifer begins to be over-drafted.

At the time, the College had not invested in a centralized irrigation system, relying on traditional timers and manual watering to manage the campus. There were no master valves or flow sensors which meant there was no systemized leak or line break detection.

Solution

A system evaluation was conducted by the staff which determined the introduction of master valves and flow sensors as a necessary step to monitor irrigation lines for leaks and breaks. Furthermore, due to the limited access to the sports fields, the nightly water window needed to be carefully managed to optimize the water source availability.

After the evaluation, a smart irrigation system was select to provide a better way of monitoring and maintaining the College's irrigation systems. Flow sensors were installed at each point of connection to automatically monitor the system's water use and shut down valves and master valves in situations when breaks occurred.

The campus also implemented daily weather adjustments using an on-site evapotranspiration (ET) gauge and tipping rain bucket. Since most of the irrigated area was turf, the maintenance staff started by allowing the system to irrigate based on 100% of ET campus-wide. After careful monitoring through the new system's cloud-based management portal, the various irrigation valves were adjusted up or down as needed, maintaining the plant's quality.

Coupling the daily weather adjustments and the controllers' ability to automatically water to the irrigation system's fullest capacity allowed for the water being applied not only as needed, but also within the allotted time each day that it was scheduled to water.

Results

In short order, the irrigation staff began to see improvements in the sports field turf. The technology the college had invested was paying off. The time previously spent on manually watering and adjusting controller programming was reduced weekly by three hours per staff member per week. The water management reports in the year following installation showed that same areas used over 40% less water the previous year (Table 1). The new technology automating the irrigation system, and the more efficient use of the less expensive well water, resulted in significant savings. In fact, the savings achieved resulted in a return on investment on the irrigation system improvements within the very first year.

Month	Days in Month	Historical ET	Actual ET	Budget (gal)	Adj. Budget (gal)	Usage (gal)	Savings (gal)	Percent Saved
Jul 2017	31	5.13	4.46	1,630,002	1,416,437	1,211,654	204,783	14%
Aug 2017	31	5.13	3.85	1,630,002	1,223,916	942,983	280,933	23%
Sep 2017	30	4.49	4.13	1,426,649	1,312,848	911,271	401,577	31%
Oct 2017	31	3.54	4.62	1,124,797	1,469,159	988,217	480,942	33%
Nov 2017	30	2.36	3.04	749,865	966,747	433,441	533,306	55%
Dec 2017	31	1.71	3.32	543,334	1,056,067	648,622	407,445	39%
Jan 2018	31	1.83	2.38	581,463	756,633	177,270	579,363	77%
Feb 2018	28	2.2	3.07	699,026	976,347	379,359	596,988	61%
Mar 2018	31	3.17	2.72	1,007,233	864,741	204,166	660,575	76%
Apr 2018	30	4.02	3.54	1,277,312	1,126,909	636,554	490,355	44%
May 2018	31	5	3.06	1,588,696	973,590	678,296	295,294	30%
Totals	**335**	**38.58**	**38.19**	**12,258,379**	**12,143,394**	**7,211,833**	**4,931,561**	44%

Table 1 Realized Savings from July 2017 through May 2018

Campus staff are cognizant of the necessity to allow technology to function as designed to maximize reduction in water use, and to shorten the time it takes to apply the necessary water. The athletic turf quality has improved greatly making the fields safer and more playable for students.

Conclusion

The effect of Allan Hancock's adoption of the latest in smart irrigation technology not only helped them conserve of the plant's most precious resource, but also allowed them to use that water as efficiently as possible.

The Benefits of Consumable Water Management Technologies

James Spindler, Director of Agronomy, CPAg, CCA, CPSS

Ecologel Solutions, LLC, 4060 SE 45th Court, Ocala, FL 34480, jim@ecologel.com

Abstract. *Water conservation has become a major concern over the last decade or more. Influenced by rising costs, recurrent drought, use restrictions, politics and social pressures, turf and landscape managers are expected to do more with less. Beyond the advancements in irrigation hardware and software, there are a variety of consumable technologies and products designed to manage water in the soil and further maximize water use efficiencies. These products and technologies include hygroscopic humectants, surfactants and superabsorbent polymers as well as mulches, compost, antitranspirants, plant growth regulators, and hormones/biostimulants. Understanding each technology and soil/water interactions will help landscape and irrigation managers decided which strategy is best suited for their specific situation. The content will include how and why to use the various technologies, how to maintain healthy turfgrass and landscapes during periods of drought or watering restrictions and the economic benefits of using water management solutions as part of a regular program.*

Keywords. Hygroscopic humectants, wetting agents, surfactants, superabsorbent polymers, drought, watering efficiency, water conservation, consumable water management technologies, mulches, compost, ground covers, anti-transpirants, plant growth regulators, plant hormone, biostimulants, watering restrictions.

Water conservation has become a major factor affecting landscape and irrigation design over the last decade or more. Influenced by rising costs, recurrent drought, use restrictions, politics and social pressures, industry professionals are expected to do more with less. In many regions, landscape have become a political or social target for water conservation and movements to reduce irrigated acreage are growing in popularity nation-wide.

The cost of water in many markets, including "water-rich" regions, has increased dramatically since the turn of the century. Surveys have revealed that the cost of water has risen by 25 to 30 percent in many municipalities, with increases reaching as high as 300 percent or more in some regions. These cost increases coupled with the threat of drought shaming have caused many property owners to reconsider landscape watering practices. Is money better spent on other budgetary items than on irrigation?

Beyond the advancements in irrigation design, hardware and software, there are a variety of consumable technologies and products designed to manage water in the soil and further maximize watering efficiencies. Combining the use of consumable technologies with modern irrigation technology provides a greater opportunity to sustain the industry without succumbing to sacrifices in landscape quality or quantities. Consumable products and technologies include hygroscopic humectants, surfactants and superabsorbent polymers as well as mulches, compost, antitranspirants, plant growth regulators, and hormones/biostimulants. Understanding each technology and soil/water interactions will help landscape and irrigation managers decided which strategy is best suited for their specific situation.

Technologies for Optimizing Soil Moisture Management

Hygroscopic Humectants

Though they are not new to the industry, hygroscopic humectants are continuing to gain favor with landscape professionals as products that are very effective at reducing overall water requirements. With a history in golf and sports turf, these products are becoming more popular for municipalities as well as residential and commericial properties, particularly those in areas with recurrant drought, watering restrictions or high water costs.

Hygroscopic humectants manage and conserve water through two modes. As the name suggests, there is a hygroscopic compontent and a humectant component. Each has a critical function in the performance of the technology. The mode of action of the hygroscopic component is to condense soil water vapor or soil humidity back into liquid droplets of water. The hygroscopic ability of these materials can be compared to condensation or "sweat" that occurs on the side of a cold drink. Rootzone humidity cannot naturally be absorbed by plant roots. Hygroscopics convert this unavilable humidity into plant usable micro-droplets of water.

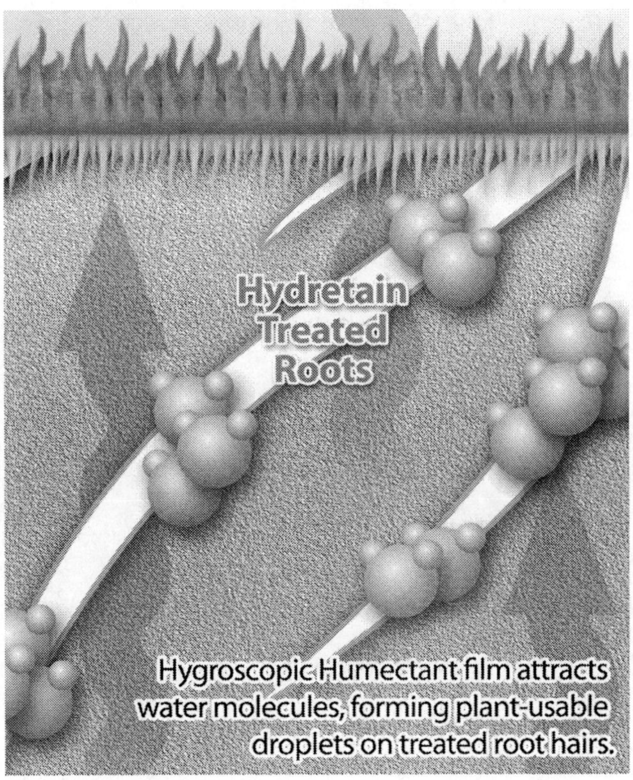

Figure 1. Diagram of the hygroscopic humectants on a soil particle. *(Courtesy of Ecologel Solutions, LLC)*

The humectant components hold the water droplets condensed by the hygroscopic components. Do not confuse a humectant with a humate. They are completely different substances with different molecular structures. The humectant component holds the droplets tightly enough to prevent it from leaving the proximity of the root, but lightly enough to allow the root to absorb the water through osmosis. The humectants in hygroscopic humectants are also utilized in cosmetics, shampoos, and other body care products where they help hold moisture in the skin and hair.

Available in both liquid and granular options, hygroscopic humectant technologies must be watered-in. Not because they will burn, but to assure that the active ingredients will coat plant roots, soil particles and organic particles in the root zone. If hygroscopic humectants dry on the plant leaves or in the thatch or mulch, they will anchor themselves at that location permanently. The active incredients will perform its function at these locations, but there are no roots present to utilize the condensed water droplets, therfore the benefits will not be realized.

Once the hygroscopic humectant molecules arrive in the rootzone they are too large to be absorbed by the roots. Once these components attach to the roots and soil particles, they remain attached and are resistant to further movement in the soil. The ingredients are primarily derived from plant byproducts (some brands are USDA BioPreferred Certified for biobased content*). Therefore, they are eventually broken down by soil microbial activity. Research and users have demonstrated that the most effective hygroscopic humectants products have been able to reduce water use by up to 50 percent or more and will typically perform for up to 90 days. In addition to providing general conservation of water, hygroscopic humectants aid in seed germination, transplant establishment and in establishing sod and sprigs. Hygroscopic humectants have also been used to suppress dust on baseball infields, horse arenas, dirt race tracks, dirt roads, etc.

Figure 2. Sports field under water restrictions. Plot treated with a hygroscopec humectant compared to adjacent untreated plot in the foreground. (Courtesy of Hydretain)

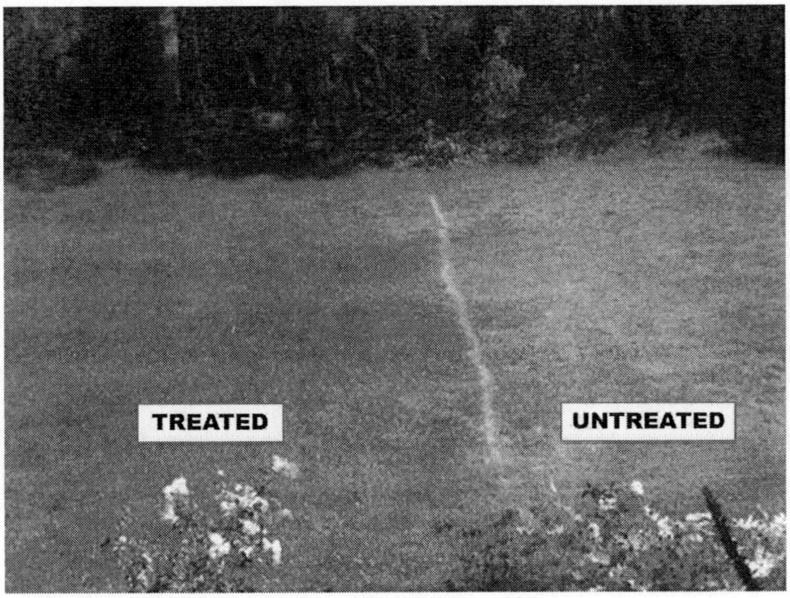

Figure 3. Golf course slope treated with hygroscopic humectant under water restrictions. (Courtesy of Lesco Moisture Manager)

Superabsorbant Polymers

Superabsorbant polymers are a technology that track their orgin to a patent by Monsanto in 1963. They described polymers as “strings of large molecules that chemists use like Tinker Toys, adding, subtracting or linking them together to create diverse uses ranging from filling for disposable diapers to dental products” (Messina, 1991).

Polymers have been adapted for use in soil to improve water availability to plants. They are utilized to “increase a soil’s water holding capacity, increase pore sizes and pore numbers in the soil, increase germination rates, and decrease or mitigate the effect of soil compaction on plant growth” (Orzolek, 1993). The five main types of soil polymers available commercially include:

- Cross-linked polyacrylamides (gel forming)
- Non-cross-linked polyacrylamides (water soluble)
- Polyacrylates
- Polyacrylontrile
- Starch-grafted copolymers

The most commonly used polymer is the cross-linked polyacrylamide. Soil polymers occur in a crystalline form. When exposed to water, they expand into a gelatin-like block. When used in soils, they function as mini-reservoirs of water. They absorb water and hold it until the plant removes the water. The literature indicates that cross-linked polyacrylamide polymers used in the field will absorb and hold 80 to 200 times their weight in water or more. Their ability to hold soil water is influenced by the amount of polymer in the soil, the type of polymer utilized and soil characteristics, such as salt content (Polhemus,

1992). The lifespan of polymers is thought to range from 2 to 10 years, depending on the type of polymer and soil conditions (Polhemus, 1992).

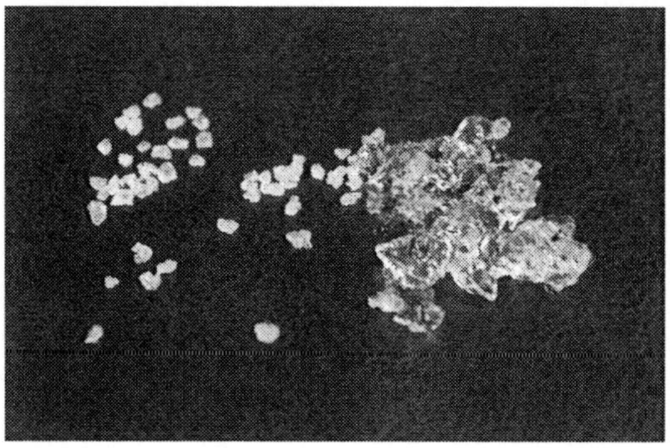

Figure 4. Comparison of dry vs. hydrated superabsorbent polymer crystals

The literature reports that the time between irrigation events can be extended with the use of polymers, but the actual water savings with use of these products is dependent on application rates and soil conditions. Cost of these products may be a limiting factor for effective application rates.

Initially, polymers were used to help reduce water use in potted plants, ornamental beds and in planting trees and shrubs. Over the years, soil-applied polymer use has expanded to turf applications. They are utilized in the establishment of sod and sprigs, improving seed germination and in general turf use. The challenge in utilizing polymers on established turfgrass is delivering the polymer crystal to the root zone. Some turf manager will aerate the turf and drag the crystal into the holes. In addition to this practice, there are now machines that will inject the polymer crystals into the soil.

Surfactants/Wetting Agents

Surfactants or wetting agents are probably the most commonly known products used to manage soil moisture. These materials are utilized for a number of applications in turf and plant management, including relief from localized dry spots, improved drainage, assisting the efficiency of various pesticides, reduced dew and frost accumulation, improved seed germination, reduced fairy ring damage, alliviation of soil compaction, improved irrigation efficiency, and more (Karnock, Xia, & Tucker, 2004).

Surfactants stand for **SURF**ace **ACT**ive Age**NTS (SURFACTANTS)**. These are agents that affect the surface of a liquid or solid. Understanding the nature of water is critical to understanding the function of surfactants. Water molecules are naturally polarized: the two hydrogen atoms are attached to one side of the oxygen atom causing each end to have a slight charge. Just as opposite charges of magnets attract to each other, the positive end of the water molecule is attracted to the negative end of another water molecule. This is called cohesion and is why we see a drop of water bead up when placed on wax paper. The waxy surface of the paper has no charge or is non-polar; therefore, the only place for the water molecules to be attracted is to each other. In soils, waxy coatings are the result of the decay of organic materials and certain species of fungi that exude waxy substances. The formation of waxy, non-polar coatings on soil particles is the primary cause of hydrophobic conditions. The non-polar soil particle surface will not attract, and may actually repel, the polar water molecule. This prevents irrigation water or rainfall from infiltrating into soils to hydrate plants.

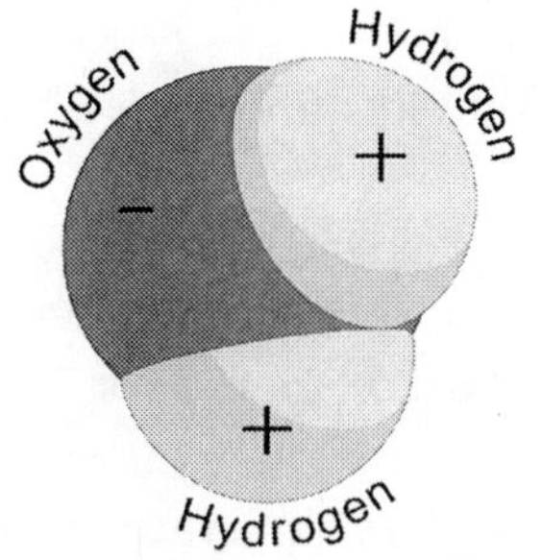

Figure 5. Graphic rendering depicts the polar nature of water molecules.

Creating a polar surface allows water molecules to enter and fill the soil. The surfactant has a non-polar and a polar end on the molecule. The non-polar end of the surfactant molecule aligns with the non-polar surface of the organic soil coating, leaving the polar end exposed outward from the soil particle. This allows the polar water molecules to be attracted to the polar surfactant molecules therefore overcoming the hydrophobic condition (Karnock, Xia, & Tucker, 2004).

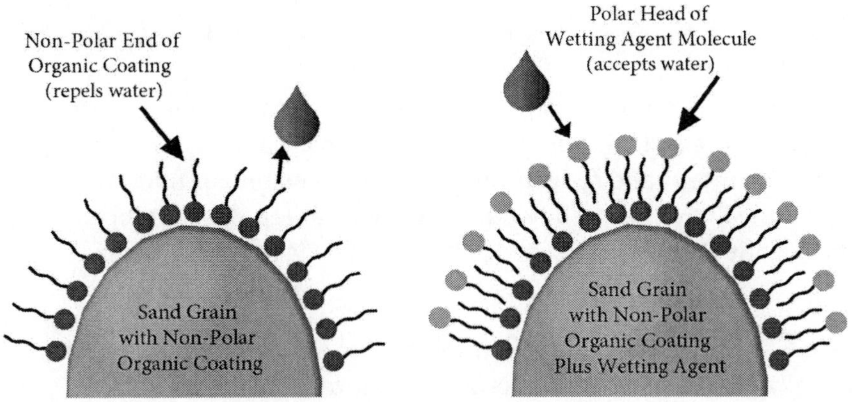

Figure 6. Diagrams illustrating wetting agent activity

There are many different kinds of surfactants, most of which fall into these four basic categories:

- Anionic—Form negatively-charged ions in water
- Cationic—Form positively-charged ions in water
- Nonionic—Does not ionize in water
- Amphoteric—Take on the ionization of the water

Non-ionic surfactants are the most common products used in the turf industry due to their safety, compatibility with other products and ease of use. As technology has improved, a number of catetories of non-ionic surfactants have been developed. These include:

- Polyoxyethylene (POE)—This is older technology originally developed to treat localized dry spots. They can be phytotoxic.
- Block Co-Polymer Surfactants—These are the most commonly used turfgrass surfactants. They are safer and are effective in treating soil water repellency, improving soil water content and plant available water. This category has two sub categories: Straight Block Co-Polymers and Reverse Co-Polymers.
- Alkyl Polyglucoside Surfactants—These are made from sugar molecules reacted with a fatty acid and are considered naturally derived. When blended with a block co-polymer, the performance appears to be better than either technology alone. These blended technologies appear to increase water infiltration, improve water availability and enhance irrigation efficiencies.
- Modified Methyl Capped Block Co-Polymer—This is a class of surfactant that is a modification of the co-polymer class. This technology forms a thinner, more continuous film around the soil particle.
- Humic Substance Redistribution Molecules—"These molecules allow water penetration through the soil profile by disrupting the hydrophobic supramolecular humic association, most prevalent in the top one to two centimeters on the soil, which lead to localized dry spots."
- Multi-branched Regenerating Wetting Agents—Most surfactants have linear molecules. These products have a much higher molecular weight and multiple branched molecules. Each branch essentially functions as wetting agent itself. (Zonteck & Kostka, 2012)

Surfactants/wetting agents have been demonstrated to possess many functions in the management of water in and around turfgrass and other plant systems. When discussing the maximization of water use efficiencies, these products tackle the barriers (non-polar coatings in the soil) that prevent water from moving into and distributing throughout the soil. Research has shown that surfactants/wetting agents can significantly improve soil moisture content and reduce variability in soil water content, improving soil moisture uniformity. In addition, they have been shown to "reduce localized dry spot incidence, allow for longer periods between irrigation events, and reduce hand watering in isolated areas" (Karchner & Richardson, 2014).

Surfactants/wetting agents are available in liquid and granular forms. The amount of water conserved, longevity of the product and cost may vary based on product type and local conditions.

Soil Barriers and Composts

Soil barriers such as mulches, pine straw, plastics, landscape fabrics, and gravel have long been used to conserve water in landscape beds. These products retain soil mositure by acting as a physical barrier to moisture loss through evaporation. They act to insulate the soil from the excessive warming drying effects of direct sun exposure. They also help reduce soil erosion and compaction, which can negatively affect water use efficiency. Furthermore, they reduce weeds, which steal water resources from desired plant materials.

Compost incorporation is also used to increase water use efficiency. Similar to the use of superabsorbent polymers, composts are known to increase soil water holding capacity. Amended soils are estimated to require up to 60% less water depending on the quality of the compost and quantity incorporated into the soil.

Antitranspirants

Antitranspirants are substances sprayed on the leaves of plants to reduce the rate of transpiration. Antitranspirants function utilizing three known modes of action. First, they reflect radiant energy way from the plant resulting in lower temperatures and transpiration rates. Second, emulsions of wax latex or other film are used to prevent the escape of moisture from the plant. Third, they prevent the stomata from opening fully and decrease the loss of water vapor from leaves. While not all experts are in agreement on the benefits of anti-transpirants, studies have shown that they can conserve water and minimize wilt and drought stress.

Growth Regulators and Biostimulants

Growth regulators and hormone biostimulants represent another class of chemistries that can be used to reduce plant water requirements. Plant growth regulators (PGRs) are designed to reduce excess vertical plant growth. While the intended benefit of these products is lower trimming and mowing maintance, slow growth has also proven to reduce plant water requirements.

Hormone biostimulants containing cytokinins have proven to increase root mass and depth. By encouraging root development, biostimulants improve water efficiency by helping plants source more water from soil. They have also provent to increase plant tolerance to a number of stresses, including drought.

Conclusions

There are a wide variety of consumable products and technologies available to help manage and conserve water. Using these technologies in conjunction with modern advancements in irrigation hardware will offer greater opportunities to maximize water use efficiency. The key to success is to recognize the value of consumables and learn which product(s) are the best fit for each specific situation. It is adviseable to remember to not think linearly. Often, there is not one single issue with one single solution. The best solution for the management and conservation of water may be to combine consumable technologies. A very common example of this is the combination of hygroscopic humectants with surfactants technologies. In this situation, the surfactant will allow water with the hygroscopic humectant to enter and disperse throughout the soil where hydrophobic non-polar organic coatings exist. Water can uniformly disperse throughout the rootzone. Then, the hygroscopic humectant can reduce evaporative loss for maximum plant water use.

Thinking outside the box and using all tools available gives landscape and irrigation managers the ability to maximize water use efficiency and optimize turf and plant performance.

*Products brands Hydretain®, LESCO Moisture Manager™, and BioPro H3O™ have been certified to contain 93% biobased contents by the USDA BioPreferred® Program.

References

Karchner, D., & Richardson, M. (2014, December). Wetting Agents Improve Root-Zone Moisture Distribution. *Golf Course Management*, pp. 80-83.

Karnock, K. J., Xia, K., & Tucker, K. A. (2004, June). Wetting agents: What are they, and how do they work? *Golf Course Management*, pp. 84-86.

Messina, F. (1991, March 18). Absorbing Issue: Polymers are Saving Grace for Large Water Unsers, but Some Question Long-Term Effects. *Los Angeles Times*, p. Internet.

Orzolek, M. (1993, January/March). Use of Hydrophylic Polymers in Horticulture. *HortTechnology*, pp. 41-44.

Polhemus, D. (1992, June). Soil Polymers for Turf Areas: A Technical Review. *Sports Turf*, pp. 14-17.Zonteck, S. J., & Kostka, S. J. (2012, July). Understanding the Different Wetting Agent Chemistries. *USGA Green Section Record, Volume 50 (15)*, pp. 1-6.

Technical Paper for 2018 Irrigation Show & Education Conference

Title: Using Persistent and Autonomous Thermal and Visual Image Data to Measure Stress and to Know; Where, When & How Much Irrigation Water to Use

James (Jim) Etro, Founder and CEO, Itri Corporation, 2507 Little River Road, Haymarket, VA 20169, jim.etro@itricorp.com

ABSTRACT

Using persistent thermal and visual image data to measure the turf canopy it is possible to enable the turf to tell you about the stress it is experiencing and its water status.

When scouting for stress one is looking for deviations in the canopy where the quality is not up to desired standard. Homogeneity of the turf's canopy is key. When assessing the need for irrigation the turf's irrigator may be using one of three operationally feasible methods of assessing the need for applying water; using an evapotranspiration equation, using a network of soil moisture sensors, or by seat-of-the-pants. Evapotranspiration calculated from observations from a weather station and applied to the turf by a turf coefficient. It is an estimation of the water used by the turf so one can know how much to irrigate. Soil moisture sensors measure the water in the soil so one can know, based on the fidelity of the network, where and how much to irrigate. Seat-of-the-pants is the art of looking at and touching the turf, then applying one's intuition and experience to know when and where to irrigate.

Persistent visual and thermal image data processed to illuminate changes in canopy vigor and canopy temperature is a tool to enhance and extend seat-of-the-pants methodologies.

Key Words: remote sensing, image data, visual, thermal, transpiration, turf quality, turf stress, scouting, irrigation, plant thermography, quality index, stress index, irrigation index,

Note: All Figures are included in Appendix A at full size.

Background

Visual Image Data and its Application for Indexing Turf Stress

Karcher and Richardson (2003) and others, found that an analysis of a digital visual image (400 – 700 nanometers) provides a reliable method to measure the reflectance of color from vegetated surfaces. The digital camera measures the hue degree of the turf (Figure 1). This hue measurement of the turf canopy can represent the homogeneity

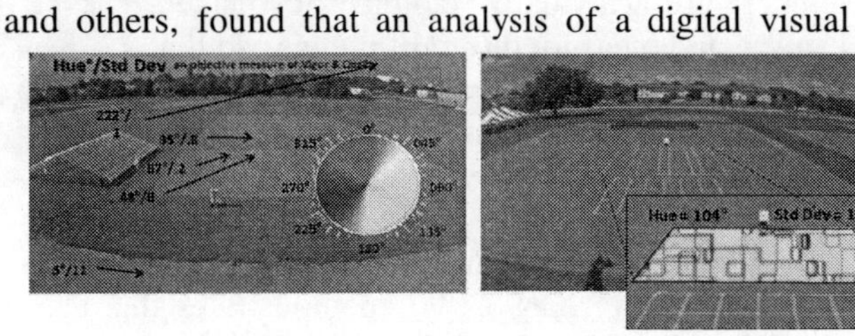

Figure 1. Hue (color) & Standard Deviation of Hue

of the turf color when an area of interest is processed to show the average hue value (the hue degree of each pixel contained in the designated area is measured) and the standard deviation of the average hue value. For that turf surface it is a representation of the turf's vigor and quality. The lower the standard deviation, the better the quality. When a 'typical' value is established a change in the deviation illuminates a change in vigor which is directly related to health. Another indicator of a decline in vigor is a change in the hue degree out of the range of green toward yellow and brown. See the color wheel shown in Figure 1.

This Hue/Std Dev value calculated at every image data collection may also be known as a Quality Index. The Daily Visual or Quality Index (QI) is the average of the standard deviation of the hue, +/- one hour of solar noon; or for 2 hours of no shade during a 'bright' part of the daylight hours.

Thermal Image Data

Using a radiometric thermal image (8,000 – 13,500 nanometers) it is possible to measure the temperature of the turf's canopy. Turf photosynthesizes during daylight hours and respires during nighttime. Both processes release water vapor as a byproduct and the evaporation of that water vapor is a cooling agent (see Figure 2). The observation of the temperature

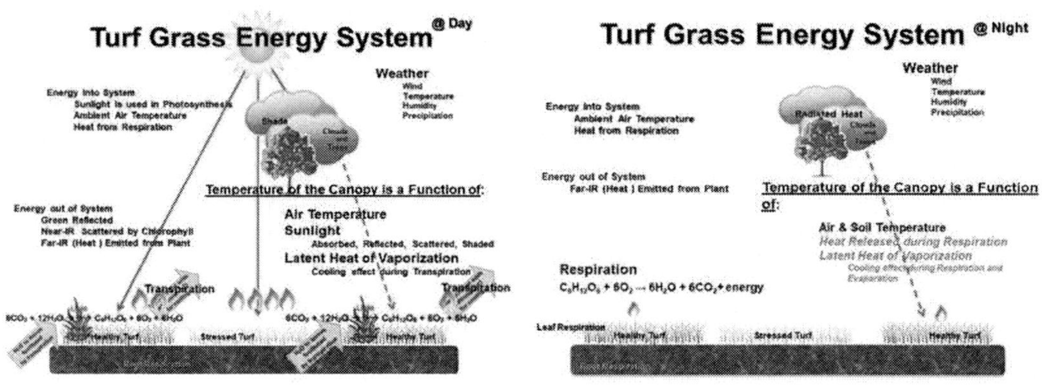

Figure 2 a&b. Turfgrass Energy - Day & Night

across the extant of the turf can indicate locations where stresses may be occurring(Figure 3). Because this cooling process, which is very evident during daylight hours, especially in direct sunlight, can highlight areas of disease, pest, and/or water status stress. This is also valuable tool when evaluated at night because although the variances of the surface temperature are small, radiometric imagers can see and measure those differences so that non-homogenous areas can be evaluated for drainage patterns and/or disease and pest issues.

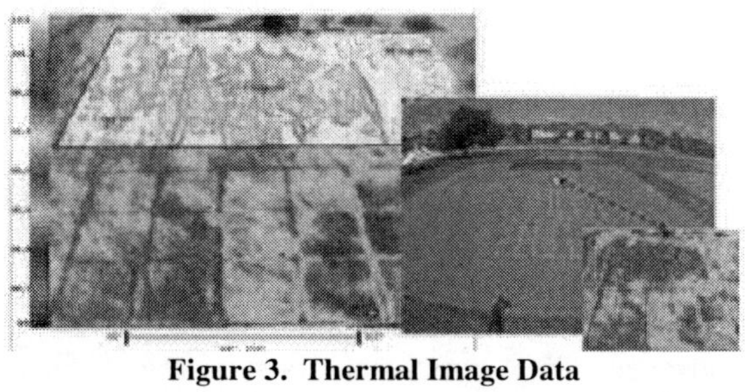

Figure 3. Thermal Image Data

The video record found at https://vimeo.com/269639967 demonstrates this use of nighttime thermal image data to locate the extant of winterkill of a warm season turf two months before green-up of the turf.

Applying Thermal Image Data for Indexing Turf Stress

Jackson *et al.* (1981) appreciated that the canopy to air temperature difference ($T_{canopy} - T_{air}$) depends on vapor pressure deficit (VPD): under non-limiting water conditions, a healthy crop transpires at the potential rate (i.e. evapotranspiration is the maximum it can be, but maximum evapotranspiration increases with increasing VPD). Thus, for several crops, when crop health and water availability is not limiting and when measured under clear sky conditions, there is a linear relationship between $T_{canopy} - T_{air}$ and VPD. Jackson called this linear relationship the theoretical 'nonwater-stressed baseline' (nwsb). For a given crop, at a given VPD, this theoretical baseline provides the minimum possible value of $(T_{canopy} - T_{air})_{nwsb}$. The $T_{canopy} - T_{air}$ for a non-transpiring crop is insensitive to VPD and can be estimated if wind speed and net solar radiation are known. This sets the 'upper limit' (ul) to $(T_{canopy} - T_{air})_{ul}$. Jackson *et al.* used the idea of 'upper and lower' baselines, to create a crop water stress index (CWSI). The $CWSI = (T_{canopy} - T_{air}) - (T_{canopy} - T_{air})_{nwsb} / (T_{canopy} - T_{air})_{ul} - (T_{canopy} - T_{air})_{nwsb}$: where $T_{canopy} - T_{air}$ is the measured difference in temperature, $(T_{canopy} - T_{air})_{nwsb}$ is the estimated difference at the same VPD under non-limiting water conditions (on-waterstressed baseline), and $(T_{canopy} - T_{air})_{ul}$ is the non-transpiring upper limit. This CWSI allows one to relate crop's temperature to the maximum and minimum values possible under similar environmental conditions. The higher the CWSI, the greater the crop stress is assumed to be.

A disadvantage of the above form of CWSI is the need to determine the non-water-stressed baseline by plotting $T_{canopy} - T_{air}$ against VPD. This requires substantial time to be spent determining the baseline for a well-watered crop, and the VPD needs to be known when measuring T_{canopy} of the crop of interest. Also, this index does not account for changes in T_{canopy} due to irradiance and wind speed, and the non-water-stressed baseline is not necessarily the same under different radiation conditions. Finally, the non-transpiring upper limit also varies, with a wide range of values (Ben-Gal *et al.*, 2009).

Establishing a Stress Index from empirical observations of the upper and lower limits is possible by understanding that transpiration is a key measurement and applying thermographic techniques to the image data. Experience gained by observing the canopy temperature shows that it is possible to make the canopy temperature an indicator of transpiration and respiration. During the day evaporation of the transpired water vapor cools the leaf/canopy. At night one can see the heat from respiration, transpiration, and evaporation of the near surface moisture. Thus, the turf's canopy temperature is the biotic integrator of the air temperature, humidity, wind, solar radiance, and the turf's health and water status.

More than six years of observation has demonstrated that an equation of the form similar to one outlined by J. Miguel Costa *et al.* (2013), addressing plant–environment interactions is a superb indicator of the stress experienced by turf. By using a thermal imaging data system co-located with a weather station to persistently measure canopy temperature and air temperature it is possible to observe/measure an upper limit and lower limit of water vapor released during transpiration. This Stress Index is used:

$$\mathbf{(SI) = (T_m - T_{LL})/(T_{UL} - T_{LL})}$$

$\mathbf{T_m}$ = canopy temperature minus air temperature measured at image data capture time.

$\mathbf{T_{LL}}$ {non-stressed condition} = early daylight canopy temperature minus air temperature

$\mathbf{T_{UL}}$ {stressed condition} = most stressed part of the day canopy temperature minus air temperature

An index value (SI) is calculated over designated areas every image and the Daily Heat Stress Index is the average of the daylight Image Indexes. When normalized by a running average of the T_{LL} and T_{UL}, it can become a reliable, disciplined, and repeatable indicator of turf health and water status; and it informs the accumulated stress of the day.

Figure 4 is a chart depicting the Image Stress recorded approximately every 10 minutes during the day at three different areas on a golf course green.

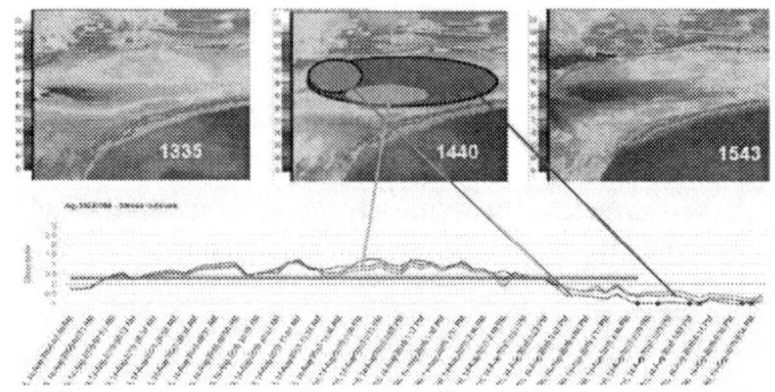

Figure 4. Image Stress at Ten Minute Intervals

Smart Irrigation Month Demonstration

To celebrate Smart Irrigation Month in July 2018, we established a demonstration to look at the application of irrigating simulated reclaimed water versus fresh well water on turf. The reclaimed water vs fresh well water demo was extended into August and an evaluation of the Irrigation Index under long rainy and mostly cloudy conditions was undertaken.

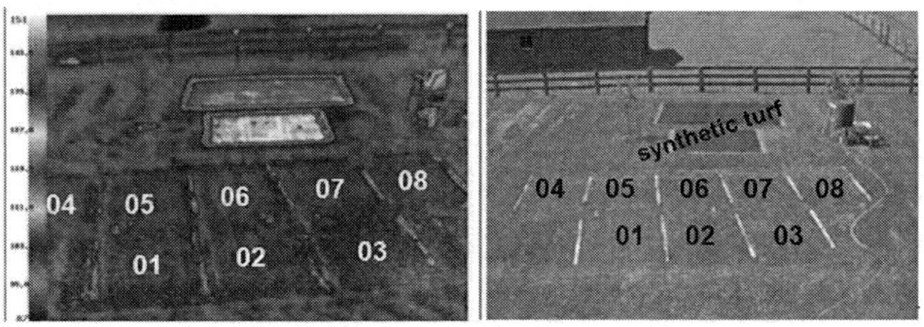

Figure 5. Demonstration Plot Lay-Out

In late June eight plots were established (Figure 5) on a predominately tall fescue area. In June, July, and August, weeds and crabgrass were pulled by hand to keep the plots weed free and ensure a homogenous canopy of turf. As the summer progressed bermudagrass progressively encroached into plots 07 and 08. By late August plot 08 was 50% tall fescue and 50% bermudagrass, plot 07 was 70% tall fescue and 30% bermudagrass[1]. The height of the grass in all the plots was maintained at 0.8" – 1.2", through the demonstration period. The longer cut area around the plots was loosely maintained at 2" – 3".

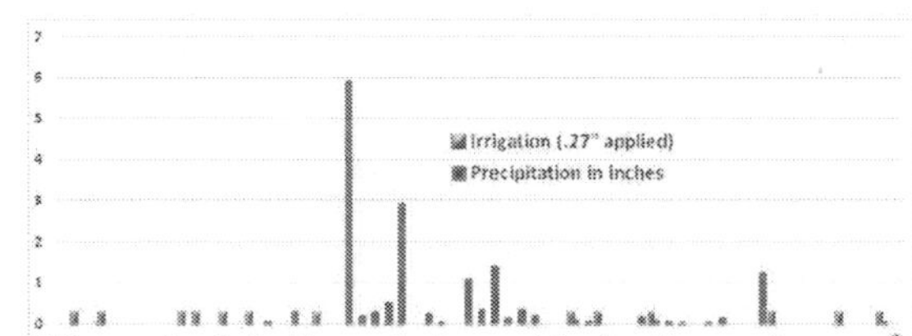

Figure 6. July through August Irrigation and Precipitation

The summer of 2018 in Northern Virginia, was unusually wet, and the solar radiance due to the cloud cover was less than usual. Typical July-August precipitation would total 8″, and the average solar radiance would be about 240 watts/meter^2/24hrs. During June the plot area received 7" of precipitation. During July rainfall was 13" and the average solar radiance was 209 watts/meter^2/24hrs. August rainfall was 5.8" and the average solar radiance was 180 watts/meter^2/24hrs. Figure 6 plots the July August irrigation and rain events.

---

[1] A study of Plot 08 from the August image data will be undertaken at a later date to investigate the difference in hue between the tall fescue and the bermudagrass.

During July, plots 02, 05, and 07 were irrigated with .27 inches of 0.3% brackish water (table salt + well water); plots 01, 03, 04, 05, 07, and 08 were irrigated with .27 inches of untreated well water. Then between July 22 through August 5th, more than 13" of rain fell. During August, plots 02, 05, and 07 were irrigated with .27 inches of 0.6% brackish water (table salt + well water); plots 03, 07, and 08 were irrigated with .27 inches of untreated well water. Irrigation was not applied to plots 01 and 04.

Results of the Visual Image Data Analysis

Results are highlighted by the charting of the data taken from Plot 02 (irrigated with brackish water to simulate reclaimed water) and Plot 06 (irrigated with fresh well water).

"Eye-balling" the plots during July, there was no evidence of declining quality or increased stress in the plots irrigated with brackish water and it was judged that .03% brackish application wasn't enough to get results in a short period of time. It was also assumed that the 22 July - 05 August, rain flushed the brackish remnants out of the root zone. Because there was no detected decline in the turf quality where the brackish water was applied in July, the brackish solution was increased to.06%. Note that during the period leading up to the application of the two different irrigation prescriptions the hue (Figure 8) and deviations of the hue (Figure 8) are very close. Then when the prescriptions, starting 02 July, are applied the data diverges until the rainy period flushes the root zone and the quality and deviation of quality converges again.

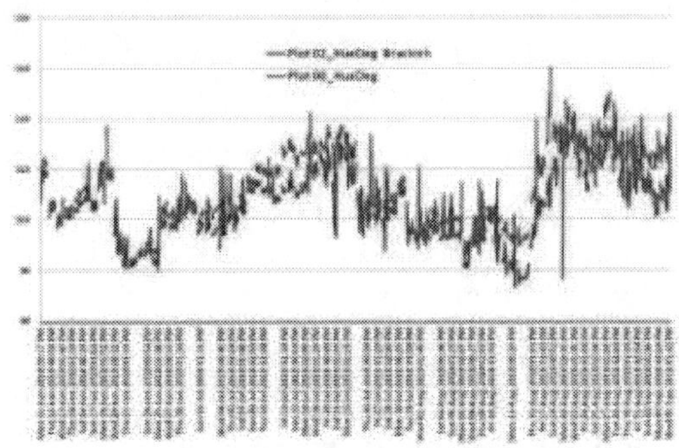

Figure 7. Hue Degree (color); Brackish versus Fresh

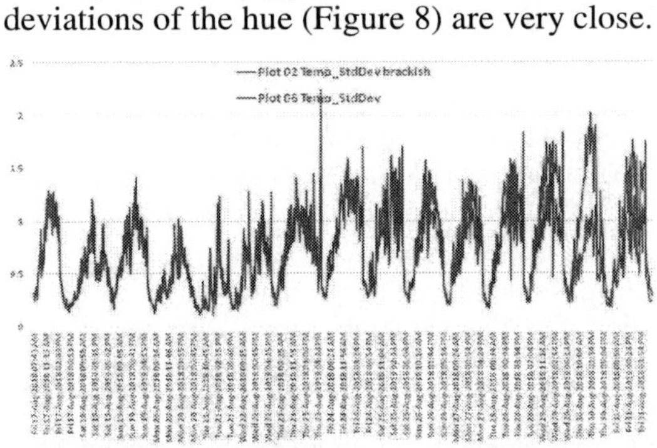

Figure 8. Standard Deviation of Hue; Brackish versus Fresh

In July, had the visual image data been closely examined and an alert for changes in hue deviation been set into the Hawk-Eye™ collection system rather than waiting until September to start organizing the data for this paper, it could have been established that the .03% brackish solution was working. Hawk-Eye™ saw what the eye missed. There was a decrease in quality as evidenced by a slight decline in hue (Figure 7) and an increase in the deviation of the hue (Figure 8).

Then in the August the plots under the .06% prescription a deeper anomaly appears. Starting on 18 August, although the hue (Figure 7) declines in Plot 02 (brackish), the standard deviation of the hue (Figure 8) in plot 06 increases significantly. Brown patch (Figure 9) was the cause of the decline in quality in Plot 06 and it was evident in all the plots irrigated with fresh

Figure 9. Brown Patch in Plot 06

water. In Plot 06 the infection covered approximately 5% of the surface area. Of interest is that the plots irrigated with brackish water and taking the precipitation, showed no signs of brown patch and all the areas near the demo plots (no other locations were irrigated at any time during the summer) showed an indication of disease.

Results of the Thermal Image Data Analysis

Due to a calibration issue and need for a firmware update in the radiometric camera no temperature retrievals where collected and archived until August. Figure 10 is a chart of the average canopy temperature in Plot 02 (brackish) and Plot 06, from 17 through 31 August; And a chart of the standard deviation of the average canopy temperature of the plot. Note that Plot 02 (brackish) is warmer than Plot 06 and it also has a slightly higher standard deviation of its average temperature. This becomes more evident after 22 August when the rain events ended for the month.

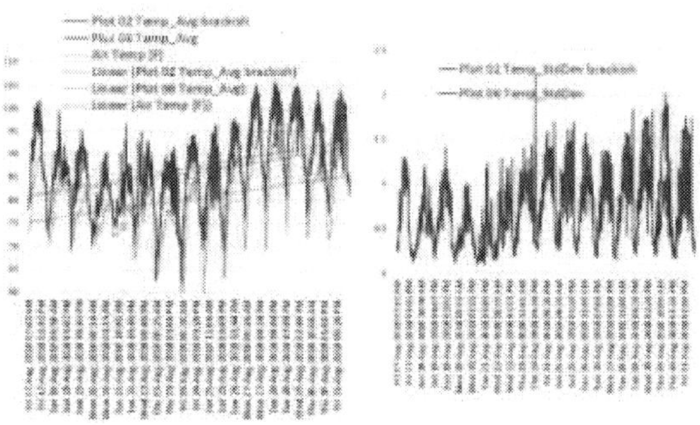

Figure 10. Canopy Temperatures vs Air Temperature; and Standard Deviation of the Canopy Temperatures

Since Plot 06 was infected with brown patch it was assumed that the canopy temperatures would have been higher than Plot 02 (brackish). However, this was not the case. It appears that the .06% saline solution irrigated over the whole Plot 02 was more effective in reducing the hue and interrupting transpiration[2]; thus, causing a loss in evaporative cooling and a relative increase in temperature. As the brown patch grew the temperature difference and the difference in standard deviations grew. But the average heat profile in the infected patch remained cooler and had lower variability from the average than the salt stressed plot.

Using Canopy and Air Temperature to Gage Turf Stress (Stress Indexing)

In calculating the stress experienced by the turf we persistently, day and night, measure the turf's canopy temperature and the local air temperature. The Stress Index equation is used:

$$(SI) = (T_m - T_{LL})/(T_{UL} - T_{LL})$$

T_m = canopy temperature minus air temperature measured at image data capture time

T_{LL} {non-stressed condition} = early daylight canopy temperature minus air temperature

T_{UL} {stressed condition} = most stressed part of the day canopy temperature minus air temperature.

From every thermal image data set (typically every 10 minutes) an "Image Index" is calculated. At the end of daylight, the Image Indexes are averaged to establish the "Daily Index". Image Indexes are used to track and report stressing events during the day and the Daily Index relates the turf's experience through the day and provides a measure of day-to-day health. The Daily Index can also inform the need for irrigation. Figure 11 is provided to show the Image Index and Daily Index of Plots 02 and 03 between 20-30 August. Figure 12 shows the upper level (UL) and lower level (LL) of stress plotted against air temperature and the water (precipitation and irrigation) introduced between 17-30 August.

---

[2] The relationship of the hue degree to the turf cultivar and the amount of chlorophyll is a question for a more detailed study.

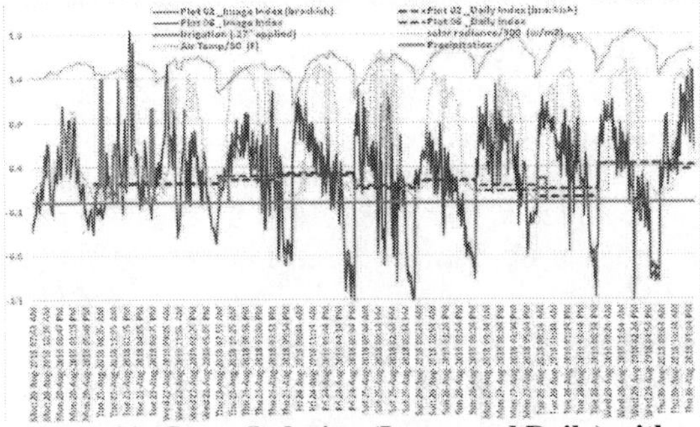

Figure 11. Stress Indexing (Image and Daily) with Air Temperature and Solar Radiance and Irrigation

In figure 11, note the response of the canopy temperature to the air temperature and the solar radiance. On several days the canopy temperature declines as the solar radiance and the air temperature increases. This is unusual and may be due to the unusually wet and partly cloudy to mostly cloudy days during the summer. Typical daily profiles in dryer summer conditions look more like 23 August than 24 August. In Figure 12, note the response of the T_{UL} and T_{LL} of the turf to the trend of the temperature and solar radiation and the water introduced. After the very wet summer and then the rain events of 18-22 August, the temperatures and the solar insolation are low. This keeps the daily stress low and because the calculation of the Upper Level of stress is a running average over several days it biases the Stress Indices to lower values than may be expected. Note the high number of Image Indexes below zero and the very low Daily Index of .06 on the 28th of August. On 29 August the Daily SI snaps back to higher values (although not exceeding an irrigation demand threshold after daylight on 28 August.)

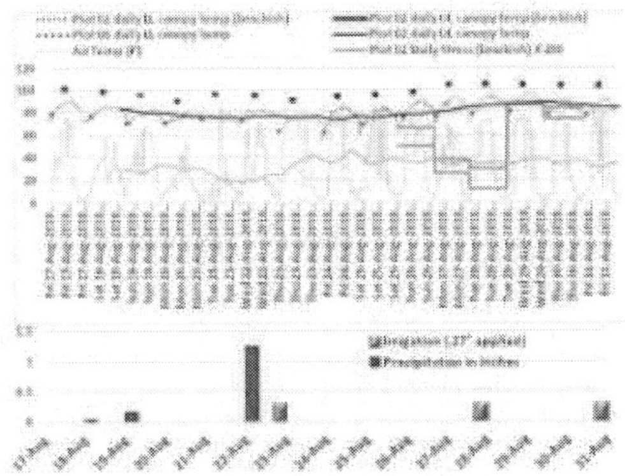

Figure 12. Daily T_{LL} and T_{UL} with Precipitation and Irrigation

Need for Persistence of Image Data

The video record summarized in Figure 13 and detailed at https://vimeo.com/269639967, demonstrating use of nighttime thermal image data to locate the extant of winterkill, also highlights an important consideration when using any types of image data for turf management. That is the need for frequent persistent image data records. It is important to have an understanding of patterns noted in the imagery because occasional (even daily) snapshots of data make it impossible to recognize persistent patterns and may lead to an incomplete understanding of the condition of the turf. Figure 14 illustrates the variability of the thermal character of over a short period (30 minutes) of time. Settling on any one image as a starting point for scouting may lead one to confusion and a poor conclusion regarding actions that may be needed, or not.

Figure 13. Using Nighttime Thermal to ID Winterkill

When using image data for assessing plant water status and guiding irrigation it is necessary to consider continuous image data measurements over the course of the daylight hours, and for several consecutive days, to achieve good results.

Figure 14. Image Data Time Series

By measuring hue (color) and the temperature of the canopy with a persistent (24/7/365) visual and thermal image data system (a Hawk-Eye™ Remote Sensing System), applying experience and intuition in intelligence algorithms, and by exploiting the Internet-of-Things; the turf can give its voice to when and where it is being stressed, and when and where it wants water.

Irrigation Guidance:

The turf's canopy temperature is the biotic integrator of the air temperature, humidity, wind, solar radiance, and the turf's health and water status. The visual Quality Index and the thermal Stress Index puts a disciplined measurement on the health and water status.

The Daily Irrigation Index threshold is the index value where the plant indicates it needs water. The Prescription is the amount of water the plant is given during the evening after it says it needs it. There are periods where the Irrigation Index may call for the Prescription two or three evenings in a row or it could go six or more days before water is called for by the plant.

The Daily "Irrigation Index" is a function of the Quality Index (QI) and the Daily Stress Index (SI) described in the Visual Image Data and the Applying Thermal Image Data for Identifying Turf Stress sections of this paper. The greatest weight is given to the Daily SI and the QI is used as a cross check when there have been long periods of rain, cool air temperatures, and frequent cloudy sky conditions.

A Hawk-Eye™ System will inform the user when the Indices exceed the threshold set for irrigation guidance. This index result is available 10 minutes after sunset and can be reported via internet in e-mail and to mobile devices by SMS text. Figure

When the Daily Irrigation Index crosses the plant's threshold that day, irrigation is applied in a predetermined amount. The amount is a constant (i.e. the same amount all season) that is specific to the location and it is based on a typical amount of irrigation that might be applied. Daily Irrigation Index measurements continue every day and the next day the Irrigation Index crosses the threshold the water is applied again.

Figure 15 shows the results of an experiment conducted in the mid-west during July 2017. Twelve plots were irrigated with four different prescriptions. Three plots received 100% replacement according to calculated ET. Three plots received 80% replacement according to calculated ET. Three plots received 0% replacement. Three plots received irrigation guidance according to the Hawk-Eye™ prescription.

Plots	17-28 July 2017 Total Inches of Water [illegible]	ET [illegible]	Turf Quality [illegible]
100% Replacement	4.065"	[illegible]	7
80% Replacement	2.641"		7
0% Replacement	0.160"		1
Hawk-Eye™ Stress Index	1.585"		7

Figure 15. Variable Irrigation of Plots and Resulting Quality

Summary

Using persistent thermal and visual image data to measure the turf canopy it is possible to enable the turf to tell you about the stress it is experiencing and its water status. This image data, processed to illuminate changes in canopy vigor and canopy temperature, is a tool to enhance and extend a person's senses and their knowledge and experience base.

The Irrigation Month Demonstration to observe and measure the impact of irrigating plots with simulated reclaimed water versus fresh well water on turf went very well despite the wet and could cover. It was shown that one can use a visual image data Quality Index to observe and quantify the impact of using reclaimed water and see when the reclaimed water has been flushed through the root zone. With respect to applying the Stress Index for irrigation guidance we saw that although it is a reliable tool in dry hot climates some work needs to be done to make it work better in more temperate climate where there may be long periods of precipitation and cloud cover.

The original development of the Hawk-Eye™ SI and Irrigation Index discussed here-in was accomplished with image data and meteorology observed and collected from fields, turf plots, and golf courses in California and the mid-west during more than ten summer seasons. That hot dry climate is very different than the mid-Atlantic climate that the demonstration described in this paper was observed in. During the Smart Irrigation Month Demonstration the weather was unusually wet, and the solar radiance due to the cloud cover was less than usual. Because of these factors the turf's demand for water was very small. During August the Hawk-Eye™ System's measure of the turf only called for irrigation (0.27") one day.

As noted in the section "Using Canopy and Air Temperature to Gage Turf Stress (Stress Indexing)" there were some surprising (to me) results and it illuminated the need to factor into the Stress Index to account for long periods of frequent precipitation and mostly cloudy sky conditions.

During this winter season we will evaluate the simple running average calculation that is used to establish the Upper Level of stress with an objective of maturing the algorithm to perform better in wet and cloudy environments.

References

Ben-Gal, Alon, Agam, Nurit, Alchanatis, Victor, Cohen, Y, Evaluating water stress in irrigated olives: correlation of soil water status, tree water status, and thermal imagery, Irrigation Science 27(5):367-376, April 2009

Costa, J. Miguel, Grant, Olga M., M. Chaves, Manuela, Thermography to explore plant–environment interactions, equation (5), Journal of Experimental Botany, Volume 64, Issue 13, 1 October 2013, Pages 3937–3949

Jackson, RD, Idso, SB, Reginato, RJ, Pinter, PJJ, Canopy temperature as a crop water stress indicator, Water Resources Research, Volume 17, Issue 4, 1981

Karcher, Douglas E. and Richardson, Michael D., Dep. of Horticulture, Univ. of Arkansas, Quantifying Turfgrass Color Using Digital Image Analysis, Crop Science, 43:943–951, 2003

Figures

Figure 2. Hue (color) & Standard Deviation of Hue

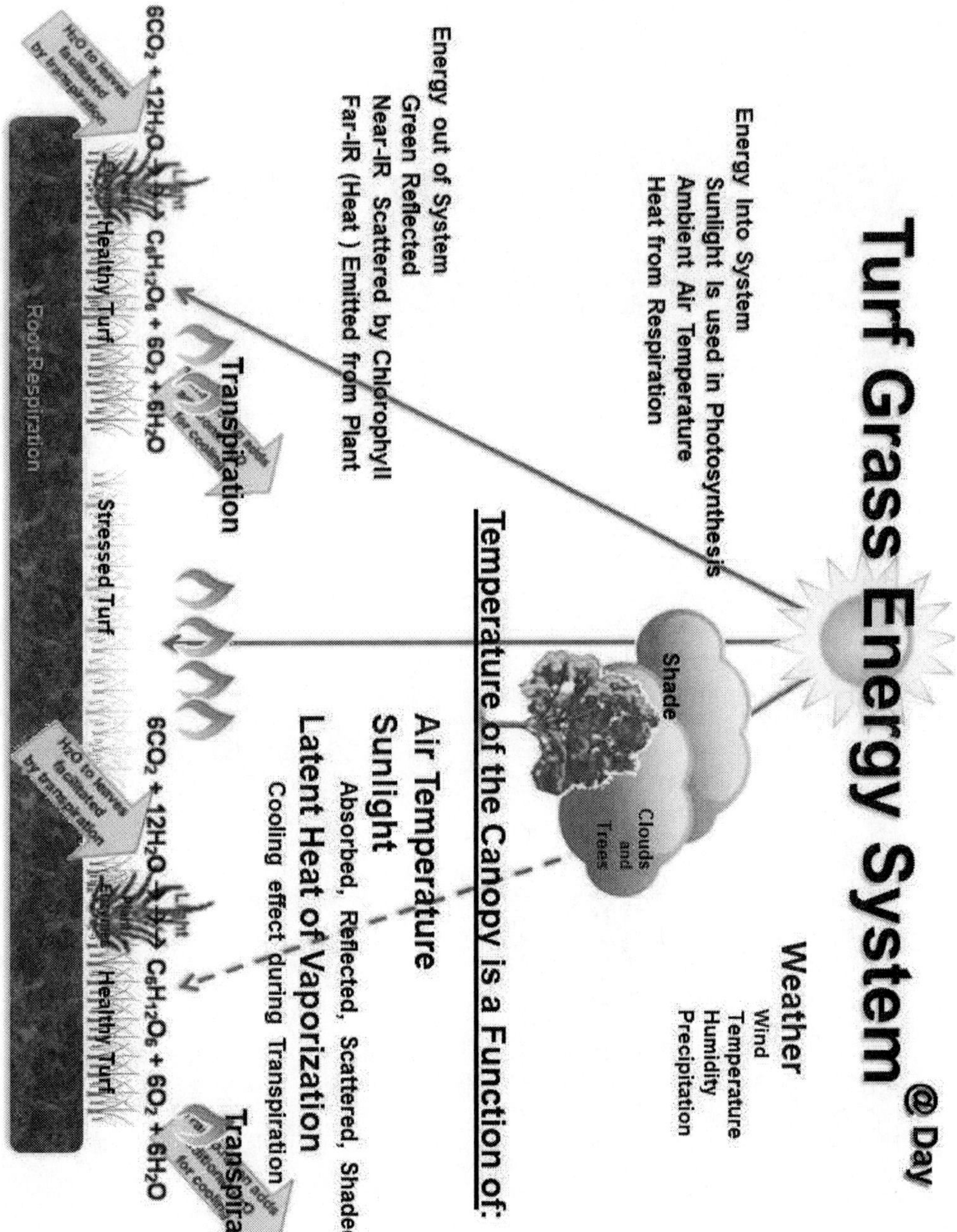

Figure 2a. Turfgrass Energy - Day & Night

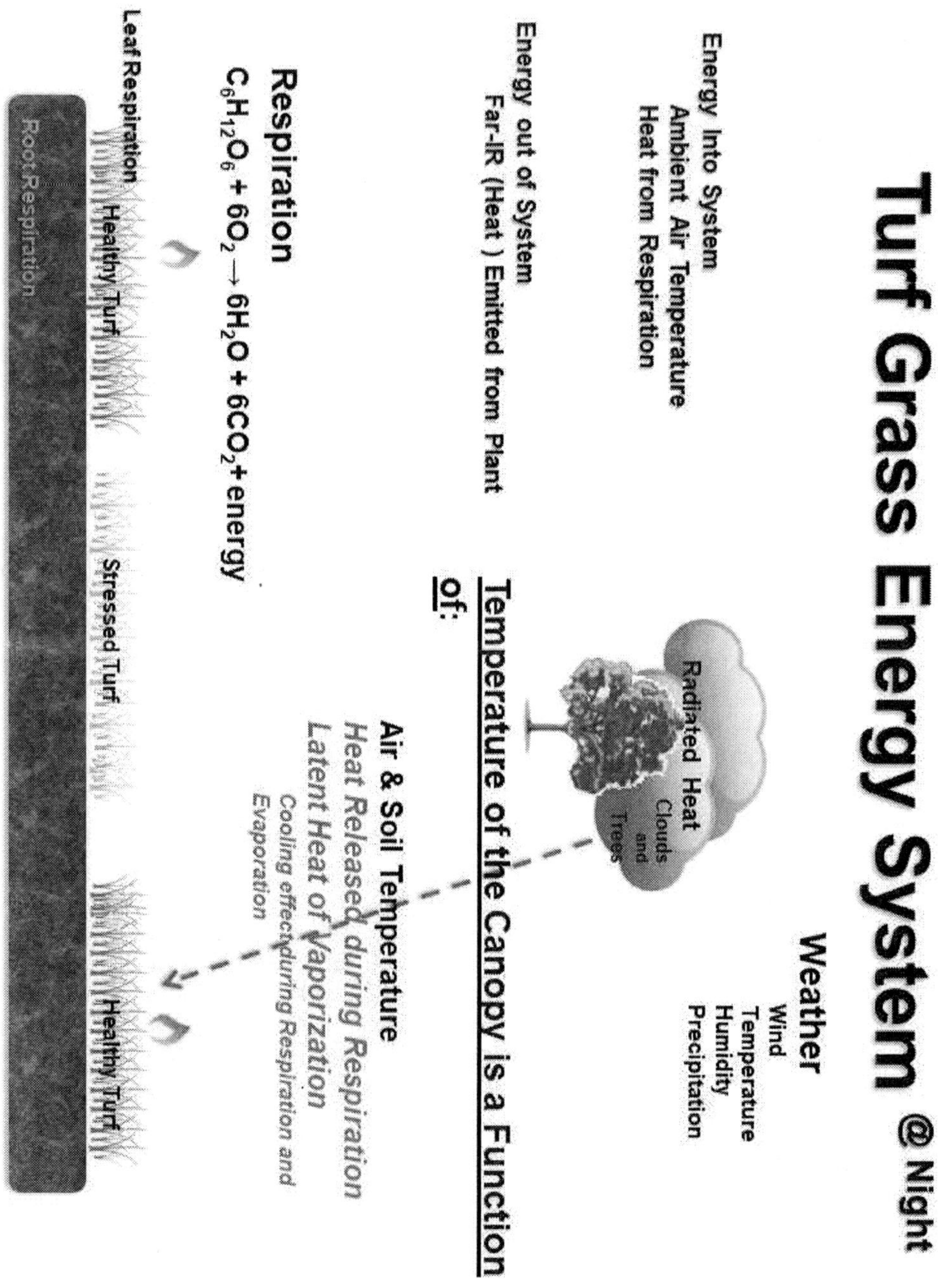

Figure 2b. Turfgrass Energy - Night

Figure 3. Thermal Image Data

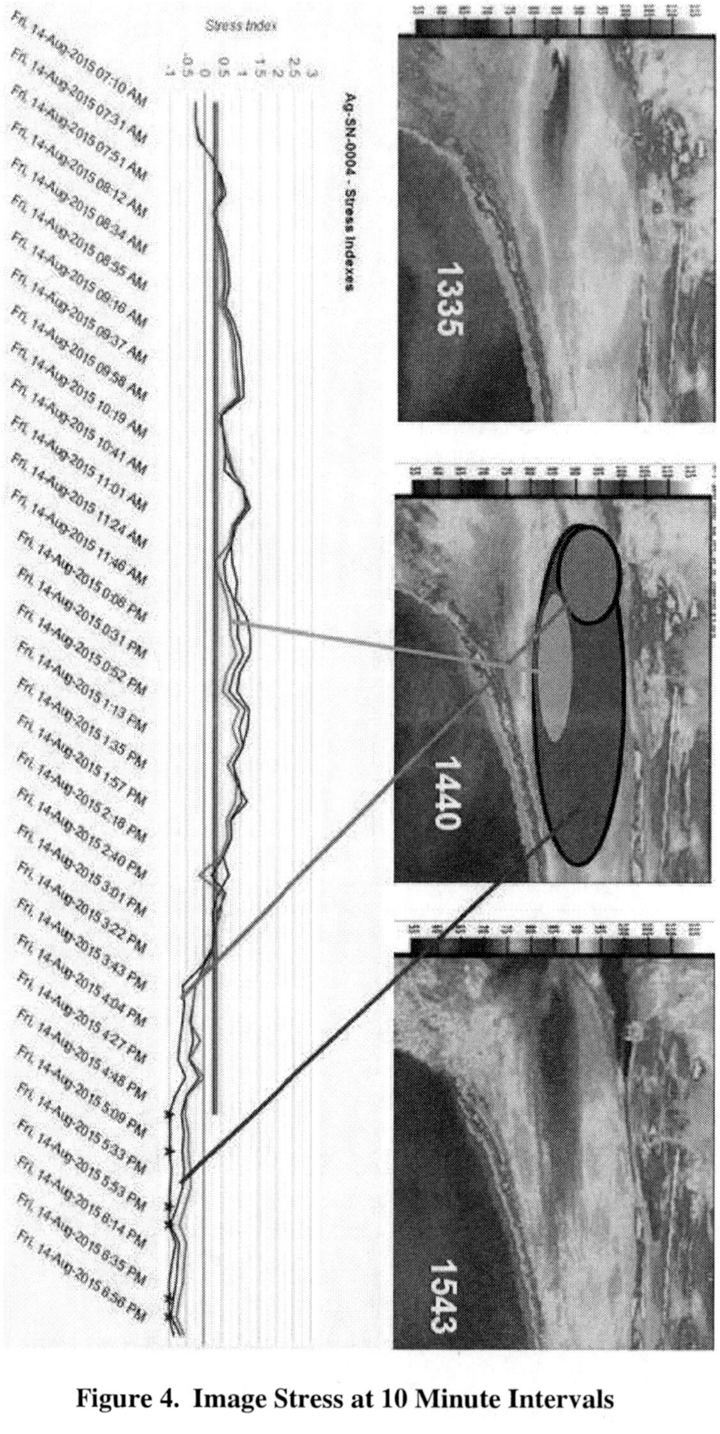

Figure 4. Image Stress at 10 Minute Intervals

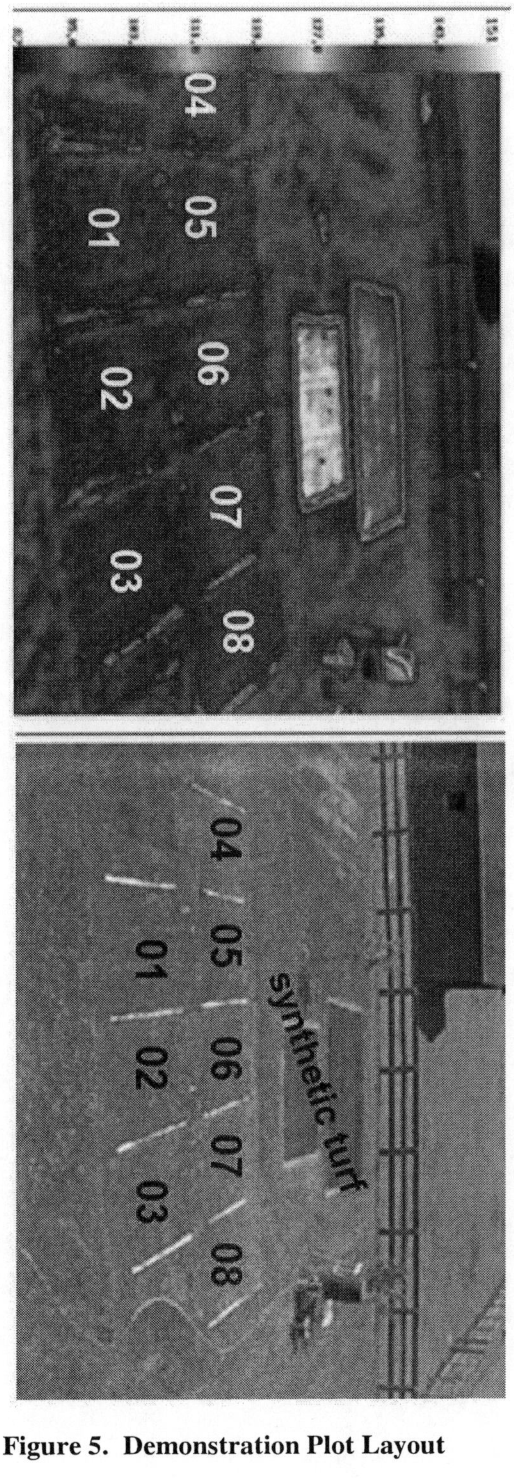

Figure 5. Demonstration Plot Layout

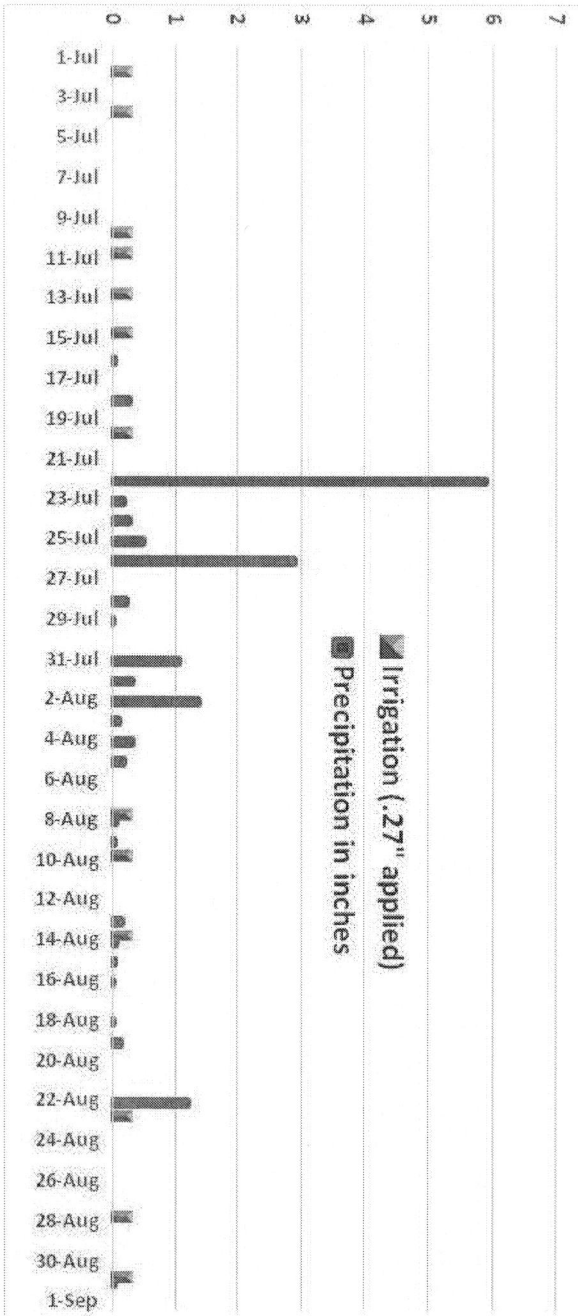

Figure 6. July through August Irrigation and Precipitation

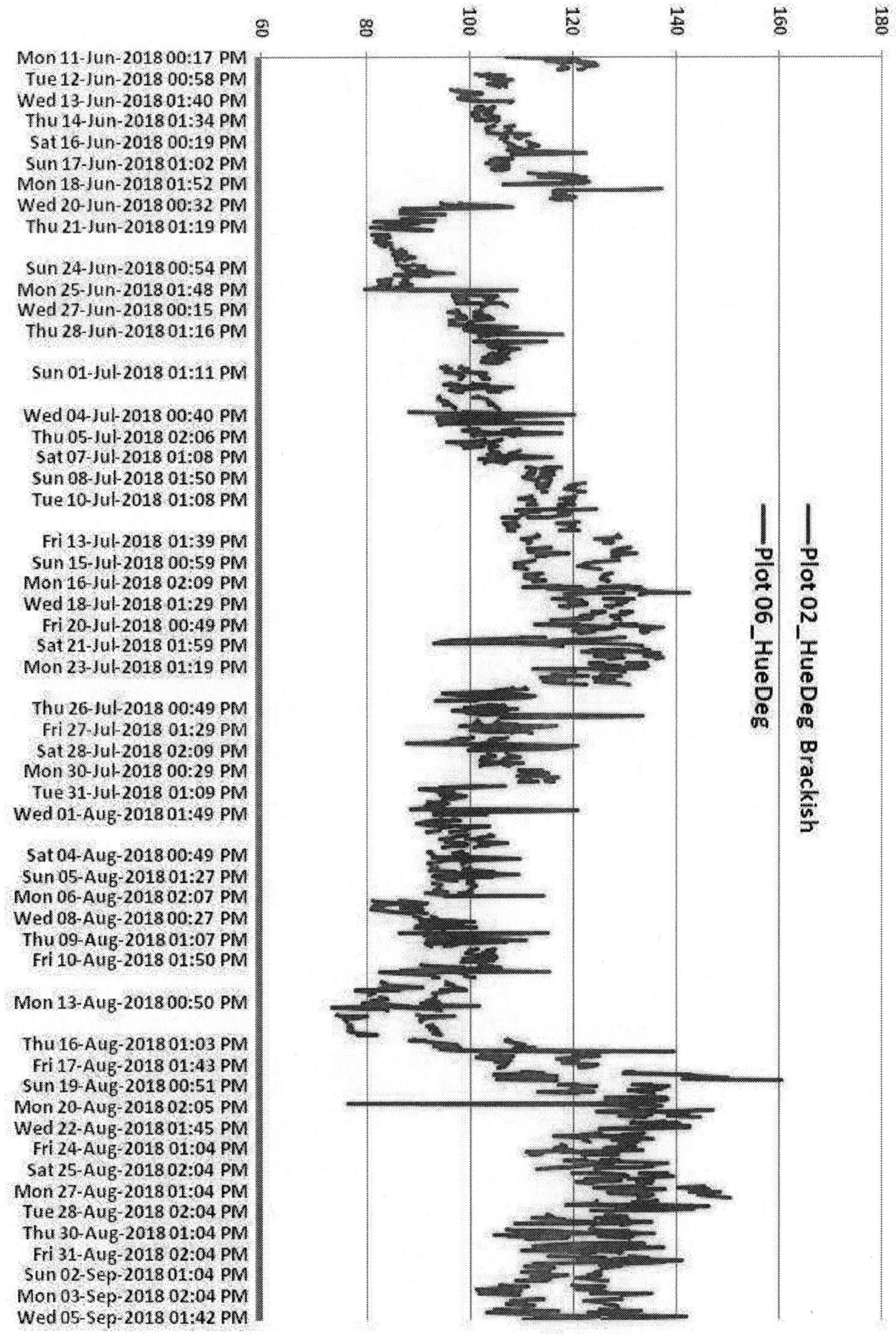

Figure 7. Hue Degree (color); Brackish versus Fresh

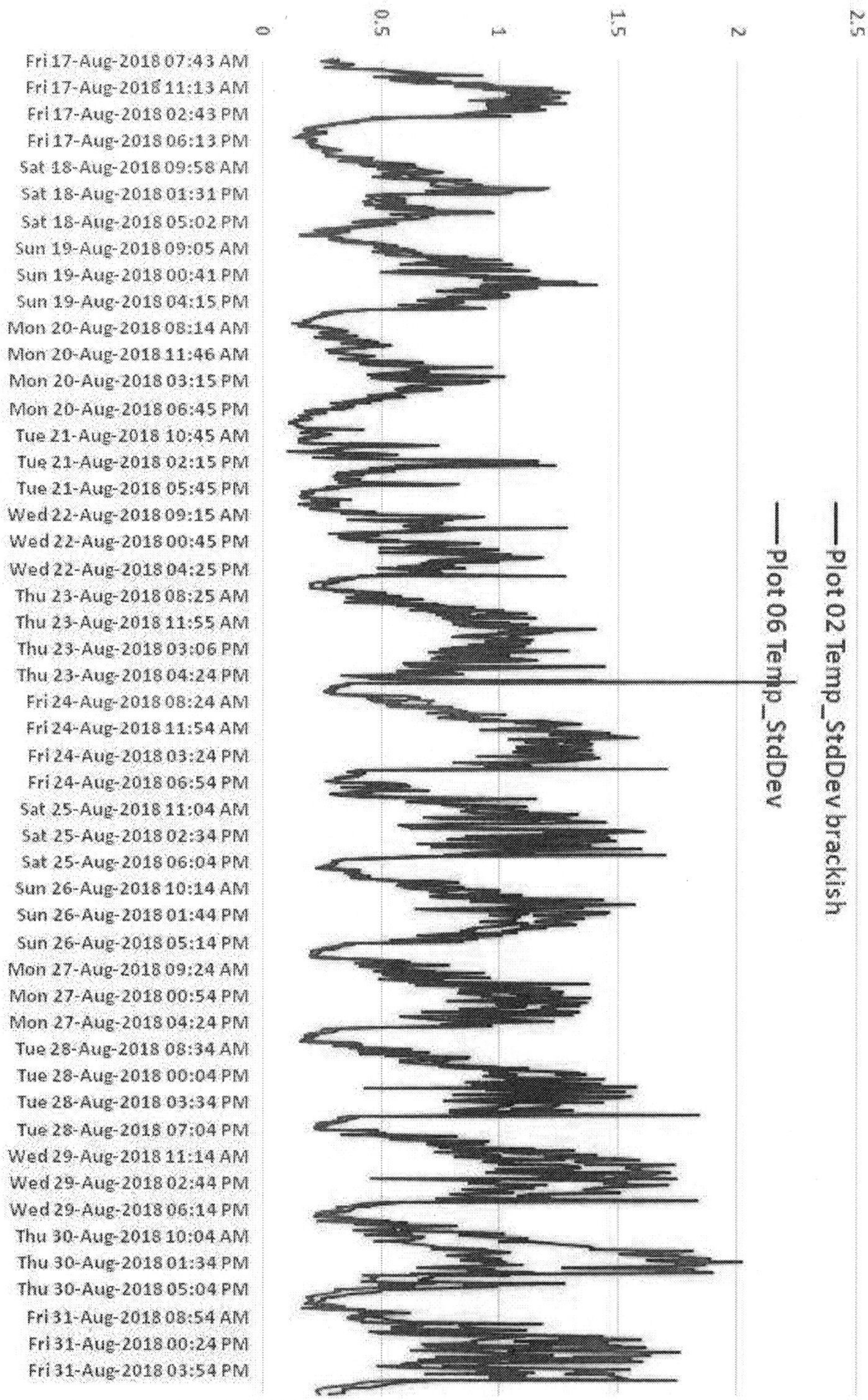

Figure 8. Standard Deviation of Hue; Brackish versus Fresh

ItriCorp

There are no boundaries when enthusiasm and excitement are applied to the task at hand.

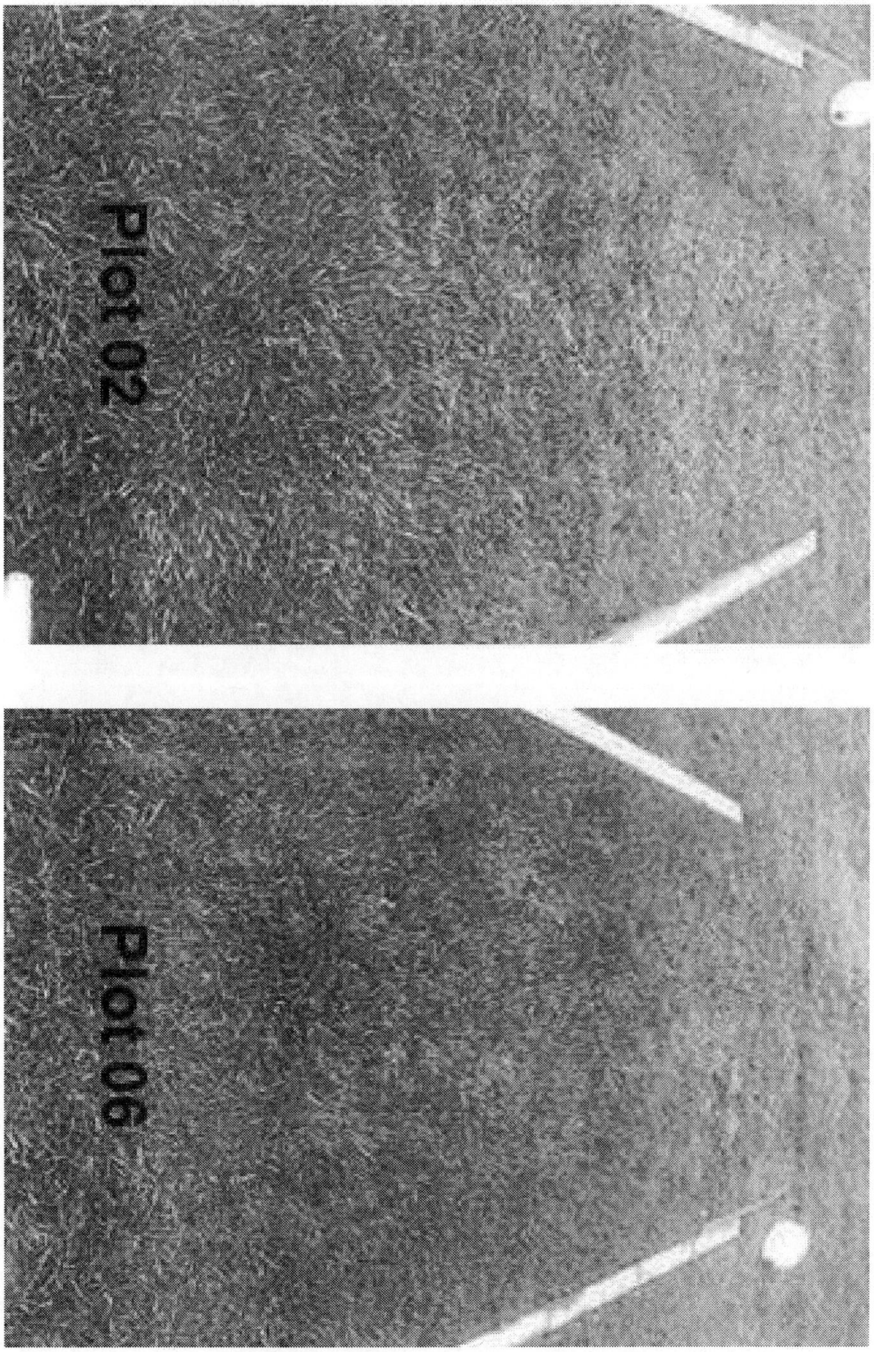

Figure 9. Brown Patch in Plot 06

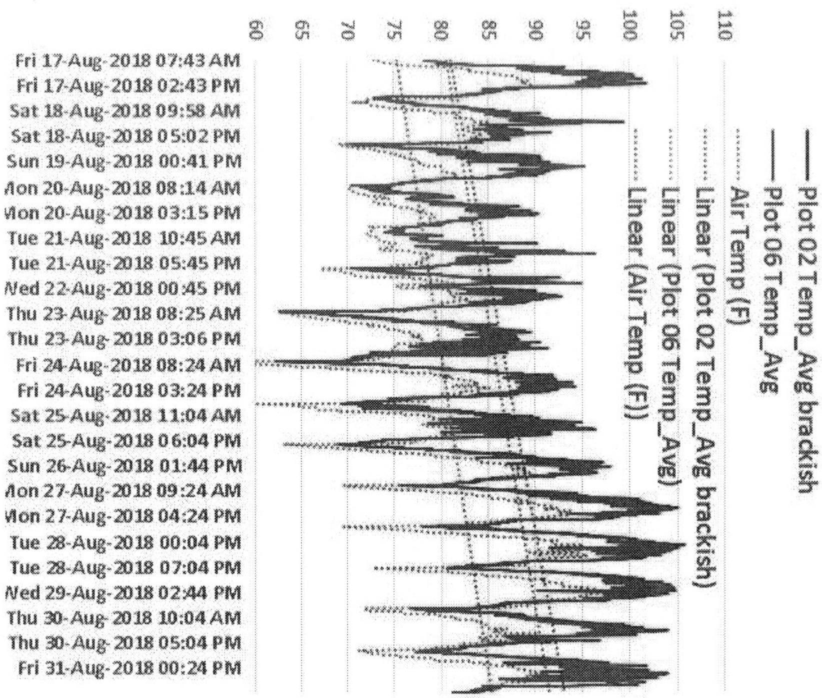

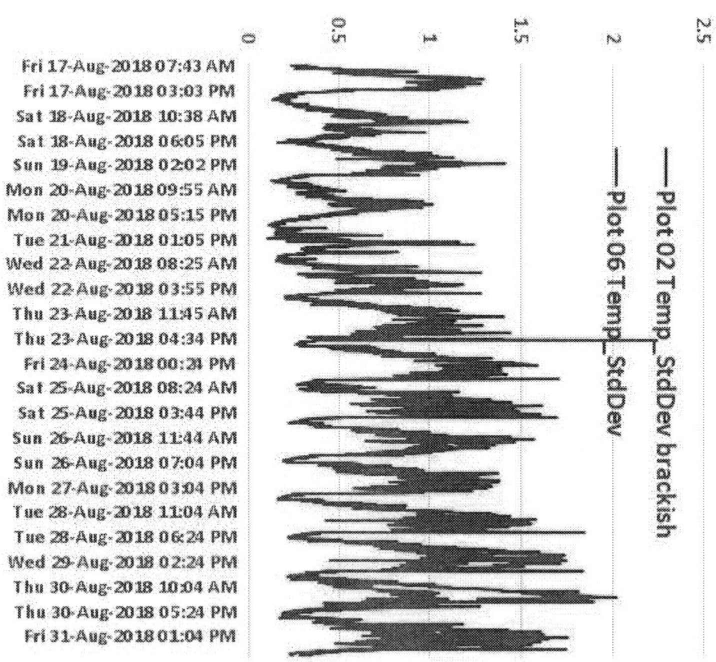

Figure 10. Canopy Temperature versus Air Temperature; and Standard Deviation of the Canopy Temperature

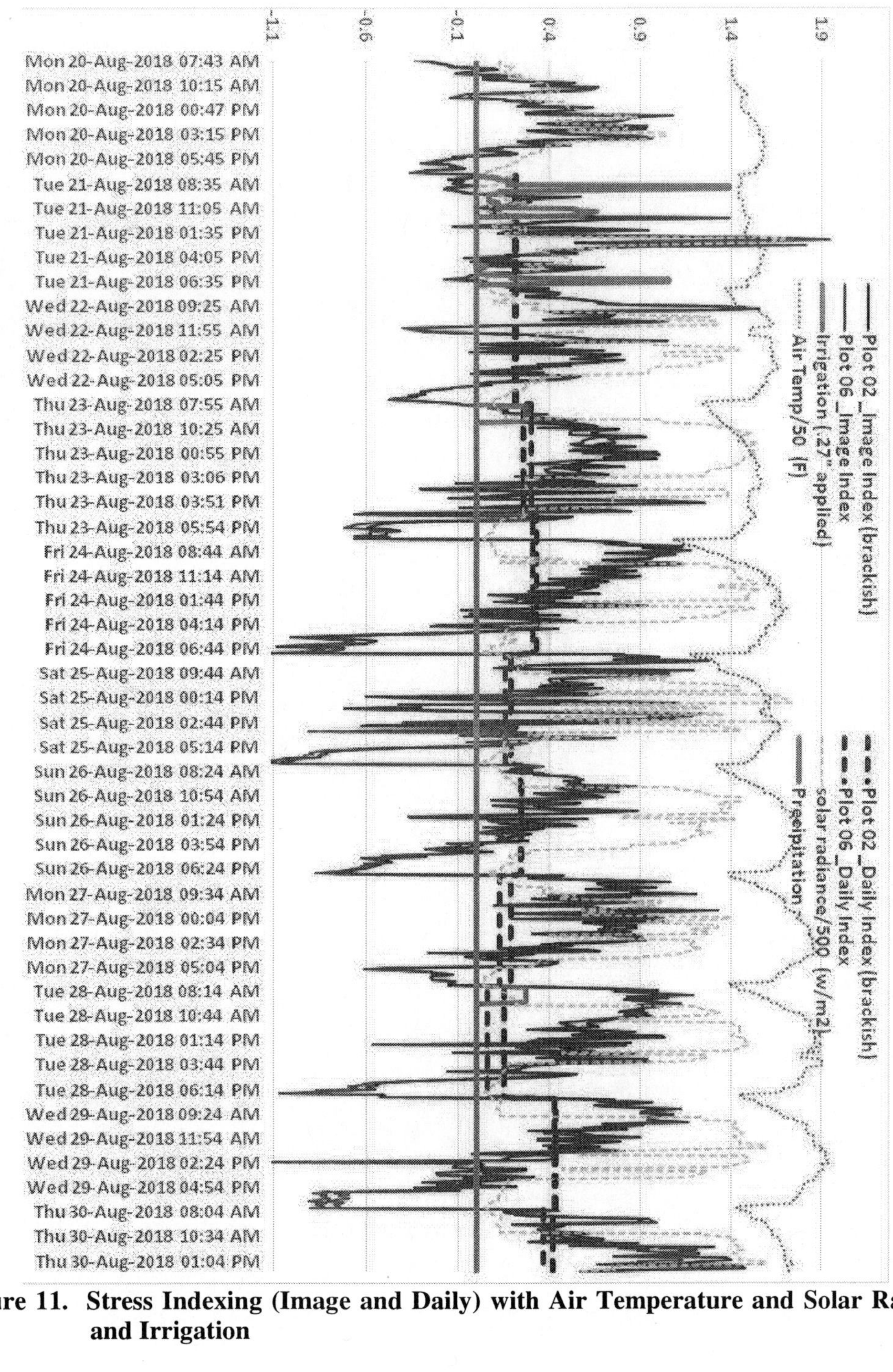

Figure 11. Stress Indexing (Image and Daily) with Air Temperature and Solar Radiance and Irrigation

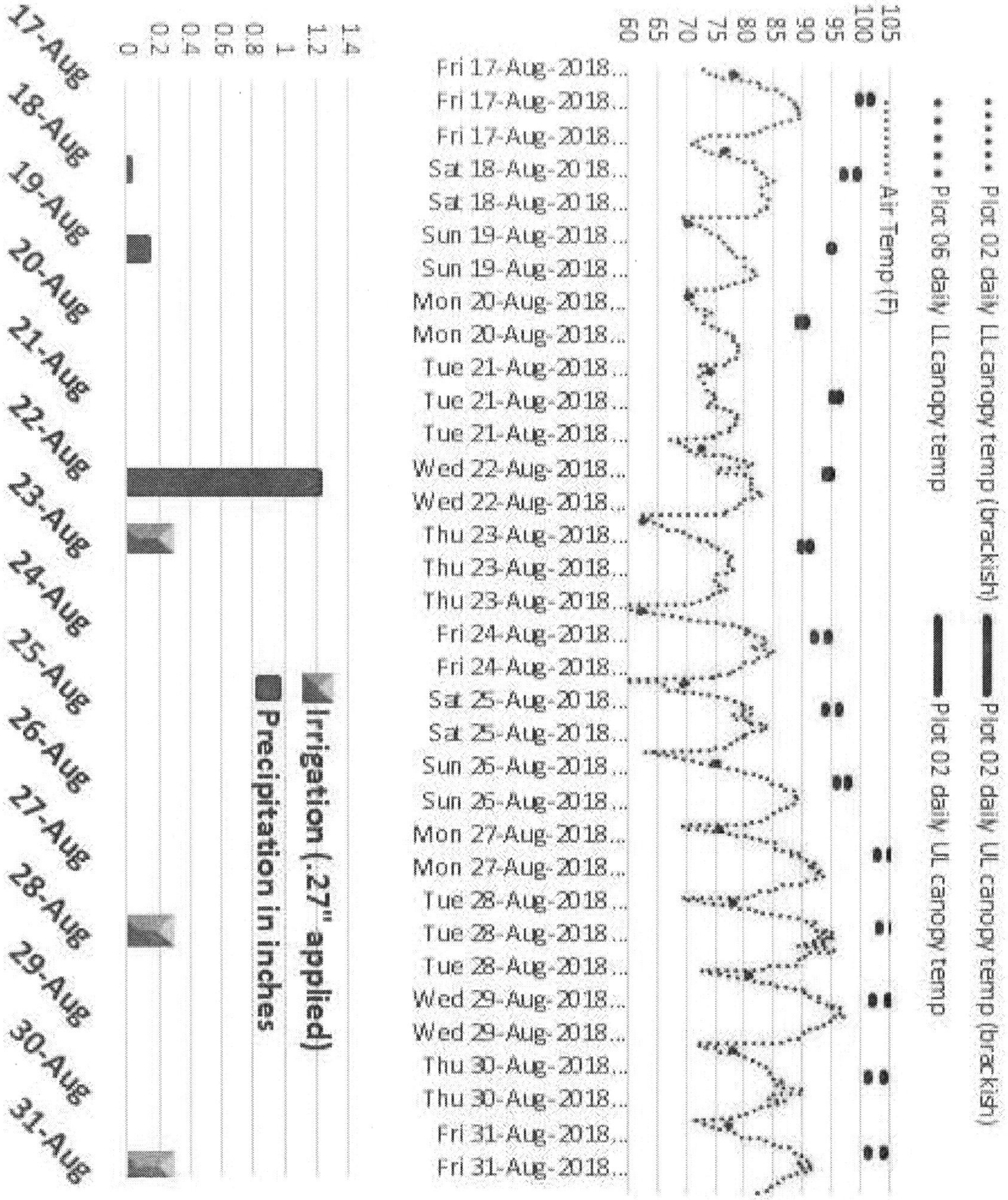

Figure 12. Daily LL and UL with Precipitation and Irrigation

Figure 13. Using Nighttime Thermal to ID Winterkill

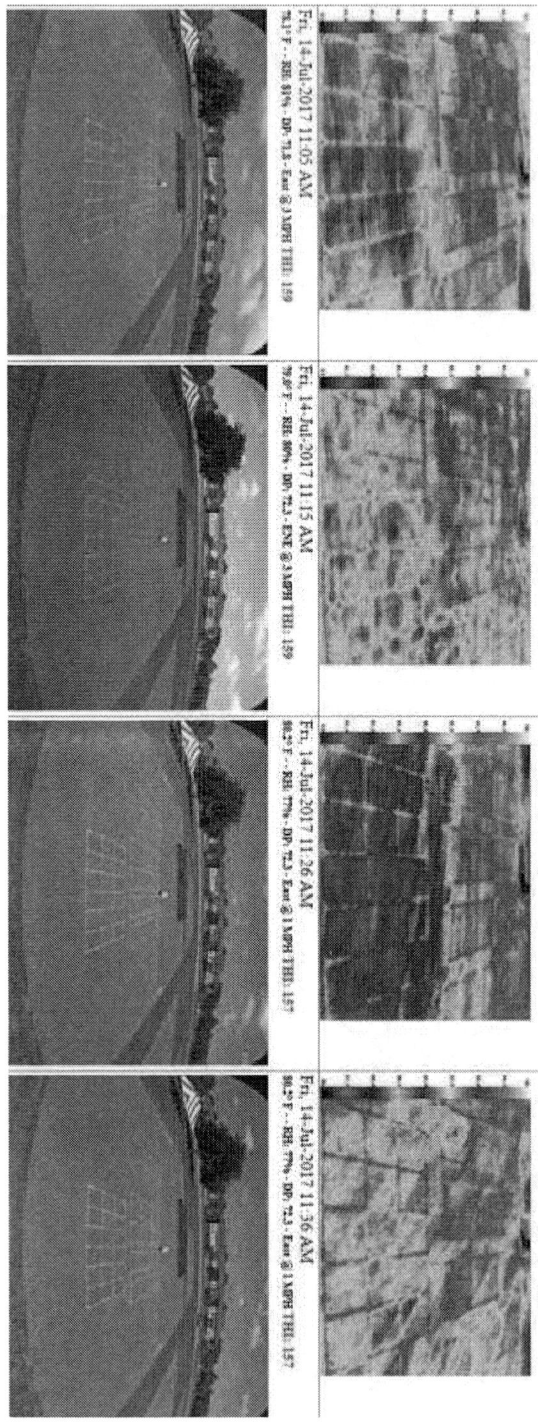

Figure 14. Image Data Time Series

Plots	17-25 July 2017 Total Inches of Water (irrigation + 0.18" precipitation)	ET $_{Kc=.8}$ recommendation	Turf Quality (from std dev Hue)
100% Replacement	4.065"	3.063" 51% Reduction over ET Recommendation	7
50% Replacement	2.641"		7
0% Replacement	0.160"		1
Hawk-Eye™ Stress Index	1.585"		7

Figure 15. Variable Irrigation of Plots and Resulting Quality

Author Index

A

B

C

D

E

F

G

H

N

O

P

R

S

T

V

W

Irrigation Association (IA)
8280 Willow Oaks Corporate Drive
Suite 400
Fairfax, Virginia 22031
USA

CURRAN ASSOCIATES INC.
proceedings.com

ISBN 978-1-5108-7677-4
9 781510 876774